SCHAUM'S OUTLINE OF

THEORY AND PROBLEM

OF

ELECTRIC CIRCUITS

Second Edition

•

BY

JOSEPH A. EDMINISTER, M.S.E.

Professor of Electrical Engineering
The University of Akron

•

SCHAUM'S OUTLINE SERIES

McGRAW-HILL BOOK COMPANY

New York St. Louis San Francisco Auckland Bogotá Guatemala Hamburg Johannesburg Lisbon London Madrid Mexico Montreal New Delhi Panama Paris San Juan São Paulo Singapore Sydney Tokyo Toronto

JOSEPH A. EDMINISTER is a Professor of Electrical Engineering at the University of Akron in Akron, Ohio, where he has also served as Assistant Dean and Acting Dean of Engineering. He has been a member of the faculty since 1957. He received his B.S.E.E. in 1957 and his M.S.E.E. in 1960 from the University of Akron. In 1974 he received his J.D., also from Akron. Dr. Edminister is a registered Professional Engineer in Ohio, a member of the bar of Ohio, and a registered patent attorney. He is the author of another Schaum's Outline, *Theory and Problems of Electromagnetics.*

Schaum's Outline of Theory and Problems of
ELECTRIC CIRCUITS

2 3 4 5 6 7 8 9 10 11 12 13 14 15 16 17 18 19 20 SHP SHP 8 7 6 5 4 3

ISBN 0-07-018984-6

Sponsoring Editor, David Beckwith
Editing Supervisor, Marthe Grice
Production Manager, Nick Monti

Library of Congress Cataloging in Publication Data

Edminister, Joseph.
Schaum's outline of theory and problems of electric circuits.

(Schaum's outline series)
Includes index.
1. Electric circuits. I. Title. II. Series.
TK454.E46 1983 621.319'2 82-18007
ISBN 0-07-018984-6

Preface

This book is designed for use as a textbook for a first course in circuit analysis or as a supplement to current standard texts. Studying from this book will help electrical engineering students and other engineering and technology students as well. Emphasis is placed on the basic laws, theorems and techniques which are common to the various approaches found in other texts.

The subject matter is divided into chapters covering duly-recognized areas of theory and study. The chapters begin with statements of pertinent definitions, principles and theorems together with illustrative examples and other descriptive material. This is followed by sets of solved and supplementary problems. The solved problems match the section of theory and serve to amplify it. They provide short, practical examples, and bring into focus the fine points which enable the student to apply the basic principles correctly and confidently. The supplementary problems match the solved problems in subject matter, are generally more numerous, and give the reader an opportunity to demonstrate problem-solving skills acquired in the theory and solved problems of the chapter. Answers are provided with each supplementary problem.

Topics covered include fundamental circuit responses, series and parallel circuit responses, power and power factor improvement and resonance phenomena. Node voltage and mesh current methods are developed first in direct current and later in alternating current phasor analysis. Theorems such as superposition, reciprocity and compensation are developed and treated in solved problems. Polyphase circuits and mutually coupled circuits are thoroughly covered. Trigonometric and exponential Fourier series are treated simultaneously, with the coefficients of one often converted to the other to show their relationship. Circuit transients are solved using classical differential equations and by Laplace transforms, which permits a convenient comparison of the two methods.

This second edition features the International System of Units (SI) throughout. Dependent voltage and current sources appear early in the text and occur throughout later chapters. Chapter 15 provides a simple, straightforward introduction to state variable methods as applied to circuit analysis.

This book is dedicated to my many former students. From them, I have learned how to teach well. To a large degree, it is they who have made possible a most satisfying and rewarding career. I want to express my gratitude to the editorial staff of McGraw-Hill, and to David Beckwith in particular. Sincere thanks are due to Thomas R. Connell, who assisted with several chapters and planned the general arrangement of topics. Many outstanding students deserve thanks for checking the problems for accuracy. And finally, I offer thanks to my wife, Nina, and to our family, for whom all of my efforts are happily made.

JOSEPH A. EDMINISTER

Contents

CONTENTS

Chapter 1

Introduction

1.1 UNITS

The International System of Units (SI) is used throughout this book. It is built on nine basic units, listed in Table 1-1, from which all other units are derived. Note that lowercase letters are used as symbols, except where the unit is named after an individual, as *ampere* (A) or *kelvin* (K). When the name is required in the plural, the usual rules of English apply; e.g., *henry* makes *henries.*

Table 1-1

Quantity	Unit	Symbol
length	meter	m
mass	kilogram	kg
time	second	s
current	ampere	A
temperature	kelvin	K
amount of substance	mole	mol
luminous intensity	candela	cd
plane angle	radian	rad
solid angle	steradian	sr

The derived units commonly used in electric circuit theory appear in Table 1-2. Some will be developed in this chapter, while the others will be explained as they are required.

Table 1-2

Quantity	Unit	Symbol
electric charge	coulomb	C
electric potential	volt	V
resistance	ohm	Ω
conductance	siemens	S
inductance	henry	H
capacitance	farad	F
frequency	hertz	Hz
force	newton	N
energy, work	joule	J
power	watt	W
magnetic flux	weber	Wb
magnetic fiux density	tesla	T

Decimal multiples and submultiples of SI units should be used whenever possible. The symbols given in Table 1-3 are prefixed to the unit symbols of Tables 1-1 and 1-2. For example, mm is used for *milli*meter, 10^{-3} m; MW for *mega*watt, 10^6 W.

Table 1-3

Factor	Prefix	Symbol
10^9	giga	G
10^6	mega	M
10^3	kilo	k
10^{-2}	centi	c
10^{-3}	milli	m
10^{-6}	micro	μ
10^{-9}	nano	n
10^{-12}	pico	p

1.2 FORCE, WORK, POWER

The derived units follow the mathematical expressions which relate the quantities. From "force equals mass times acceleration," the *newton force* (N) is defined as the unbalanced force which imparts to a one-kilogram mass an acceleration of one meter per second squared. Thus, $1\ \text{N} = 1\ \text{kg} \cdot \text{m/s}^2$.

Work results when a force acts over a distance. A *joule* of work is equivalent to a newton-meter: $1\ \text{J} = 1\ \text{N} \cdot \text{m}$. Work and energy have the same units; thus, the energy stored in a capacitor will be expressed in joules.

Power is the rate at which the work is done, or the rate at which energy is changed from one form to another. The derived unit of power, the *watt* (W), is one joule per second (J/s). For example, to lift a 3-newton weight vertically through 2 meters in two minutes requires (3 N)(2 m) = 6 J of work (increased potential energy) and hence an average power of

$$\frac{6\ \text{J}}{120\ \text{s}} = 0.05\ \text{W}$$

1.3 CURRENT AND ELECTRIC CHARGE

The basic unit of current is the *ampere* (A), defined as that constant current in two infinite, parallel conductors of negligible circular cross section, one meter apart in vacuum, which produces a force between the conductors of 2×10^{-7} newtons per meter length. Of course, current results from charges in motion, and one ampere is one coulomb of charge moving past a fixed location in one second. In this way, the derived unit of charge, the *coulomb* (C), is equivalent to an ampere-second: $1\ \text{A} = 1\ \text{C/s}$ or $1\ \text{C} = 1\ \text{A} \cdot \text{s}$. The moving charges may be positive or negative. A liquid may contain positive ions, as suggested in Fig. 1-1(a). If these positive ions move to the left so that they pass the fixed plane B at the rate of *one coulomb per second*, then the current is one ampere, directed to the left. Negative ions moving to the right at the rate of one coulomb per second would also constitute a current of one ampere directed to the left, as shown in Fig. 1-1(b). Of more interest in electric circuit theory is the current in a metallic conductor, Fig. 1-1(c). The atoms of such a conductor are fixed in a crystal structure. The positively charged protons in the atomic nucleus are surrounded by the negatively charged electrons. All good conductors will have one or two electrons in the outermost shell, with all the other electrons locked within inner shells closer to the nucleus. These outer electrons are free to move about from one atom to the next. However, Coulomb forces prevent them from piling up in one location. A reasonably accurate picture of conduction in a copper wire, for example, is that approximately 8.5×10^{28} electrons per cubic meter are free to move. The electron charge is $-e = -1.602 \times 10^{-19}$ C, so that for a current of 1 A some 6.24×10^{18} electrons per second would have to pass a fixed cross section of the wire.

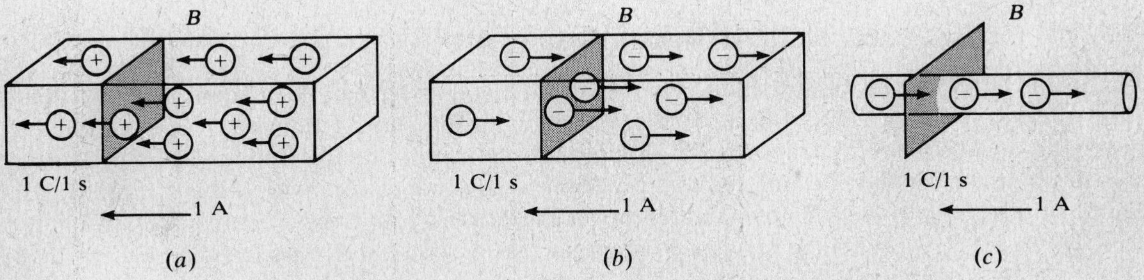

Fig. 1-1

1.4 ELECTRIC POTENTIAL

Electric charges experience forces in electric fields which, if unopposed, can accelerate the particle containing the charge. Of more interest here is the work done to move such a charge against the field. Consider the charge Q in Fig. 1-2, acted on by the field force $\mathbf{F}$ and an equal and opposite applied force $\mathbf{F}_a$. To move the charge to position 1, work is required. If Q is *one coulomb* and the work done by $\mathbf{F}_a$ is *one joule*, then position 1 is said to be at a potential of *one volt positive* with respect to the original point 0. That is, 1 V = 1 J/C.

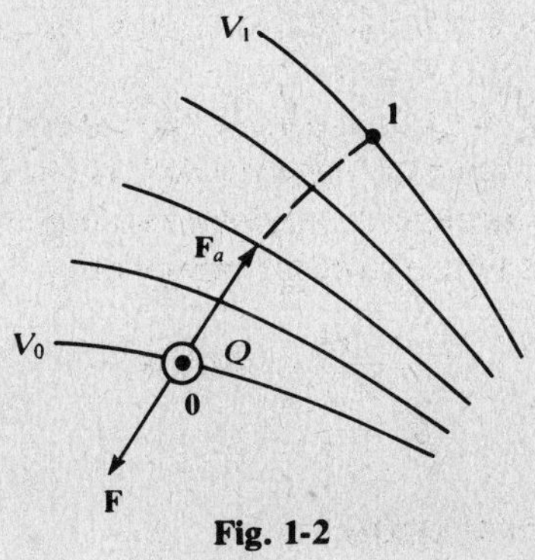

Fig. 1-2

The volt may also be defined (and, in fact, officially is defined) as the difference in electric potential between two points along a conductor carrying a constant current of one ampere when the power dissipated between the two points is one watt. The power p is the product of the current and the voltage difference, $p = vi$ or $v = p/i$. Hence

$$1\text{ V} = \frac{1\text{ W}}{1\text{ A}} = \frac{1\text{ J/s}}{1\text{ C/s}} = \frac{1\text{ J}}{1\text{ C}}$$

1.5 NOTATION

It is common in circuit analysis to distinguish between constant and time-varying quantities by employing capital letters for the constant and lowercase for the variable. For example, a constant current of ten amperes would be written $I = 10$ A, while a sinusoidal current of amplitude ten amperes would be written

$$i = (10\ \text{A}) \sin \omega t \qquad \text{or} \qquad i = 10 \sin \omega t \quad (\text{A})$$

Normally, the second form will be used in this book; it indicates that i will be in A provided the constant 10 is in A, ω is in rad/s, and t is in s.

1.6 CONTINUOUS AND DISCONTINUOUS FUNCTIONS

In an electric circuit, the current in a branch is the first derivative of the amount of charge that is free to move, $i = dq/dt$. (This will be further examined in Chapter 2.) But the charge is carried by electrons, and electrons cannot appear or disappear abruptly. For example, in Fig. 1-3 branch current i is shown in the conductor connected to a capacitor whose charge is q. Now, if the charge increases in the time interval $0 < t < t_1$ in the manner shown in Fig. 1-4(*a*), then the current is as shown in Fig. 1-4(*b*). From t_1 to t_2 the charge remains constant and the current is zero. As the charge q returns to zero at a constant rate in the interval $t_2 < t < t_3$, the current is constant and negative. The current function is *discontinuous* at $t = 0, t_2, t_3$, because at those instants $i(t^-) \neq i(t^+)$. In contrast, the charge function is always *continuous*, since a discontinuity would require that a quantity of charge instantaneously appeared or disappeared. The electrons may at times move rapidly, but never with infinite speed.

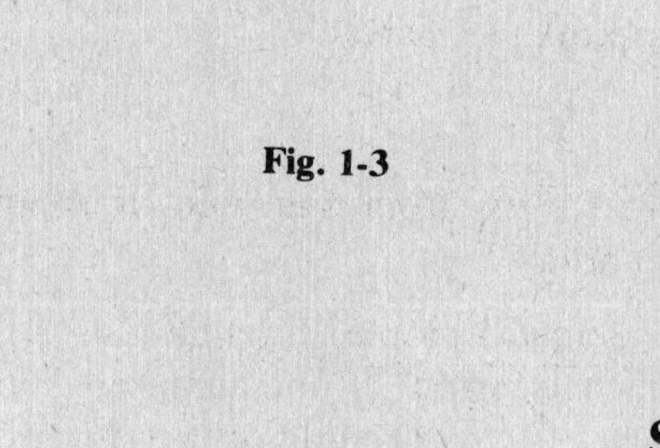

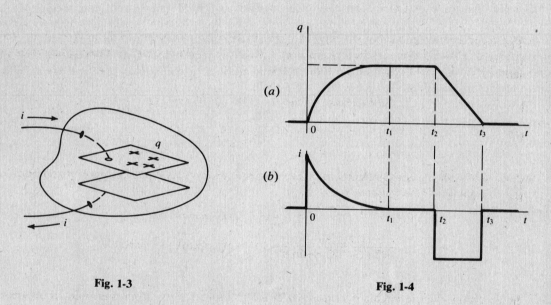

Fig. 1-3

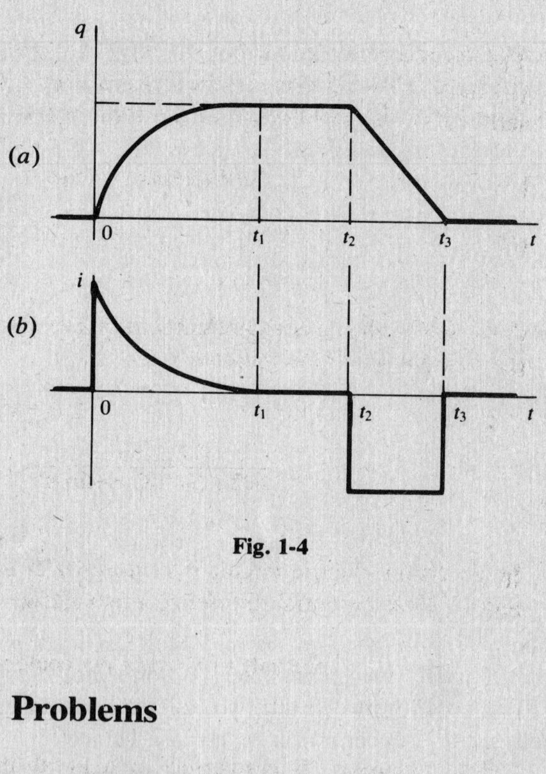

Fig. 1-4

Solved Problems

1.1 Obtain the time, in nanoseconds and in microseconds, at which the function

$$y(t) = Ye^{-8\times10^5 t}$$

has a value which is 20% of $y(0)$.

Since $y(0) = Y$,

$$0.20\,Y = Ye^{-8\times10^5 t}$$

Solving by natural logarithms,

$$t = 2.01 \times 10^{-6}\ \text{s} = 2010\ \text{ns} = 2.01\ \mu\text{s}$$

1.2 A steady-state impedance Z is given by

$$Z = [(10.5)^2 + (2\pi f 4 \times 10^{-5})^2]^{1/2} \quad (\Omega)$$

where f is the frequency in hertz. Obtain Z for frequencies 9.95 kHz and 3.20 MHz.

$$\{(10.5)^2 + [2\pi(9.95 \times 10^3)4 \times 10^{-5}]^2\}^{1/2} = 10.79\ \Omega$$

$$\{(10.5)^2 + [2\pi(3.20 \times 10^6)4 \times 10^{-5}]^2\}^{1/2} = 30.24\ \Omega$$

1.3 In simple rectilinear motion a 2.5-kg mass is given a constant acceleration of 25 m/s^2. (*a*) Find the acting force F. (*b*) Find the position and kinetic energy of the body after four seconds, if it was initially at rest at $x = 0$.

(*a*) $$F = ma = (2.5\text{ kg})(25\text{ m/s}^2) = 62.5\text{ kg}\cdot\text{m/s}^2 = 62.5\text{ N}$$

(*b*) $$x = \tfrac{1}{2}at^2 = \tfrac{1}{2}(25\text{ m/s}^2)(4\text{ s})^2 = 200\text{ m}$$

Then, by the work-energy relation,

$$\text{KE} = Fx = (62.5\text{ N})(200\text{ m}) = 12\,500\text{ N}\cdot\text{m} = 12\,500\text{ J}$$

or 12.5 kJ.

1.4 The force applied to an object moving in the x-direction varies according to $F = 12/x^2$ (N). (*a*) Find the work done in the interval $1\text{ m} \le x \le 3\text{ m}$. (*b*) What constant force acting over the same interval would result in the same work?

(*a*) $$dW = F\,dx \qquad \text{so} \qquad W = \int_1^3 \frac{12}{x^2}\,dx = 12\left[\frac{-1}{x}\right]_1^3 = 8\text{ J}$$

(*b*) $$8\text{ J} = F_c(2\text{ m}) \qquad \text{or} \qquad F_c = 4\text{ N}$$

1.5 A conductor has a constant current of five amperes. How many electrons pass a fixed point on the conductor in one minute?

$$5\text{ A} = (5\text{ C/s})(60\text{ s/min}) = 300\text{ C/min}$$

$$\frac{300\text{ C/min}}{1.602 \times 10^{-19}\text{ C/electron}} = 1.87 \times 10^{21}\text{ electrons per minute}$$

1.6 In a certain electric circuit it requires 9.25 μJ to move 0.5 μC from one point to a second point. What potential difference exists between the two points?

one volt = one joule per coulomb $$V = \frac{9.25 \times 10^{-6}\text{ J}}{0.5 \times 10^{-6}\text{ C}} = 18.5\text{ V}$$

1.7 Electrical energy is converted to heat at the rate of 7.56 kJ/min in a resistor which has 270 C/min passing through. What is the voltage difference across the resistor terminals?

From $P = VI$,

$$V = \frac{P}{I} = \frac{7.56 \times 10^3\text{ J/min}}{270\text{ C/min}} = 28\text{ J/C} = 28\text{ V}$$

1.8 A certain circuit element has a current $i = 2.5 \sin \omega t$ (mA), where ω is the angular frequency in rad/s, and a voltage difference between terminals $v = 45 \sin \omega t$ (V). Find the average power P_{avg} and the energy W_T transferred in one period of the sine function.

Energy is the time-integral of instantaneous power:

$$W_T = \int_0^{2\pi/\omega} vi\,dt = 112.5 \int_0^{2\pi/\omega} \sin^2 \omega t\,dt = \frac{112.5\,\pi}{\omega} \quad \text{(mJ)}$$

The average power is then

$$P_{avg} = \frac{W_T}{2\pi/\omega} = 56.25 \text{ mW}$$

Note that P_{avg} is independent of ω.

1.9 The unit of energy used by electric utility companies is the kilowatt-hour (kWh). (*a*) How many joules are in 1 kWh? (*b*) A color television set rated at 650 W is turned on from 7 p.m. to 11:30 p.m. What is the energy consumption, in megajoules and in kilowatt-hours?

(*a*) $$1 \text{ kWh} = (1000 \text{ W})(1 \text{ h}) = (1000 \text{ J/s})(3600 \text{ s}) = 3.6 \times 10^6 \text{ J}$$

or 3.6 MJ.

(*b*) $$(650 \times 10^{-3} \text{ kW})(4.5 \text{ h}) = 2.93 \text{ kWh}$$

and $$(2.93 \text{ kWh})(3.6 \text{ MJ/kWh}) = 10.5 \text{ MJ}$$

1.10 Figure 1-5(*a*) is the graph of the current into a circuit element. Find the charge transferred by this current, assuming that the element is initially uncharged.

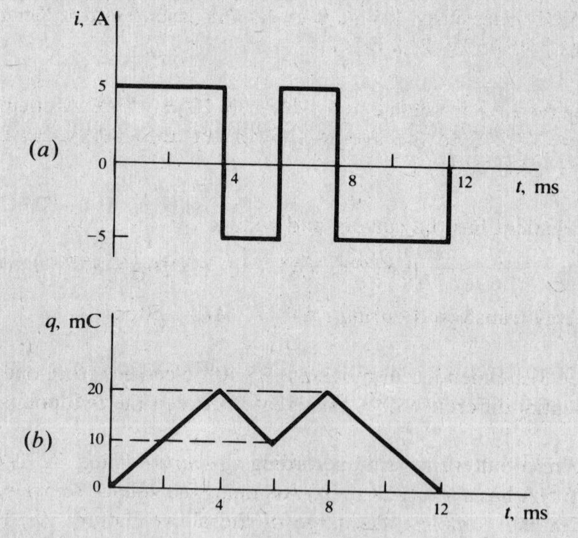

Fig. 1-5

From $i = dq/dt$ and $q(0) = 0$,

$$q = \int_0^t i\,dt$$

The integral of i can be found graphically, by calculating the area under the curve of Fig. 1-5(*a*) between 0 and t. The result is shown in Fig. 1-5(*b*). Note that q is a continuous function of time, even though i is not. Note also that because the graph of i is antisymmetric about $t = 6$ ms, the graph of q is symmetric about that point.

Supplementary Problems

1.11 Obtain the time t, in microseconds and in milliseconds, at which

$$v = 250(1 - e^{-2\times 10^3 t}) \quad \text{(V)}$$

has the value 97.5 volts. *Ans.* 247 μs, 0.247 ms

1.12 Obtain the time t, in microseconds and in nanoseconds, of the first minimum (most negative) value of the sinusoid $v = V_m \sin \omega t$, where $V_m > 0$ and $\omega = 2\pi \times 10^6$ rad/s. *Ans.* 0.75 μs, 750 ns

1.13 Obtain the work and power associated with the movement of a 3.5×10^{-2}-kg mass by a 7.5×10^{-4}-N force over a two-meter distance in an elapsed time of fourteen seconds. *Ans.* 1.5 mJ, 0.107 mW

1.14 Obtain the work and power required to move a 5.0-kg mass up a frictionless plane inclined at an angle of 30° with the horizontal for a distance of 2.0 m along the plane in a time of 3.5 s.
Ans. 49.0 J, 14.0 W

1.15 A unit of power used for electric motors is the *horsepower* (hp), equal to 746 watts. How much energy does a 5-hp motor deliver in two hours? Express the answer in MJ. *Ans.* 26.9 MJ

1.16 An AWG #12 copper wire contains approximately 2.77×10^{23} free electrons per meter length, assuming one conduction electron per atom. For a current of 25 A, find the percentage of these electrons that will pass a given point in one minute. *Ans.* 3.38%

1.17 Find the charge, in coulombs, and the number of electrons that must pass a fixed point in one hour in the filament of a 100-watt light bulb at 120 volts. Assume direct current. *Ans.* 3000 C/h, 1.87×10^{22} h^{-1}

1.18 A pulse of electricity measures 305 V, 0.15 A, and lasts 500 μs. What power and energy does this represent? *Ans.* 45.75 W, 22.9 mJ

1.19 Households are served by a voltage $v = 120\sqrt{2} \sin 120\pi t$ (V). A circuit rated at 20 A has a current $i = 20\sqrt{2} \sin 120\pi t$ (A). Find the average power and the energy delivered in one period of the sine function. *Ans.* 2400 W, 40 J

1.20 A certain circuit element has the current and voltage

$$i = 10e^{-5000t} \quad \text{(A)} \qquad v = 50(1 - e^{-5000t}) \quad \text{(V)}$$

Find the total energy transferred during $t \geq 0$. *Ans.* 50 mJ

1.21 Work equal to 136 J is expended in moving 8.5×10^{18} electrons from one point to another in an electric circuit. What potential difference does this create between the two points? *Ans.* 100 V

1.22 A typical 12-volt auto battery is rated according to *ampere-hours*. A 70-A · h battery, for example, at a discharge rate of 3.5 A has a life of 20 h. (*a*) Assuming the voltage remains constant, obtain the energy and power delivered in the complete discharge of the above battery. (*b*) Repeat for a discharge rate of 7.0 A. *Ans.* (*a*) 3.02 MJ, 42 W; (*b*) 3.02 MJ, 84 W

1.23 For $t \geq 0$, $q = 4 \times 10^{-4}(1 - e^{-250t})$ (C). Obtain the current at (*a*) $t = 0$, (*b*) $t = 3$ ms.
Ans. (*a*) 0.10 A, (*b*) 0.0472 A

1.24 The *capacitance* of a circuit element is defined as q/V, where q is the magnitude of charge stored in the element and V is the magnitude of the voltage difference across the element. The SI derived unit of capacitance is the *farad* (F). Express the farad in terms of the basic units.
Ans. $1\ \text{F} = 1\ \text{A}^2 \cdot \text{s}^4/\text{kg} \cdot \text{m}^2$

Chapter 2

Circuit Concepts

2.1 CIRCUIT ELEMENTS

A *circuit diagram* or *network* is constructed from series and parallel combinations of two-terminal elements to represent an electrical device. Analysis of the circuit diagram predicts the performance of the actual device. The basic building block is the two-terminal element shown in general form in Fig. 2-1, with a single function or single device contained within the symbol and two perfectly conducting leads. *Active* elements are voltage or current sources which are able to supply energy to the network. Resistors, inductors, and capacitors are *passive* elements which either absorb or store the energy from the sources.

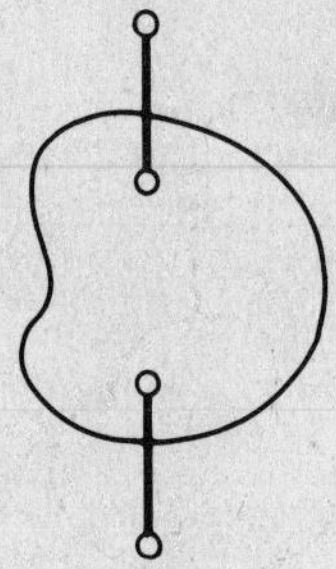

Fig. 2-1

Figure 2-2 illustrates seven basic circuit elements. *Independent* active circuit elements are shown at (*a*) and (*c*), where the specified voltage or current is not altered by changes in the connected network. Voltage or current sources which change according to variables in the connected network are *dependent* sources, with diamond-shaped symbols as in (*b*) and (*d*). The three passive circuit elements are shown at (*e*), (*f*), and (*g*).

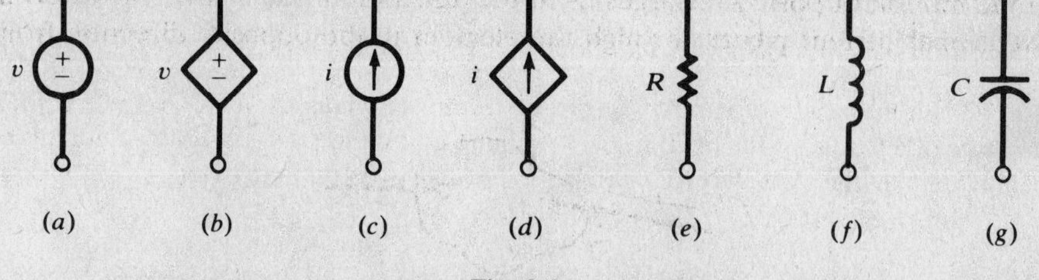

Fig. 2-2

The circuit diagrams discussed here are termed *lumped-parameter* circuits since a single element at one location in the diagram is used to represent a distributed resistance, inductance, or capacitance. For example, a coil consisting of many turns of insulated wire has resistance and inductance distributed along the entire length of the wire. Nevertheless, a single resistance *lumped* at one place, as shown in Fig. 2-3(*b*), is used to represent the distributed resistance of the coil. The inductance is likewise lumped at one place, in series with the resistance.

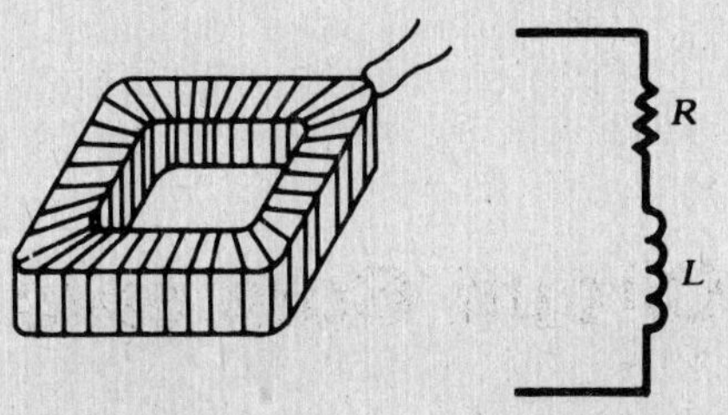

(a) (b)

Fig. 2-3

2.2 ELECTRIC POTENTIAL

A magnitude and a polarity must be specified to describe completely the potential or voltage. The polarity marks in the circuit diagram are placed near the two conductors where the voltage is defined. If, for instance, $v = 15$ V in Fig. 2-4, terminal A is positive with respect to terminal B. If $v = 10 \sin \omega t$, terminal A is positive with respect to B for $0 < \omega t < \pi$ and B is positive with respect to A for $\pi < \omega t < 2\pi$.

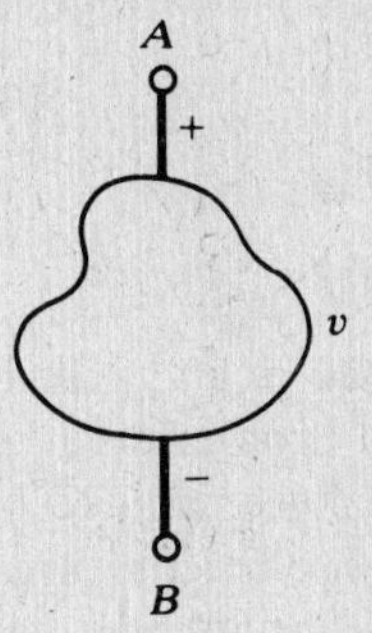

Fig. 2-4

2.3 CURRENT

To be completely defined, a current must be given a magnitude and a direction. In Fig. 2-5, current i is shown entering a generalized circuit element. If $i = 2.5$ A, then the conventional current (attributed to the motion of positive charges) is in the direction of the arrow. However, if $i = -4.0$ A, then the conventional current passes through the element in the opposite direction from that of the arrow.

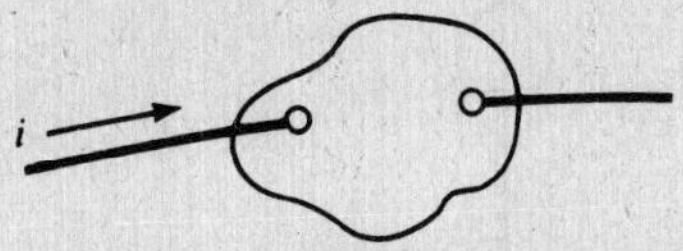

Fig. 2-5

2.4 SIGN CONVENTION

When current enters a circuit element at the +-marked terminal for the voltage v across the element, the power absorbed is $p = vi$. In Fig. 2-6(a), $v_a = 20$ V, $v_b = -15$ V, $v_c = 5$ V. At element A

$$p = -v_a i = -(20)(3) = -60 \text{ W}$$

Negative absorption is positive emission; consequently, element A must be a source. Elements B and C have 45 W and 15 W absorbed power, respectively. The circuit of Fig. 2-6(*b*), with two batteries and one resistor in series, corresponds exactly to the general circuit of Fig. 2-6(*a*). The 5-V battery is being charged at the rate of 15 joules per second.

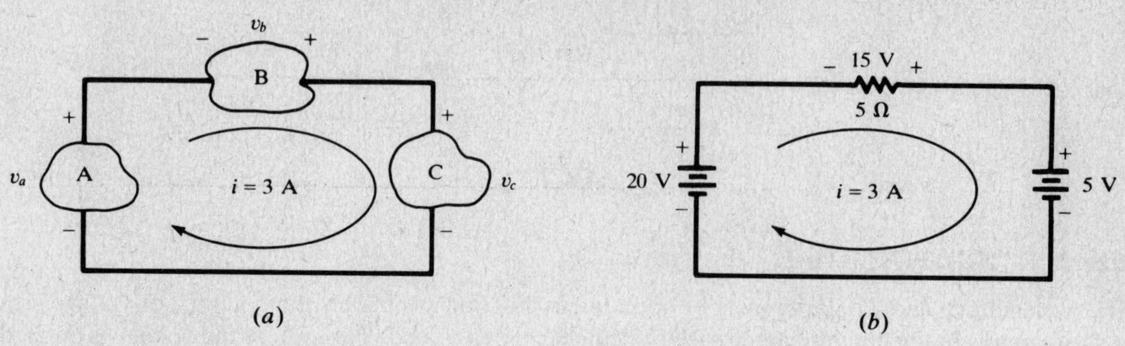

Fig. 2-6

At times, it is convenient to use the term *power delivered* rather than power absorbed. Power delivered by an active circuit element is the negative of the power absorbed.

2.5 CIRCUIT DIAGRAMS

Every circuit diagram can be constructed in a variety of ways which may look different but are in fact identical. The diagram presented in a problem may not suggest the best of several methods of solution. Consequently, a diagram should be examined before a solution is started and redrawn if necessary to show more clearly how the elements are interconnected. An extreme example is illustrated in Fig. 2-7, where the three circuits are actually identical. In Fig. 2-7(*a*) the three "junctions" labeled *A* are shown as two "junctions" in (*b*). However, resistor R_4 is bypassed by a short circuit and may be removed for purposes of analysis. Then, in Fig. 2-7(*c*) the single junction *A* is shown with its three meeting branches.

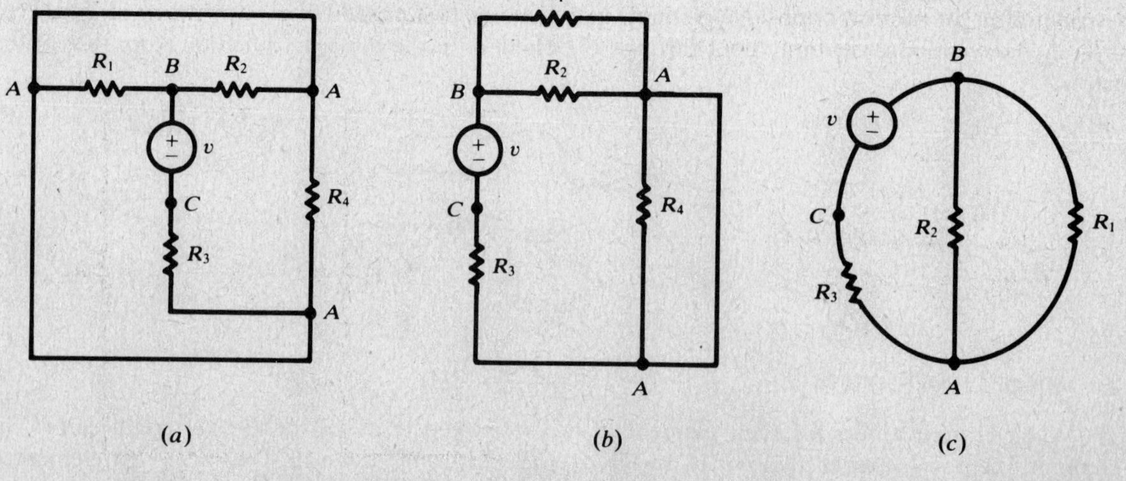

Fig. 2-7

2.6 VOLTAGE-CURRENT RELATIONS

The passive circuit elements resistance, inductance, and capacitance are conveniently defined by the manner in which voltage and current are related for the individual element. In Table 2-1, current is assumed to enter the element at the positive-marked voltage terminal, following the sign convention of Section 2.4.

Table 2-1

Circuit element	Units	Voltage	Current	Power
 **Resistance**	ohms (Ω)	$v = Ri$ **(Ohm's law)**	$i = \dfrac{v}{R}$	$p = vi = i^2R$
 **Inductance**	henries (H)	$v = L\dfrac{di}{dt}$	$i = \dfrac{1}{L}\int v\,dt + k_1$	$p = vi = Li\dfrac{di}{dt}$
 **Capacitance**	farads (F)	$v = \dfrac{1}{C}\int i\,dt + k_2$	$i = C\dfrac{dv}{dt}$	$p = vi = Cv\dfrac{dv}{dt}$

2.7 SERIES AND PARALLEL ELEMENTS

The three resistors in the series connection shown in Fig. 2-8(a) have the same current i. The voltage across the entire circuit is the sum of the individual voltages. A single *equivalent* resistor R_{eq} replaces the three resistors [Fig. 2-8(b)].

$$v = v_1 + v_2 + v_3$$
$$iR_{eq} = i(R_1 + R_2 + R_3)$$
$$R_{eq} = R_1 + R_2 + R_3$$

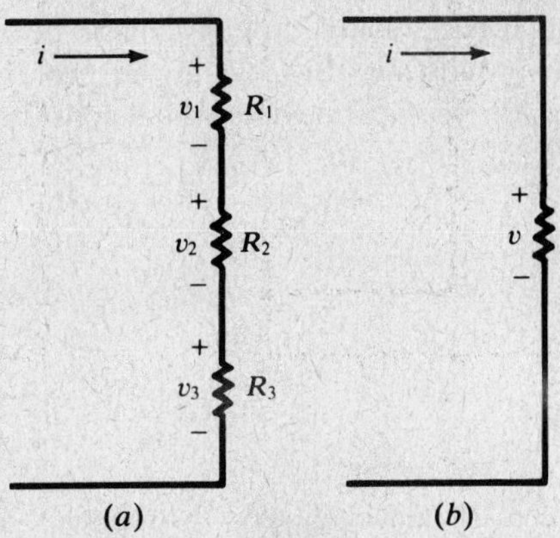

Fig. 2-8

In a parallel connection of three resistors [Fig. 2-9(*a*)], the total current is the sum of the three individual currents.

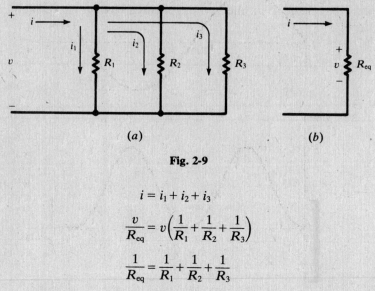

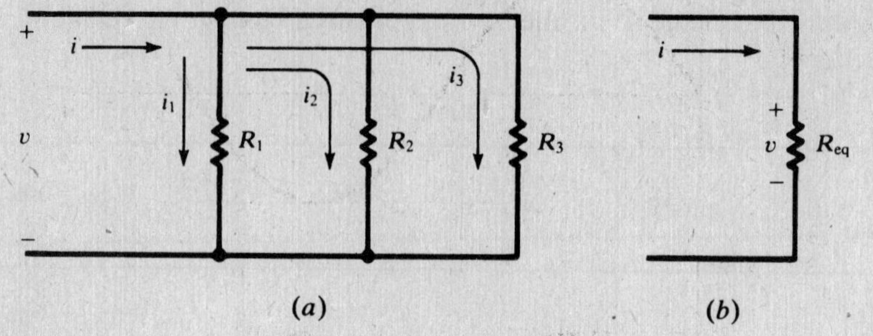

Fig. 2-9

$$i = i_1 + i_2 + i_3$$

$$\frac{v}{R_{eq}} = v\left(\frac{1}{R_1} + \frac{1}{R_2} + \frac{1}{R_3}\right)$$

$$\frac{1}{R_{eq}} = \frac{1}{R_1} + \frac{1}{R_2} + \frac{1}{R_3}$$

For two resistors in parallel,

$$R_{eq} = R_1R_2/(R_1 + R_2)$$

From Table 2-1, $v = Ri$ and $v = L\,di/dt$. Therefore, series and parallel combinations of inductances will have identical L_{eq}-expressions to those for resistance. Combinations of capacitances, because of the inverse relationship of v and C as shown in Table 2-1, will have expressions for series and parallel equivalents which are reversed. That is, two capacitors in parallel have $C_{eq} = C_1 + C_2$, which is of the same form as resistors in series, $R_{eq} = R_1 + R_2$. See Problems 2.10 and 2.12.

All the above results follow more fundamentally from a single condition: namely, that the power in the equivalent element be equal to the total power in the elements it replaces.

2.8 RESISTANCE

Energy can be stored in the electric field of a capacitor (see Problem 2.18) or the magnetic field of an inductor and be returned at a later time when the capacitor is discharged or the field of the inductor is allowed to collapse. In direct contrast, a resistor (also called a *resistance*) takes energy from the driving source which cannot be returned. Thus the average power for inductance and capacitance must be zero, while the power expression for a resistance, $p = i^2R$, is clearly always positive. All electrical devices which consume energy must have resistance in their circuit models.

EXAMPLE 2.1 A resistor of 4 Ω has a current $i = 2.5 \sin 500\pi t$ (A). Determine the instantaneous voltage, the instantaneous power, and the energy over one cycle.

$$v_R = Ri = 10 \sin 500\pi t \quad \text{(V)}$$

$$p = Ri^2 = 25 \sin^2 500\pi t \quad \text{(W)}$$

$$w = \int_0^t p\,dt = 25\left(\frac{t}{2} - \frac{\sin 1000\pi t}{2000\pi}\right) \quad \text{(J)}$$

The plots of i, p, and w in Fig. 2-10 illustrate that p is always positive. It follows that the energy increases at all times; this is the energy absorbed by the resistor.

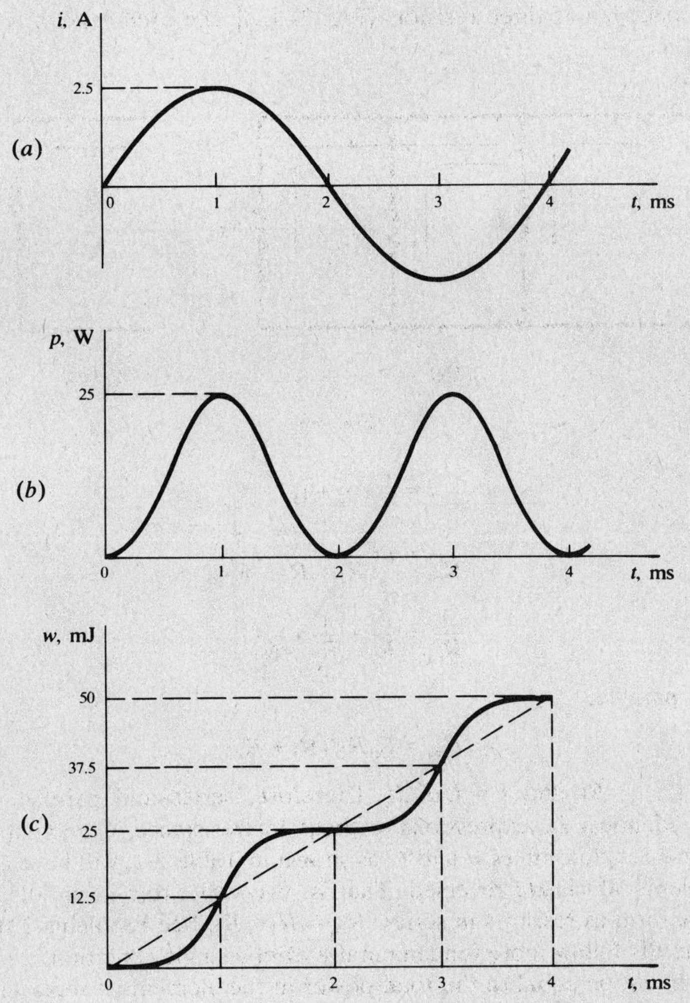

Fig. 2-10

2.9 INDUCTANCE

An inductor (also called an *inductance*) is a circuit element that stores energy during some periods of time and returns it during others, such that the average power is zero. The coil windings in electric motors, transformers, and similar devices have inductances in their circuit models. From Table 2-1, the current in an inductance is given by a time-integral of the applied voltage; this current must therefore be a continuous function of time. In particular, if $i = 0$ at $t = 0^-$, it is not possible for i at $t = 0^+$ to have any value other than zero.

EXAMPLE 2.2 In the interval $0 \le t \le (\pi/50)$ s an inductance of 30 mH has a current $i = 10 \sin 50t$ (A). At all other times the current is zero. Obtain the voltage, power, and energy for the inductance.

$$v_L = L\frac{di}{dt} = 15 \cos 50t \quad (\text{V}) \qquad p = v_L i = 75 \sin 100t \quad (\text{W}) \qquad w = \int_0^t p\,dt + w(0)$$

If $w(0) = 0$,

$$w = 0.75(1 - \cos 100t) \quad (\text{J})$$

As is shown by Fig. 2-11, the given current results in a maximum stored energy of 1.50 J at $t = (\pi/100)$ s. During the next $(\pi/100)$ s this energy returns to the source, and the stored energy is zero again at $t = (\pi/50)$ s.

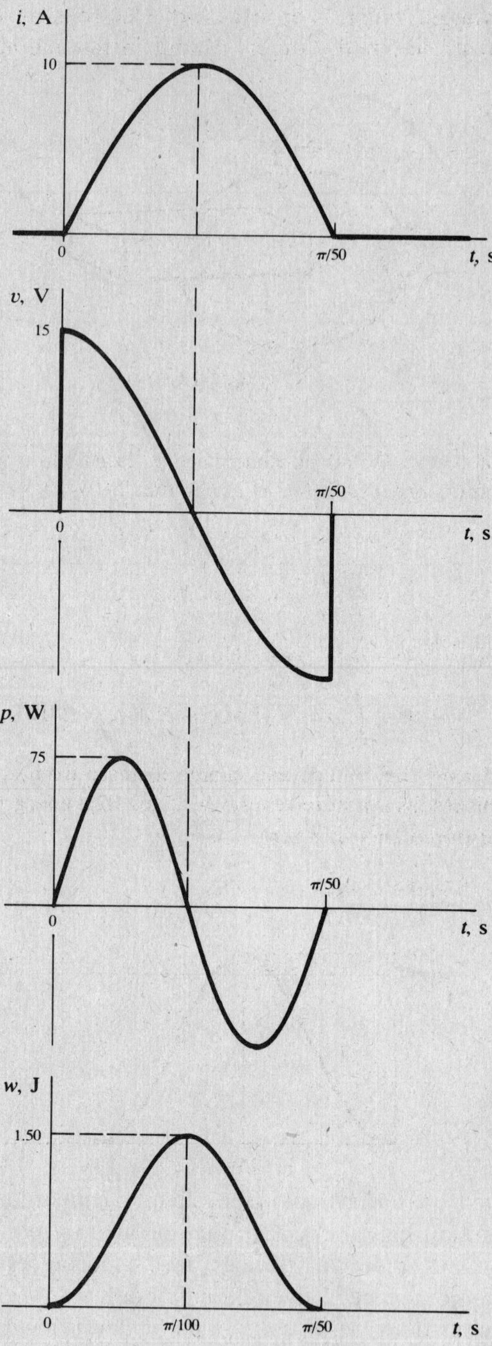

Fig. 2-11

2.10 CAPACITANCE

A capacitor (also called a *capacitance*) is a circuit element that, like the inductor, stores and returns energy. In the capacitor, storage takes place in an electric field, whereas the storage in the inductor is in a magnetic field. The capacitor voltage is a time-integral of the current (Table 2-1); thus v_C must be a continuous function of time. Charge on the capacitor is directly proportional to the voltage: $q = Cv_C$. This provides the best evidence that the voltage cannot change abruptly from one value to another; for the charge, which is an excess of electrons on the negative conductor in Fig. 2-12 and a deficiency of electrons on the other, cannot instantaneously change from one value to another.

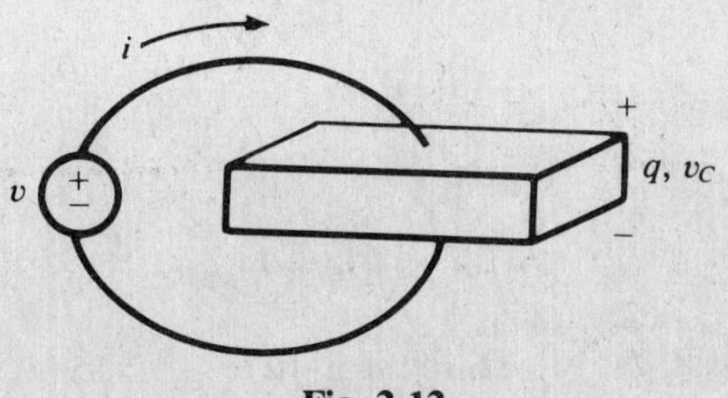

Fig. 2-12

EXAMPLE 2.3 In the interval $0 \le t \le 5\pi$ ms, a capacitance of 20 μF has a voltage $v_C = 50 \sin 200t$ (V). Find the charge, current, power, and energy. Plot $w(t)$, assuming that $w(0) = 0$.

$$q = Cv_C = 1000 \sin 200t \quad (\mu\text{C})$$

$$i = C\frac{dv_C}{dt} = 0.20 \cos 200t \quad (\text{A})$$

$$p = v_C i = 5 \sin 400t \quad (\text{W})$$

$$w = \int_0^t p\,dt = 12.5(1 - \cos 400t) \quad (\text{mJ})$$

In the interval $0 \le t \le 2.5\pi$ ms the voltage and charge increase from zero to 50 V and 1000 μC, respectively. Figure 2-13 shows that this results in a stored energy of 25 mJ. This energy is returned to the source in the next 2.5π-ms interval, and the final stored energy is zero.

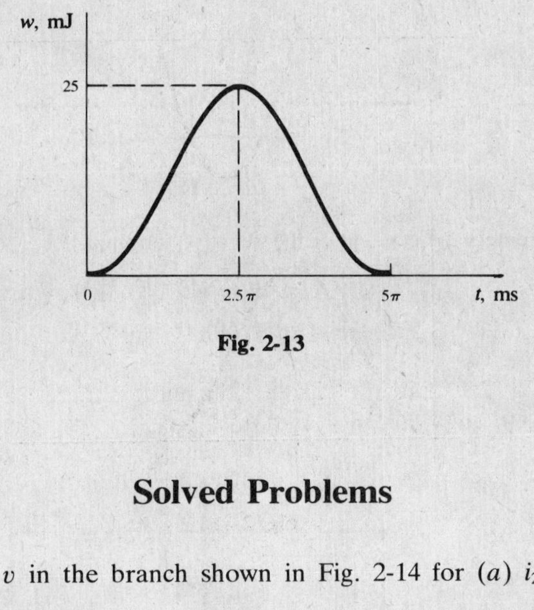

Fig. 2-13

Solved Problems

2.1 Obtain the voltage v in the branch shown in Fig. 2-14 for (*a*) $i_2 = 1$ A, (*b*) $i_2 = -2$ A, (*c*) $i_2 = 0$ A.

Voltage v is the sum of the current-independent 10-V source and the current-dependent voltage source v_x. Note that the factor 15 multiplying the control current carries the units Ω.

(*a*) $$v = 10 + v_x = 10 + 15(1) = 25 \text{ V}$$

(*b*) $$v = 10 + v_x = 10 + 15(-2) = -20 \text{ V}$$

(*c*) $$v = 10 + 15(0) = 10 \text{ V}$$

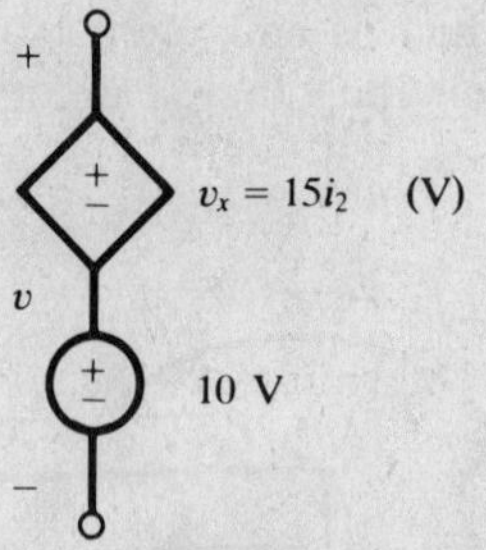

Fig. 2-14

2.2 Find the voltage v across the 10-Ω resistor in Fig. 2-15, if the control current i_1 in the dependent current source is (*a*) 2 A, (*b*) −1 A.

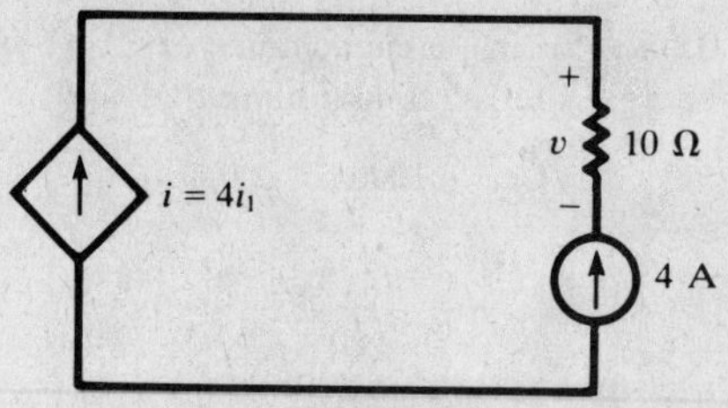

Fig. 2-15

(*a*) $$v = (i - 4)10 = [4(2) - 4]10 = 40 \text{ V}$$

(*b*) $$v = (i - 4)10 = [4(-1) - 4]10 = -80 \text{ V}$$

2.3 Find the power absorbed by the generalized circuit element in Fig. 2-16, for (*a*) $v = 50$ V, (*b*) $v = -50$ V.

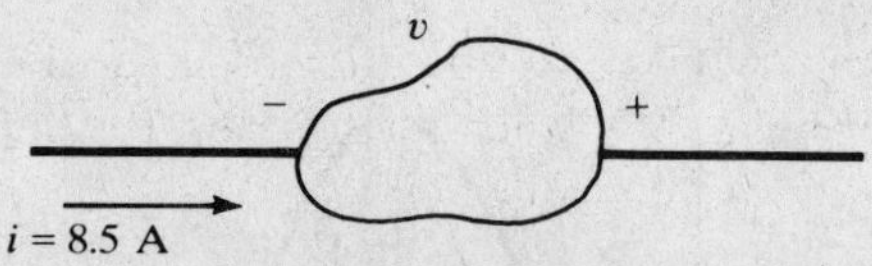

Fig. 2-16

Since the current enters the element at the negative terminal,

(*a*) $$p = -vi = -(50)(8.5) = -425 \text{ W}$$

(*b*) $$p = -vi = -(-50)(8.5) = 425 \text{ W}$$

2.4 For the circuit of Fig. 2-17 find the currents, in microamperes, if the powers absorbed are: $p_1 = 27.75$ nW, $p_2 = -0.3\ \mu$W, and $p_3 = 1.20\ \mu$W.

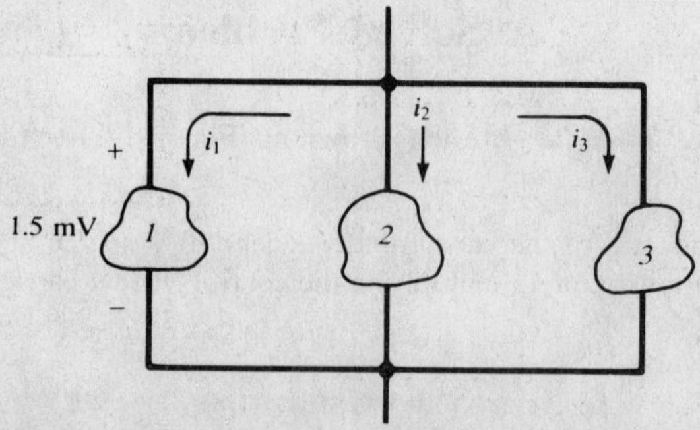

Fig. 2-17

$$p_1 = vi_1 \quad \text{or} \quad 27.75 \times 10^{-9} = 1.5 \times 10^{-3} i_1 \quad \text{or} \quad i_1 = 1.85 \times 10^{-5}\ \text{A} = 18.5\ \mu\text{A}$$

$$p_2 = vi_2 \quad \text{or} \quad -0.30 \times 10^{-6} = 1.5 \times 10^{-3} i_2 \quad \text{or} \quad i_2 = -2 \times 10^{-4}\ \text{A} = -200\ \mu\text{A}$$

$$p_3 = vi_3 \quad \text{or} \quad 1.20 \times 10^{-6} = 1.5 \times 10^{-3} i_3 \quad \text{or} \quad i_3 = 0.8 \times 10^{-3}\ \text{A} = 800\ \mu\text{A}$$

2.5 Find the power delivered by the sources in the circuit of Fig. 2-18.

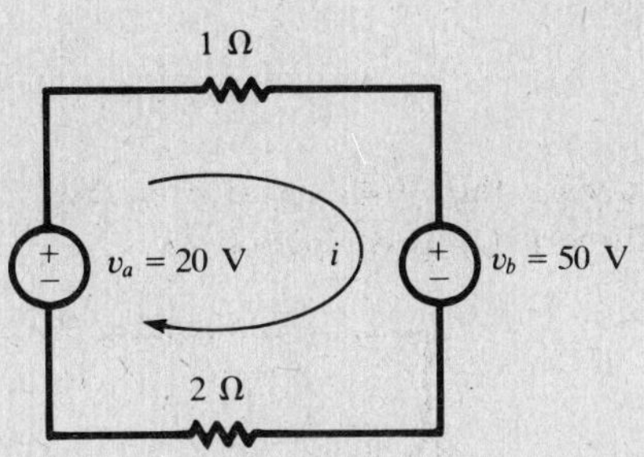

Fig. 2-18

$$i = \frac{20 - 50}{3} = -10\ \text{A}$$

The powers *absorbed by* the sources are:

$$p_a = -v_a i = -(20)(-10) = 200\ \text{W}$$

$$p_b = v_b i = (50)(-10) = -500\ \text{W}$$

Since power delivered is the negative of power absorbed, source v_a delivers -200 W and v_b delivers 500 W.

2.6 (*a*) Obtain the equivalent resistance for the circuit shown in Fig. 2-19(*a*). (*b*) With a constant voltage applied to terminals *ab*, which resistor absorbs the greatest power?

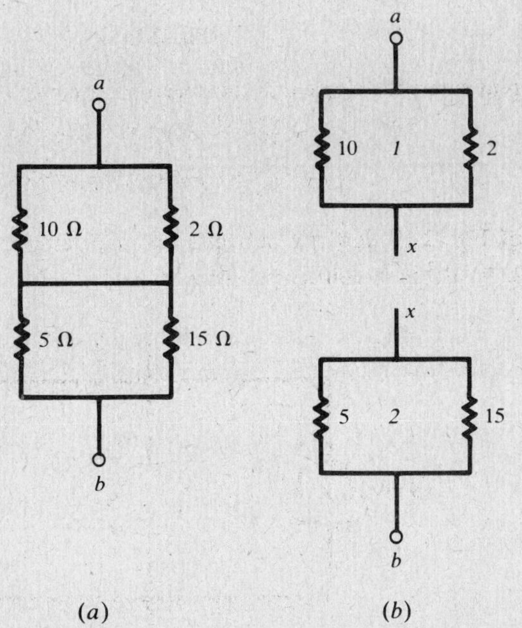

Fig. 2-19

(*a*) Decompose the circuit into series subcircuits *1* and *2*, as shown in Fig. 2-19(*b*). Then,

$$R_{eq1} = \frac{2(10)}{12} = \frac{20}{12}\ \Omega$$

$$R_{eq2} = \frac{5(15)}{20} = \frac{45}{12}\ \Omega$$

$$R_{eq} = R_{eq1} + R_{eq2} = \frac{65}{12}\ \Omega$$

(*b*) There is a single voltage difference across subcircuit *1*, and a single voltage difference across subcircuit *2*. Hence,

$$\frac{P_{10}}{P_2} = \frac{2}{10} \qquad \text{and} \qquad \frac{P_5}{P_{15}} = \frac{15}{5}$$

from which the maximum power must be the larger of P_2 and P_5. But, since the same current flows through the two subcircuits and $P = I^2R_{eq}$,

$$\frac{P_{10}+P_2}{P_5+P_{15}} = \frac{R_{eq1}}{R_{eq2}}$$

$$\frac{P_2}{P_5}\frac{(P_{10}/P_2)+1}{1+(P_{15}/P_5)} = \frac{20}{45}$$

$$\frac{P_2}{P_5}\frac{(2/10)+1}{1+(5/15)} = \frac{20}{45}$$

$$\frac{P_2}{P_5} = \frac{40}{81}$$

The 5-Ω resistor absorbs the greatest power.

2.7 Find the current I supplied by the 50-V source in Fig. 2-20(*a*).

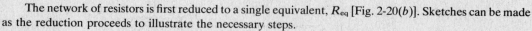

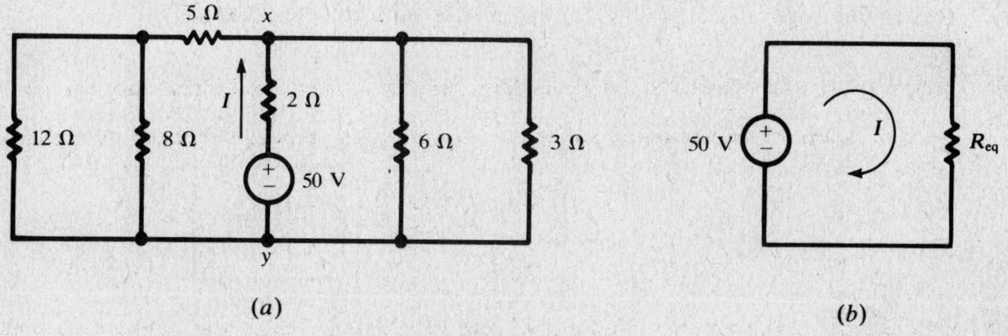

Fig. 2-20

The network of resistors is first reduced to a single equivalent, R_{eq} [Fig. 2-20(*b*)]. Sketches can be made as the reduction proceeds to illustrate the necessary steps.

The resistance to the left of branch xy is

$$R_{eq1} = 5 + \frac{8(12)}{20} = 9.80\ \Omega$$

and to the right, $R_{eq2} = 6(3)/9 = 2\ \Omega$. These two parts are in parallel with each other, and in series with the 2 Ω. Thus,

$$R_{eq} = 2 + \frac{(9.80)(2)}{11.80} = 3.66\ \Omega$$

and $I = 50/3.66 = 13.7$ A.

2.8 A resistance of 25 Ω has a voltage $v = 150 \sin 377t$ (V). Obtain the current i and the power p.

From Table 2-1,

$$i = \frac{v}{R} = 6 \sin 377t \quad \text{(A)} \qquad p = vi = 900 \sin^2 377t \quad \text{(W)}$$

In a resistance the voltage and current are proportional; expressed as functions of time, they differ only in the amplitude. The waveforms shown in Fig. 2-10(a) and (b) are similar to those for this problem.

2.9 The current in a 5-Ω resistor increases linearly from zero to 10 A in 2 ms. At $t = 2^+$ ms the current is again zero, and it increases linearly to 10 A at $t = 4$ ms. This pattern repeats each 2 ms. Sketch the corresponding v.

Since $v = Ri$, the voltage maximum must be (5)(10) = 50 V. In Fig. 2-21 the plots of i and v are shown. The identical nature of the functions is evident.

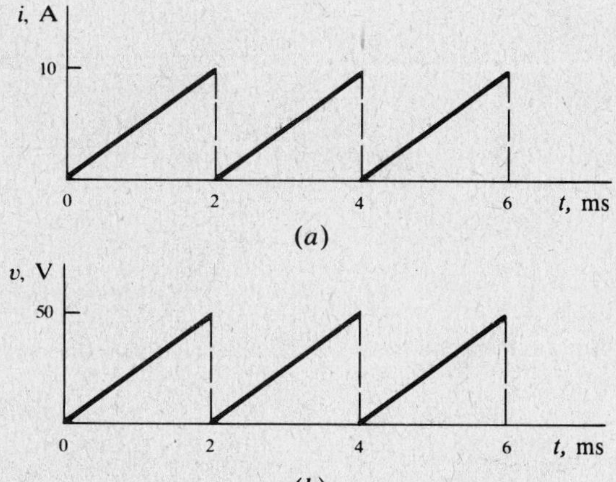

Fig. 2-21

2.10 Develop the expression for the equivalent inductance of several inductances in series.

The power in the equivalent inductance must equal the sum of the powers in the inductances in series.

$$L_{eq} i \frac{di}{dt} = L_1 i \frac{di}{dt} + L_2 i \frac{di}{dt} + \cdots + L_n i \frac{di}{dt}$$

whence $L_{eq} = L_1 + L_2 + \cdots + L_n$.

2.11 In Fig. 2-22 there are two sources, v_a and v_b. What is L_{eq} of the network as seen by each source?

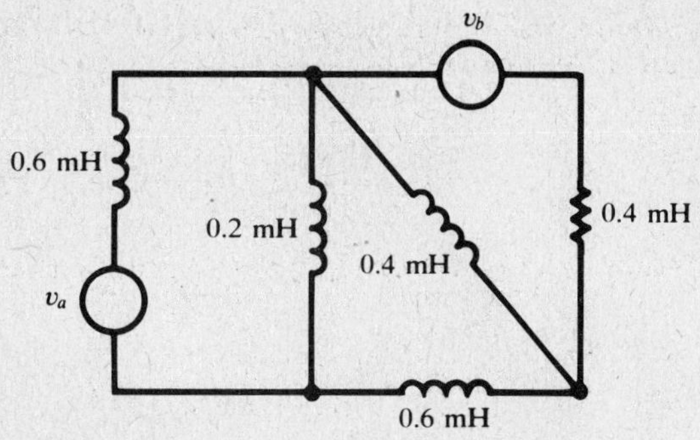

Fig. 2-22

To find L_{eq} from source v_a, remove v_b and replace it with a short circuit. Then

$$L_{eqa} = 0.6 + \frac{\left[\frac{(0.4)(0.4)}{0.8} + 0.6\right](0.2)}{\left[\frac{(0.4)(0.4)}{0.8} + 0.6\right] + 0.2} = 0.760 \text{ mH}$$

Similarly, $L_{eqb} = 0.661$ mH.

2.12 Develop the expression for the equivalent capacitance of several capacitances in series.

Assuming there are no initial charges on the capacitors, the voltages are:

$$v_1 = \frac{1}{C_1}\int i\,dt \qquad v_2 = \frac{1}{C_2}\int i\,dt \qquad \cdots$$

and so the powers are:

$$p_1 = \frac{1}{C_1} i \int i\,dt \qquad p_2 = \frac{1}{C_2} i \int i\,dt \qquad \cdots$$

Therefore, the total power is

$$\frac{1}{C_1} i \int i\,dt + \frac{1}{C_2} i \int i\,dt + \cdots = \frac{1}{C_{eq}} i \int i\,dt$$

whence
$$\frac{1}{C_{eq}} = \frac{1}{C_1} + \frac{1}{C_2} + \cdots$$

2.13 Determine the equivalent capacitance of the four capacitances shown in Fig. 2-23.

$$C_{eq} = \frac{(0.2+0.8)(0.7+0.3)}{(0.2+0.8)+(0.7+0.3)} = 0.5\ \mu\text{F}$$

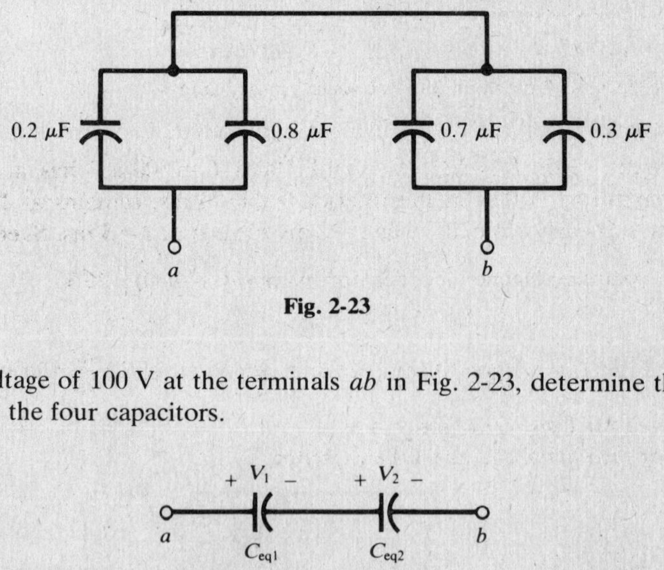

Fig. 2-23

2.14 For a constant voltage of 100 V at the terminals *ab* in Fig. 2-23, determine the corresponding charge on each of the four capacitors.

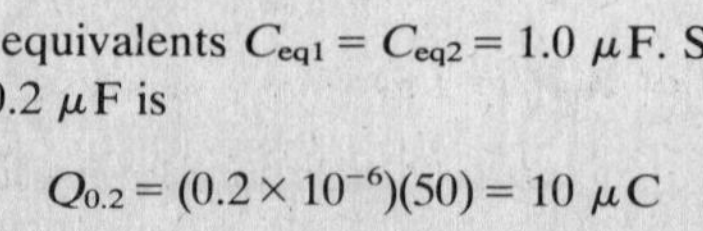

Fig. 2-24

The two parallel circuits have equivalents $C_{eq1} = C_{eq2} = 1.0\ \mu$F. See Fig. 2-24. Thus $V_1 = V_2 = 50$ V. Since $Q = CV$, the charge on the 0.2 μF is

$$Q_{0.2} = (0.2 \times 10^{-6})(50) = 10\ \mu\text{C}$$

Similarly, $Q_{0.8} = 40\ \mu$C, $Q_{0.3} = 15\ \mu$C, and $Q_{0.7} = 35\ \mu$C.

2.15 A single circuit element has the current and voltage functions graphed in Fig. 2-25. Determine the element.

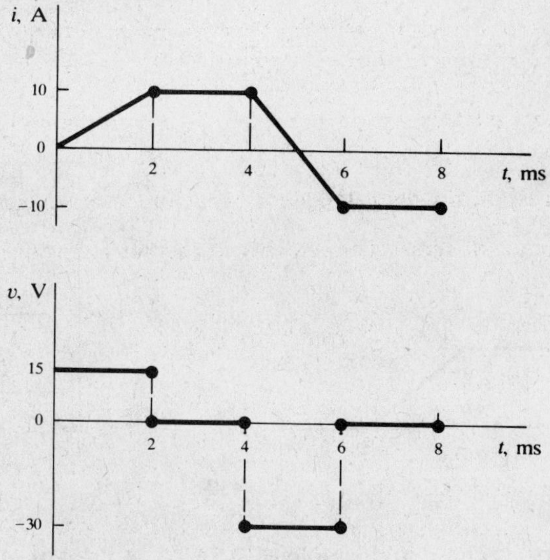

Fig. 2-25

The element cannot be a resistor since v and i are not proportional. v is an integral of i. For $2\text{ ms} < t < 4\text{ ms}$, $i \neq 0$ but v is constant (zero); hence the element cannot be a capacitor. For $0 < t < 2$ ms,

$$\frac{di}{dt} = 5 \times 10^3 \text{ A/s} \qquad \text{and} \qquad v = 15 \text{ V}$$

Consequently,

$$L = v \Big/ \frac{di}{dt} = 3 \text{ mH}$$

(Examine the interval $4\text{ ms} < t < 6$ ms: L must be the same.)

2.16 A series circuit with $R = 2\ \Omega$, $L = 2$ mH, and $C = 500\ \mu$F has a current which increases linearly from zero to 10 A in the interval $0 \leq t \leq 1$ ms, remains at 10 A for $1\text{ ms} \leq t \leq 2$ ms, and decreases linearly from 10 A at $t = 2$ ms to zero at $t = 3$ ms. Sketch v_R, v_L, and v_C.

v_R must be a time function identical to i, with $V_{max} = 2(10) = 20$ V.
For $0 < t < 1$ ms,

$$\frac{di}{dt} = 10 \times 10^3 \text{ A/s} \qquad \text{and} \qquad v_L = L\frac{di}{dt} = 20 \text{ V}$$

When $di/dt = 0$, for $1\text{ ms} < t < 2$ ms, $v_L = 0$.

Assuming zero initial charge on the capacitor,

$$v_C = \frac{1}{C}\int i\,dt$$

For $0 \leq t \leq 1$ ms,

$$v_C = \frac{1}{5 \times 10^{-4}}\int_0^t 10^4 t\,dt = 10^7 t^2 \quad \text{(V)}$$

This voltage reaches a value of 10 V at 1 ms. For $1\text{ ms} < t < 2$ ms,

$$v_C = (20 \times 10^3)(t - 10^{-3}) + 10 \quad \text{(V)}$$

See Fig. 2-26.

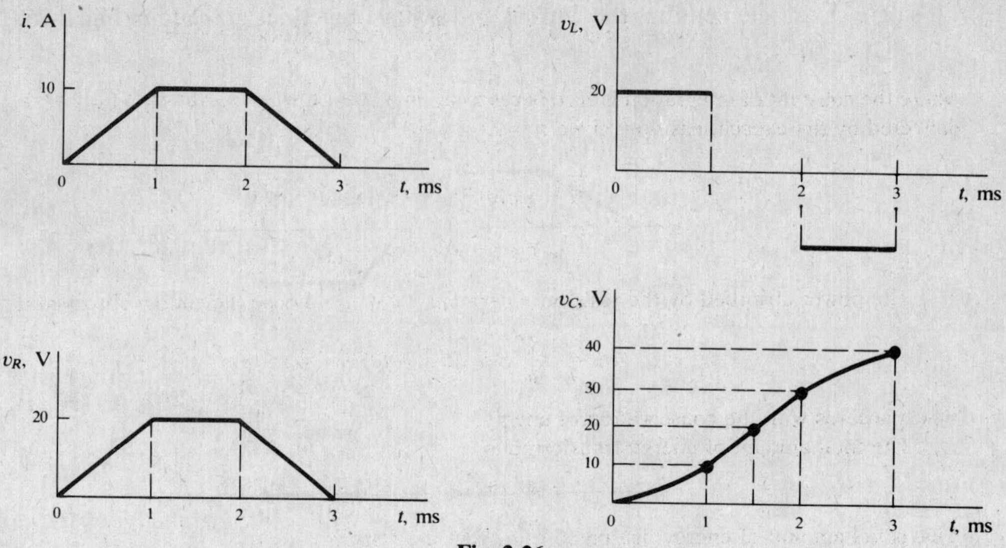

Fig. 2-26

2.17 A 25-μF capacitor has an initial charge of 2.5 μC. In the interval $0 < t < 1\ \mu$s,

$$i = 5 \times 10^6 t - 5 \quad \text{(A)}$$

Obtain the voltage and the charge at (*a*) $t = 0.5\ \mu$s, (*b*) $t = 0.9\ \mu$s.

$$q = \int_0^t i\,dt + k = \int_0^t (5 \times 10^6 t - 5)\,dt + 2.5 \times 10^{-6} = 2.5 \times 10^6 t^2 - 5t + 2.5 \times 10^{-6} \quad \text{(C)}$$

and

$$v = \frac{q}{25 \times 10^{-6}} \quad \text{(V)}$$

(*a*) At $t = 0.5\ \mu$s: $q = 0.625\ \mu$C, $v = 25$ mV.

(*b*) At $t = 0.9\ \mu$s: $q = 25$ nC, $v = 1$ mV.

2.18 The capacitor in the circuit shown in Fig. 2-27 has an initial voltage of 100 V which decreases for $t > 0$ as

$$v_C = 100e^{-t/0.015} \quad \text{(V)}$$

Find the energy delivered by the capacitor, as a function of time. Compare with the energy absorbed by the resistor, and determine the total energy transferred.

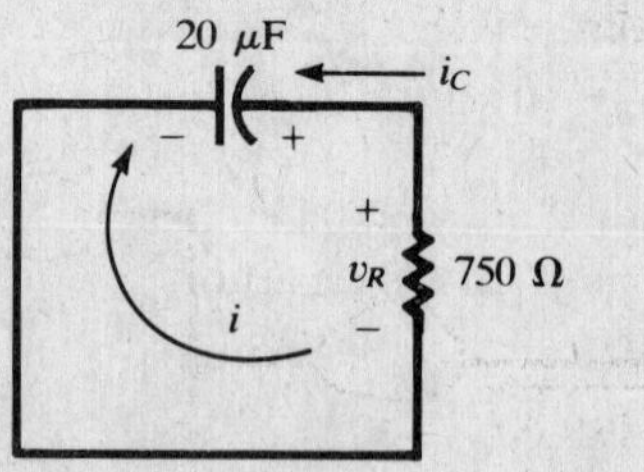

Fig. 2-27

Current i_C enters the capacitor by the + terminal, while clockwise i is of opposite sign. We have:

$$i_C = C\frac{dv_C}{dt}$$

and so the energy absorbed by the capacitor—i.e., the energy stored in its electric field—is

$$w_C = \int v_C i_C \, dt = \int C v_C \frac{dv_C}{dt} \, dt = C \int d(v_C^2/2) = \frac{C v_C^2}{2}$$

where the constant of integration is zero because w_C must vanish when v_C vanishes (at $t = \infty$). The energy delivered by the capacitor is then given by

$$w'_C = w_C(0) - w_C = \frac{C}{2}[v_C^2(0) - v_C^2]$$

$$= 0.10(1 - e^{-t/0.0075}) \quad \text{(J)}$$

The power absorbed by the resistor is $p_R = v_R i = v_C(-i_C)$; hence the energy absorbed is

$$w_R = -\int_0^t v_C i_C \, dt = w_C(0) - w_C(t) = w'_C$$

which accords with the conservation of energy.

The total amount of energy transferred is

$$w'_C(\infty) = w_C(0) = 0.10 \text{ J}$$

This is the amount of energy dissipated as heat in the resistor.

Supplementary Problems

2.19 Find voltages v_1, v_2, and v_3 in Fig. 2-28, given that the generalized circuit elements have absorbed powers $p_1 = 250$ W, $p_2 = -125$ W and $p_3 = 100$ W. *Ans.* 50 V, −25 V, −20 V

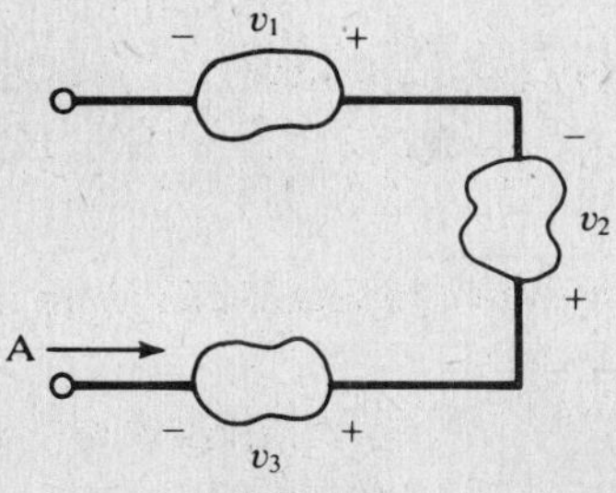

Fig. 2-28

2.20 In the circuit shown in Fig. 2-29, $v_1 = 20$ V and $v_2 = v_3 = 15$ V. What is the total power absorbed in the three circuit elements? *Ans.* 50 W

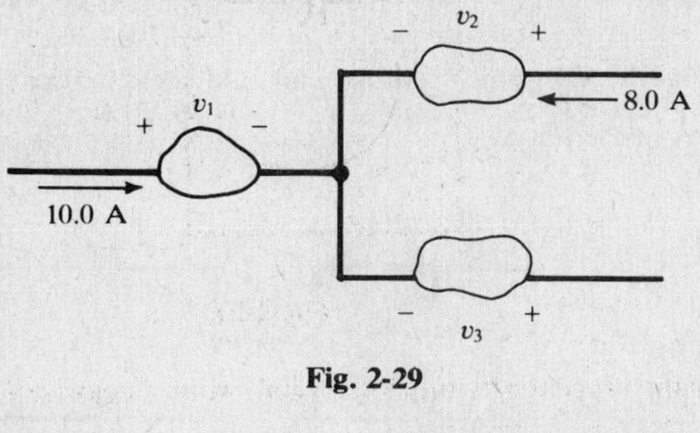

Fig. 2-29

2.21 Consider a current i entering a generalized circuit element at the negative terminal, with v the voltage across the element. If the absorbed power is −25 mW and the current magnitude 6.4 mA, obtain the possible combinations of constant v and i. *Ans.* ±3.91 V, ±6.4 mA

2.22 Find the current i in the circuit shown in Fig. 2-30, if the control v_2 of the dependent voltage source has the value (*a*) 4 V, (*b*) 5 V, (*c*) 10 V. *Ans.* 1 A, 0 A, −5 A

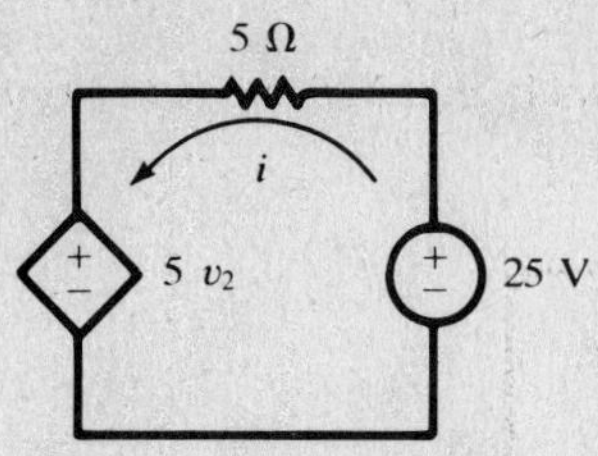

Fig. 2-30

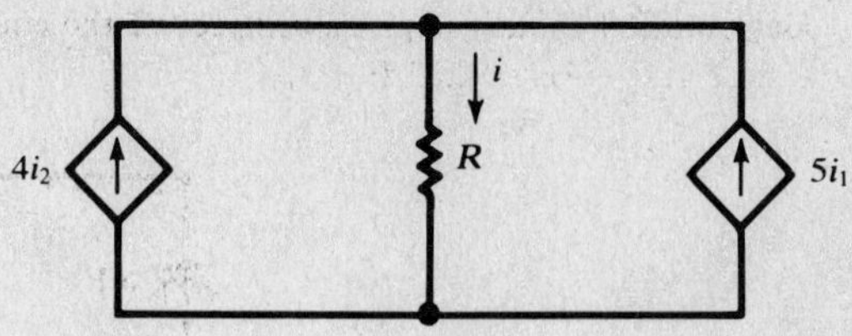

Fig. 2-31

2.23 In the circuit shown in Fig. 2-31, find the current i, given (*a*) $i_1 = 2$ A, $i_2 = 0$; (*b*) $i_1 = -1$ A, $i_2 = 4$ A; (*c*) $i_1 = i_2 = 1$ A. *Ans.* (*a*) 10 A; (*b*) 11 A; (*c*) 9 A

2.24 Three resistors are in series and have a total constant voltage V_T. R_1 has a voltage of 20 V, R_2 has a power of 25 W, and $R_3 = 2\ \Omega$. If the constant current is 5 A, find V_T. *Ans.* 35 V

2.25 A 2-Ω resistor is in series with a parallel combination of three resistors, 10 Ω, 10 Ω, and 5 Ω. (*a*) If the 5-Ω resistor has a constant current of 14 A, what is the total voltage V_T across the entire circuit? (*b*) Compare the ratio V_T/I_2 to the equivalent resistance. *Ans.* (*a*) 126 V; (*b*) both 4.50 Ω

2.26 In the circuit shown in Fig. 2-32, the adjustable resistor R is set such that the power in the 5-Ω resistor is 20 W. (*a*) Find R. (*b*) Find the maximum current and power delivered by the source as R is adjusted. *Ans.* (*a*) 16 Ω; (*b*) 12.5 A, 625 W

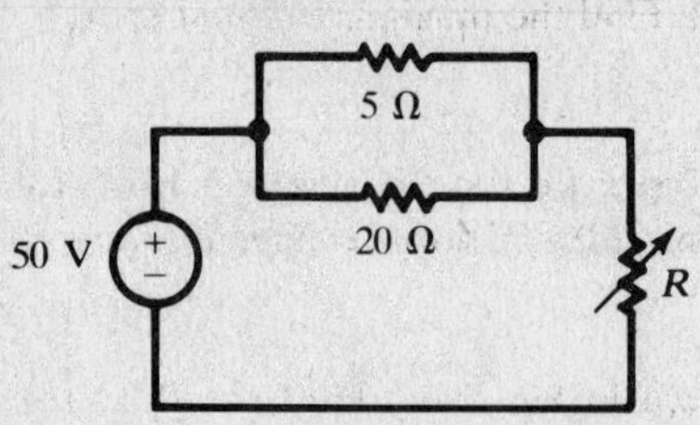

Fig. 2-32

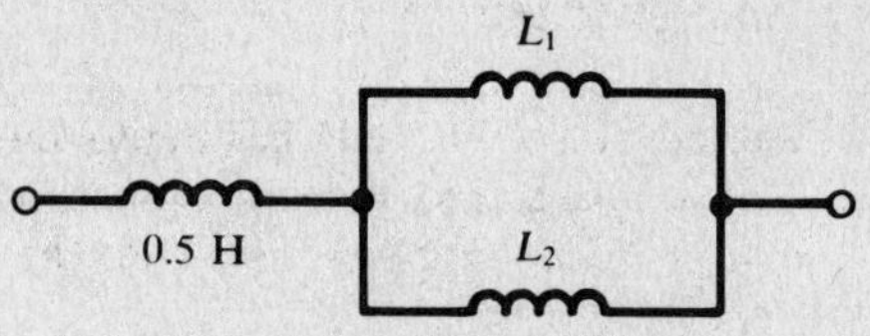

Fig. 2-33

2.27 In Fig. 2-33, $L_1 = 2L_2$. Find L_1 and L_2 if the equivalent inductance of the three inductances is 0.70 H. *Ans.* 0.60 H, 0.30 H

2.28 Three inductances are in parallel: 0.50 H, 0.80 H and L. (*a*) Find L if the equivalent inductance is 75.5 mH. (*b*) Is there a value of L for which $L_{eq} = 0.50$ H? (*c*) If L is adjustable without limit, what is the maximum possible value of L_{eq}? *Ans.* (*a*) 0.10 H; (*b*) no; (*c*) 0.308 H

2.29 Find the value of C in Fig. 2-34 which results in an equivalent capacitance of 0.5 μF. *Ans.* 0.4 μF

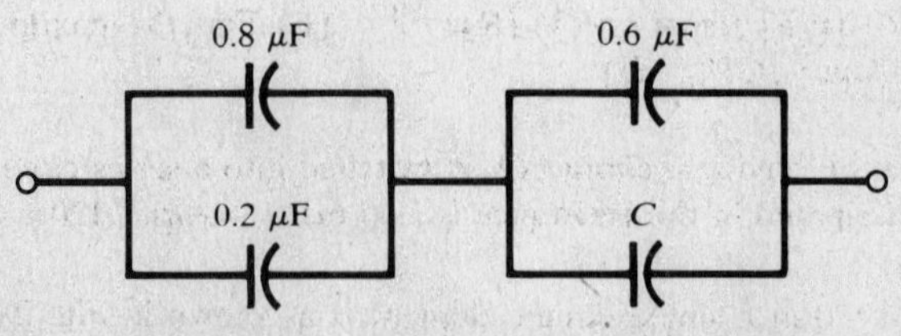

Fig. 2-34

2.30 A series combination of the two capacitances $C_1 = 20\ \mu F$ and $C_2 = 40\ \mu F$ is charged by a momentary connection to a source of 50 V. Find the voltages V_1 and V_2 and the charges Q_1 and Q_2 on the capacitors. *Ans.* 33.3 V, 16.7 V; 667 μC, 667 μC

2.31 Determine the equivalent capacitance of the combination shown in Fig. 2-35, as viewed from the voltage source v. *Ans.* 2.85 F

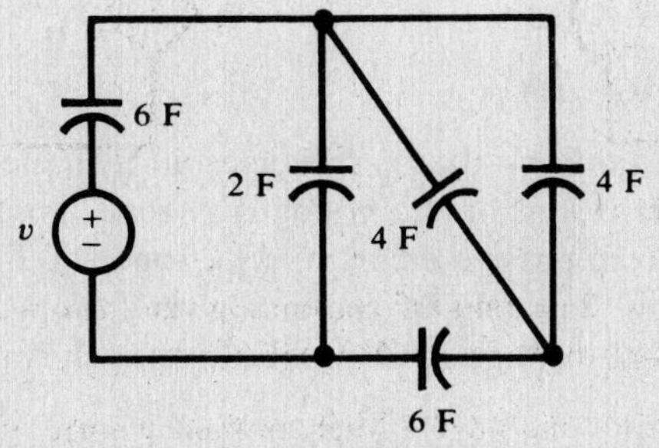

Fig. 2-35

2.32 A resistance of 25 Ω has a voltage $v = 150 \sin 377t$ (V). Find the instantaneous and the average power. *Ans.* $900 \sin^2 377t$ (W), 450 W

2.33 A resistance of 5 Ω has, for $0 < t < 2$ ms, a current $i = 5 \times 10^3 t$ (A). Find the instantaneous and the average power. *Ans.* $125t^2$ (W), for t in ms; 167 W

2.34 An inductance L (H) has a current $i = I(1 - e^{-Rt/L})$ (A). Find the maximum stored energy. *Ans.* $\frac{1}{2}LI^2$ (J)

2.35 An inductance of 3 mH has the following voltage waveform: for $0 < t < 2$ ms, $v = 15$ V; for 2 ms $< t <$ 4 ms, $v = 0$; and for 4 ms $< t < 6$ ms, $v = -30$ V. Assuming $i(0) = 0$, find the current at time (*a*) 1 ms, (*b*) 4 ms, (*c*) 5 ms. *Ans.* (*a*) 5 A; (*b*) 10 A; (*c*) 0

2.36 A capacitance of 60 μF has the voltage waveform shown in Fig. 2-36. Find (*a*) P_{max}, (*b*) i and p at $t = 3$ ms. *Ans.* (*a*) 75 W; (*b*) −1.5 A, −37.5 W

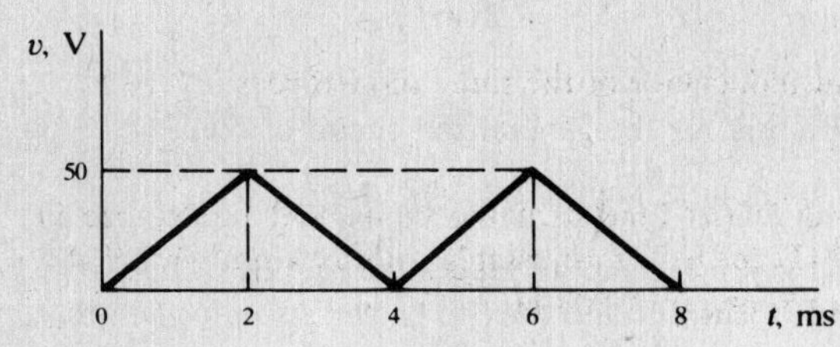

Fig. 2-36

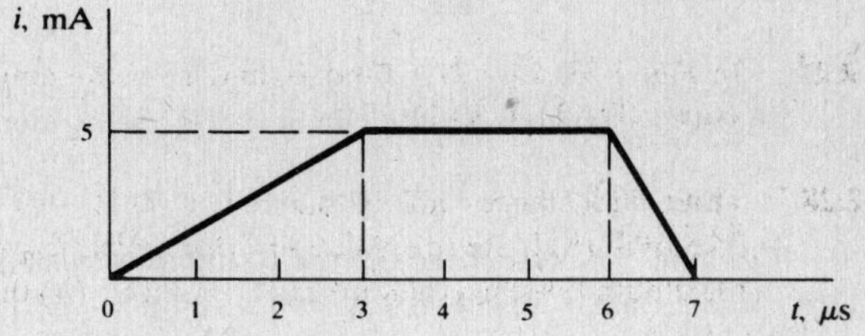

Fig. 2-37

2.37 A capacitance C (F) has a current $i = (V_m/R)e^{-t/RC}$ (A). Find the maximum stored energy, assuming zero initial charge. *Ans.* $\frac{1}{2}CV_m^2$ (J)

2.38 A capacitance of 2 μF having a charge Q_0 is switched into a series circuit of 10 Ω resistance. Find Q_0 if the total energy dissipated in the resistance is 3.60 mJ. *Ans.* 120 μC

2.39 The current after $t = 0$ in a single circuit element is as shown in Fig. 2-37. Find the voltage across the element at $t = 6.5\ \mu$s, if the element is (*a*) 10 kΩ, (*b*) 15 mH, (*c*) 0.3 nF with $Q(0) = 0$. *Ans.* (*a*) 25 V; (*b*) −75 V; (*c*) 81.3 V

Chapter 3

DC Resistive Circuits

3.1 KIRCHHOFF'S VOLTAGE LAW

For any closed path in a network that is traversed in a single direction, *Kirchhoff's voltage law* (*KVL*) states that the algebraic sum of the voltages is zero. Some of the voltages may be voltage sources, while others result from passive elements (Section 2.1). For dc resistive circuits, these latter voltages will be of the form $V = IR$. In traversing the loop, if an element is entered at the negative-potential end, then the voltage is taken negative in the sum.

EXAMPLE 3.1 Starting at the lower left corner of the single-loop circuit of Fig. 3-1 and applying KVL to a clockwise path of elements results in the following equation:

$$-V_a + V_1 + V_b + V_2 + V_3 = 0$$

An equation can be written for a closed path such as *mnom* in Fig. 3-1 by introducing voltage V_{om}, where *o* is assumed positive with respect to *m*. Again starting at the lower left corner,

$$-V_a + V_1 + V_{om} = 0$$

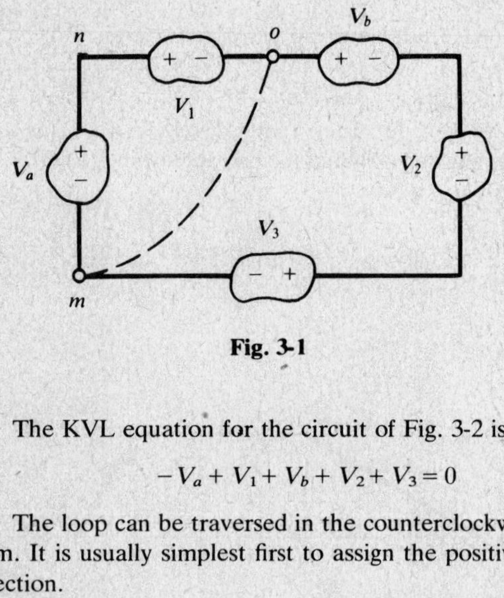

Fig. 3-1

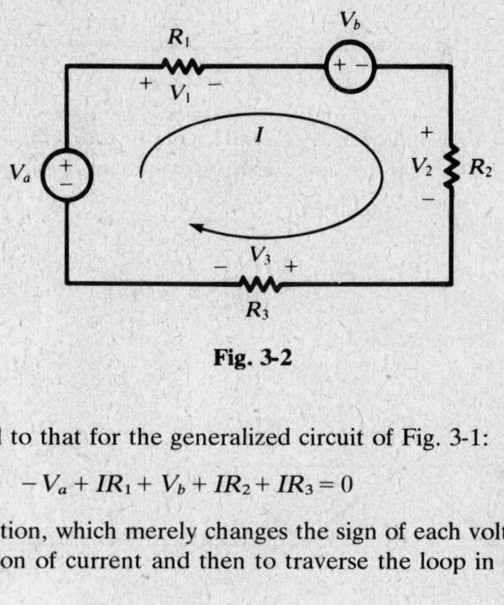

Fig. 3-2

The KVL equation for the circuit of Fig. 3-2 is identical to that for the generalized circuit of Fig. 3-1:

$$-V_a + V_1 + V_b + V_2 + V_3 = 0 \qquad \text{or} \qquad -V_a + IR_1 + V_b + IR_2 + IR_3 = 0$$

The loop can be traversed in the counterclockwise direction, which merely changes the sign of each voltage term. It is usually simplest first to assign the positive direction of current and then to traverse the loop in that direction.

3.2 KIRCHHOFF'S CURRENT LAW

The connection of two or more circuit elements creates a junction called a *node*. A junction of two elements is a *simple node*; a junction of three or more elements is a *principal node*. In the node voltage method of circuit analysis (Section 4.6), equations will be obtained at the principal nodes by applying *Kirchhoff's current law* (*KCL*). This law states that at any node (principal or not) the sum of the currents entering equals the sum of the currents leaving. The conservation of electric charge is the basis of the law. Alternative statements of KCL are (i) the total current into a node is zero; (ii) the total current out of a node is zero.

EXAMPLE 3.2 In Fig. 3-3, five branches connect at a common junction, forming a principal node. The total current into the node is

$$I_1 - I_2 + I_3 - I_4 - I_5 = 0$$

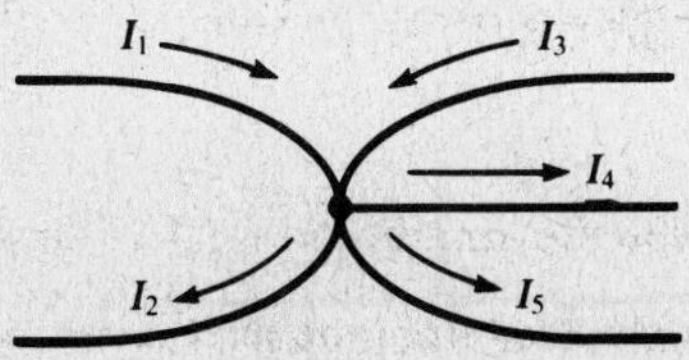

Fig. 3-3

The same equation is obtained when the sum of currents entering is set equal to the sum leaving:

$$I_1 + I_3 = I_2 + I_4 + I_5$$

3.3 VOLTAGE DIVISION AND CURRENT DIVISION

A set of two or more series-connected resistors (Fig. 3-4) is frequently referred to as a *voltage divider.* From Ohm's law,

$$\frac{V_j}{V_k} = \frac{IR_j}{IR_k} = \frac{R_j}{R_k}$$

$$\frac{P_j}{P_k} = \frac{I^2R_j}{I^2R_k} = \frac{R_j}{R_k}$$

that is, the total voltage V_T and the total absorbed power P_T are divided in the ratio of the resistances.

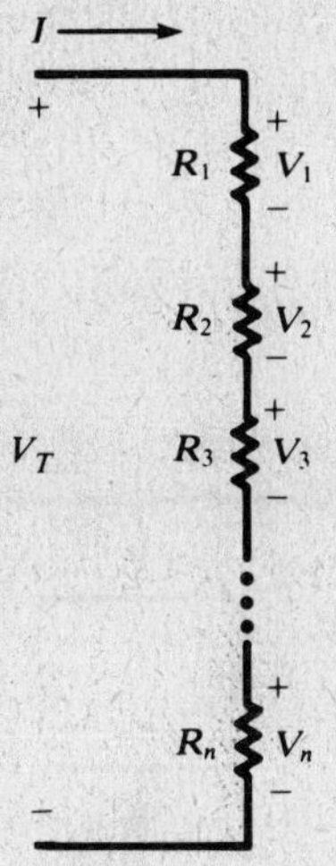

Fig. 3-4

Two or more resistors in a parallel connection (Fig. 3-5) will divide the total current I_T and the total absorbed power P_T in the *inverse* ratio of the resistances:

$$\frac{I_j}{I_k} = \frac{V/R_j}{V/R_k} = \frac{R_k}{R_j}$$

$$\frac{P_j}{P_k} = \frac{V^2/R_j}{V^2/R_k} = \frac{R_k}{R_j}$$

In particular, for $n = 2$,

$$I_1 = \frac{R_2}{R_1 + R_2} I_T \qquad P_1 = \frac{R_2}{R_1 + R_2} P_T$$

$$I_2 = \frac{R_1}{R_1 + R_2} I_T \qquad P_2 = \frac{R_1}{R_1 + R_2} P_T$$

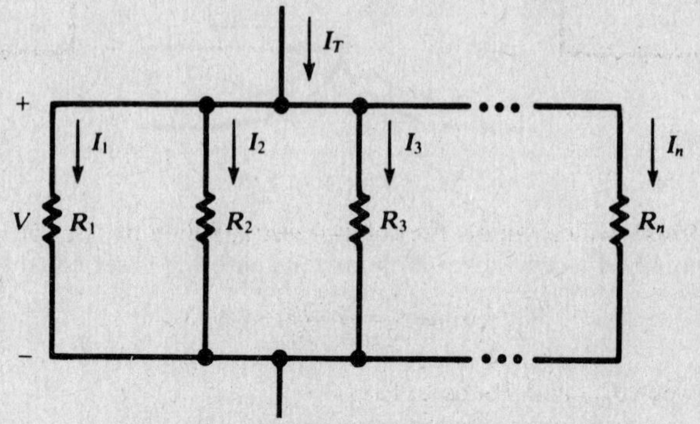

Fig. 3-5

3.4 SERIES-PARALLEL NETWORK REDUCTION

The mesh current and node voltage methods of Chapter 4 are the principal techniques of dc resistive circuit analysis. However, the equivalent resistance of series and parallel branches (Section 2.7), combined with the voltage and current division rules, provide another method of analyzing a network. This method is tedious and usually requires the drawing of several additional circuits. Even so, the process of reducing the network provides a very clear picture of the overall functioning of the network in terms of voltages, currents, and power. The reduction begins with a scan of the network to pick out series and parallel combinations of resistors.

EXAMPLE 3.3 Obtain the total power supplied by the 60-V source and the power absorbed in each resistor in the network of Fig. 3-6.

$$R_{ab} = 7 + 5 = 12\ \Omega$$

$$R_{cd} = \frac{(12)(6)}{12 + 6} = 4\ \Omega$$

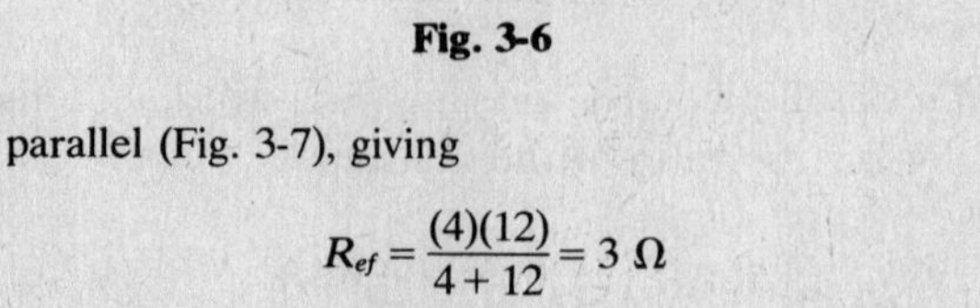

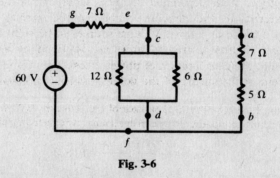

Fig. 3-6

These two equivalents are in parallel (Fig. 3-7), giving

$$R_{ef} = \frac{(4)(12)}{4 + 12} = 3\ \Omega$$

Then this 3-Ω equivalent is in series with the 7 Ω (Fig. 3-8), so that for the entire circuit,

$$R_{eq} = 7 + 3 = 10\ \Omega$$

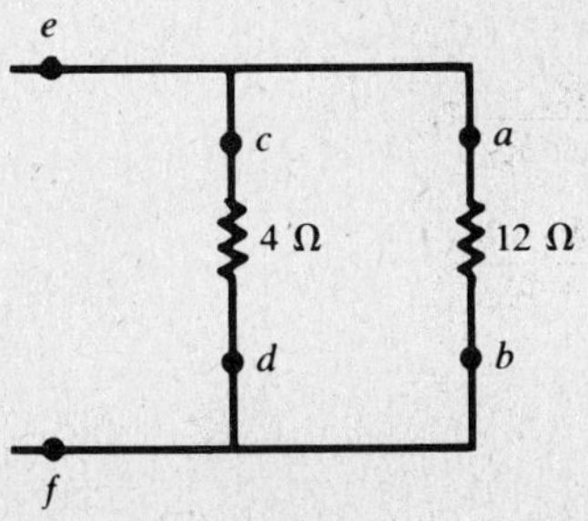

Fig. 3-7

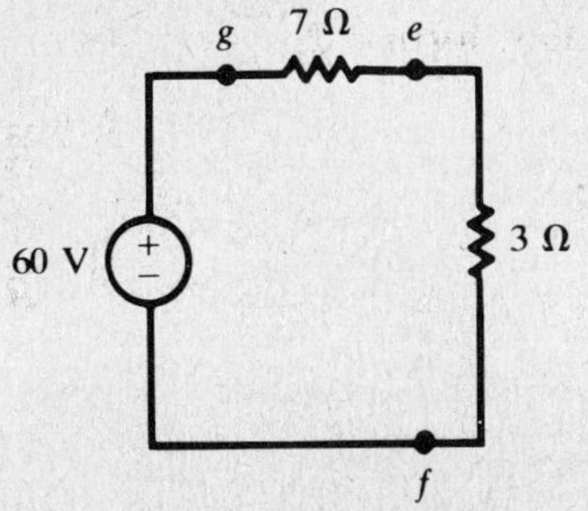

Fig. 3-8

The total power absorbed, which equals the total power supplied by the source, can now be calculated as

$$P_T = \frac{V^2}{R_{eq}} = \frac{(60)^2}{10} = 360 \text{ W}$$

This power is divided between R_{ge} and R_{ef} as follows:

$$P_{ge} = P_{7\Omega} = \frac{7}{7+3}(360) = 252 \text{ W} \qquad P_{ef} = \frac{3}{7+3}(360) = 108 \text{ W}$$

Power P_{ef} is further divided between R_{cd} and R_{ab} as follows:

$$P_{cd} = \frac{12}{4+12}(108) = 81 \text{ W} \qquad P_{ab} = \frac{4}{4+12}(108) = 27 \text{ W}$$

Finally, these powers are divided between the individual resistances as follows:

$$P_{12\Omega} = \frac{6}{12+6}(81) = 27 \text{ W} \qquad P_{7\Omega} = \frac{7}{7+5}(27) = 15.75 \text{ W}$$

$$P_{6\Omega} = \frac{12}{12+6}(81) = 54 \text{ W} \qquad P_{5\Omega} = \frac{5}{7+5}(27) = 11.25 \text{ W}$$

3.5 SUPERPOSITION

A linear network (e.g., a dc resistive network) which contains two or more *independent* sources can be analyzed to obtain the various voltages and branch currents by allowing the sources to act one at a time, then superposing the results. This principle applies because of the linear relationship between current and voltage. With *dependent* sources, superposition can be used only when the control functions are external to the network containing the sources, so that the controls are unchanged as the sources act one at a time. Voltage sources to be suppressed while a single source acts are replaced by short circuits; current sources are replaced by open circuits. Superposition cannot be directly applied to the computation of power, because power in an element is proportional to the square of the current or the square of the voltage, which is nonlinear.

EXAMPLE 3.4 Compute the current in the 23-Ω resistor of Fig. 3-9(*a*) by applying the superposition principle. With the 200-V source acting alone, the 20-A current source is replaced by an open circuit, Fig. 3-9(*b*).

$$R_{eq} = 47 + \frac{(27)(4+23)}{54} = 60.5 \ \Omega$$

$$I_T = \frac{200}{60.5} = 3.31 \text{ A}$$

$$I'_{23\Omega} = \left(\frac{27}{54}\right)(3.31) = 1.65 \text{ A}$$

When the 20-A source acts alone, the 200-V source is replaced by a short circuit, Fig. 3-9(*c*). The equivalent resistance to the left of the source is

$$R_{eq} = 4 + \frac{(27)(47)}{74} = 21.15\ \Omega$$

Then

$$I''_{23\Omega} = \left(\frac{21.15}{21.15 + 23}\right)(20) = 9.58\ \text{A}$$

The total current in the 23-Ω resistor is

$$I_{23\Omega} = I'_{23\Omega} + I''_{23\Omega} = 11.23\ \text{A}$$

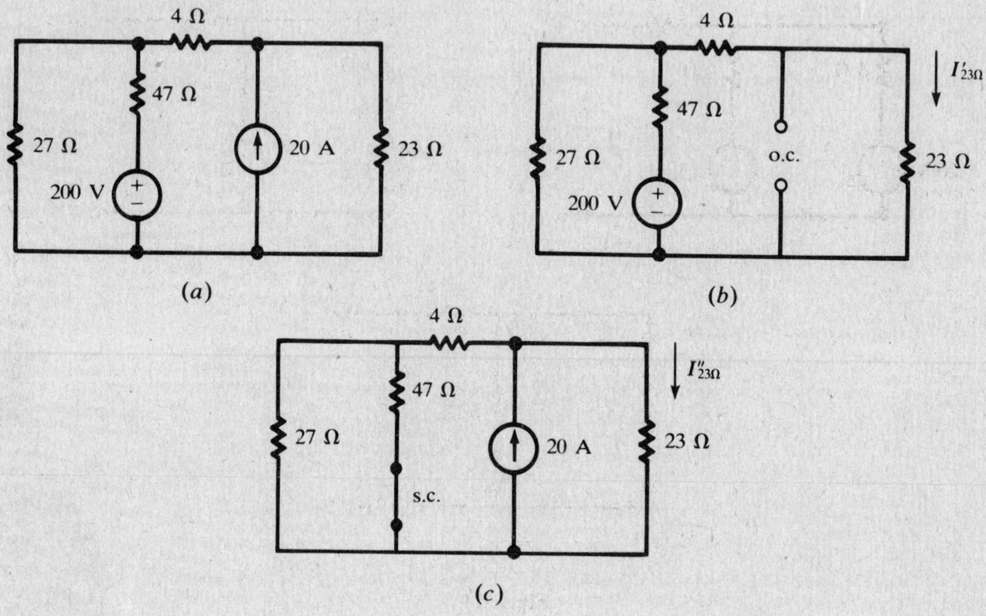

Fig. 3-9

3.6 THEVENIN'S AND NORTON'S THEOREMS

A linear, active, resistive network which contains one or more voltage or current sources can be replaced by a single voltage source and a series resistance (*Thévenin's theorem*), or by a single current source and a parallel resistance (*Norton's theorem*). The voltage is called the *Thévenin equivalent voltage*, V', and the current the *Norton equivalent current*, I'. The two resistances are the same, R'. When terminals *ab* in Fig. 3-10(*a*) are *open-circuited*, a voltage will appear between them.

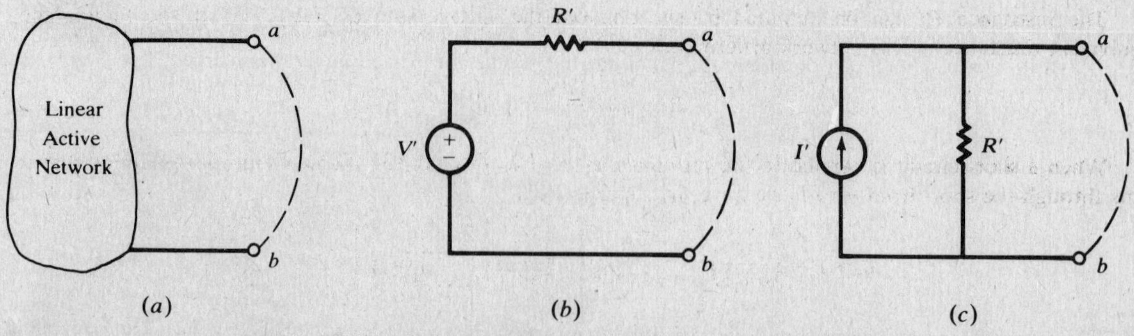

Fig. 3-10

From Fig. 3-10(*b*) it is evident that this must be the voltage V' of the Thévenin equivalent circuit. If a *short circuit* is applied to the terminals, as suggested by the dashed line in Fig. 3-10(*a*), a current will result. From Fig. 3-10(*c*) it is evident that this current must be I' of the Norton equivalent circuit. Now, if the circuits in (*b*) and (*c*) are equivalents of the same active network, they are equivalent to each other. It follows that $I' = V'/R'$. If both V' and I' have been determined from the active network, then $R' = V'/I'$.

EXAMPLE 3.5 Obtain the Thévenin and Norton equivalent circuits for the active network in Fig. 3-11(*a*).

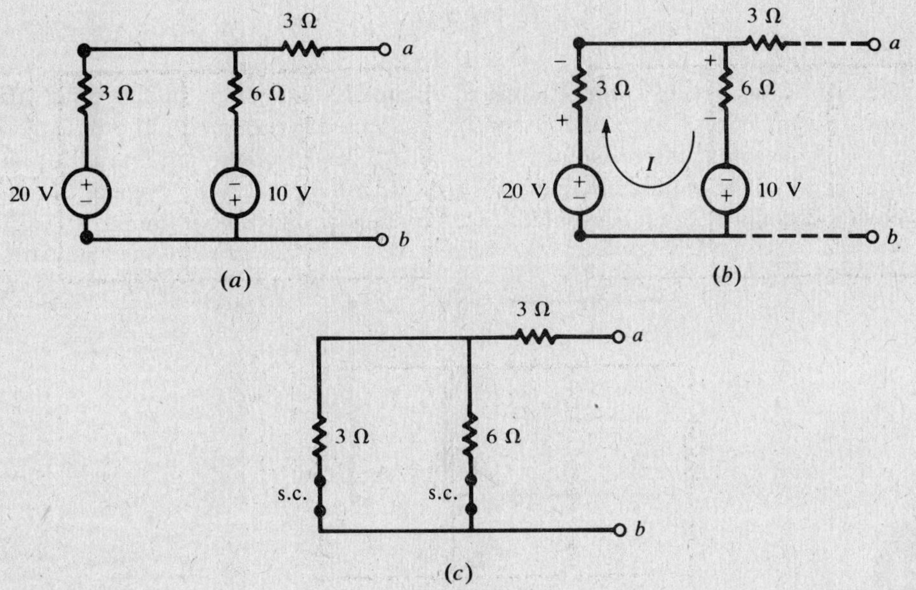

Fig. 3-11

With terminals *ab* open, the two sources drive a clockwise current through the 3-Ω and 6-Ω resistors [Fig. 3-11(*b*)].

$$I = \frac{20+10}{3+6} = \frac{30}{9}\text{ A}$$

Since no current passes through the upper right 3-Ω resistor, the Thévenin voltage can be taken from either active branch:

$$V_{ab} = V' = 20 - \left(\frac{30}{9}\right)(3) = 10\text{ V}$$

or

$$V_{ab} = V' = \left(\frac{30}{9}\right)6 - 10 = 10\text{ V}$$

The resistance R' can be obtained by shorting out the voltage sources [Fig. 3-11(*c*)] and finding the equivalent resistance of this network at terminals *ab*:

$$R' = 3 + \frac{(3)(6)}{9} = 5\ \Omega$$

When a short circuit is applied to the terminals, current I_{sc} results from the two sources. Assuming that it runs through the short from *a* to *b*, we have, by superposition,

$$I_{sc} = I' = \left(\frac{6}{6+3}\right)\left[\frac{20}{3+\frac{(3)(6)}{9}}\right] - \left(\frac{3}{3+3}\right)\left[\frac{10}{6+\frac{(3)(3)}{6}}\right] = 2\text{ A}$$

Figure 3-12 shows the two equivalent circuits. In the present case, V', R', and I' were obtained independently. Since they are related by Ohm's law, any two may be used to obtain the third.

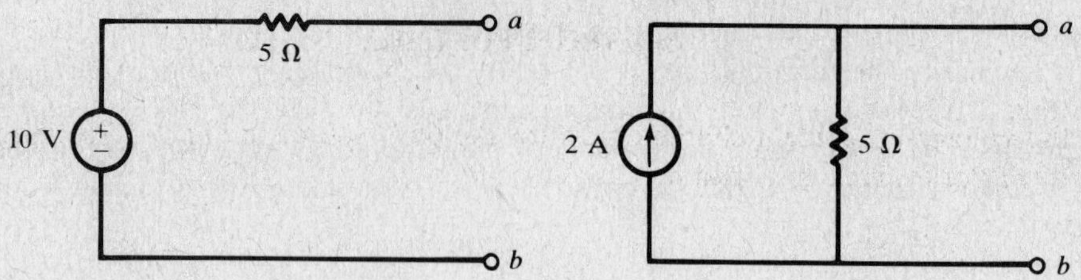

(a) Thévenin Equivalent (b) Norton Equivalent

Fig. 3-12

The value of Thévenin and Norton equivalent circuits is clear when an active network is to be examined under a number of load conditions, each represented by a resistor. This is suggested in Fig. 3-13, where it is evident that the resistors $R_1, R_2, \ldots, R_n$ can be connected one at a time, and the resulting current and power readily obtained. If this were attempted in the original circuit using, for example, network reduction, the task would be very tedious and time-consuming.

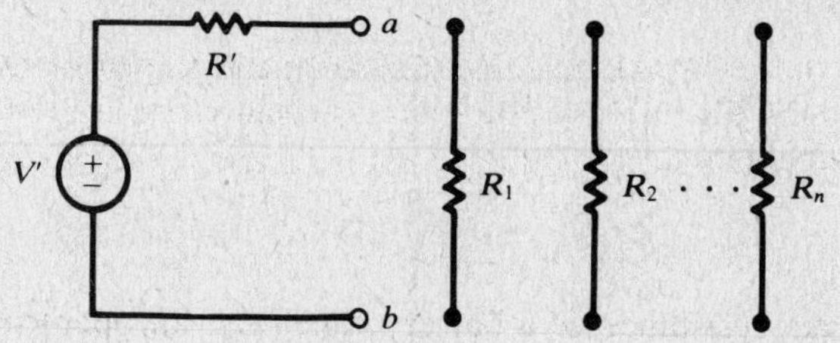

Fig. 3-13

3.7 MAXIMUM POWER TRANSFER THEOREM

At times it is desired to obtain the maximum power transfer from an active network to an external load resistor R_L. Assuming that the network is linear, it can be reduced to an equivalent circuit as in Fig. 3-14. Then

$$I = \frac{V'}{R' + R_L}$$

and so the power absorbed by the load is

$$P_L = \frac{V'^2 R_L}{(R' + R_L)^2} = \frac{V'^2}{4R'}\left[1 - \left(\frac{R' - R_L}{R' + R_L}\right)^2\right]$$

It is seen that P_L attains its maximum value, $V'^2/4R'$, when $R_L = R'$, in which case the power in R' is also $V'^2/4R'$. Consequently, when the power transferred is a maximum, the efficiency is 50%.

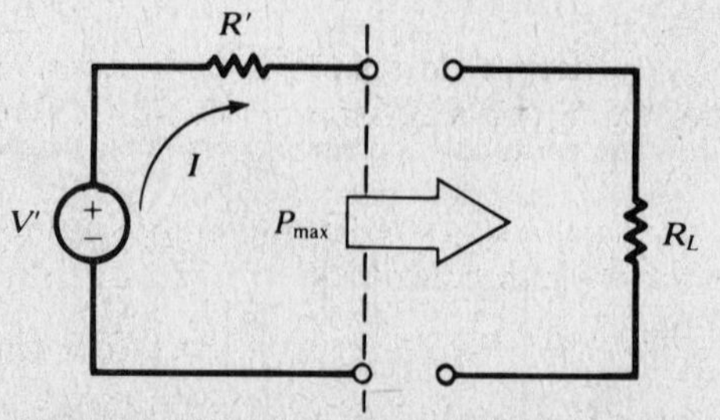

Fig. 3-14

Solved Problems

3.1 Plot the voltage-vs.-current characteristic for the 60-V source in Fig. 3-15(*a*). Show the points for switch positions *a*, *b*, *c*, and *d*.

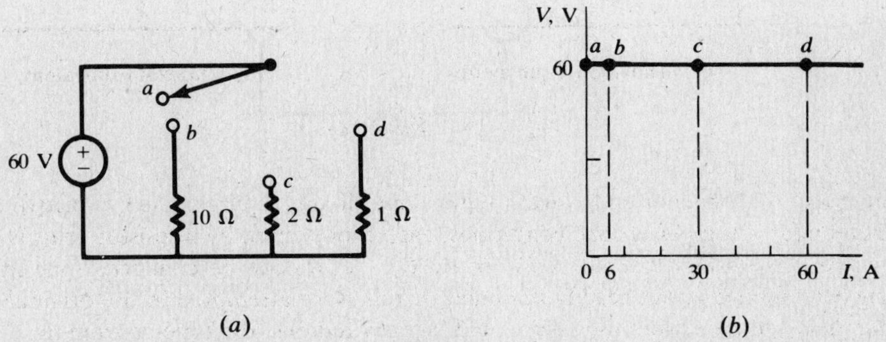

(*a*) (*b*)

Fig. 3-15

$I_a = 60/\infty = 0$ A; $I_b = 60/10 = 6$ A; $I_c = 60/2 = 30$ A; and $I_d = 60/1 = 60$ A.

The plot is shown in Fig. 3-15(*b*). The source remains constant at 60 V for all currents. However, zero resistance is not permitted.

3.2 Calculate the internal resistance of a battery which has an open-circuit voltage of 12.0 V and delivers 100 A to a resistance of 0.10 Ω.

The battery with its internal resistance R is modeled in Fig. 3-16.

$$I = 100 = \frac{12}{R + 0.10}$$

from which $R = 0.02$ Ω.

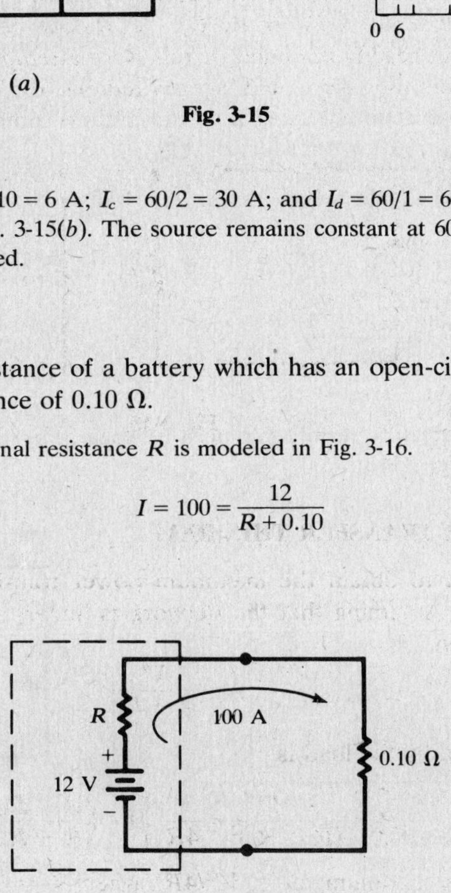

Fig. 3-16

3.3 Measurements made on a practical dc source show a terminal voltage of 100 V for a load resistance of 100 Ω, and 105 V for a resistance of 210 Ω. Obtain the circuit model for this source.

A constant-voltage source and a series resistance can be used to model the practical source, as in Fig. 3-16. The following equations may be written:

$$I_1 = 100/100 = 1.0 \text{ A} \qquad V = 1.0R + 100 \quad \text{(V)}$$

$$I_2 = 105/210 = 0.5 \text{ A} \qquad V = 0.5R + 105 \quad \text{(V)}$$

Solving the two equations simultaneously gives $V = 110$ V, $R = 10$ Ω.

3.4 Calculate the power delivered by the dependent source in Fig. 3-17.

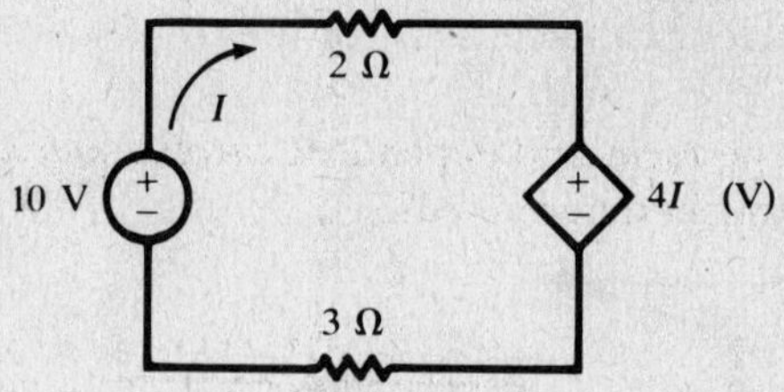

Fig. 3-17

Using Kirchhoff's voltage law, $10 = 2I + 4I + 3I$, or $I = 1.11$ A. The current enters the positive terminal. Thus power *absorbed* is $1.11 \times 4(1.11) = 4.93$ W and power delivered is -4.93 W.

3.5 Design a 10-mA current supply using a constant 20-V source and a series resistance R. Plot current vs. load resistance for $0 \le R_L \le 100\ \Omega$.

Assuming that 10 mA is the maximum current, it will occur at $R_L = 0$. Then

$$10 \times 10^{-3} = \frac{20}{R} \qquad R = 2000\ \Omega$$

For $R_L = 100\ \Omega$,

$$I = \frac{20}{2000 + 100} = 9.52 \text{ mA}$$

Since the current-vs.-resistance relationship must be linear, the characteristic over the range $0 \le R_L \le 100\ \Omega$ is as shown in Fig. 3-18.

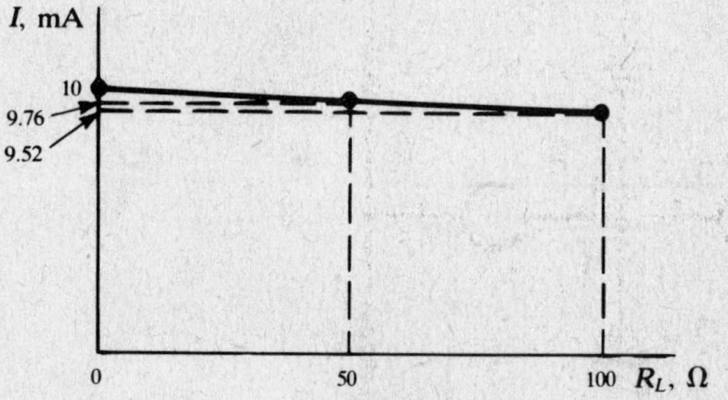

Fig. 3-18

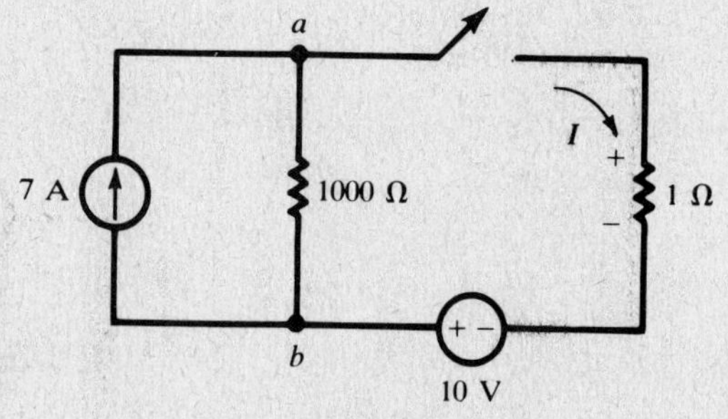

Fig. 3-19

3.6 Two 1-Ω resistors, rated 25 W and 50 W, are available for use in the circuit shown in Fig. 3-19. (*a*) Can either resistor be used? (*b*) Find the power absorbed by each element.

The 1000-Ω resistor makes the 7-A current source practical by providing a current path when the switch is open. When the switch is closed,

$$I = \left(\frac{1000}{1000 + 1}\right)(7) = 6.99 \approx 7 \text{ A}$$

$$P_{1\Omega} = (7)^2(1) = 49 \text{ W}$$

(*a*) The 25-W resistor may not be used, since 49 W far exceeds its rating.

(*b*)

$$P_{10V} = -(7)(10) = -70 \text{ W}$$

$$V_{ab} = 7(1) - 10 = -3 \text{ V} \qquad P_{7A} = (3)(7) = 21 \text{ W}$$

Current enters the 7-A source at the + terminal. Consequently, this *source* is absorbing power of 21 W, while the 10-V source *delivers* power of 70 W. (Assuming 7 A in the 1-Ω resistor makes both the current and power zero in the 1000-Ω resistor.)

3.7 In Fig. 3-20 both independent and dependent current sources drive current through resistor R. Is the value of R uniquely determined?

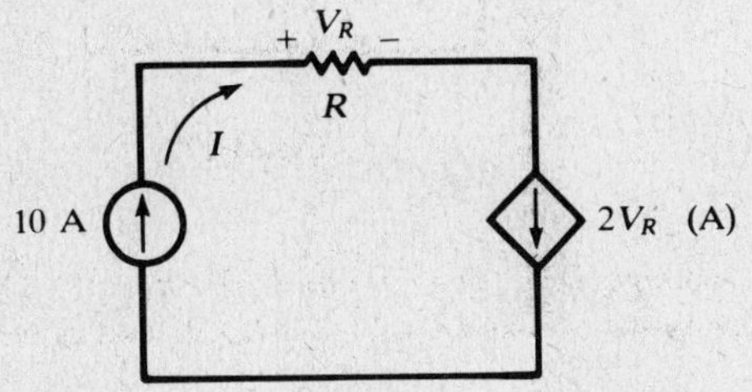

Fig. 3-20

The current I must be 10 A, by definition of an independent source. Then,

$$I = 10\text{ A} = 2V_R \qquad V_R = 5\text{ V} \qquad 5\text{ V} = (10)(R) \qquad R = 0.5\ \Omega$$

No other value for R is possible.

3.8 In the circuit shown in Fig. 3-21, find the power absorbed by the 5-V battery.

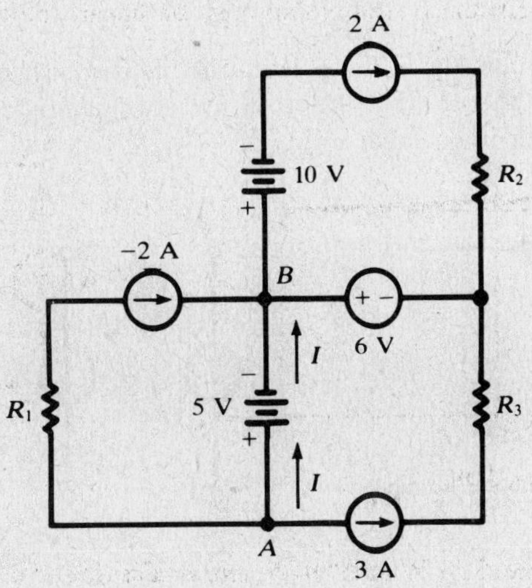

Fig. 3-21

Current I entering the 5-V battery at the + terminal may be found using KCL either at node A or B. At A,

$$3 + I + (-2) = 0 \qquad I = -1\text{ A}$$

Then power *absorbed* is $P_{5V} = (-1)(5) = -5$ W; the 5-V battery *supplies* or *delivers* 5 W to the circuit.

3.9 Refer to Fig. 3-22. Calculate the energy dissipated in the 5-Ω resistor in the interval $0 < t < 5$ ms, if $v_g = 5000t$ (V).

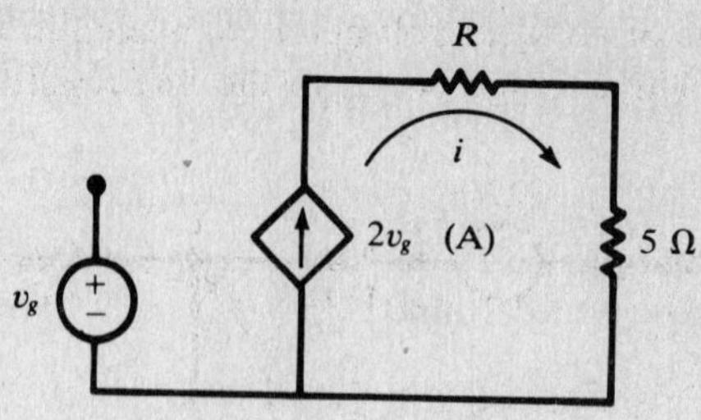

Fig. 3-22

$$i = 2v_g = 10^4 t \quad \text{(A)}$$

$$p = i^2 5 = 5 \times 10^8 t^2 \quad \text{(W)}$$

$$W = \int_0^{5\times10^{-3}} 5 \times 10^8 t^2 \, dt = (5 \times 10^8)\left[\frac{t^3}{3}\right]_0^{5\times10^{-3}} = 20.8 \text{ J}$$

3.10 Calculate the output power of a 250-V dc motor if the efficiency is 92% when the input current is 12 A.

By definition,

$$\text{percent efficiency} = \frac{P_{\text{out}}}{P_{\text{in}}}(100\%)$$

For the data, $P_{\text{in}} = (250)(12) = 3000$ W, and so

$$P_{\text{out}} = \frac{92\%}{100\%}(3000) = 2760 \text{ W}$$

3.11 For the circuit shown in Fig. 3-23, imagine the elements shown on the right connected one at a time to the terminals *ab*. The control for the dependent sources is I_x. Determine the dependent parameter in each case.

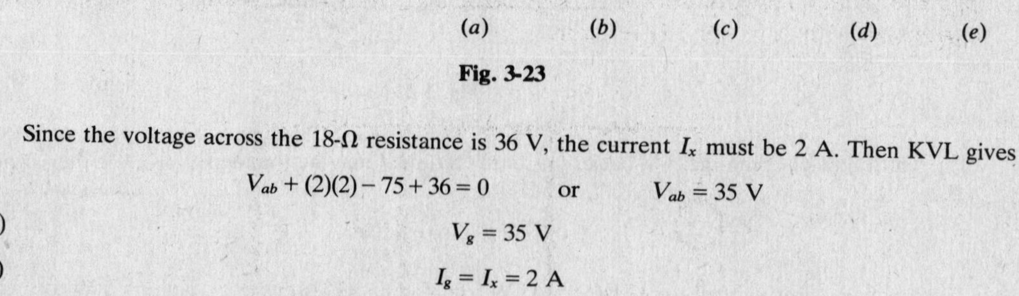

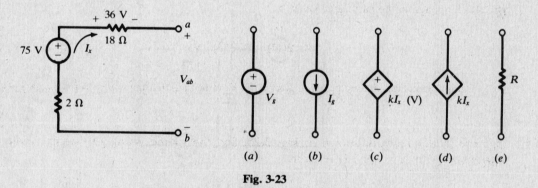

Fig. 3-23

Since the voltage across the 18-Ω resistance is 36 V, the current I_x must be 2 A. Then KVL gives

$$V_{ab} + (2)(2) - 75 + 36 = 0 \qquad \text{or} \qquad V_{ab} = 35 \text{ V}$$

(*a*) $V_g = 35$ V

(*b*) $I_g = I_x = 2$ A

(*c*) $kI_x = 35$ V $\qquad k = 17.5\ \Omega$

(*d*) $kI_x = -I_x \qquad k = -1$

(*e*) $V_R = I_x R = 35$ V $\qquad R = 17.5\ \Omega$

3.12 Determine the readings of an ideal voltmeter connected in Fig. 3-24 to (*a*) terminals *a* and *b*, (*b*) terminals *c* and *g*. The average power in the 5-Ω resistor is 20 W.

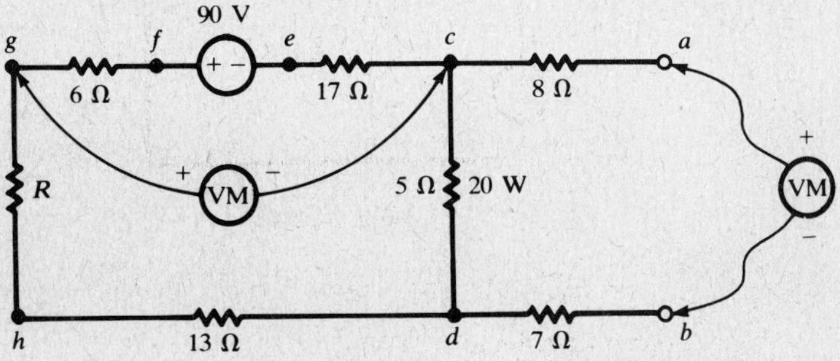

Fig. 3-24

$$P = I^2(5) \qquad I = \sqrt{\frac{20}{5}} = \pm 2 \text{ A}$$

The direction of I through the 5-Ω resistor is determined by noting that the polarity of the 90-V source requires that the current pass from d to c. Thus d is positive with respect to c and $V_{dc} = (2)(5) = 10$ V.

An *ideal* voltmeter indicates the voltage without drawing any current. It may be considered as having an infinite resistance.

(*a*) KVL applied to the closed loop *acdba* results in

$$V_{ac} + V_{cd} + V_{db} + V_{ba} = 0$$
$$0 - 10 + 0 - \text{VM} = 0$$
$$\text{VM} = -10 \text{ V}$$

If the meter is of the digital type, it will indicate −10 V. A moving-coil galvanometer will try to go downscale, with the pointer stopping at the pin. If the leads are reversed, it will indicate 10 V. (And with its + lead at point b, it is known that b is 10 volts positive with respect to a.)

(*b*) KVL applied to the path *cefgc* gives

$$V_{ce} + V_{ef} + V_{fg} + V_{gc} = 0$$
$$2(17) - 90 + 2(6) + \text{VM} = 0$$
$$\text{VM} = 44 \text{ V}$$

In this connection, the meter reads positive 44 V, indicating that point g is 44 volts above point c.

3.13 For the ladder network shown in Fig. 3-25, find the source voltage V_s which results in a current of 7.5 mA in the 3-Ω resistor.

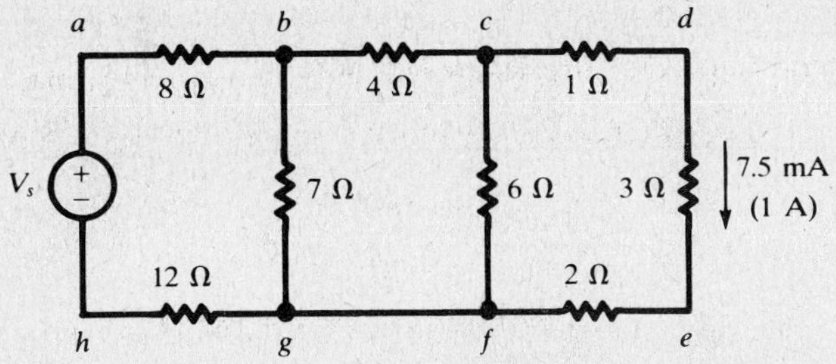

Fig. 3-25

A current of 1 A will be assumed: the voltage necessary to produce 1 A is in the same ratio to 1 A as V_s is to 7.5 mA, because of the linearity of the network.

$$V_{cf} = 1(1+3+2) = 6 \text{ V} \qquad I_{cf} = \frac{6}{6} = 1 \text{ A}$$

Then, by KCL, $I_{bc} = 1 + 1 = 2$ A, and

$$v_{bg} = 2(4) + 6 = 14 \text{ V} \qquad I_{bg} = \frac{14}{7} = 2 \text{ A}$$

Again from KCL, $I_{ab} = 2+2 = 4$ A; and so $V_{ah} = 4(8) + 14 + 4(12) = 94$ V. Now, scaling down,

$$\frac{V_{ah}}{1 \text{ A}} = \frac{V_s}{7.5 \text{ mA}} \qquad \text{whence} \qquad V_s = 0.705 \text{ V}$$

3.14 Determine the current I for the circuit shown in Fig. 3-26.

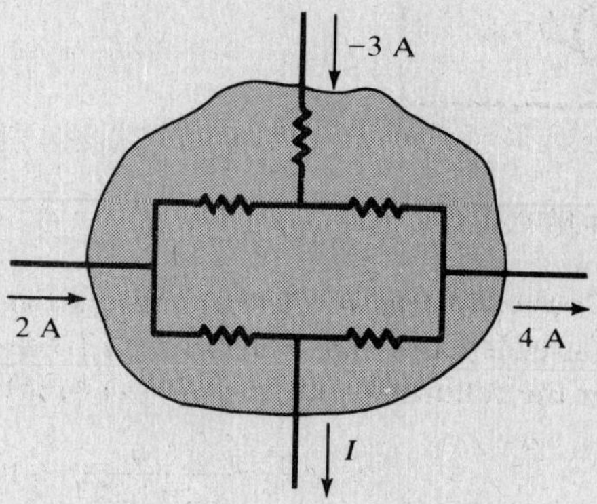

Fig. 3-26

Without the values of the resistors, it is not possible to calculate the branch currents. Even so, the network within the shaded area may be viewed as a single node, at which KCL gives

$$2 - 3 - I - 4 = 0 \qquad \text{or} \qquad I = -5 \text{ A}$$

3.15 In Problem 2.7 a 50-V source was found to deliver 13.7 A to the network shown in Fig. 3-27. Obtain the currents in all the network branches.

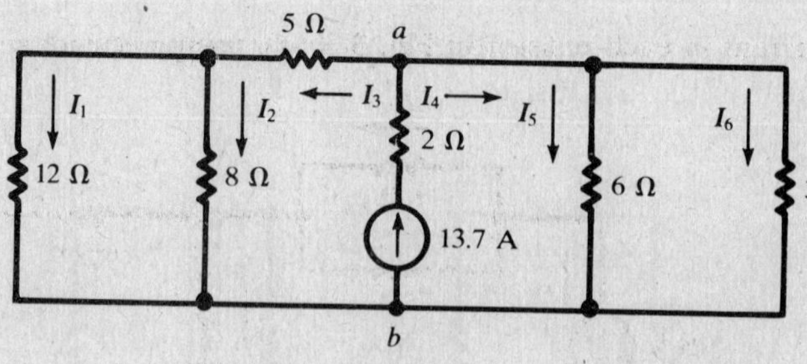

Fig. 3-27

The equivalent resistances to the left and right of branch ab are computed as follows:

$$R_{eq}^{(\text{left})} = 5 + \frac{(12)(8)}{20} = 9.80 \ \Omega \qquad R_{eq}^{(\text{right})} = \frac{(6)(3)}{9} = 2.00 \ \Omega$$

Referring to the reduced network, Fig. 3-28,

$$I_3 = \left(\frac{2.0}{11.8}\right)(13.7) = 2.32 \text{ A}$$

$$I_4 = \left(\frac{9.8}{11.8}\right)(13.7) = 11.38 \text{ A}$$

The further division of these currents is obtained by referring to the original network.

$$I_1 = \left(\frac{8}{20}\right)(2.32) = 0.93 \text{ A} \qquad I_2 = \left(\frac{12}{20}\right)(2.32) = 1.39 \text{ A}$$

$$I_5 = \left(\frac{3}{9}\right)(11.38) = 3.79 \text{ A} \qquad I_6 = \left(\frac{6}{9}\right)(11.38) = 7.59 \text{ A}$$

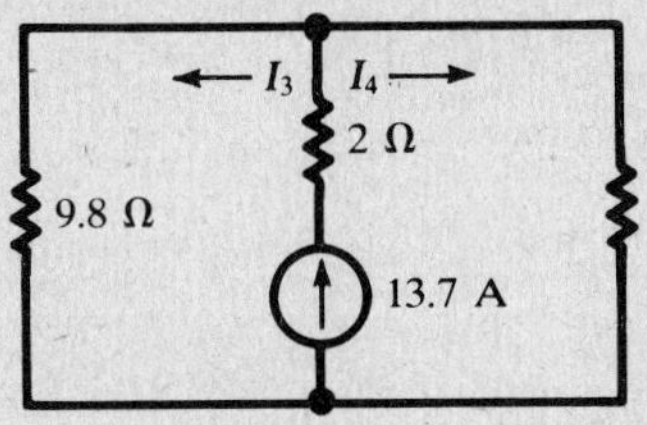

Fig. 3-28

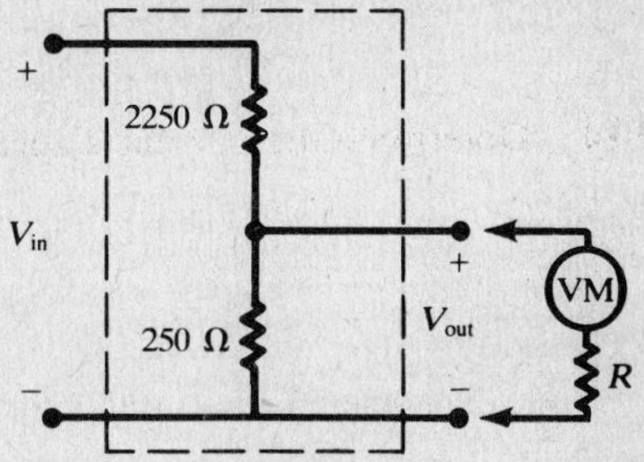

Fig. 3-29

3.16 The voltage divider shown in Fig. 3-29 is also called an *attenuator.* (When a single resistor has an adjustable tap, it is called a *potentiometer* or *pot.*) To discover the effect of loading, compute the ratio V_{out}/V_{in} for the following values of R: (*a*) ∞, (*b*) 1 MΩ, (*c*) 10 kΩ, (*d*) 1 kΩ.

(*a*)
$$V_{out}/V_{in} = \frac{250}{2250 + 250} = 0.100$$

(*b*) With $R = 10^6$ Ω, the parallel equivalent with the 250 Ω must be calculated first.

$$R_{eq} = \frac{(250)(10^6)}{250 + 10^6} = 249.9\ \Omega \qquad V_{out}/V_{in} = \frac{249.9}{2250 + 249.9} = 0.100$$

(*c*)
$$R_{eq} = \frac{(250)(10\,000)}{250 + 10\,000} = 243.9\ \Omega \qquad V_{out}/V_{in} = \frac{243.9}{2250 + 243.9} = 0.098$$

(*d*)
$$R_{eq} = \frac{(250)(1000)}{250 + 1000} = 200.0\ \Omega \qquad V_{out}/V_{in} = \frac{200}{2250 + 200} = 0.082$$

A voltage divider may appear to be constructed to result in a ratio of 1 : 10. However, the loading can change this considerably.

3.17 Obtain the current in each resistor in Fig. 3-30(*a*), using network reduction methods.

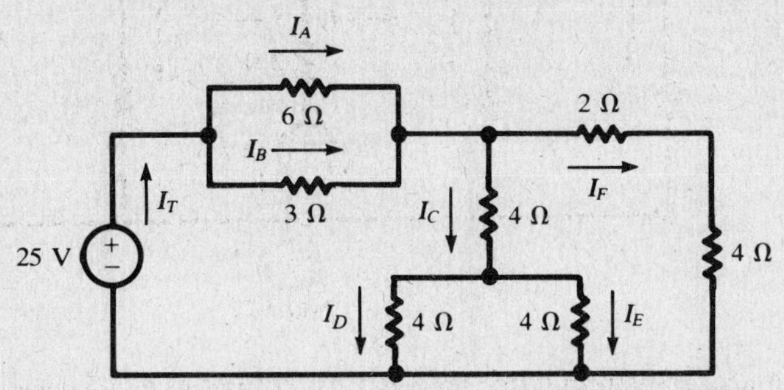

(*a*)

Fig. 3-30

As a first step, two-resistor parallel combinations are converted to their equivalents. For the 6 Ω and 3 Ω, $R_{eq} = (6)(3)/(6+3) = 2\ \Omega$. For the two 4-Ω resistors, $R_{eq} = 2\ \Omega$. The circuit is redrawn with series resistors added [Fig. 3-30(*b*)]. Now the two 6-Ω resistors in parallel have the equivalent $R_{eq} = 3\ \Omega$, and this is in series with the 2 Ω. Hence, $R_T = 5\ \Omega$, as shown in Fig. 3-30(*c*). The resulting total current is

$$I_T = \frac{25}{5} = 5\ \text{A}$$

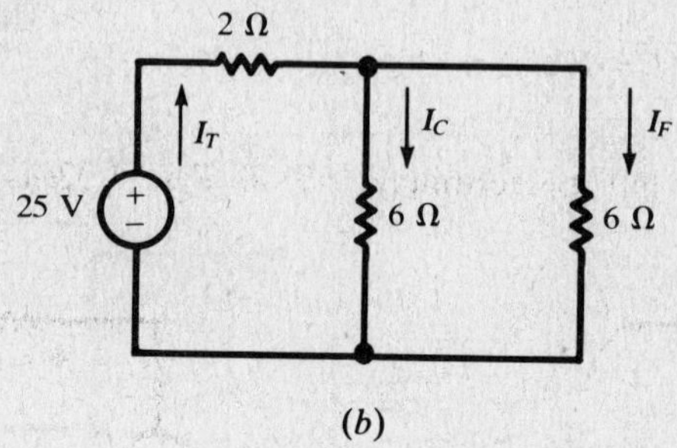

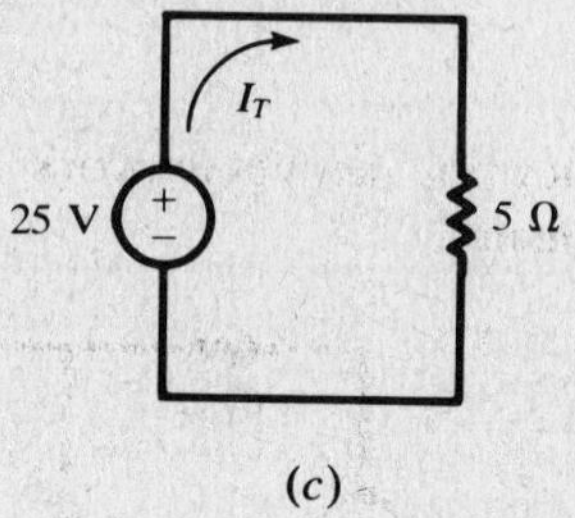

Fig. 3-30 (*cont.*)

Now the branch currents can be obtained by working back through the circuits of Fig. 3-30(*b*) and 3-30(*a*).

$$I_C = I_F = \tfrac{1}{2} I_T = 2.5\ \text{A}$$

$$I_D = I_E = \tfrac{1}{2} I_C = 1.25\ \text{A}$$

$$I_A = \frac{3}{6+3} I_T = \frac{5}{3}\ \text{A}$$

$$I_B = \frac{6}{6+3} I_T = \frac{10}{3}\ \text{A}$$

3.18 Obtain the current I_x in the 10-Ω resistor in Fig. 3-31(*a*), using superposition.

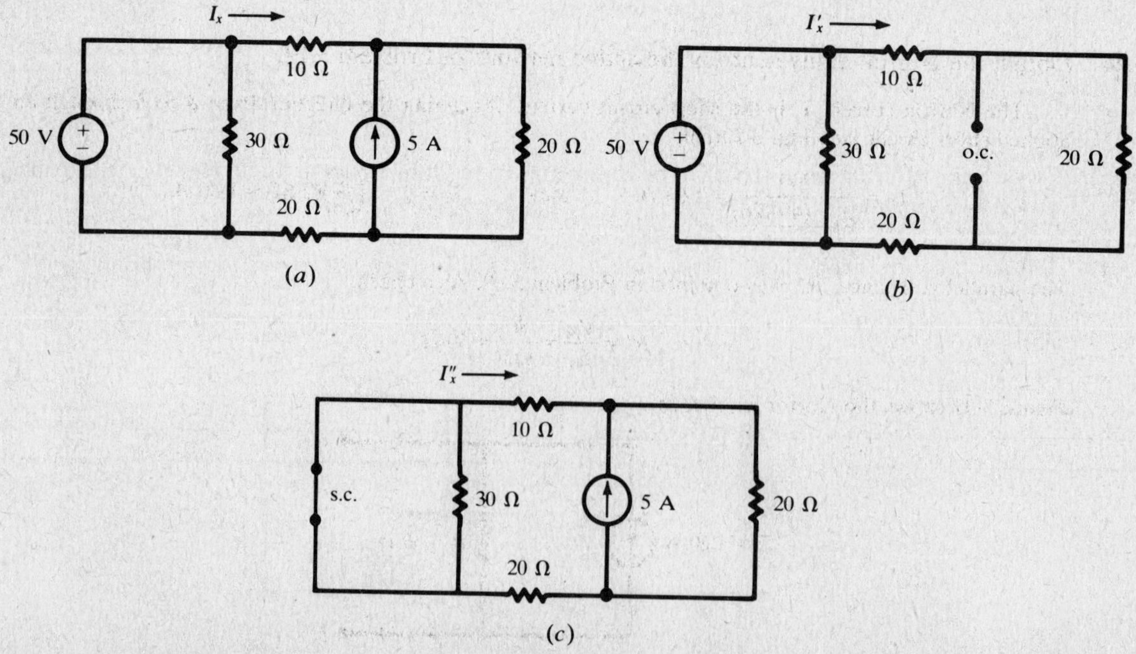

Fig. 3-31

To permit the 50-V source to act alone, the 5-A current source is replaced by an open circuit [Fig. 3-31(*b*)]. Then

$$I'_x = \frac{50}{10+20+20} = 1 \text{ A}$$

Next, the voltage source is removed and replaced by a short circuit [Fig. 3-31(*c*)].

$$I''_x = \left(\frac{20}{50}\right)(-5) = -2 \text{ A}$$

Now $$I_x = I'_x + I''_x = -1 \text{ A}$$

3.19 Replace the active network to the left of terminals *ab* in Fig. 3-32(*a*) by a Thévenin equivalent.

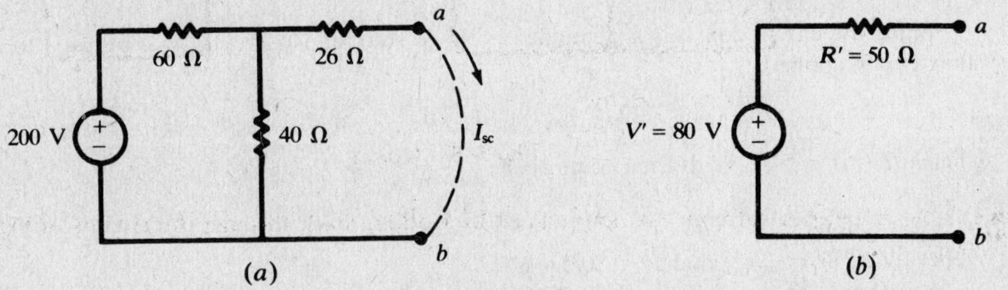

Fig. 3-32

The open-circuit voltage V_{ab} is the voltage across the 40-Ω resistor:

$$V_{ab} = \left(\frac{40}{60+40}\right)(200) = 80 \text{ V}$$

Resistance R' may be found by looking into the circuit from *ab* with the voltage source shorted.

$$R' = 26 + \frac{(40)(60)}{100} = 50\ \Omega$$

See Fig. 3-32(*b*) for the Thévenin equivalent.

3.20 Obtain the Norton equivalent for the active network of Problem 3.19.

The Norton current I' is the short-circuit current. Assuming the direction from *a* to *b* through an applied short circuit [see Fig. 3-32(*a*)],

$$I_T = \frac{200}{60 + \dfrac{(40)(26)}{66}} = 2.64 \text{ A} \qquad \text{and} \qquad I_{sc} = I' = \left(\frac{40}{66}\right)(2.64) = 1.60 \text{ A}$$

The parallel resistance, R', was obtained in Problem 3.19. As a check,

$$R' = \frac{V_{oc}}{I_{sc}} = \frac{80}{1.60} = 50\ \Omega$$

Figure 3-33 shows the Norton equivalent.

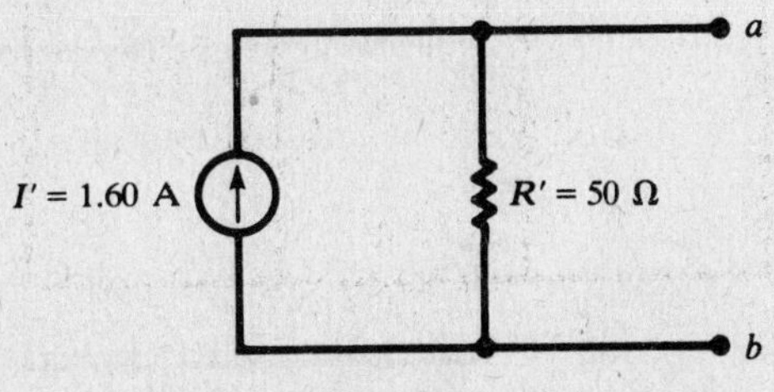

Fig. 3-33

3.21 Obtain a Thévenin equivalent for the active network shown in Fig. 3-34.

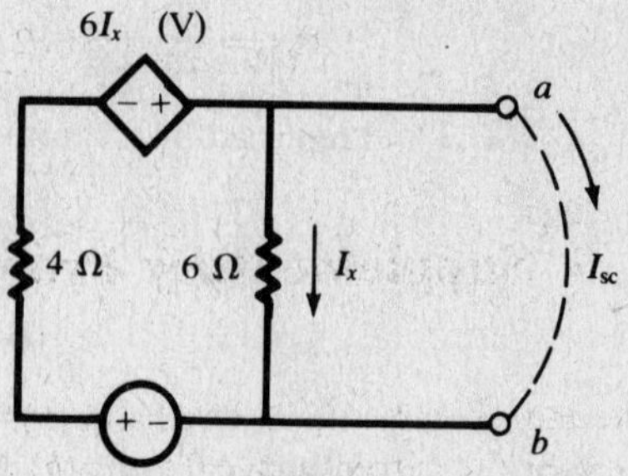

Fig. 3-34

Since the circuit contains a dependent voltage source, V_{oc} and I_{sc} will be used to find R'. With a short circuit applied,

$$-20 + I_{sc}(4) - 6I_x = 0 \qquad \text{and} \qquad I_x = 0$$

whence $I_{sc} = I' = 5$ A. With the circuit open,

$$-20 + 4I_x - 6I_x + 6I_x = 0 \qquad \text{or} \qquad I_x = 5 \text{ A}$$

Hence, $V' = 5(6) = 30$ V and $R' = 30/5 = 6\ \Omega$.

See Fig. 3-35(a) and (b) for both Thévenin and Norton equivalent circuits.

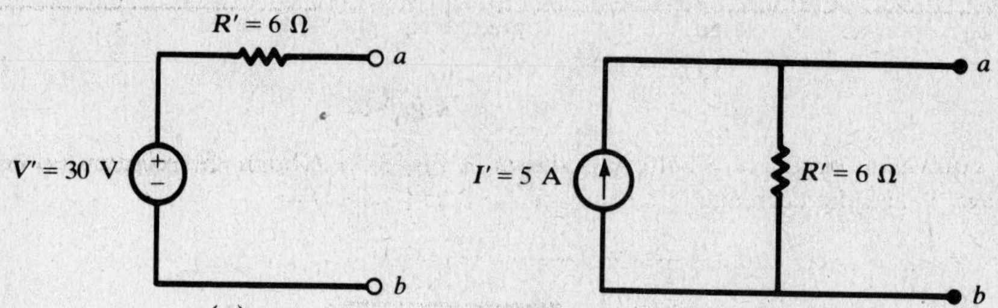

(a) (b)

Fig. 3-35

3.22 A Thévenin equivalent circuit can be converted into a Norton equivalent circuit, under certain limitations on the series resistance R'. State these limitations.

Since $I' = V'/R'$, the value $R' = 0$ is not permitted, since it would imply infinite current I'. At the other extreme, $R' = \infty$ would lead to $I' = 0$. A *practical* voltage source must include a series resistance that is neither zero nor infinity. Similarly, a *practical* current source must include a shunting resistance that is neither zero nor infinity.

3.23 Find the value of the adjustable resistance R which results in maximum power transfer across the terminals ab of the circuit shown in Fig. 3-36.

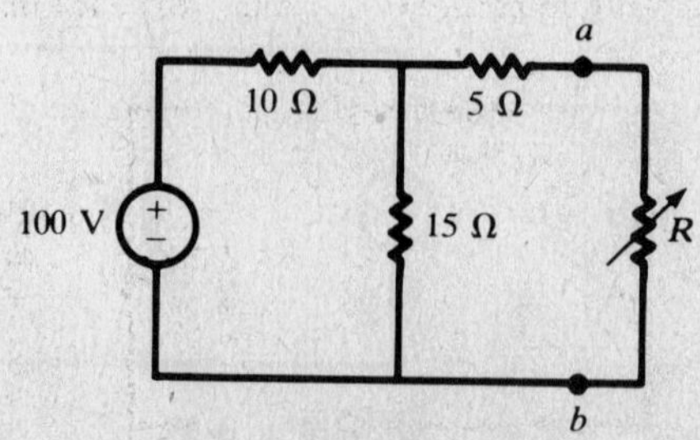

Fig. 3-36

First a Thévenin equivalent is obtained, with $V' = 60$ V and $R' = 11\ \Omega$. By Section 3.7, maximum power transfer occurs for $R = R' = 11\ \Omega$, with

$$P_{max} = \frac{V'^2}{4R'} = 81.82\ \text{W}$$

Supplementary Problems

3.24 What circuit elements are indicated by Fig. 3-37(a) and (b)?
Ans. (a) A 10-V source over the current range 0 to I_0. (b) A 3-A source over the voltage range 0 to V_0.

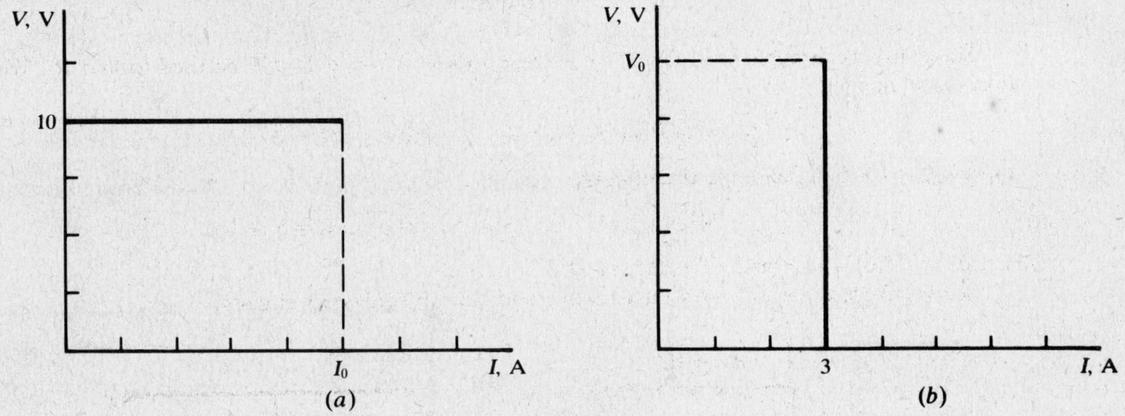

Fig. 3-37

3.25 The equivalent circuit of a battery is shown in Fig. 3-38. Obtain the relationship between the terminal voltage V and the current I. *Ans.* $V = -0.05I + 45$ (V)

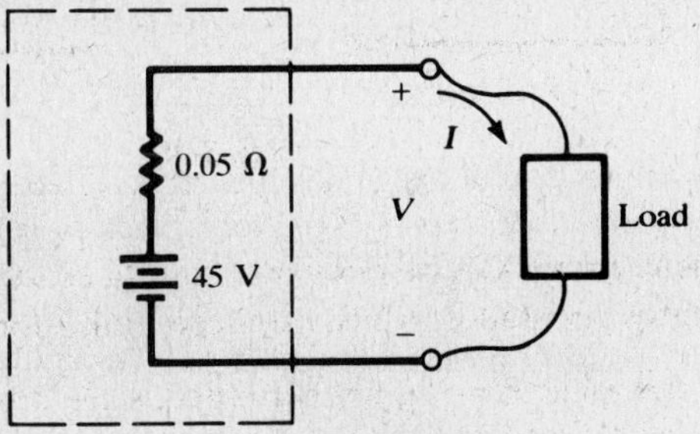

Fig. 3-38

3.26 A practical current source has a current of 22.0 A. Loading the source with 50.0 Ω results in a terminal voltage of 390.3 V. Obtain the source constants, I and R. *Ans.* 22.0 A, 27.5 Ω

3.27 In Fig. 3-39, are the power ratings of the resistors adequate if the resistors are connected to terminals ab (a) one at a time? (b) all three in parallel? *Ans.* (a) each inadequate; (b) adequate

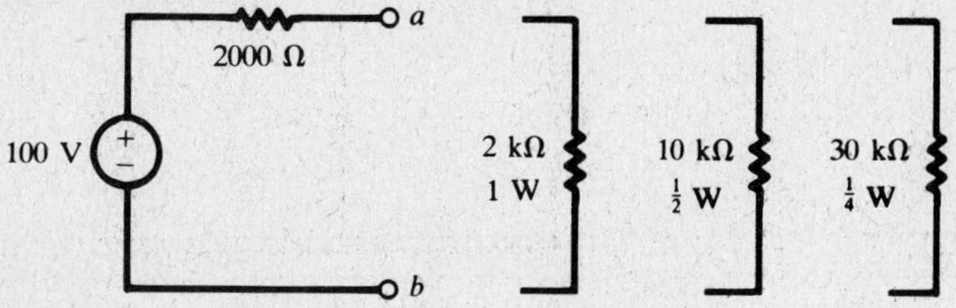

Fig. 3-39

3.28 Determine the power delivered by the current source in Fig. 3-40. *Ans.* 228 W

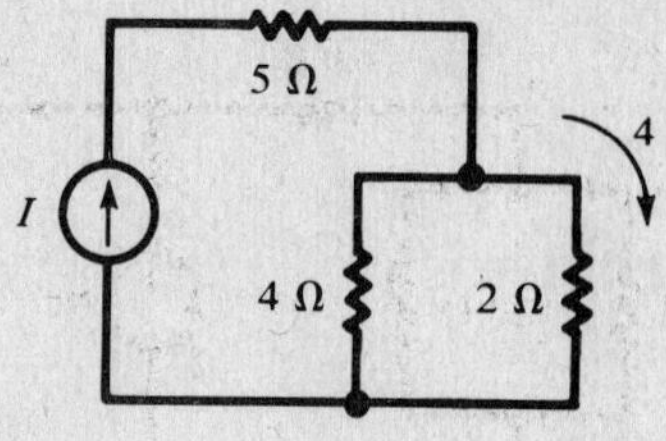

Fig. 3-40

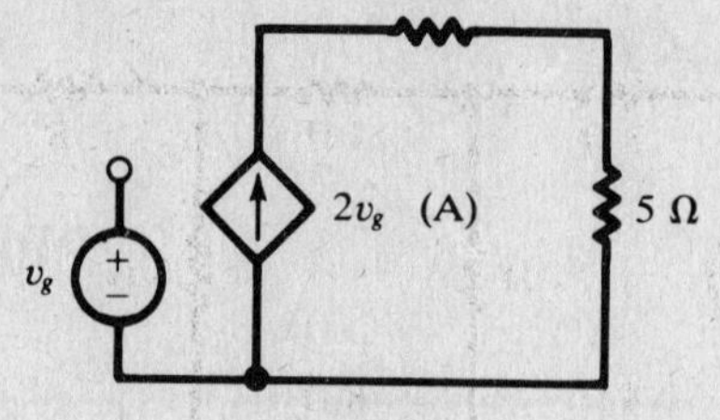

Fig. 3-41

3.29 Calculate the energy dissipated in the 5-Ω resistor of Fig. 3-41 during the interval $0 \leq t \leq 5$ s, if

$$v_g = (10 + 10e^{-t/10}) \quad \text{(V)}$$

Ans. 32.1 kJ

3.30 For the circuit of Fig. 3-42, show that the power supplied by the sources is equal to the power dissipated in the resistors. *Ans.* $P_T = 940.3$ W

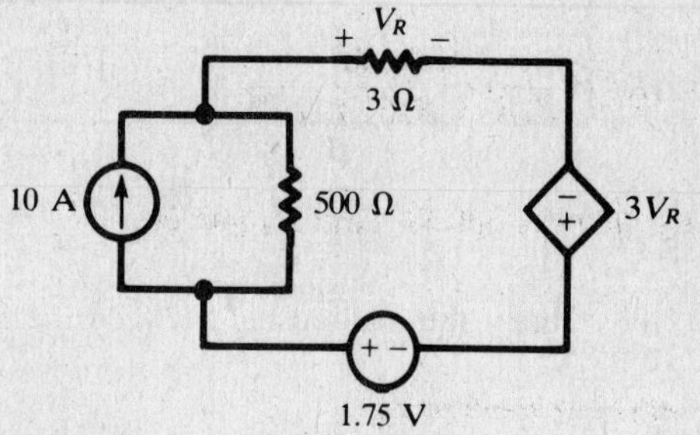

Fig. 3-42

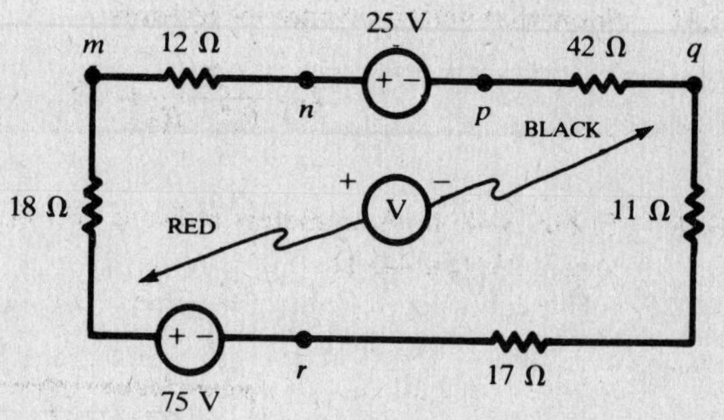

Fig. 3-43

3.31 In Fig. 3-43, determine the reading of a digital voltmeter when connected (*a*) red to *m*, black to *r*; (*b*) red to *q*, black to *r*; (*c*) red to *q*, black to *m*; (*d*) red to *p*, black to *n*.
Ans. (*a*) 66 V; (*b*) 14 V; (*c*) −52 V; (*d*) −25 V

3.32 Determine the currents I_1 and I_2 in Fig. 3-44. *Ans.* −6 A, 9 A

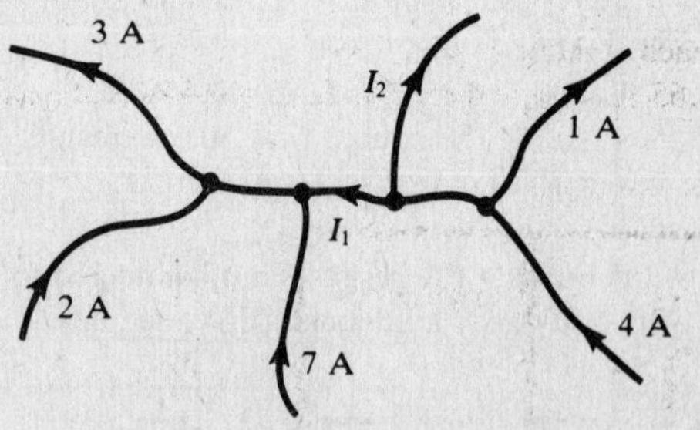

Fig. 3-44

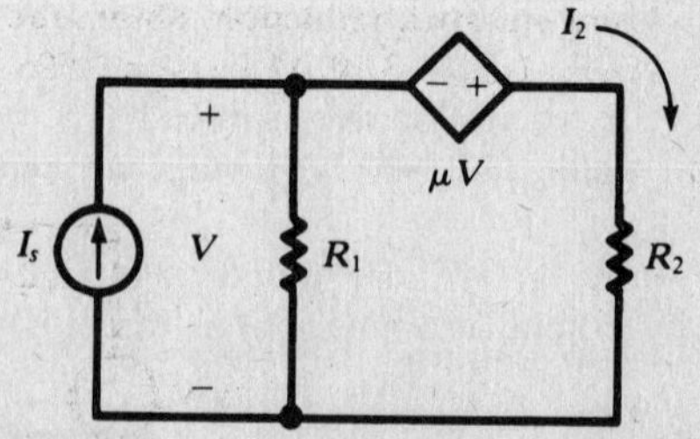

Fig. 3-45

3.33 Find I_2/I_s for the circuit of Fig. 3-45.

Ans. $\dfrac{R_1(1+\mu)}{R_2+(1+\mu)R_1}$

3.34 Find the value of the source voltage in Fig. 3-46 if the power dissipated in the 3-Ω resistor is 0.75 W. *Ans.* $V_s = 0.49R + 3.0$ (V)

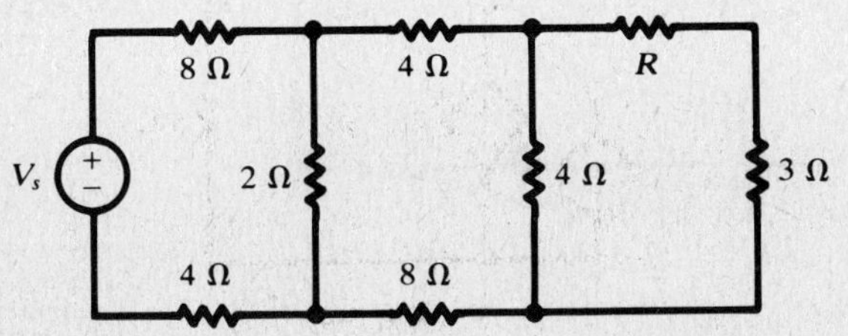

Fig. 3-46

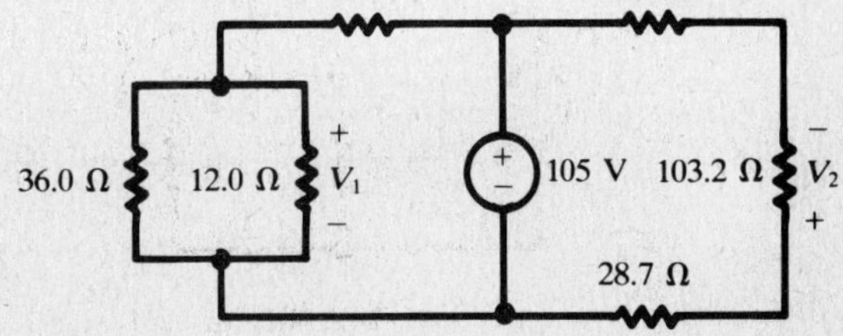

Fig. 3-47

3.35 Using the voltage division rule, calculate V_1 and V_2 in Fig. 3-47. *Ans.* 11.39 V, −73.07 V

3.36 Design a high-voltage divider circuit that reduces 1 MV to 100 V and limits the current to a maximum of 0.5 A. *Ans.* $R_H = 2$ MΩ, $R_L = 200$ Ω

3.37 Show that with four parallel resistors:

$$I_4 = \frac{R^*}{R^* + R_4} I_T \qquad \text{where} \qquad R^* \equiv \frac{R_1R_2R_3}{R_1R_3 + R_2R_3 + R_1R_2}$$

3.38 In Fig. 3-48, both ammeters indicate 1.70 A. The source supplies 300 W. Find R_1 and R_2. *Ans.* 23.9 Ω, 443 Ω

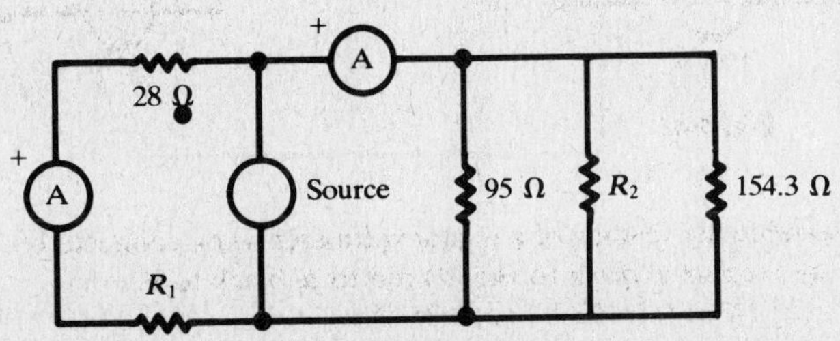

Fig. 3-48

3.39 Using network reduction, obtain the current in each resistor in Fig. 3.49.
Ans. $I_{2.45\Omega} = 3.10$ A, $I_{17.47\Omega} = 0.595$ A, $I_{6.30\Omega} = 1.65$ A, $I_{6.7\Omega} = 0.855$ A, $I_{10.0\Omega} = 0.466$ A, $I_{12.0\Omega} = 0.389$ A

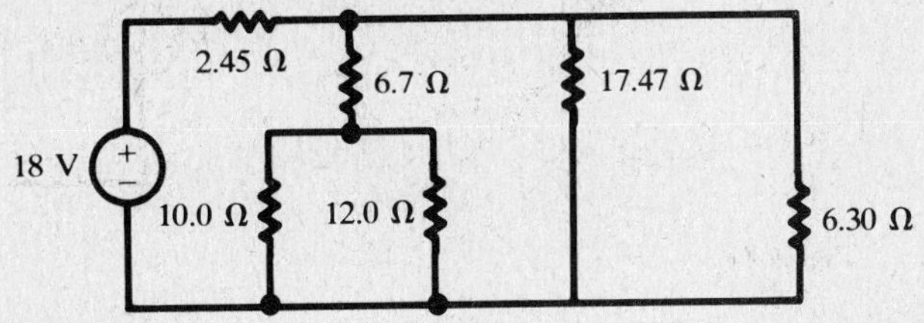

Fig. 3-49

3.40 The 12-Ω resistor in Fig. 3-50 dissipates 147 W. Find the source voltage V_s. *Ans.* 100 V

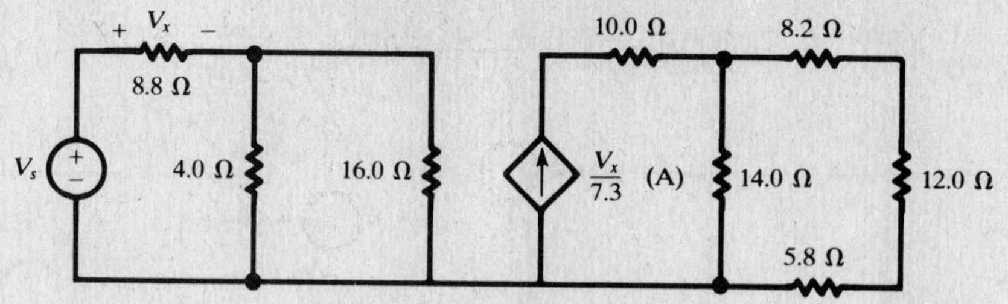

Fig. 3-50

3.41 Using superposition, find the current I in Fig. 3-51. *Ans.* 16.2 A

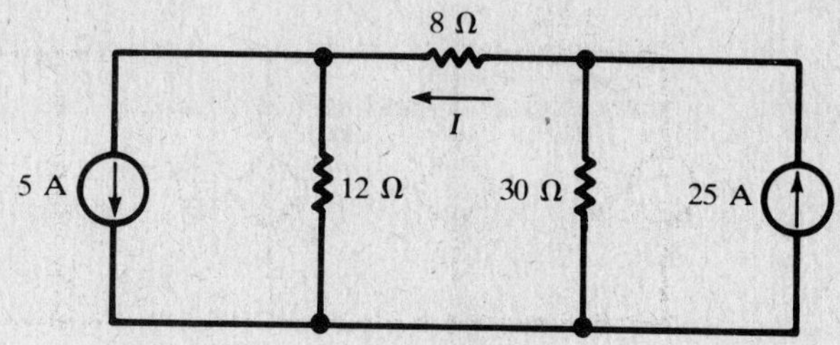

Fig. 3-51

3.42 Using superposition, find the current I in Fig. 3-52. *Ans.* −12 A

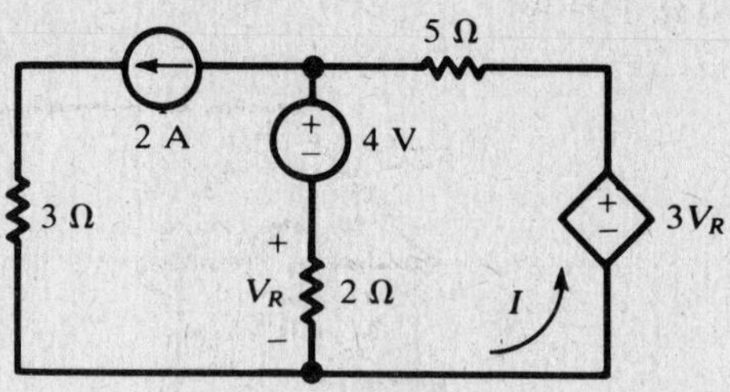

Fig. 3-52

3.43 At no-load, the terminal voltage of a dc generator is 120 V. When delivering its rated current of 40 A, its terminal voltage drops to 112 V. Represent the generator by its Thévenin equivalent.
Ans. $V' = 120$ V, $R' = 0.2\ \Omega$

3.44 In Fig. 3-53, replace the network to the left of the terminals ab with its Thévenin equivalent. Also find the Norton equivalent of the network.
Ans. $V' = \dfrac{100R}{25+R}$ (V), $R' = \dfrac{50R+625}{R+25}$ (Ω), $I' = \dfrac{2R}{R+12.5}$ (A)

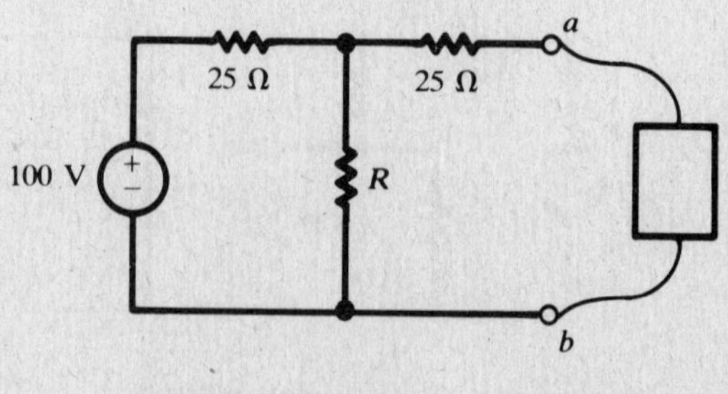

Fig. 3-53

3.45 Replace the circuit in Fig. 3-54 with its Norton equivalent. *Ans.* $I' = 0.8$ A, $R' = 10.0\ \Omega$

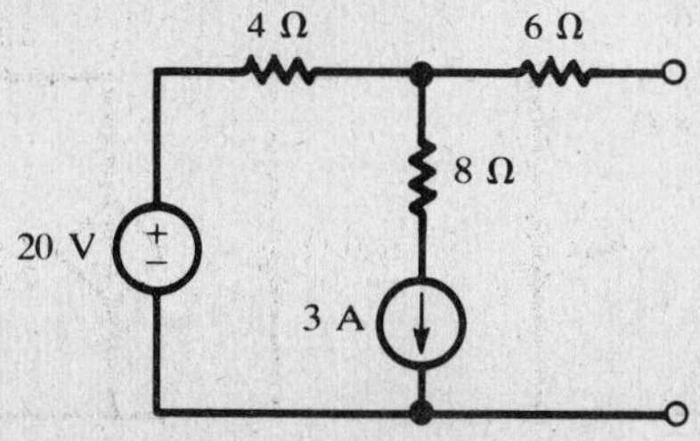

Fig. 3-54

3.46 Replace the circuit in Fig. 3-55 with its Thévenin equivalent. *Ans.* $V' = -2903V_{in}$, $R' = 38.707\ k\Omega$

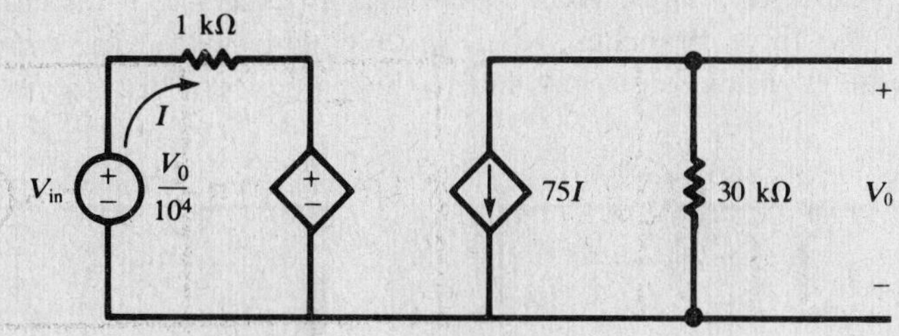

Fig. 3-55

3.47 For the circuit in Fig. 3-56, find the value of R that will receive maximum power. Determine this power. *Ans.* 10.0 Ω, 1.09 W

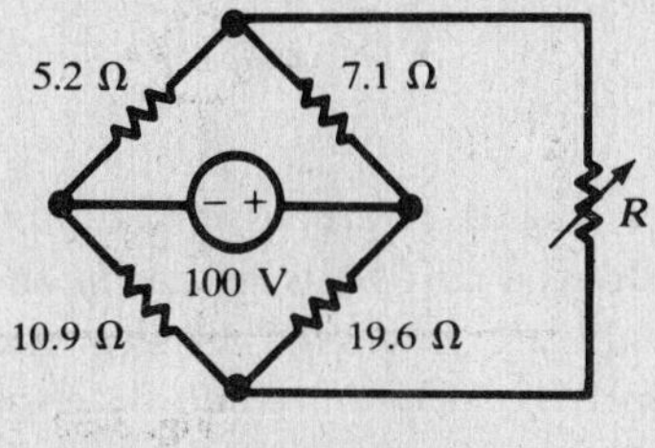

Fig. 3-56

Chapter 4

DC Mesh and Node Analysis

4.1 BRANCH AND MESH CURRENTS

A three-branched active network is shown in Fig. 4-1(*a*). A solution for the *branch currents* I_1, I_2, I_3 can be obtained by applying Kirchhoff's current law at principal nodes (Section 3.2) and equating voltages of parallel branches. Thus, KCL at node a gives $I_1 = I_2 + I_3$; and the voltage V_{ab} can be written for each of the three branches, when the elements within the branches are specified. The result will be three independent equations in the three unknown branch currents.

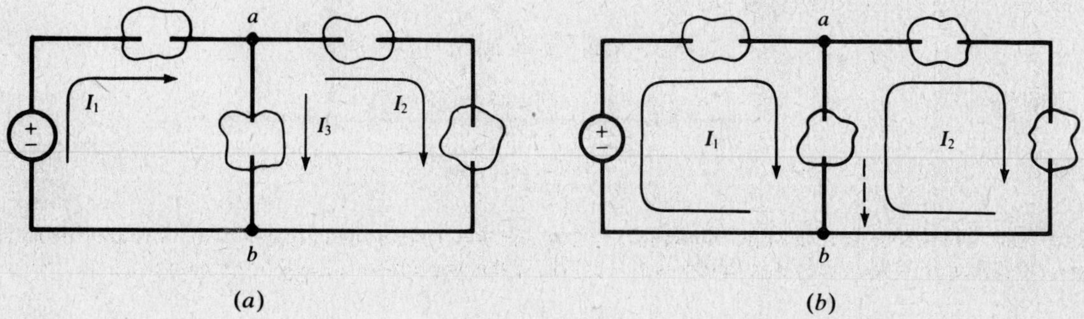

Fig. 4-1

In Fig. 4-1(*b*), *mesh currents* I_1 and I_2 are assigned to the same three-branched network. The current in the center branch (broken arrow) is given by the difference of the two mesh currents, $I_1 - I_2$. In this way KCL is implicitly included in the mesh current method. Two independent equations in the unknowns I_1 and I_2 can be obtained by applying Kirchhoff's voltage law to the two loops in which the mesh currents flow. It is generally best to traverse the loops in the same direction as the currents so that the signs of the voltage terms follow a simple pattern. For a general planar network, a mesh current is assigned to the perimeter of each of the bounded regions into which the network divides the plane. It is also possible to use other sets of loops (see Problem 4.5), in which case one usually speaks of the currents as *loop currents.*

EXAMPLE 4.1 Solve the network of Fig. 4-2 by the branch current method.

Applying KCL at node a,

$$I_1 = I_2 + I_3 \tag{1}$$

Writing V_{ab} for the three branches connected to a and b (i.e., applying KVL to the left loop and the outside loop),

$$20 - I_1(5) = I_3(10) \tag{2}$$

$$20 - I_1(5) = I_2(2) + 8 \tag{3}$$

Solving (*1*), (*2*), and (*3*) simultaneously,

$$I_1 = 2\text{ A} \qquad I_2 = 1\text{ A} \qquad I_3 = 1\text{ A}$$

The branch current method can be difficult to apply to extensive networks, because it does not suggest a starting point in the network and a logical progression through the branches until all currents are obtained.

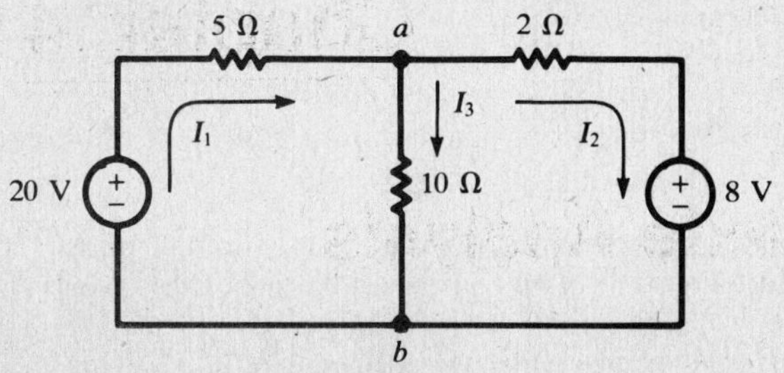

Fig. 4-2

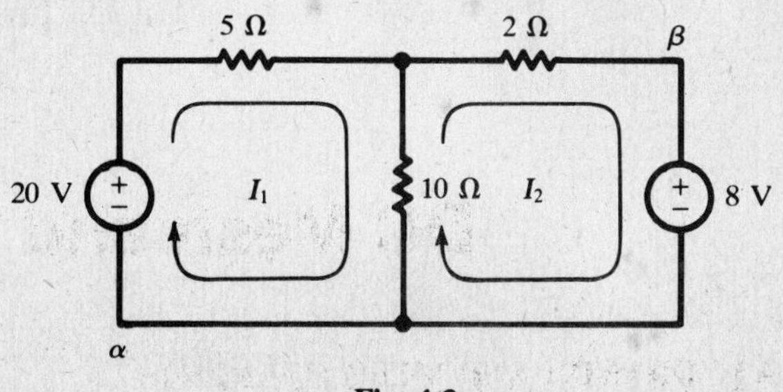

Fig. 4-3

EXAMPLE 4.2 Solve the network of Fig. 4-3 (the same as that of Fig. 4-2) by the mesh current method.

The mesh currents are shown on the circuit diagram. Applying KVL around the left loop, starting at point α,

$$-20 + 5I_1 + 10(I_1 - I_2) = 0$$

and around the right loop, starting at point β,

$$8 + 10(I_2 - I_1) + 2I_2 = 0$$

Rearranging terms,

$$15I_1 - 10I_2 = 20 \qquad (1)$$

$$-10I_1 + 12I_2 = -8 \qquad (2)$$

Solving these simultaneous equations results in $I_1 = 2$ A and $I_2 = 1$ A. If the current in the 10-Ω resistor in the downward direction is required, it is found as $I_1 - I_2 = 1$ A; this was identified as I_3 in Example 4.1.

4.2 MATRICES AND MESH CURRENTS

The n simultaneous equations of an n-mesh network can be written in matrix form. (Refer to Appendix C for an introduction to matrices and determinants.)

EXAMPLE 4.3 When KVL is applied to the three-mesh network of Fig. 4-4, the following three equations are obtained.

$$\begin{aligned}(R_A + R_B)I_1 \quad - R_BI_2 \quad &= V_a \\ -R_BI_1 + (R_B + R_C + R_D)I_2 \quad - R_DI_3 &= 0 \\ -R_DI_2 + (R_D + R_E)I_3 &= -V_b\end{aligned}$$

Placing the equations in matrix form,

$$\begin{bmatrix} R_A + R_B & -R_B & 0 \\ -R_B & R_B + R_C + R_D & -R_D \\ 0 & -R_D & R_D + R_E \end{bmatrix} \begin{bmatrix} I_1 \\ I_2 \\ I_3 \end{bmatrix} = \begin{bmatrix} V_a \\ 0 \\ -V_b \end{bmatrix}$$

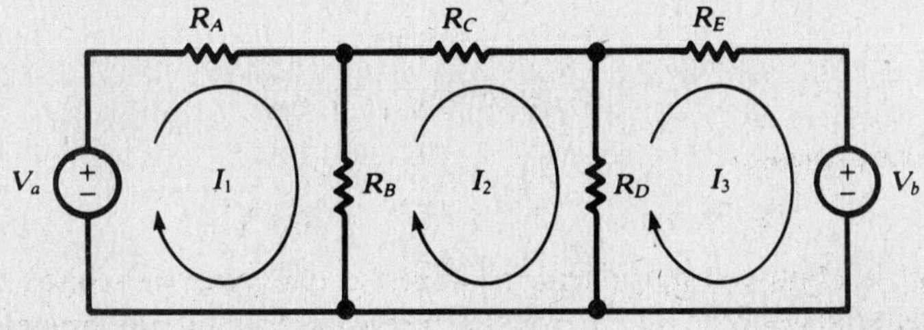

Fig. 4-4

The elements of the matrices can be indicated in general form as follows:

$$\begin{bmatrix} R_{11} & R_{12} & R_{13} \\ R_{21} & R_{22} & R_{23} \\ R_{31} & R_{32} & R_{33} \end{bmatrix} \begin{bmatrix} I_1 \\ I_2 \\ I_3 \end{bmatrix} = \begin{bmatrix} V_1 \\ V_2 \\ V_3 \end{bmatrix} \tag{1}$$

Now element R_{11} (row 1, column 1) is the sum of all resistances through which mesh current I_1 passes. In Fig. 4-4, this is $R_A + R_B$. Similarly, elements R_{22} and R_{33} are the sums of all resistances through which I_2 and I_3, respectively, pass.

Element R_{12} (row 1, column 2) is the sum of all resistances through which mesh currents I_1 and I_2 pass. The sign of R_{12} is + if the two currents are in the same direction through each resistance, and − if they are in opposite directions. In Fig. 4-4, R_B is the only resistance common to I_1 and I_2; and the current directions are opposite in R_B, so that the sign is negative. Similarly, elements R_{21}, R_{23}, R_{13}, and R_{31} are the sums of the resistances common to the two mesh currents indicated by the subscripts, with the signs determined as described above for R_{12}. It should be noted that for all i and j, $R_{ij} = R_{ji}$. As a result, the resistance matrix is symmetric about the principal diagonal.

The current matrix requires no explanation, since the elements are in a single column with subscripts 1, 2, 3 . . . to identify the current with the corresponding mesh. These are the unknowns in the mesh current method of network analysis.

Element V_1 in the voltage matrix is the sum of all source voltages driving mesh current I_1. A voltage is counted positive in the sum if I_1 passes from the − to the + terminal of the source; otherwise it is counted negative. In other words, a voltage is positive if the source drives in the direction of the mesh current. In Fig. 4-4, mesh 1 has a source V_a driving in the direction of I_1; mesh 2 has no source; and mesh 3 has a source V_b driving opposite to the direction of I_3, making V_3 negative.

4.3 DETERMINANTS AND THE MESH CURRENT METHOD

The matrix equation arising from the mesh current method may be solved by various techniques. One of these, the *method of determinants* (Cramer's rule), will be presented here. It should be stated, however, that other techniques are far more efficient for large networks.

EXAMPLE 4.4 Solve matrix equation (*1*) of Example 4.3 by the method of determinants.

The unknown current I_1 is obtained as the ratio of two determinants. The denominator determinant has the elements of the resistance matrix. This may be referred to as the *determinant of the coefficients* and given the symbol Δ_R. The numerator determinant has the same elements as Δ_R except in the first column, where the elements of the voltage matrix replace those of the determinant of the coefficients. Thus

$$I_1 = \begin{vmatrix} V_1 & R_{12} & R_{13} \\ V_2 & R_{22} & R_{23} \\ V_3 & R_{32} & R_{33} \end{vmatrix} \Bigg/ \begin{vmatrix} R_{11} & R_{12} & R_{13} \\ R_{21} & R_{22} & R_{23} \\ R_{31} & R_{32} & R_{33} \end{vmatrix} \equiv \frac{1}{\Delta_R} \begin{vmatrix} V_1 & R_{12} & R_{13} \\ V_2 & R_{22} & R_{23} \\ V_3 & R_{32} & R_{33} \end{vmatrix}$$

Similarly,

$$I_2 = \frac{1}{\Delta_R} \begin{vmatrix} R_{11} & V_1 & R_{13} \\ R_{21} & V_2 & R_{23} \\ R_{31} & V_3 & R_{33} \end{vmatrix} \qquad I_3 = \frac{1}{\Delta_R} \begin{vmatrix} R_{11} & R_{12} & V_1 \\ R_{21} & R_{22} & V_2 \\ R_{31} & R_{32} & V_3 \end{vmatrix}$$

An expansion of the numerator determinants by cofactors of the voltage terms results in a set of equations which can be helpful in understanding the network, particularly in terms of its driving-point and transfer resistances:

$$I_1 = V_1\left(\frac{\Delta_{11}}{\Delta_R}\right) + V_2\left(\frac{\Delta_{21}}{\Delta_R}\right) + V_3\left(\frac{\Delta_{31}}{\Delta_R}\right) \tag{1}$$

$$I_2 = V_1\left(\frac{\Delta_{12}}{\Delta_R}\right) + V_2\left(\frac{\Delta_{22}}{\Delta_R}\right) + V_3\left(\frac{\Delta_{32}}{\Delta_R}\right) \tag{2}$$

$$I_3 = V_1\left(\frac{\Delta_{13}}{\Delta_R}\right) + V_2\left(\frac{\Delta_{23}}{\Delta_R}\right) + V_3\left(\frac{\Delta_{33}}{\Delta_R}\right) \tag{3}$$

Here, Δ_{ij} stands for the cofactor of R_{ij} (the element in row i, column j) in Δ_R. Care must be taken with the signs of the cofactors—see Appendix C.

4.4 INPUT RESISTANCE

In single-source networks, the *input* or *driving-point resistance* is often of interest. Such a network is suggested in Fig. 4-5, where the driving voltage has been designated as V_1 and the corresponding current as I_1. Since the only source is V_1, the equation for I_1 is [see (*1*) of Example 4.4]:

$$I_1 = V_1\left(\frac{\Delta_{11}}{\Delta_R}\right)$$

The input resistance is the ratio of V_1 to I_1:

$$R_{\text{input},1} = \frac{\Delta_R}{\Delta_{11}}$$

The reader should verify that Δ_R/Δ_{11} actually carries the units Ω.

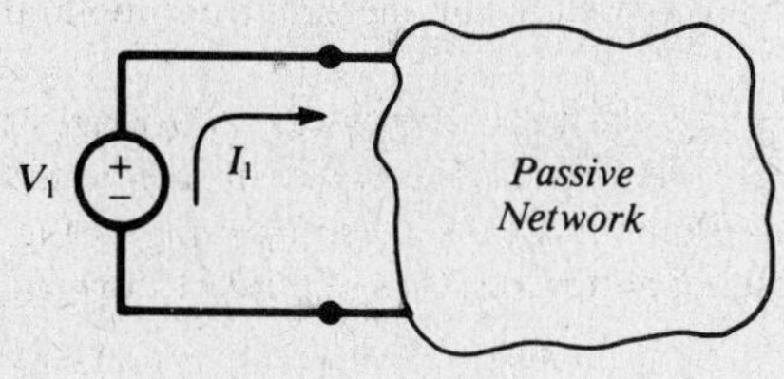

Fig. 4-5

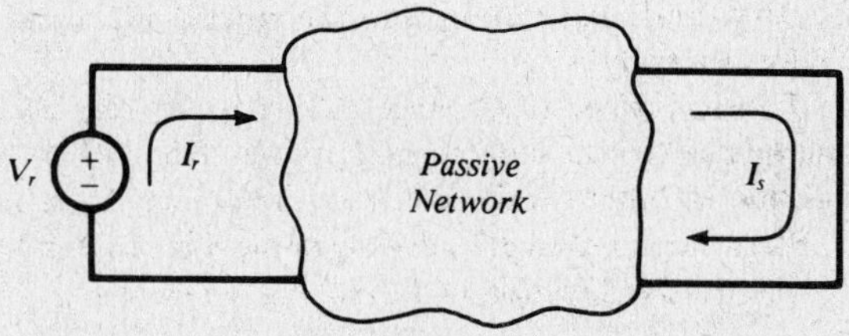

Fig. 4-6

4.5 TRANSFER RESISTANCE

A driving voltage in one part of a network results in currents in all the network branches. For example, a voltage source applied to a passive network results in an output current in that part of the network where a load resistance has been connected. In such a case the network has an overall *transfer resistance*. Consider the passive network suggested in Fig. 4-6, where the voltage source has been designated as V_r and the output current as I_s. The mesh current equation for I_s contains only one term, the one resulting from V_r in the numerator determinant:

$$I_s = (0)\left(\frac{\Delta_{1s}}{\Delta_R}\right) + \cdots + 0 + V_r\left(\frac{\Delta_{rs}}{\Delta_R}\right) + 0 + \cdots$$

The network transfer resistance is the ratio of V_r to I_s:

$$R_{\text{transfer},rs} = \frac{\Delta_R}{\Delta_{rs}}$$

Because the resistance matrix is symmetric, $\Delta_{rs} = \Delta_{sr}$, and so

$$R_{\text{transfer},rs} = R_{\text{transfer},sr}$$

This expresses an important property of linear networks: If a certain voltage in mesh r gives rise to a certain current in mesh s, then the same voltage in mesh s produces the same current in mesh r.

Consider now the more general situation of an n-mesh network containing a number of voltage sources. The solution for the current in mesh k can be rewritten in terms of input and transfer resistances [refer to (*1*), (*2*), and (*3*) of Example 4.4]:

$$I_k = \frac{V_1}{R_{\text{transfer},1k}} + \cdots + \frac{V_{k-1}}{R_{\text{transfer},(k-1)k}} + \frac{V_k}{R_{\text{input},k}} + \frac{V_{k+1}}{R_{\text{transfer},(k+1)k}} + \cdots + \frac{V_n}{R_{\text{transfer},nk}}$$

There is nothing new here mathematically, but in this form the current equation does illustrate the superposition principle very clearly, showing how the resistances control the effects which the voltage sources have on a particular mesh current. A source far removed from mesh k will have a high transfer resistance into that mesh and will therefore contribute very little to I_k. Source V_k, and others in meshes adjacent to mesh k, will provide the greater part of I_k.

4.6 NODE VOLTAGE METHOD

The network shown in Fig. 4-7(a) contains five nodes, where *4* and *5* are simple nodes and *1*, *2*, and *3* are principal nodes. In the node voltage method, one of the principal nodes is selected as the reference and equations based on KCL are written at the other principal nodes. At each of these other principal nodes, a voltage is assigned, where it is understood that this is a voltage *with respect to the reference node.* These voltages are the unknowns and, when determined by a suitable method, result in the network solution.

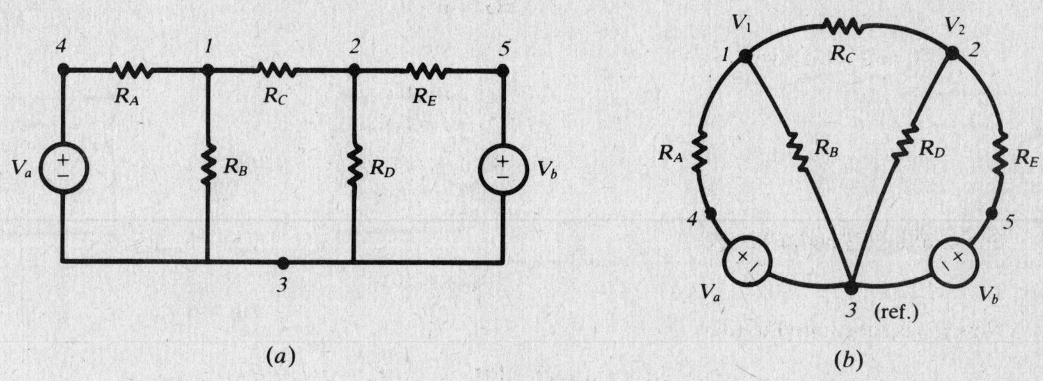

Fig. 4-7

The network is redrawn in Fig. 4-7(b) and node *3* selected as the reference for voltages V_1 and V_2. KCL requires that the total current out of node *1* be zero:

$$\frac{V_1 - V_a}{R_A} + \frac{V_1}{R_B} + \frac{V_1 - V_2}{R_C} = 0$$

Similarly, the total current out of node *2* must be zero:

$$\frac{V_2 - V_1}{R_C} + \frac{V_2}{R_D} + \frac{V_2 - V_b}{R_E} = 0$$

(Applying KCL in this form does not imply that the actual branch currents all are directed out of either node. Indeed, the current in branch *12* is necessarily directed *out of* one node and *into* the other.) Putting the two equations for V_1 and V_2 in matrix form,

$$\begin{bmatrix} \dfrac{1}{R_A} + \dfrac{1}{R_B} + \dfrac{1}{R_C} & -\dfrac{1}{R_C} \\ -\dfrac{1}{R_C} & \dfrac{1}{R_C} + \dfrac{1}{R_D} + \dfrac{1}{R_E} \end{bmatrix} \begin{bmatrix} V_1 \\ V_2 \end{bmatrix} = \begin{bmatrix} V_a/R_A \\ V_b/R_E \end{bmatrix}$$

Note the symmetry of the coefficient matrix. The 1,1-element contains the reciprocals of all resistances connected to node *1*; the 2,2-element contains the reciprocals of all resistances connected to node *2*. The 1,2- and 2,1-elements are each equal to the *negative* of the sum of the reciprocals of the resistances of all branches joining nodes *1* and *2*. (There is just one such branch in the present circuit.)

On the right-hand side, the current matrix contains V_a/R_A and V_b/R_E, the driving currents. Both these terms are taken positive because they both drive a current *into* a node. Further discussion of the elements in the matrix representation of the node voltage equations is given in Chapter 8, where the networks are treated in the sinusoidal steady state.

Solved Problems

4.1 Use branch currents in the network shown in Fig. 4-8 to find the current supplied by the 60-V source.

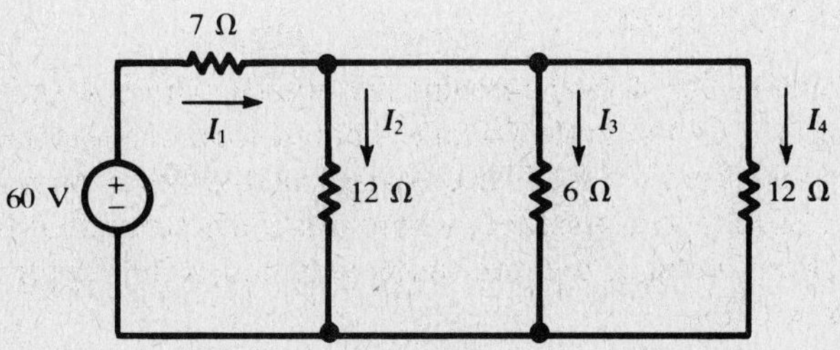

Fig. 4-8

KVL and KCL give:

$$I_2(12) = I_3(6) \quad (1)$$
$$I_2(12) = I_4(12) \quad (2)$$
$$60 = I_1(7) + I_2(12) \quad (3)$$
$$I_1 = I_2 + I_3 + I_4 \quad (4)$$

Substituting (*1*) and (*2*) in (*4*),

$$I_1 = I_2 + 2I_2 + I_2 = 4I_2 \quad (5)$$

Now (*5*) is substituted in (*3*):

$$60 = I_1(7) + \tfrac{1}{4}I_1(12) = 10I_1 \qquad \text{or} \qquad I_1 = 6\text{ A}$$

4.2 Solve Problem 4.1 by the mesh current method.

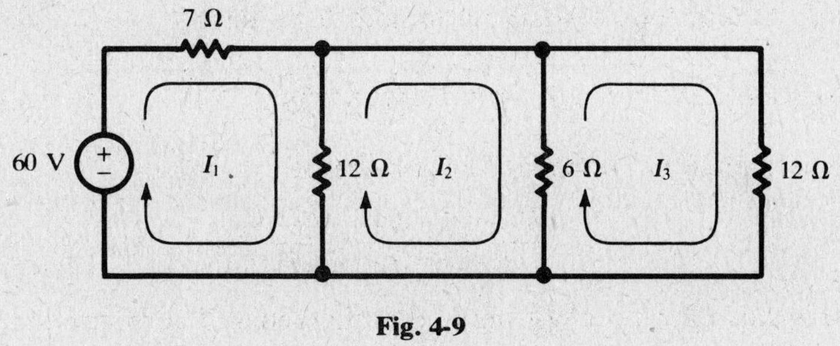

Fig. 4-9

Applying KVL to each mesh results in

$$60 = 7I_1 + 12(I_1 - I_2)$$
$$0 = 12(I_2 - I_1) + 6(I_2 - I_3)$$
$$0 = 6(I_3 - I_2) + 12I_3$$

Rearranging terms and putting the equations in matrix form,

$$\begin{aligned} 19I_1 - 12I_2 \qquad &= 60 \\ -12I_1 + 18I_2 - 6I_3 &= 0 \\ -\ 6I_2 + 18I_3 &= 0 \end{aligned} \qquad \text{or} \qquad \begin{bmatrix} 19 & -12 & 0 \\ -12 & 18 & -6 \\ 0 & -6 & 18 \end{bmatrix} \begin{bmatrix} I_1 \\ I_2 \\ I_3 \end{bmatrix} = \begin{bmatrix} 60 \\ 0 \\ 0 \end{bmatrix}$$

Using Cramer's rule to find I_1,

$$I_1 = \begin{vmatrix} 60 & -12 & 0 \\ 0 & 18 & -6 \\ 0 & -6 & 18 \end{vmatrix} \div \begin{vmatrix} 19 & -12 & 0 \\ -12 & 18 & -6 \\ 0 & -6 & 18 \end{vmatrix} = 17\,280 \div 2880 = 6\text{ A}$$

4.3 In Problem 4.2, obtain $R_{\text{input},1}$ and use it to calculate I_1.

$$R_{\text{input},1} = \frac{\Delta_R}{\Delta_{11}} = \frac{2880}{\begin{vmatrix} 18 & -6 \\ -6 & 18 \end{vmatrix}} = \frac{2880}{288} = 10\ \Omega$$

Then $$I_1 = \frac{60}{R_{\text{input},1}} = \frac{60}{10} = 6\ \text{A}$$

4.4 Obtain $R_{\text{transfer},12}$ and $R_{\text{transfer},13}$ for the network of Problem 4.2 and use them to calculate I_2 and I_3.

The cofactor of the 1,2-element in Δ_R must include a negative sign:

$$\Delta_{12} = (-1)^{1+2}\begin{vmatrix} -12 & -6 \\ 0 & 18 \end{vmatrix} = 216 \qquad R_{\text{transfer},12} = \frac{2880}{216} = 13.33\ \Omega$$

Then, $I_2 = 60/13.33 = 4.50$ A.

$$\Delta_{13} = (-1)^{1+3}\begin{vmatrix} -12 & 18 \\ 0 & -6 \end{vmatrix} = 72 \qquad R_{\text{transfer},13} = \frac{2880}{72} = 40\ \Omega$$

Then, $I_3 = 60/40 = 1.50$ A.

4.5 Solve Problem 4.1 by use of the loop currents indicated in Fig. 4-10.

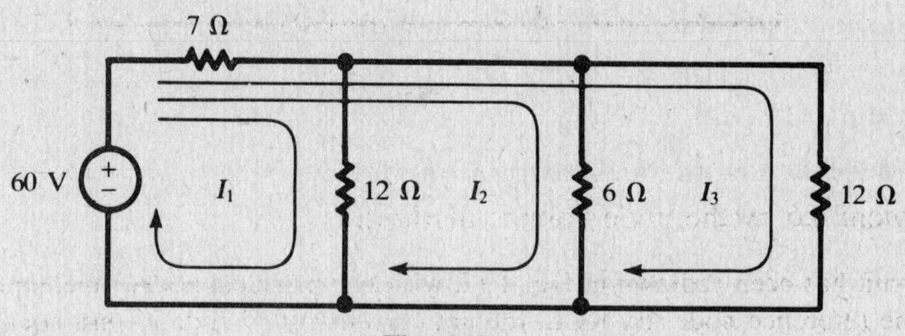

Fig. 4-10

The elements in the matrix form of the equations are obtained by inspection, following the rules of Section 4.2.

$$\begin{bmatrix} 19 & 7 & 7 \\ 7 & 13 & 7 \\ 7 & 7 & 19 \end{bmatrix}\begin{bmatrix} I_1 \\ I_2 \\ I_3 \end{bmatrix} = \begin{bmatrix} 60 \\ 60 \\ 60 \end{bmatrix}$$

Thus $$\Delta_R = \begin{bmatrix} 19 & 7 & 7 \\ 7 & 13 & 7 \\ 7 & 7 & 19 \end{bmatrix} = 2880$$

Notice that in Problem 4.2, too, $\Delta_R = 2880$, although the elements in the determinant were different. *All valid sets of meshes or loops yield the same numerical value for* Δ_R. The three numerator determinants are:

$$N_1 = \begin{vmatrix} 60 & 7 & 7 \\ 60 & 13 & 7 \\ 60 & 7 & 19 \end{vmatrix} = 4320 \qquad N_2 = 8640 \qquad N_3 = 4320$$

Consequently,

$$I_1 = \frac{N_1}{\Delta_R} = \frac{4320}{2880} = 1.5\ \text{A} \qquad I_2 = \frac{N_2}{\Delta_R} = 3\ \text{A} \qquad I_3 = \frac{N_3}{\Delta_R} = 1.5\ \text{A}$$

The current supplied by the 60-V source is the sum of the three loop currents, $I_1 + I_2 + I_3 = 6$ A.

4.6 Write the mesh current matrix equation for the network of Fig. 4-11 by inspection, and solve for the currents.

$$\begin{bmatrix} 7 & -5 & 0 \\ -5 & 19 & -4 \\ 0 & -4 & 6 \end{bmatrix}\begin{bmatrix} I_1 \\ I_2 \\ I_3 \end{bmatrix} = \begin{bmatrix} -25 \\ 25 \\ 50 \end{bmatrix}$$

Solving,

$$I_1 = \begin{vmatrix} -25 & -5 & 0 \\ 25 & 19 & -4 \\ 50 & -4 & 6 \end{vmatrix} \div \begin{vmatrix} 7 & -5 & 0 \\ -5 & 19 & -4 \\ 0 & -4 & 6 \end{vmatrix} = (-700) \div 536 = -1.31 \text{ A}$$

Similarly,

$$I_2 = \frac{N_2}{\Delta_R} = \frac{1700}{536} = 3.17 \text{ A} \qquad I_3 = \frac{N_3}{\Delta_R} = \frac{5600}{536} = 10.45 \text{ A}$$

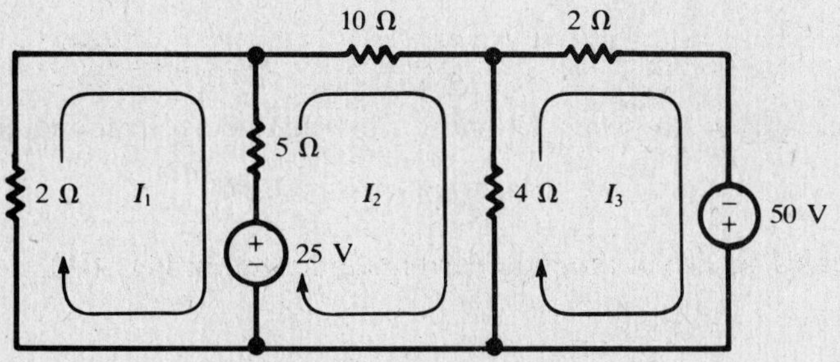

Fig. 4-11

4.7 Solve Problem 4.6 by the node voltage method.

The circuit has been redrawn in Fig. 4-12, with two principal nodes numbered *1* and *2* and the third chosen as the reference node. By KCL, the net current out of node *1* must equal zero.

$$\frac{V_1}{2} + \frac{V_1 - 25}{5} + \frac{V_1 - V_2}{10} = 0$$

Similarly, at node *2*,

$$\frac{V_2 - V_1}{10} + \frac{V_2}{4} + \frac{V_2 + 50}{2} = 0$$

Putting the two equations in matrix form,

$$\begin{bmatrix} \frac{1}{2} + \frac{1}{5} + \frac{1}{10} & -\frac{1}{10} \\ -\frac{1}{10} & \frac{1}{10} + \frac{1}{4} + \frac{1}{2} \end{bmatrix}\begin{bmatrix} V_1 \\ V_2 \end{bmatrix} = \begin{bmatrix} 5 \\ -25 \end{bmatrix}$$

The determinant of coefficients and the numerator determinants are

$$\Delta = \begin{vmatrix} 0.80 & -0.10 \\ -0.10 & 0.85 \end{vmatrix} = 0.670$$

$$N_1 = \begin{vmatrix} 5 & -0.10 \\ -25 & 0.85 \end{vmatrix} = 1.75 \qquad N_2 = \begin{vmatrix} 0.80 & 5 \\ -0.10 & -25 \end{vmatrix} = -19.5$$

From these,

$$V_1 = \frac{1.75}{0.670} = 2.61 \text{ V} \qquad V_2 = \frac{-19.5}{0.670} = -29.1 \text{ V}$$

In terms of these voltages, the currents in Fig. 4-11 are determined as follows:

$$I_1 = \frac{-V_1}{2} = -1.31 \text{ A} \qquad I_2 = \frac{V_1 - V_2}{10} = 3.17 \text{ A} \qquad I_3 = \frac{V_2 + 50}{2} = 10.45 \text{ A}$$

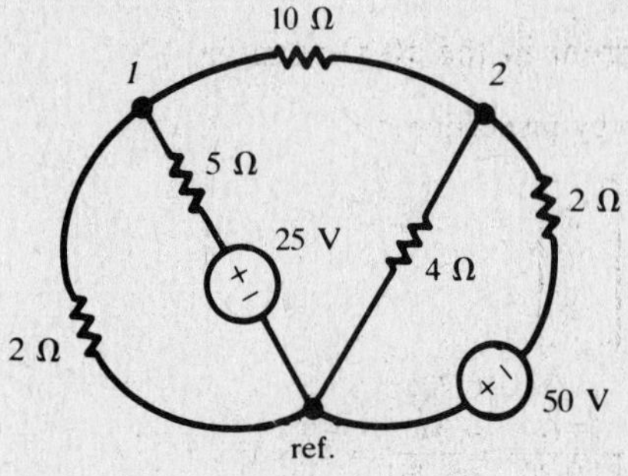

Fig. 4-12

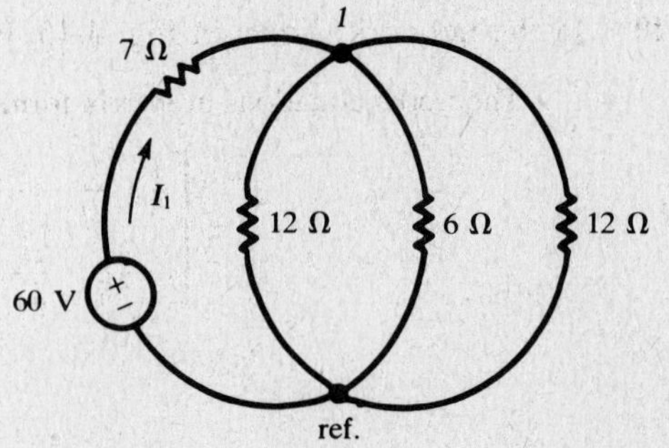

Fig. 4-13

4.8 Solve the network of Problems 4.1 and 4.2 by the node voltage method. See Fig. 4-13.

With two principal nodes, only one equation is necessary.

$$\frac{V_1 - 60}{7} + \frac{V_1}{12} + \frac{V_1}{6} + \frac{V_1}{12} = 0$$

from which $V_1 = 18$ V. Then,

$$I_1 = \frac{60 - V_1}{7} = 6 \text{ A}$$

4.9 For the network shown in Fig. 4-14, find V_s which makes $I_0 = 7.5$ mA.

The node voltage method will be used and the matrix form of the equations written by inspection.

$$\begin{bmatrix} \frac{1}{20}+\frac{1}{7}+\frac{1}{4} & -\frac{1}{4} \\ -\frac{1}{4} & \frac{1}{4}+\frac{1}{6}+\frac{1}{6} \end{bmatrix} \begin{bmatrix} V_1 \\ V_2 \end{bmatrix} = \begin{bmatrix} V_s/20 \\ 0 \end{bmatrix}$$

Solving for V_2,

$$V_2 = \frac{\begin{vmatrix} 0.443 & V_s/20 \\ -0.250 & 0 \end{vmatrix}}{\begin{vmatrix} 0.443 & -0.250 \\ -0.250 & 0.583 \end{vmatrix}} = 0.0638\, V_s$$

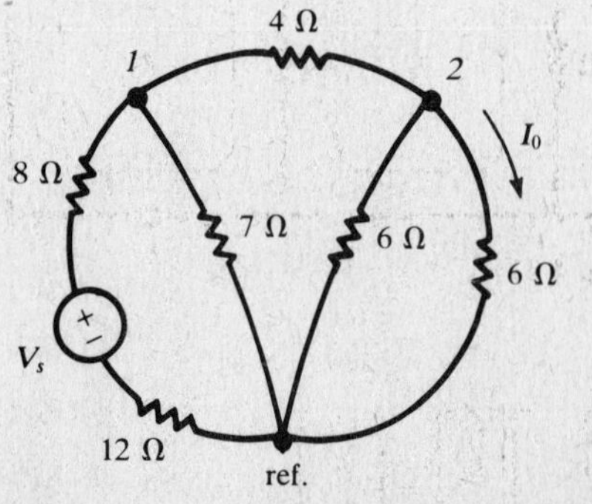

Fig. 4-14

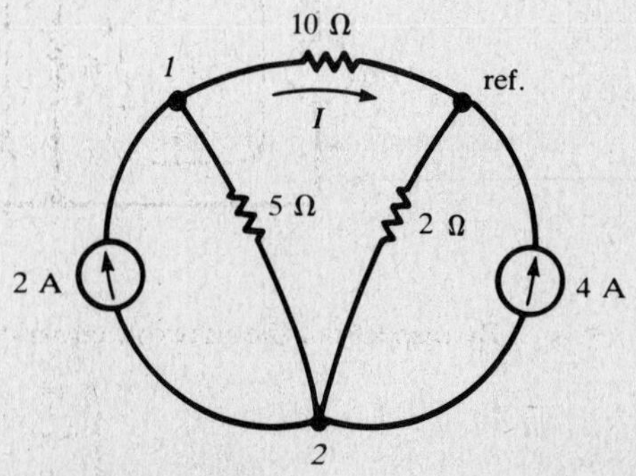

Fig. 4-15

Then

$$7.5 \times 10^{-3} = I_0 = \frac{V_2}{6} = \frac{0.0638 V_s}{6}$$

from which $V_s = 0.705$ V.

4.10 In the network shown in Fig. 4-15, find the current in the 10-Ω resistor.

The nodal equations in matrix form are written by inspection.

$$\begin{bmatrix} \frac{1}{5}+\frac{1}{10} & -\frac{1}{5} \\ -\frac{1}{5} & \frac{1}{5}+\frac{1}{2} \end{bmatrix} \begin{bmatrix} V_1 \\ V_2 \end{bmatrix} = \begin{bmatrix} 2 \\ -6 \end{bmatrix}$$

$$V_1 = \frac{\begin{vmatrix} 2 & -0.20 \\ -6 & 0.70 \end{vmatrix}}{\begin{vmatrix} 0.30 & -0.20 \\ -0.20 & 0.70 \end{vmatrix}} = 1.18 \text{ V}$$

Then, $I = V_1/10 = 0.118$ A.

4.11 Find the voltage V_{ab} in the network shown in Fig. 4-16.

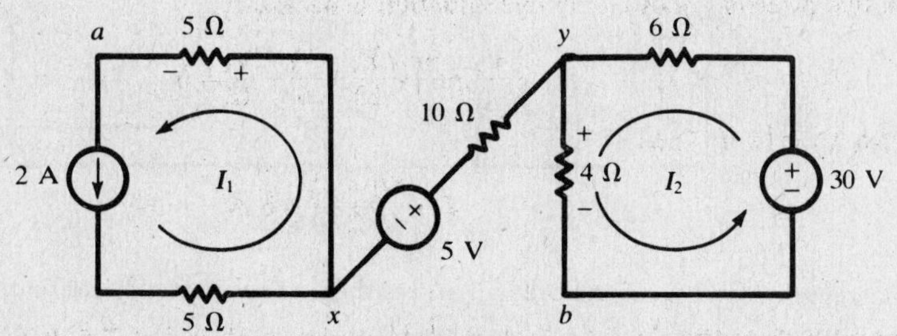

Fig. 4-16

The two closed loops are independent and no current can pass through the connecting branch.

$$I_1 = 2 \text{ A} \qquad I_2 = \frac{30}{10} = 3 \text{ A}$$

$$V_{ab} = V_{ax} + V_{xy} + V_{yb} = -I_1(5) - 5 + I_2(4) = -3 \text{ V}$$

4.12 For the ladder network of Fig. 4-17, obtain the transfer resistance as expressed by the ratio of V_{in} to I_4.

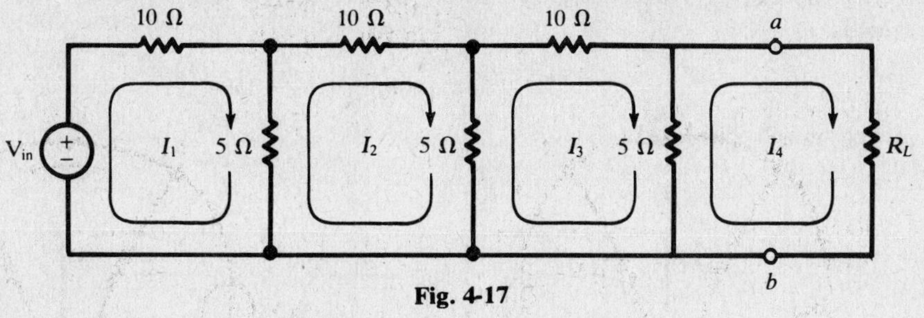

Fig. 4-17

By inspection, the network equation is

$$\begin{bmatrix} 15 & -5 & 0 & 0 \\ -5 & 20 & -5 & 0 \\ 0 & -5 & 20 & -5 \\ 0 & 0 & -5 & 5+R_L \end{bmatrix} \begin{bmatrix} I_1 \\ I_2 \\ I_3 \\ I_4 \end{bmatrix} = \begin{bmatrix} V_{\text{in}} \\ 0 \\ 0 \\ 0 \end{bmatrix}$$

$$\Delta_R = 5125R_L + 18\,750 \qquad N_4 = 125V_{in}$$

$$I_4 = \frac{N_4}{\Delta_R} = \frac{V_{in}}{41R_L + 150} \quad \text{(A)}$$

and

$$R_{\text{transfer},14} = \frac{V_{in}}{I_4} = 41R_L + 150 \quad (\Omega)$$

4.13 Obtain a Thévenin equivalent for the circuit of Fig. 4-17 to the left of terminals ab.

The short-circuit current I_{sc} is obtained from the three-mesh circuit shown in Fig. 4-18.

$$\begin{bmatrix} 15 & -5 & 0 \\ -5 & 20 & -5 \\ 0 & -5 & 15 \end{bmatrix} \begin{bmatrix} I_1 \\ I_2 \\ I_{sc} \end{bmatrix} = \begin{bmatrix} V_{in} \\ 0 \\ 0 \end{bmatrix}$$

$$I_{sc} = \frac{V_{in}\begin{vmatrix} -5 & 20 \\ 0 & -5 \end{vmatrix}}{\Delta_R} = \frac{V_{in}}{150}$$

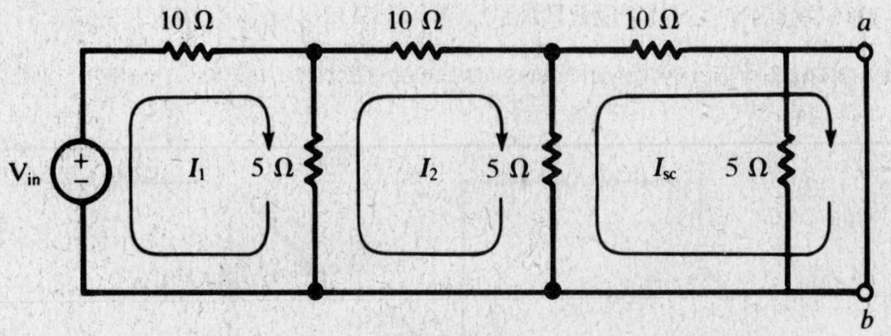

Fig. 4-18

The open-circuit voltage V_{oc} is the voltage across the 5-Ω resistor indicated in Fig. 4-19.

$$\begin{bmatrix} 15 & -5 & 0 \\ -5 & 20 & -5 \\ 0 & -5 & 20 \end{bmatrix} \begin{bmatrix} I_1 \\ I_2 \\ I_3 \end{bmatrix} = \begin{bmatrix} V_{in} \\ 0 \\ 0 \end{bmatrix}$$

$$I_3 = \frac{25V_{in}}{5125} = \frac{V_{in}}{205} \quad \text{(A)}$$

Then, $V_{oc} = I_3(5) = V_{in}/41$, and

$$R_{Th} = \frac{V_{oc}}{I_{sc}} = \frac{150}{41} \ \Omega$$

The Thévenin equivalent circuit is shown in Fig. 4-20. With R_L connected to terminals ab, the output current is

$$I_4 = \frac{V_{in}/41}{(150/41) + R_L} = \frac{V_{in}}{41R_L + 150} \quad \text{(A)}$$

agreeing with Problem 4.12.

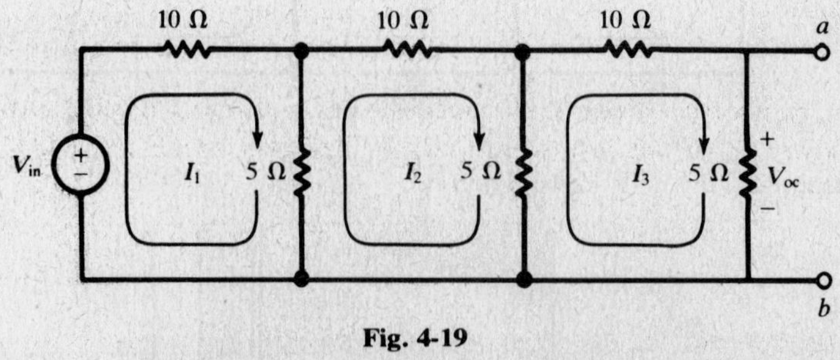

Fig. 4-19

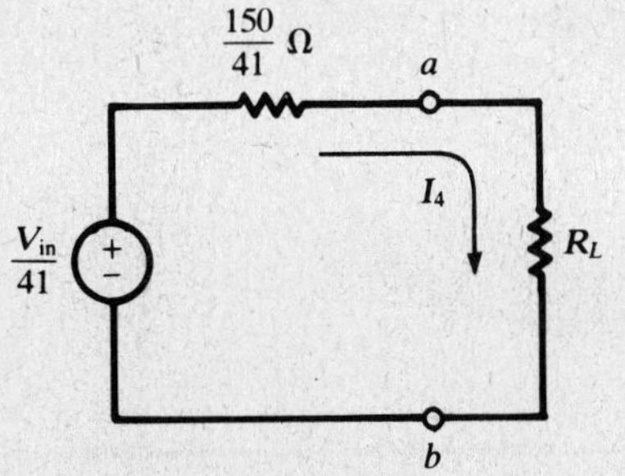

Fig. 4-20

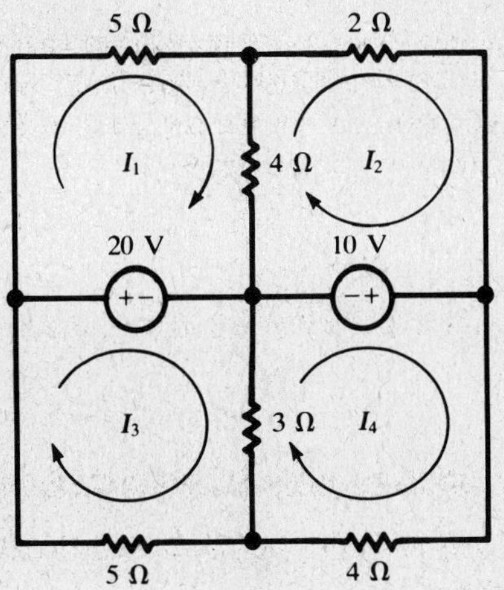

Fig. 4-21

4.14 Obtain the four mesh currents in Fig. 4-21.

The matrix form of the equations is written by inspection:

$$\left[\begin{array}{cc|cc} 9 & -4 & 0 & 0 \\ -4 & 6 & 0 & 0 \\ \hline 0 & 0 & 8 & -3 \\ 0 & 0 & -3 & 7 \end{array}\right]\begin{bmatrix} I_1 \\ I_2 \\ I_3 \\ I_4 \end{bmatrix} = \begin{bmatrix} 20 \\ -10 \\ -20 \\ 10 \end{bmatrix}$$

It should be noted that with only voltage sources in the common branches between loops, the equations for I_1 and I_2 are independent from the equations for I_3 and I_4. This allows the following evaluations of the various determinants:

$$\Delta_R = \begin{vmatrix} 9 & -4 \\ -4 & 6 \end{vmatrix}\begin{vmatrix} 8 & -3 \\ -3 & 7 \end{vmatrix} = (38)(47) = 1786$$

$$N_1 = \begin{vmatrix} 20 & -4 \\ -10 & 6 \end{vmatrix}(47) = 3760 \qquad N_2 = -470 \qquad N_3 = -4180 \qquad N_4 = 760$$

Then $\qquad I_1 = \dfrac{N_1}{\Delta_R} = 2.11\text{ A} \qquad I_2 = -0.263\text{ A} \qquad I_3 = -2.34\text{ A} \qquad I_4 = 0.426\text{ A}$

Supplementary Problems

4.15 (*a*) Find the current I in the network of Fig. 4-22 using the branch current method. (*b*) Repeat using the mesh current method. (*c*) Choose a set of loop currents and find current I. *Ans.* -8.77 A

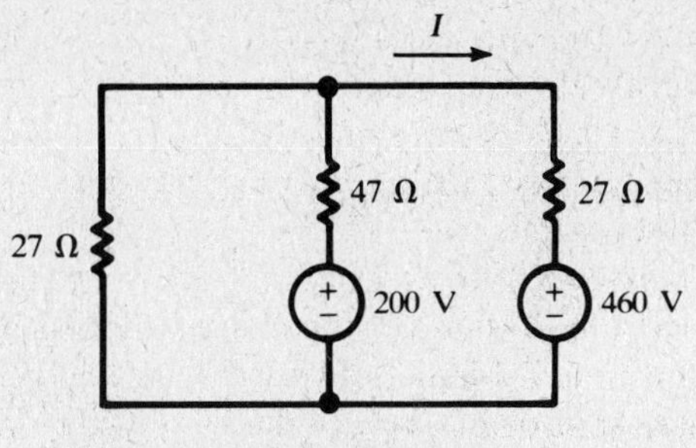

Fig. 4-22

4.16 The circuit of Problem 4.15 has been redrawn in Fig. 4-23. Using the node voltage method, find V_1 and use it to check the value of I in Problem 4.15. *Ans.* 223.3 V

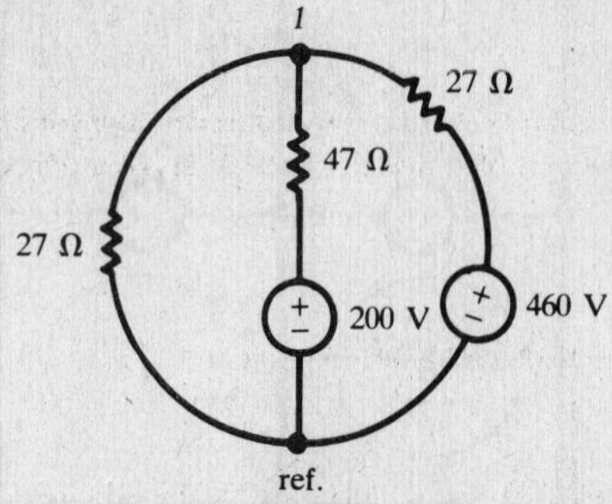

Fig. 4-23

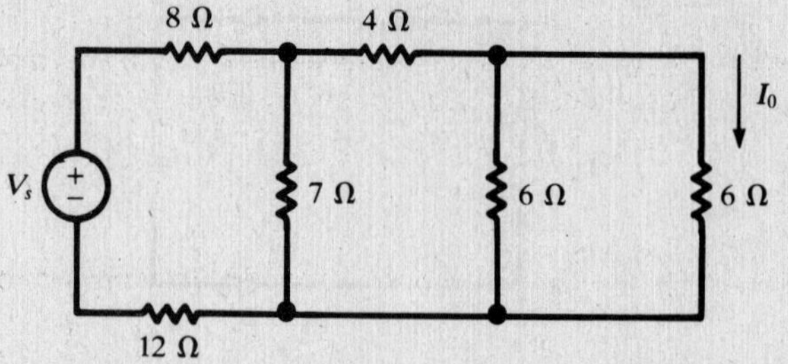

Fig. 4-24

4.17 In Fig. 4-24, $I_0 = 7.5$ mA. Use mesh currents to find the required value of V_s. *Ans.* 0.705 V

4.18 For the network of Fig. 4-24, find the input resistance as seen by the source, using appropriate determinants from Problem 4.17. Check the result by network reduction. *Ans.* 23.5 Ω

4.19 For the network of Fig. 4-24, obtain the transfer resistance which relates current I_0 to the source voltage V_s. *Ans.* 94.0 Ω

4.20 For the network shown in Fig. 4-25, obtain the mesh currents I_1, I_2, and I_3. *Ans.* 5.0 A, 1.0 A, 0.5 A

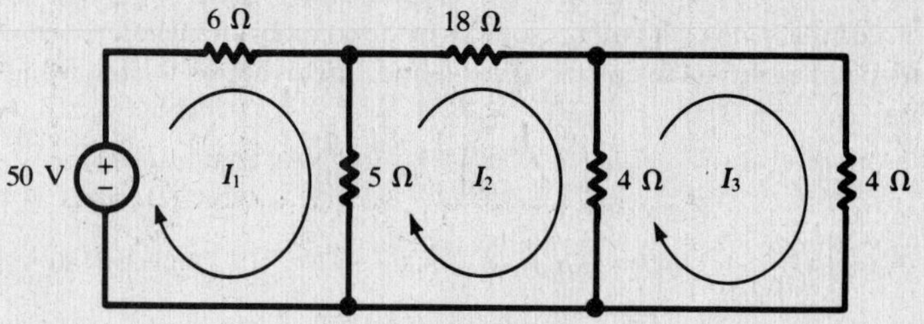

Fig. 4-25

4.21 From the resistance matrix of Problem 4.20, compute $R_{\text{input},1}$, $R_{\text{transfer},12}$, and $R_{\text{transfer},13}$. Use these resistances to obtain the currents I_1, I_2, and I_3. *Ans.* 10 Ω, 50 Ω, 100 Ω

4.22 Loop currents are shown in the network of Fig. 4-26. Write the matrix equation by inspection and solve for I_1, I_2, and I_3. *Ans.* 3.55 A, −1.98 A, −2.98 A

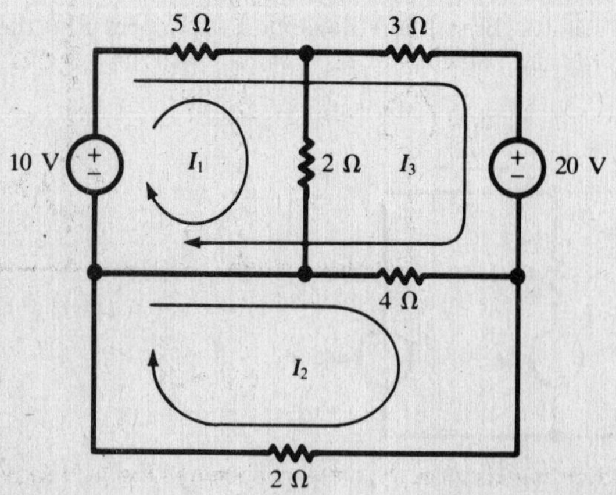

Fig. 4-26

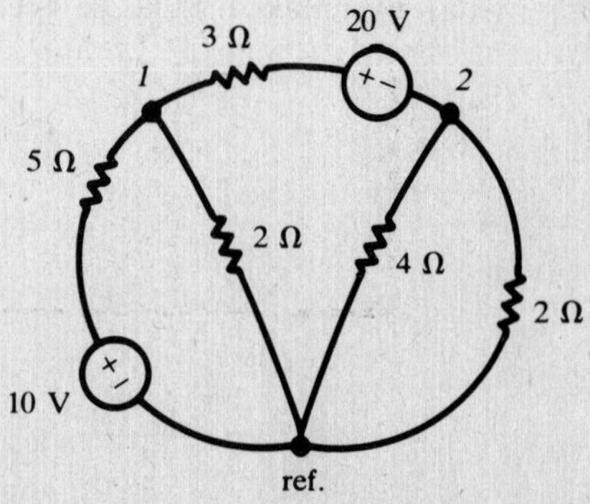

Fig. 4-27

4.23 The network of Problem 4.22 has been redrawn in Fig. 4-27 for solution by the node voltage method. Find the node voltages V_1 and V_2, and with them verify the three currents obtained in Problem 4.22. *Ans.* 7.11 V, −3.96 V

4.24 Apply the mesh current method to the network of Fig. 4-28. Expand the numerator determinant for current I_1 about the column containing the voltage sources to show the part of I_1 due to the 10 V and the part due to the 27 V. *Ans.* 2.13 A, 2.13 A

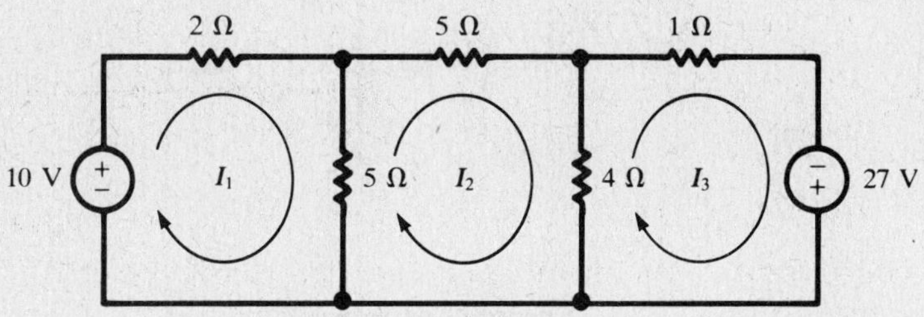

Fig. 4-28

4.25 In the network shown in Fig. 4-29, find the voltage between points a and b, V_{ab}. *Ans.* 17.65 V

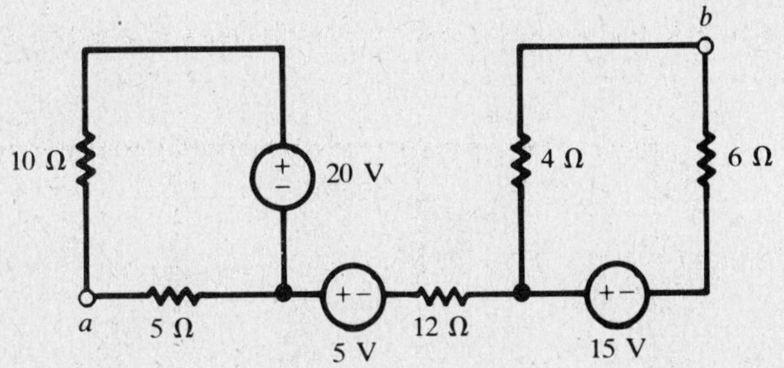

Fig. 4-29

4.26 For the circuit shown in Fig. 4-30, use mesh current or node voltage methods to obtain V_{oc}, I_{sc}, and R' at terminals ab. Consider a positive with respect to b. *Ans.* −6.29 V, −0.667 A, 9.44 Ω

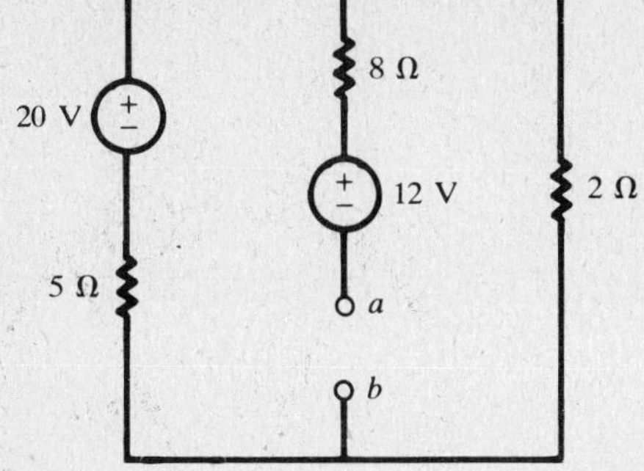

Fig. 4-30

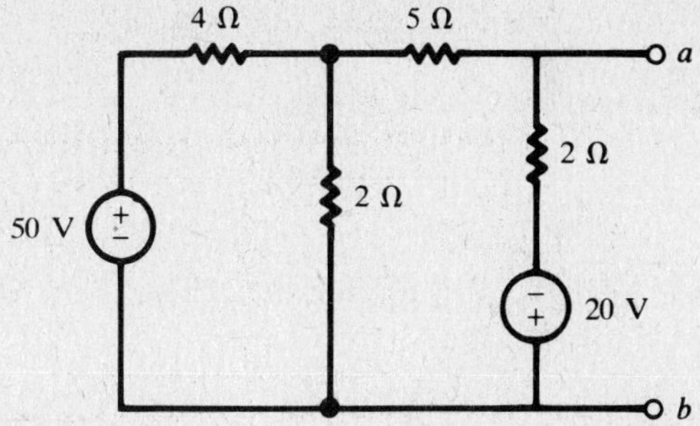

Fig. 4-31

4.27 Use the node voltage method to find V_{oc} and I_{sc} at terminals ab of the network shown in Fig. 4-31 (a positive with respect to b). *Ans.* −11.2 V, −7.37 A

4.28 The network of Problem 4.14 has been duplicated in Fig. 4-32. Find the node voltages V_1, V_2, V_3, and V_4. *Ans.* 20 V, 9.47 V, 10 V, 8.30 V

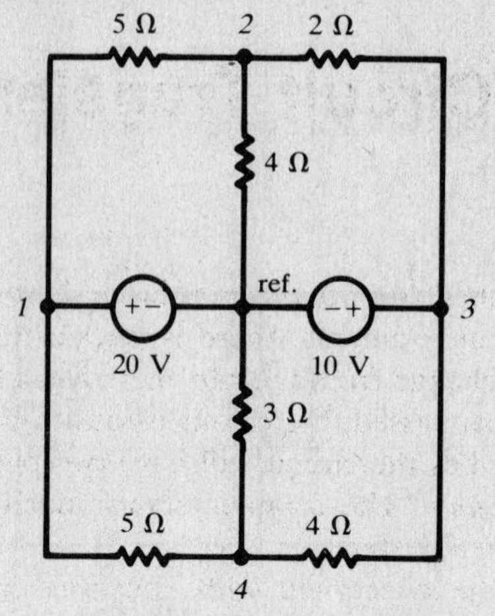

Fig. 4-32

Chapter 5

Circuit Transients

5.1 INTRODUCTION

Whenever a circuit is switched from one condition to another, either by a change in the applied source or a change in the circuit elements, there is a transitional period during which the branch currents and element voltages change from their former values to new ones. This period is called the *transient.* After the transient has passed, the circuit is said to be in the *steady state.* Now, the linear differential equation that describes the circuit will have two parts to its solution, the *complementary function* and the *particular solution.* The complementary function corresponds to the transient, and the particular solution to the steady state.

The present chapter will be concerned with situations in which the transient is of primary interest; the steady-state component, if not zero, will have some other constant value.

5.2 INITIALLY CHARGED *RC* CIRCUIT

As was seen in Problem 2.18, a capacitor with an initial charge Q_0 has an initial stored energy

$$W_0 = \tfrac{1}{2}CV_0^2 = \frac{Q_0^2}{2C}$$

where V_0 is the initial potential difference across the capacitor. When a conducting path is provided, for example by connecting at $t = 0$ a resistance R across the capacitor terminals, as in Fig. 5-1, the charge leaves the capacitor plates and the stored energy drops to zero. Kirchhoff's voltage law for the closed path of Fig. 5-1 is, for the period $t > 0$,

$$v_C + v_R = 0$$

$$\frac{q}{C} + R\frac{dq}{dt} = 0$$

$$\frac{dq}{dt} + \frac{1}{RC}q = 0$$

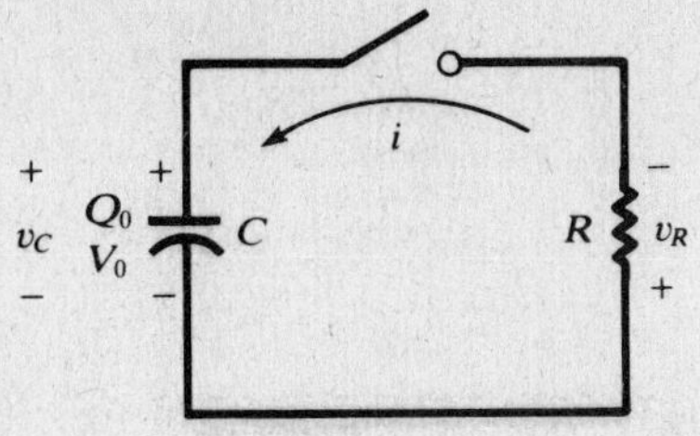

Fig. 5-1

This is a first-order, linear differential equation. Because it is homogeneous (the right-hand side is zero), the particular solution is zero, and the equation has only the complementary function as its solution. Consequently, we expect the charge function q to consist only of a transient, the steady-state charge being zero.

Any differential equations text will provide several methods of solving such an equation. The solution which satisfies the initial condition $q(0) = Q_0$ is

$$q = Q_0 e^{-t/RC}$$

From this,

$$v_C = \frac{q}{C} = V_0 e^{-t/RC}$$

$$v_R = -v_C = -V_0 e^{-t/RC}$$

$$i = \frac{v_R}{R} = -\frac{V_0}{R} e^{-t/RC}$$

The current magnitude at $t = 0^+$ is V_0/R and the direction of i is opposite to that shown on Fig. 5-1. Note also that the polarities on the diagram agree with the positive direction of i, which results in a negative sign on v_R. This shows the actual polarity on the resistor to be positive at the upper terminal, which is evident from the sign on the initial charge Q_0.

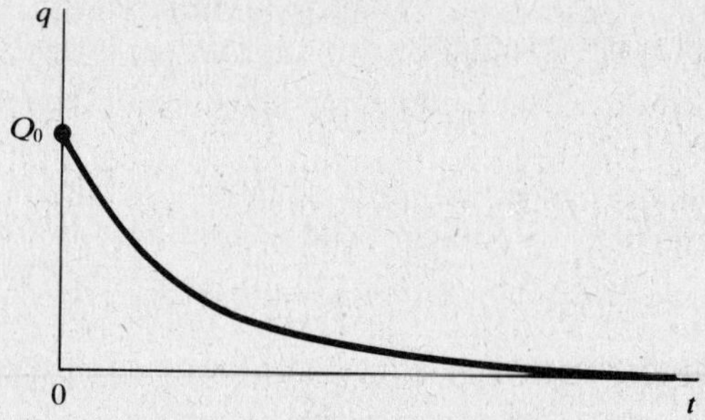

Fig. 5-2

The charge transient, plotted in Fig. 5-2, shows an exponential decay from $q(0) = Q_0$ to $q(\infty) = 0$. The voltage transients, Fig. 5-3, confirm that KVL is satisfied for all t.

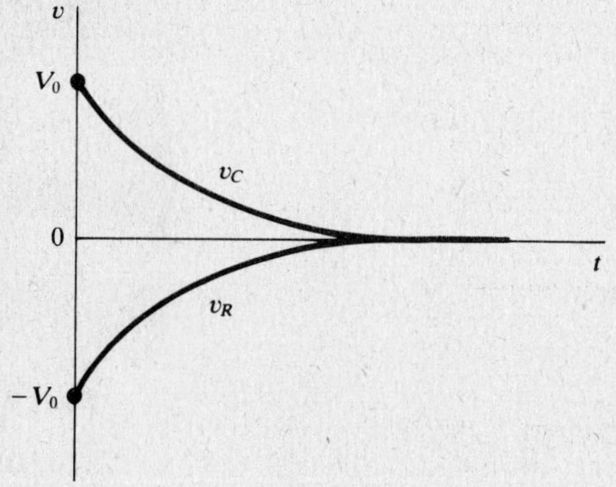

Fig. 5-3

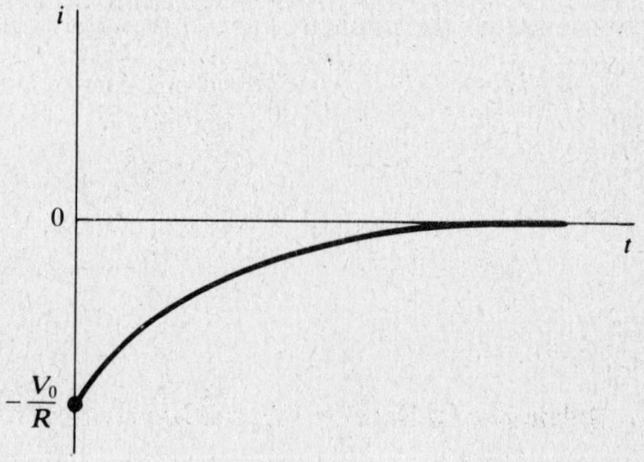

Fig. 5-4

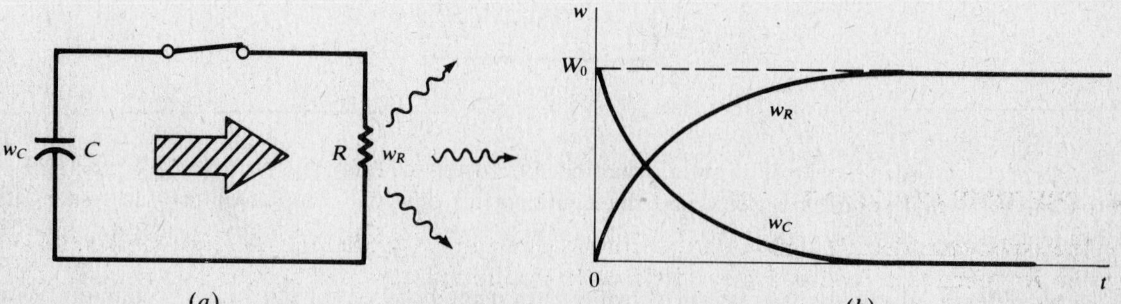

(a) (b)

Fig. 5-5

The transient of the energy stored in the capacitor can be obtained as

$$w_C = \frac{1}{2} C v_C^2 = W_0 e^{-2t/RC}$$

and the energy dissipated in the resistor is (see Problem 2.18)

$$w_R = W_0 - w_C = W_0(1 - e^{-2t/RC})$$

The transfer of energy from capacitor to resistor is suggested in Fig. 5-5(*a*); the two energy functions are plotted in Fig. 5-5(*b*). Note that the stored energy decays as $e^{-2t/RC}$, whereas the charge, voltages, and current decay as $e^{-t/RC}$.

5.3 *RL* CIRCUIT WITH INITIAL CURRENT

A differential equation of the same form as for the *RC* circuit results when KVL is applied to an *RL* circuit such as shown in Fig. 5-6. In order to maintain an initial stored energy in the inductance, it is necessary to stipulate a current I_0 for the period $t < 0$. Then, for $t \geq 0$, when the switch is in position *2*,

$$v_L + v_R = 0$$

$$L\frac{di}{dt} + Ri = 0$$

$$\frac{di}{dt} + \frac{R}{L} i = 0$$

and continuity of the inductor current at $t = 0$ gives the initial condition $i(0^+) = i(0^-) = I_0$.

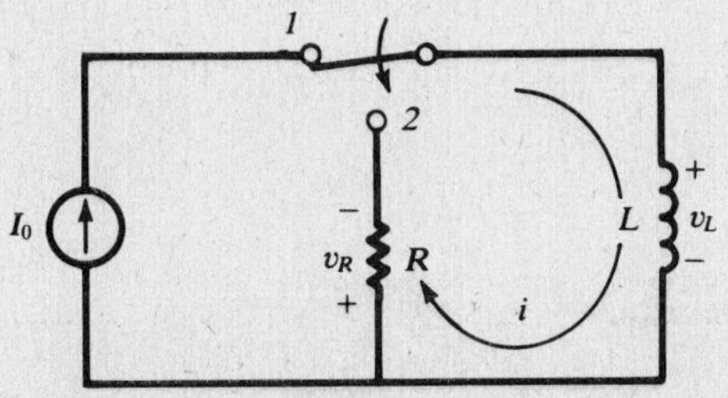

Fig. 5-6

Solving as in Section 5.2, we have for the various transients:

$$i = I_0 e^{-Rt/L}$$

$$v_R = I_0 R e^{-Rt/L}$$

$$v_L = -I_0 R e^{-Rt/L}$$

$$w_L = W_0 e^{-2Rt/L} \qquad (W_0 \equiv \tfrac{1}{2} L I_0^2)$$

$$w_R = W_0(1 - e^{-2Rt/L})$$

5.4 THE TIME CONSTANT

The exponential decay function may be written in the form $e^{-t/\tau}$, where τ is the *time constant* (in s). For the *RC* circuit of Section 5.2, $\tau = RC$; while for the *RL* circuit of Section 5.3, $\tau = L/R$. The general decay function

$$f(t) = Ae^{-t/\tau} \qquad (t > 0)$$

is plotted in Fig. 5-7, with time measured in multiples of τ. It is seen that

$$f(\tau) = Ae^{-1} = 0.368\,A$$

i.e., at $t = \tau$ the function is 36.8% of the initial value. It may also be said that the function has undergone 63.2% of the change from $f(0^+)$ to $f(\infty)$. At $t = 5\tau$, the function has the value $0.0067\,A$, which is less than 1% of the initial value. From a practical standpoint, the transient is often regarded as over after $t = 5\tau$.

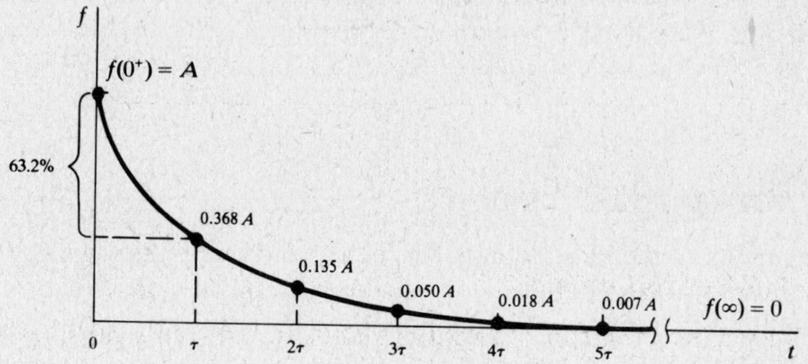

Fig. 5-7

The tangent to the exponential curve at $t = 0^+$ can be used to estimate the time constant. In fact, since

$$\text{slope} = f'(0^+) = -\frac{A}{\tau}$$

the tangent line must cut the horizontal axis at $t = \tau$ (see Fig. 5-8). More generally, the tangent at $t = t_0$ has horizontal intercept $t_0 + \tau$. Thus, if the two values $f(t_0)$ and $f'(t_0)$ are known, the entire curve can be constructed.

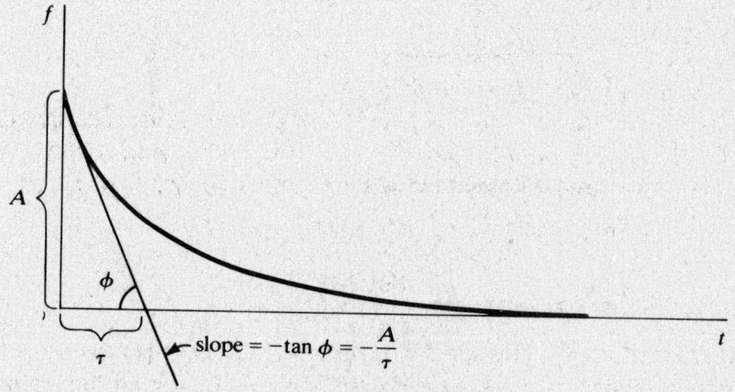

Fig. 5-8

At times a transient is only partially displayed (on chart paper or on the face of an oscilloscope), and the simultaneous values of function and slope needed in the above method are not available. In that case, any pair of data points, perhaps read from instruments, may be used to find the equation of the transient. Thus, referring to Fig. 5-9,

$$f_1 = Ae^{-t_1/\tau} \qquad f_2 = Ae^{-t_2/\tau}$$

which may be solved simultaneously to give

$$\tau = \frac{t_2 - t_1}{\ln f_1 - \ln f_2}$$

and then A in terms of τ and either f_1 or f_2.

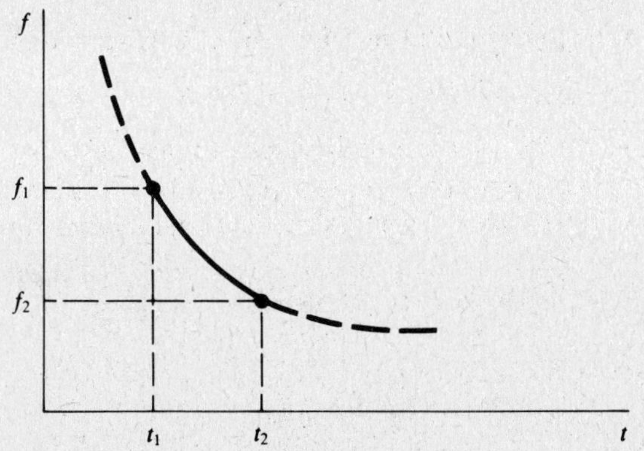

Fig. 5-9

5.5 EQUIVALENT *RC* OR *RL* CIRCUITS

A circuit which contains a number of resistors and capacitors can sometimes be reduced to one having a single R_{eq} and C_{eq}. For example, the circuit shown in Fig. 5-10(*a*) is reducible to that shown in Fig. 5-10(*b*). All transient currents in Fig. 5-10(*a*) will have time constant $\tau = R_{eq}C_{eq}$. Thus, if $i_1 = I_1 e^{-t/\tau}$, then current division (Section 3.3) gives

$$i_2 = \frac{R_3}{R_2 + R_3} I_1 e^{-t/\tau} \qquad i_3 = \frac{R_2}{R_2 + R_3} I_1 e^{-t/\tau}$$

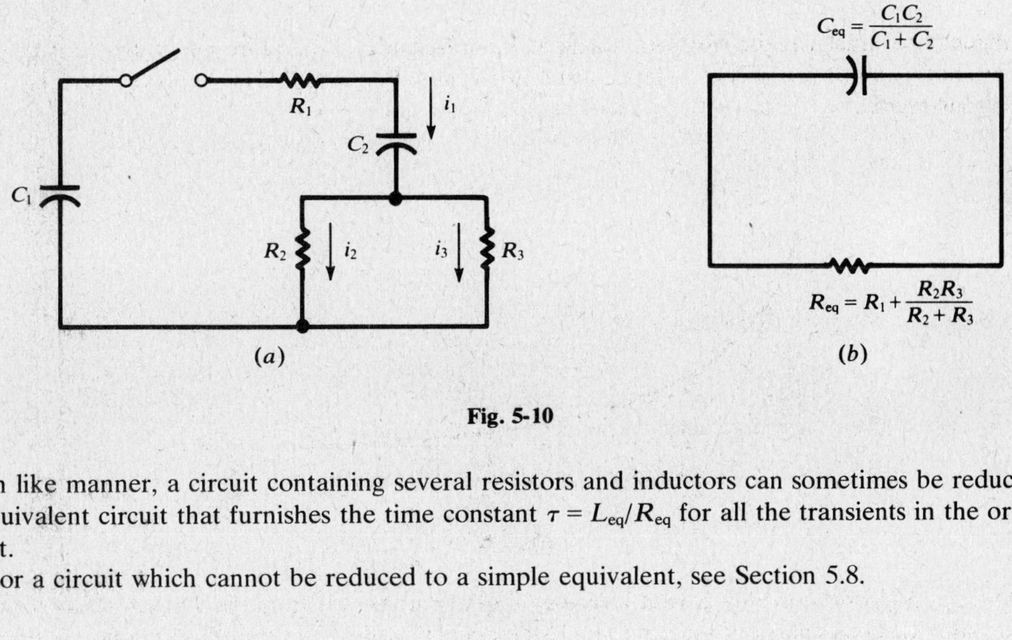

Fig. 5-10

In like manner, a circuit containing several resistors and inductors can sometimes be reduced to an equivalent circuit that furnishes the time constant $\tau = L_{eq}/R_{eq}$ for all the transients in the original circuit.

For a circuit which cannot be reduced to a simple equivalent, see Section 5.8.

5.6 *RL* AND *RC* CIRCUITS WITH SOURCES

An *RC* or *RL* circuit which was given some initial charge or current, and which was then allowed to transfer the stored energy to a passive element, exhibited *natural response.* This term *natural* underscores the fact that the circuit contained no voltage or current sources for $t > 0$. In contrast, the term *forced response* applies to those circuits which do contain voltage or current sources after $t = 0$. To describe the forced response, the complete solution (transient plus steady-state) of the non-homogeneous differential equation for $t > 0$ is required. Assuming time-independent sources, this complete solution will be of the form

$$f(t) = Ae^{-t/\tau} + B$$

The constant B, the steady-state solution, is determined by $B = f(\infty)$; then τ and A are obtained by applying one of the methods of Section 5.4 to the new function $F(t) \equiv f(t) - B$.

EXAMPLE 5.1 The switch in Fig. 5-11 has been in position *1* for a long time; it is moved to *2* at $t = 0$. Obtain the expression for i, for $t > 0$.

With the switch on *1*, $i(0^-) = 50/40 = 1.25$ A. With an inductance in the circuit, $i(0^-) = i(0^+)$. Long after the switch has been moved to *2*, $i(\infty) = 10/40 = 0.25$ A. In the above notation,

$$B = i(\infty) = 0.25 \text{ A} \qquad A = i(0^+) - B = 1.00 \text{ A}$$

and the time constant is $\tau = L/R = (1/2000)$ s. Then, for $t > 0$,

$$i = 1.00\,e^{-2000t} + 0.25 \quad \text{(A)}$$

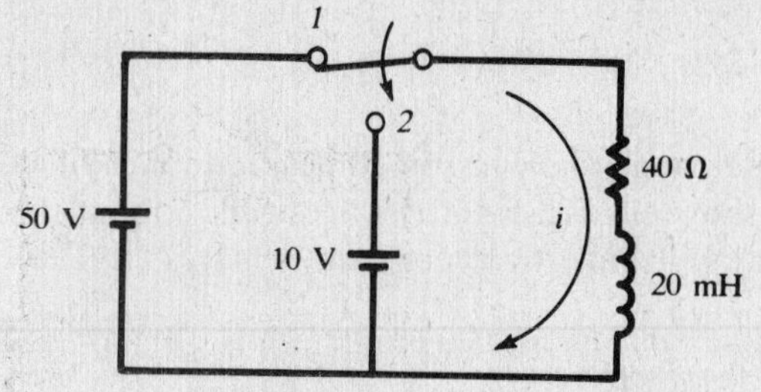

Fig. 5-11

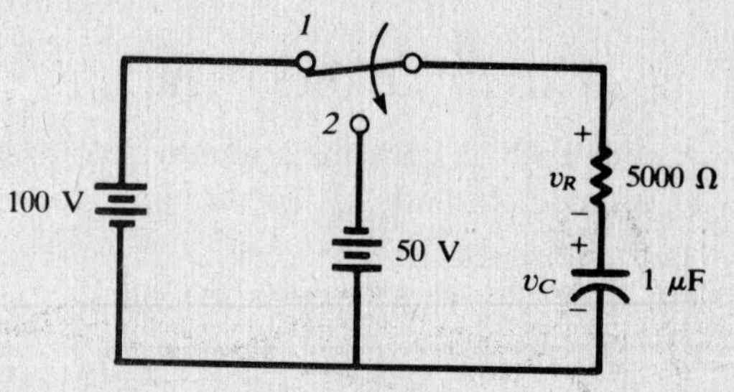

Fig. 5-12

EXAMPLE 5.2 The switch in the circuit shown in Fig. 5-12 is moved from *1* to *2* at $t = 0$. Find v_C and v_R, for $t > 0$.

With the switch on *1*, the 100-V source results in $v_C(0^-) = 100$ V; and, by continuity of charge, $v_C(0^+) = v_C(0^-)$. In position *2*, with the 50-V source of opposite polarity, $v_C(\infty) = -50$ V. Thus,

$$B = v_C(\infty) = -50 \text{ V} \qquad A = v_C(0^+) - B = 150 \text{ V}$$

$$\tau = RC = \frac{1}{200} \text{ s}$$

and

$$v_C = 150e^{-200t} - 50 \quad \text{(V)}$$

Finally, KVL gives $v_R + v_C + 50 = 0$, or

$$v_R = -150e^{-200t} \quad \text{(V)}$$

EXAMPLE 5.3 Obtain the energy functions for the circuit of Example 5.2.

$$w_C = \tfrac{1}{2}Cv_C^2 = 1.25(3e^{-200t} - 1)^2 \quad \text{(mJ)}$$

$$w_R = \int_0^t \frac{v_R^2}{R}\,dt = 11.25(1 - e^{-400t}) \quad \text{(mJ)}$$

5.7 SERIES *RLC* CIRCUIT

The differential equation of such a circuit is of second order; therefore its solution contains two constants, which are determined by two conditions, usually imposed at $t = 0$. Depending upon the relative values of the circuit parameters, the solution will be *overdamped, critically damped,* or *underdamped* (*oscillatory*). The model from mechanics shown in Fig. 5-13 helps one to visualize the three possibilities. The mass M has two adjustable damping vanes D which restrict the vertical motion. If the mass is displaced from the rest position ($z = 0$) and released (at $t = 0$), it will ultimately return to rest at $z = 0$. However, it may do so in an overdamped motion (curve *1* in Fig. 5-14), a critically damped motion (curve *2*), or an oscillatory motion (curve *3*).

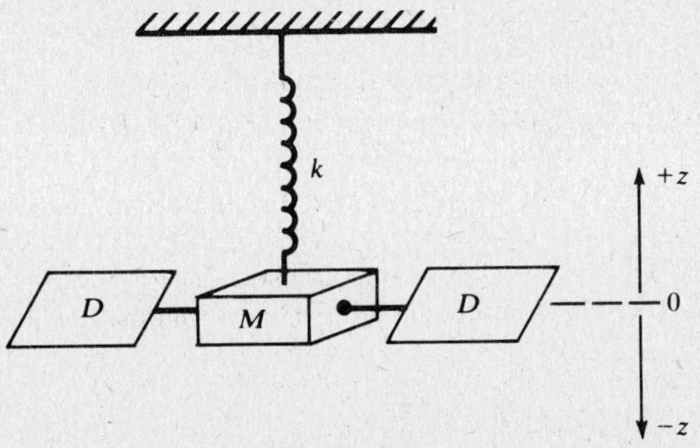

Fig. 5-13

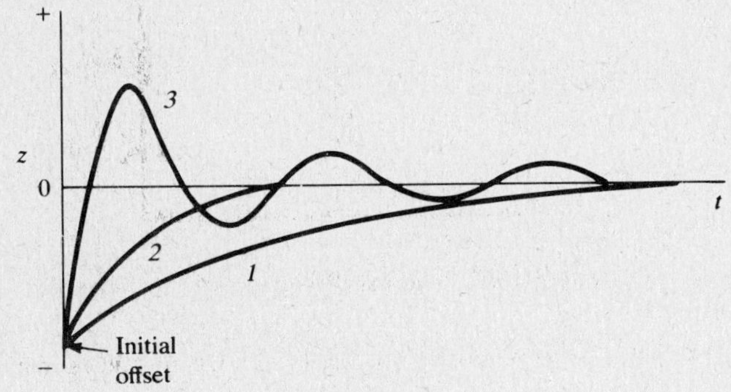

Fig. 5-14

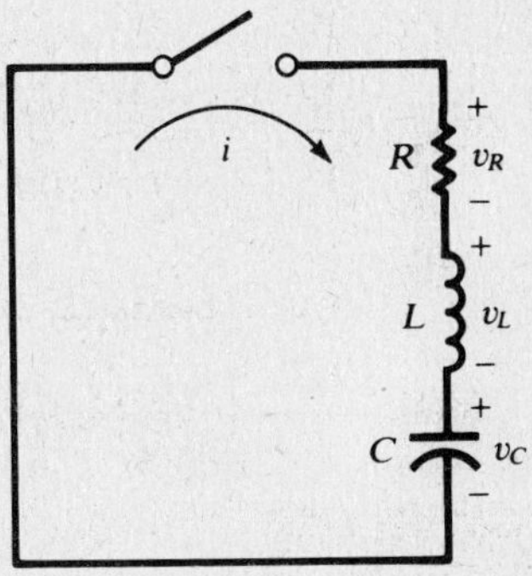

Fig. 5-15

The series *RLC* circuit shown in Fig. 5-15 contains no source. Kirchhoff's voltage law for the closed loop after the switch is closed is

$$v_R + v_L + v_C = 0$$

or

$$Ri + L\frac{di}{dt} + \frac{1}{C}\int i\,dt = 0$$

Differentiating and dividing by L,

$$\frac{d^2i}{dt^2} + \frac{R}{L}\frac{di}{dt} + \frac{1}{LC}i = 0$$

It is important to notice that this same *homogeneous* differential equation for the current would have been obtained if, after the switch were closed, the circuit had included a constant-voltage source. In either case, therefore, the current is a pure transient, with steady-state value zero. (This does not hold true if $C = 0$; see Example 5.1.)

An expression of the form $i = A_1e^{s_1t} + A_2e^{s_2t}$ will be a solution to the differential equation if

$$A_1e^{s_1t}\left(s_1^2 + \frac{R}{L}s_1 + \frac{1}{LC}\right) + A_2e^{s_2t}\left(s_2^2 + \frac{R}{L}s_2 + \frac{1}{LC}\right) = 0$$

that is, if s_1 and s_2 are the two roots of $s^2 + (R/L)s + (1/LC) = 0$:

$$s_1 = -\frac{R}{2L} + \sqrt{\left(\frac{R}{2L}\right)^2 - \frac{1}{LC}} \equiv -\alpha + \beta \qquad s_2 = -\frac{R}{2L} - \sqrt{\left(\frac{R}{2L}\right)^2 - \frac{1}{LC}} \equiv -\alpha - \beta$$

where $\alpha \equiv R/2L$, $\omega_0 \equiv 1/\sqrt{LC}$, and $\beta \equiv \sqrt{\alpha^2 - \omega_0^2}$.

Overdamped Case ($\alpha > \omega_0$)

Here, $i = e^{-\alpha t}(A_1 e^{\beta t} + A_2 e^{-\beta t})$, where α and β are real positive numbers.

EXAMPLE 5.4 In Fig. 5-15, $R = 200\ \Omega$, $L = 0.10$ H, $C = 13.33\ \mu$F, and $v_C(0^-) = 200$ V. Obtain the current transient, if the switch is closed at $t = 0$.

For this circuit,

$$\alpha = \frac{R}{2L} = 10^3\ \text{s}^{-1} \qquad \omega_0^2 = \frac{1}{LC} = 7.5 \times 10^5\ \text{s}^{-2} \qquad \beta = \sqrt{\alpha^2 - \omega_0^2} = 500\ \text{s}^{-1}$$

Noting that $\alpha^2 > \omega_0^2$, we have, for $t > 0$,

$$i = e^{-1000t}(A_1 e^{500t} + A_2 e^{-500t}) \tag{1}$$

With inductance in the circuit, $i(0^+) = i(0^-) = 0$, which is one condition for establishing the values of the constants A_1 and A_2. The other condition is provided by the continuity of the capacitor charge, and hence of the capacitor voltage, at $t = 0$. Thus, at $t = 0^+$, KVL reads:

$$(0)R + L\left.\frac{di}{dt}\right|_{0^+} + v_C(0^-) = 0 \qquad \text{or} \qquad \left.\frac{di}{dt}\right|_{0^+} = -2000\ \text{A/s}$$

Applying these two conditions to (1) and its time derivative,

$$0 = A_1 + A_2 \qquad \text{and} \qquad -2000 = -500A_1 - 1500A_2$$

from which $A_1 = -2$ A, $A_2 = 2$ A, and

$$i = -2e^{-500t} + 2e^{-1500t} \quad \text{(A)}$$

This transient current is plotted in Fig. 5-16. Because of the polarity of the initial voltage on the capacitor, the current is opposite in direction to the assumed current i in Fig. 5-15.

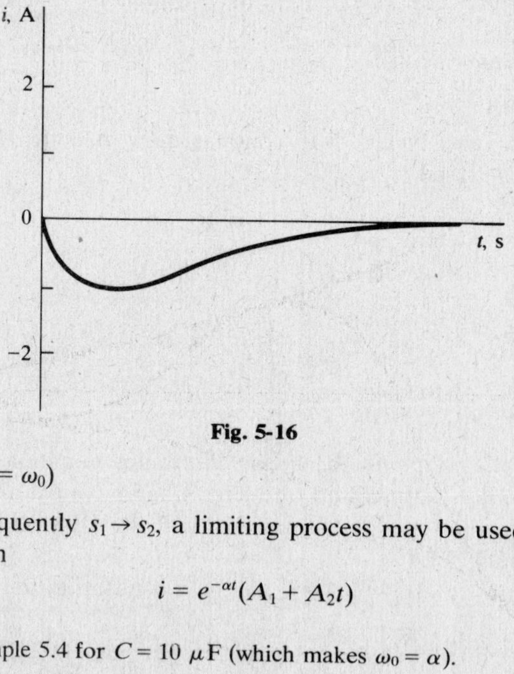

Fig. 5-16

Critically Damped Case ($\alpha = \omega_0$)

When $\beta \to 0$ and consequently $s_1 \to s_2$, a limiting process may be used to show that the solution for the current has the form

$$i = e^{-\alpha t}(A_1 + A_2 t)$$

EXAMPLE 5.5 Repeat Example 5.4 for $C = 10\ \mu$F (which makes $\omega_0 = \alpha$).

As in Example 5.4,

$$i(0^+) = 0 \qquad \left.\frac{di}{dt}\right|_{0^+} = -2000\ \text{A/s} \qquad \alpha = 1000\ \text{s}^{-1}$$

These determine the unknown constants as $A_1 = 0$, $A_2 = -2000$ A; hence

$$i = -2000te^{-1000t} \quad \text{(A)}$$

See Fig. 5-17.

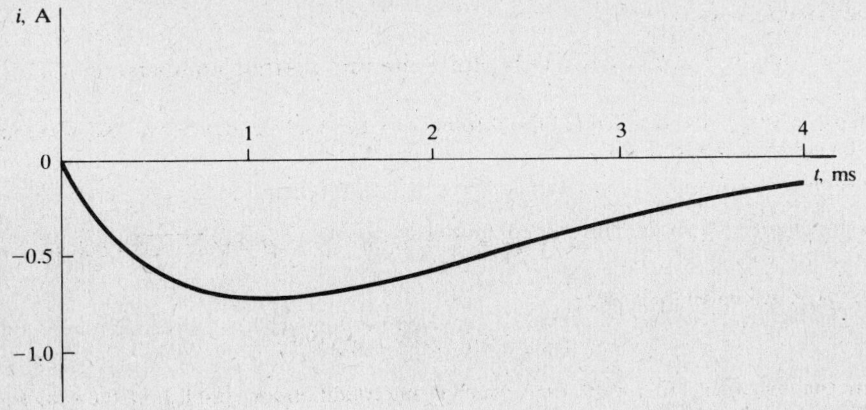

Fig. 5-17

Underdamped Case ($\alpha < \omega_0$)

In this case, β is a pure imaginary, $\beta = j|\beta|$, and s_1 and s_2 are complex conjugates. The general form of the current is

$$i = e^{-\alpha t}(A_1 \cos|\beta|t + A_2 \sin|\beta|t) \qquad \text{or} \qquad i = e^{-\alpha t}A_3 \sin(|\beta|t + \phi)$$

where A_3 and ϕ are two new constants.

EXAMPLE 5.6 Repeat Example 5.4 for $C = 1\ \mu\text{F}$.

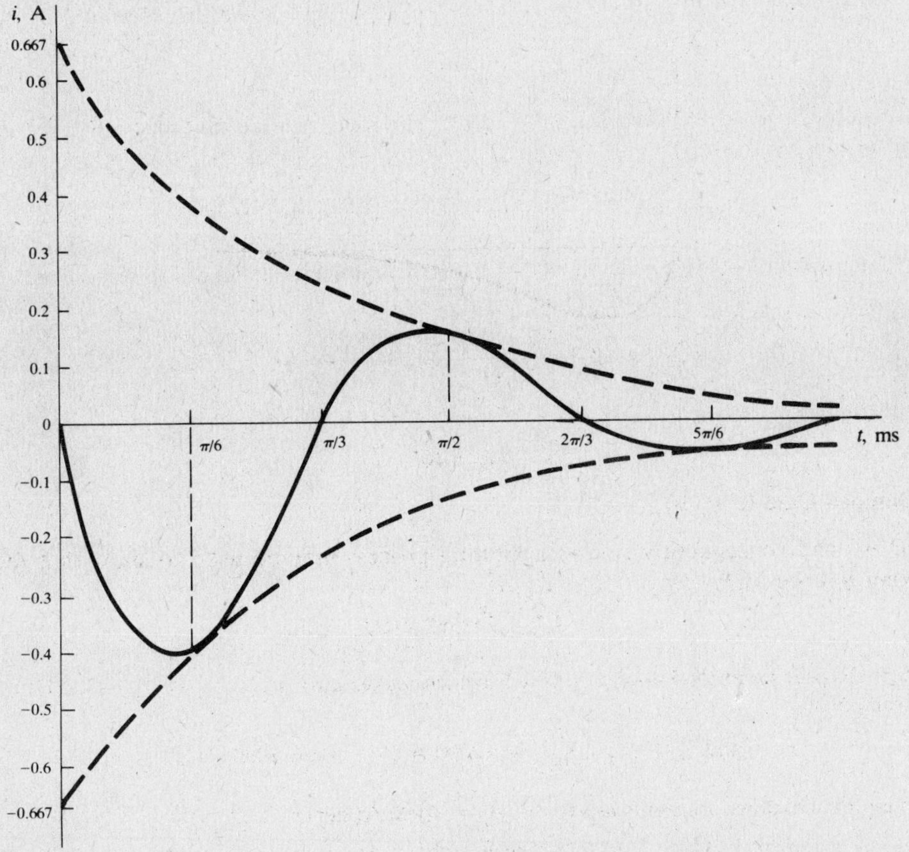

Fig. 5-18

As before,

$$i(0^+) = 0 \qquad \left.\frac{di}{dt}\right|_{0^+} = -2000 \text{ A/s} \qquad \alpha = 1000 \text{ s}^{-1}$$

but now

$$\beta = \sqrt{10^6 - 10^7} = j3000 \text{ rad/s}$$

Then

$$i = e^{-1000t} A_3 \sin(3000t + \phi)$$

The constants are obtained from the initial conditions as $\phi = 0$, $A_3 = -0.667$ A; hence,

$$i = -0.667 e^{-1000t} \sin 3000t \quad \text{(A)}$$

See Fig. 5-18. The factors $\pm 0.667 e^{-1000t}$ provide an envelope within which the sine function is confined.

5.8 TWO-MESH CIRCUITS

Example 5.7 below shows how the response may be obtained for a circuit that does not lend itself to the reduction techniques of Section 5.5. It will be seen that the method is rather cumbersome: better methods (state variables, Laplace transforms) will be treated in later chapters.

EXAMPLE 5.7 For the circuit of Fig. 5-19, choose mesh currents i_1 and i_2, as indicated. KVL yields the two first-order differential equations

$$R_1 i_1 + L_1 \frac{di_1}{dt} + R_1 i_2 = V \tag{1}$$

$$R_1 i_1 + (R_1 + R_2) i_2 + L_2 \frac{di_2}{dt} = V \tag{2}$$

which must be solved simultaneously. To accomplish this, take the time derivative of (*1*),

$$R_1 \frac{di_1}{dt} + L_1 \frac{d^2 i_1}{dt^2} + R_1 \frac{di_2}{dt} = 0 \tag{3}$$

and then eliminate i_2 and di_2/dt between (*1*), (*2*), and (*3*). The result is a second-order equation for i_1, of the type treated in Section 5.7:

$$\frac{d^2 i_1}{dt^2} + \frac{R_1 L_1 + R_2 L_1 + R_1 L_2}{L_1 L_2} \frac{di_1}{dt} + \frac{R_1 R_2}{L_1 L_2} i_1 = \frac{R_2 V}{L_1 L_2} \tag{4}$$

The steady-state solution of (*4*) is evidently $i_1(\infty) = V/R_1$; the transient solution will be determined by the roots s_1 and s_2 of

$$s^2 + \frac{R_1 L_1 + R_2 L_1 + R_1 L_2}{L_1 L_2} s + \frac{R_1 R_2}{L_1 L_2} = 0$$

together with the initial conditions

$$i_1(0^+) = 0 \qquad \left.\frac{di_1}{dt}\right|_{0^+} = \frac{V}{L_1}$$

(both i_1 and i_2 must be continuous at $t = 0$). Once the expression for i_1 is known, that for i_2 follows from (*1*).

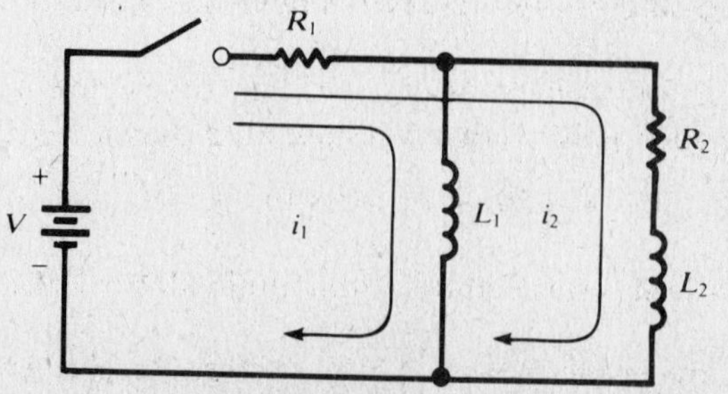

Fig. 5-19

5.9 THE UNIT STEP FUNCTION

Closing or opening a switch appears to be a simple, unambiguous operation, and in the examples and problems of this chapter no attention is paid to the switch itself. However, consider the circuit shown in Fig. 5-20(a) and the voltage v. At the moment the switch contacts meet, $v = 50$ V. Before the closing, v has a value determined by the active elements in the network. Even if the network is passive, static charges can result in a nonzero value for v. The switch in Fig. 5-20(b) moves from a position which applies a short circuit across the network terminals to one which applies voltage V. Even so, there could be a hiatus in the switching operation. Arcing, contact resistance, contact bounce, and contact capacitance are a few of the problems faced in practical switching.

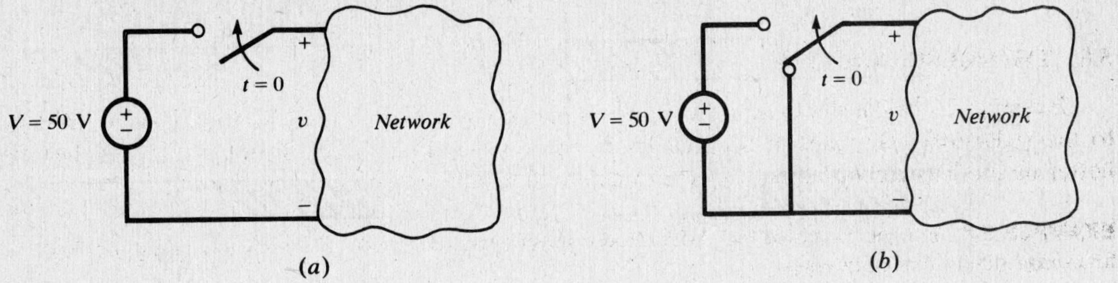

Fig. 5-20

A nondimensional *unit step function*, defined by

$$u(x) = \begin{cases} 0 & (x < 0) \\ +1 & (x > 0) \end{cases}$$

may be used to represent the effect of an ideal switch. Thus, in Fig. 5-21, $v = 0$ for $t < 0$ and $v = 50$ V for $t > 0$.

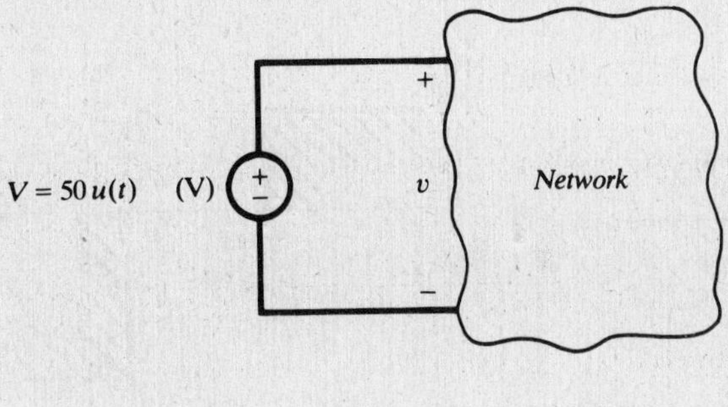

Fig. 5-21

An instantaneous change at some moment other than $t = 0$ can be represented by means of a unit step function having shifted and/or sign-reversed argument. Table 5-1 indicates the possibilities ($t' > 0$).

A square-wave pulse of any desired width can be constructed by combining two unit step functions. For example, the 10-V pulse with a 1-μs duration shown in Fig. 5-22 can be written

$$v = 10[u(t) - u(t - 10^{-6})] \quad \text{(V)}$$

EXAMPLE 5.8 Evaluate $v = 25\,u(-t-5) - 50\,u(t) + 50\,u(t-10)$ (V) at (a) $t = -6$ s, (b) -3 s, (c) 3 s, and (d) 11 s.

Each term may be sketched separately, as in Fig. 5-23. Then, by pointwise combination of the graphs: (a) 25 V, (b) 0, (c) -50 V, (d) 0.

Table 5-1

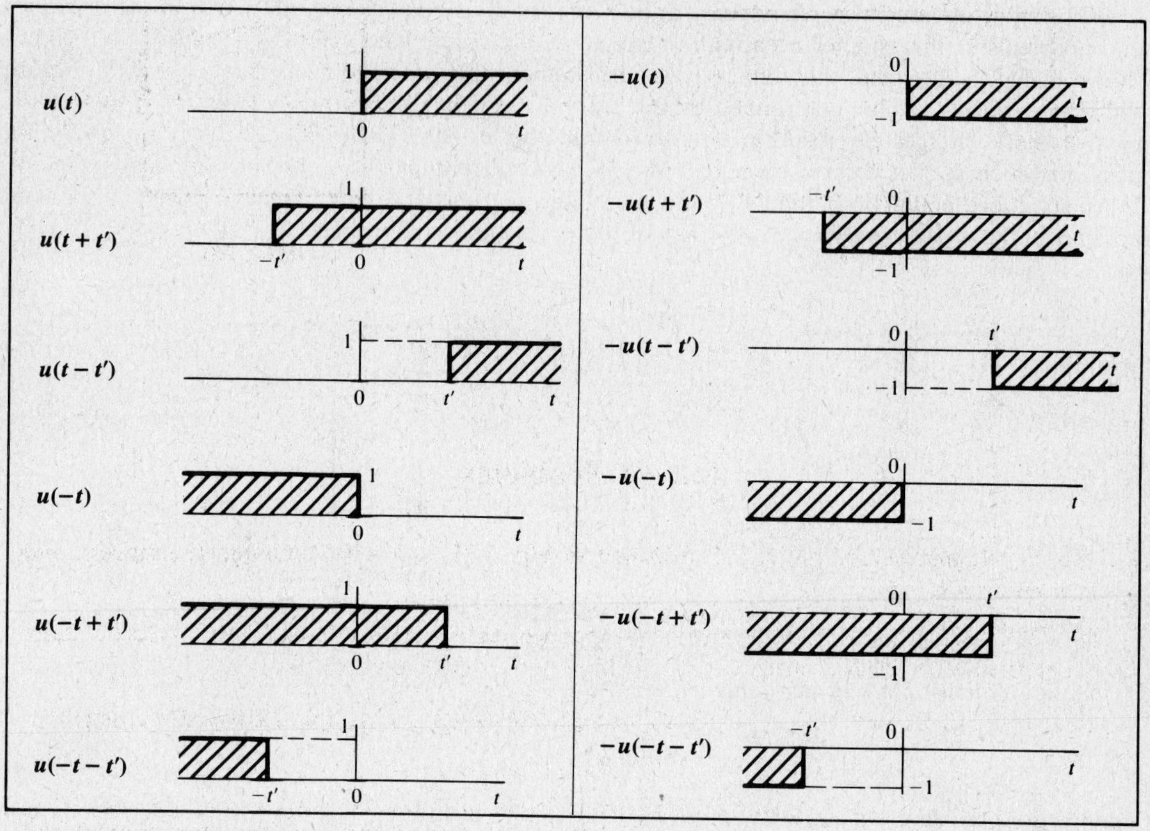

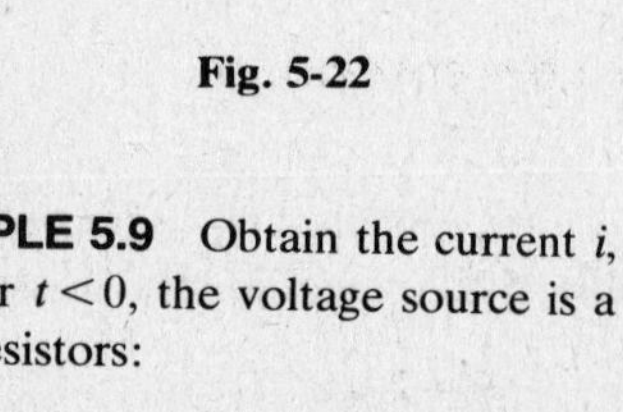

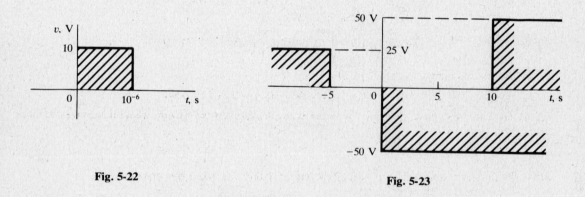

Fig. 5-22

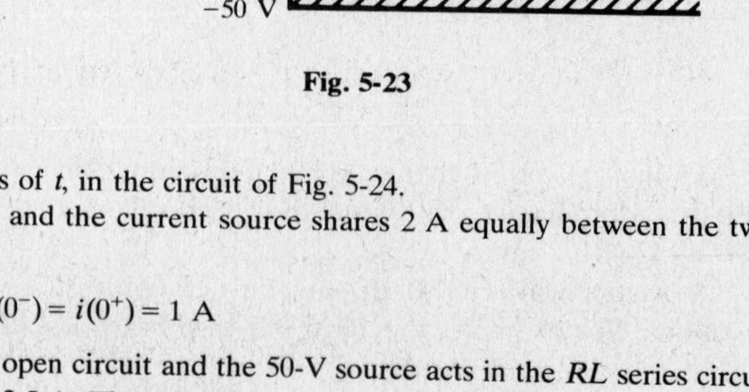

Fig. 5-23

EXAMPLE 5.9 Obtain the current i, for all values of t, in the circuit of Fig. 5-24.

For $t<0$, the voltage source is a short circuit and the current source shares 2 A equally between the two 10-Ω resistors:

$$i(t)=i(0^-)=i(0^+)=1\ \text{A}$$

For $t>0$, the current source is replaced by an open circuit and the 50-V source acts in the RL series circuit ($R=20\ \Omega$). Consequently, as $t\to\infty$, $i\to -50/20=-2.5$ A. Then, by Section 5.6,

$$i(t)=[i(0^+)-i(\infty)]e^{-Rt/L}+i(\infty)=3.5e^{-100t}-2.5 \quad \text{(A)}$$

By means of unit step functions, the two formulas may be combined into a single formula valid for all t:

$$i(t)=1\,u(-t)+(3.5e^{-100t}-2.5)\,u(t) \quad \text{(A)}$$

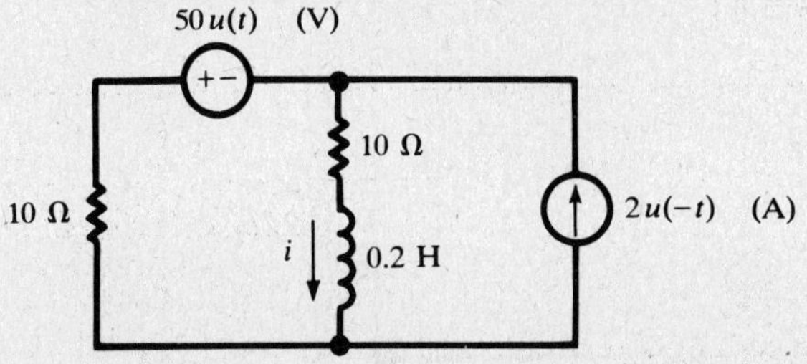

Fig. 5-24

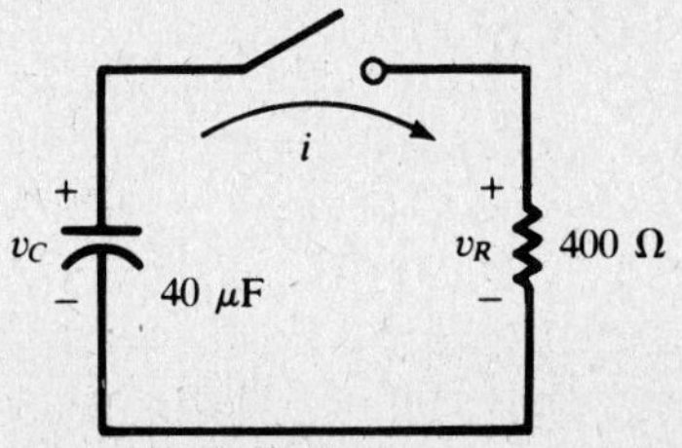

Fig. 5-25

Solved Problems

5.1 At $t = 0^-$, just before the switch is closed in Fig. 5-25, $v_C = 100$ V. Obtain the current and charge transients.

With the polarities as indicated on the diagram, $v_R = v_C$ for $t > 0$, and $1/RC = 62.5\ \text{s}^{-1}$. Also, $v_C(0^+) = v_C(0^-) = 100$ V. Thus,

$$v_R = v_C = 100e^{-62.5t} \quad (\text{V}) \qquad i = \frac{v_R}{R} = 0.25e^{-62.5t} \quad (\text{A}) \qquad q = Cv_C = 4000e^{-62.5t} \quad (\mu\text{C})$$

5.2 In Problem 5.1, obtain the power and energy in the resistor, and compare the latter with the initial energy stored in the capacitor.

$$p_R = v_R i = 25e^{-125t} \quad (\text{W})$$

$$w_R = \int_0^t p_R\,dt = \int_0^t 25e^{-125t}\,dt = 0.20(1 - e^{-125t}) \quad (\text{J})$$

The initial stored energy is

$$W_0 = \tfrac{1}{2}CV_0^2 = \tfrac{1}{2}(40 \times 10^{-6})(100)^2 = 0.20\ \text{J} = w_R(\infty)$$

i.e., all the stored energy in the capacitor is eventually delivered to the resistor, where it is converted into heat.

5.3 An RC transient identical to that in Problems 5.1 and 5.2 has a power transient

$$p_R = 360e^{-t/0.00001} \quad (\text{W})$$

Obtain the initial charge Q_0, if $R = 10\ \Omega$.

$$p_R = P_0e^{-2t/RC} \qquad \text{or} \qquad \frac{2}{RC} = 10^5 \qquad \text{or} \qquad C = 2\ \mu\text{F}$$

$$w_R = \int_0^t p_R\,dt = 3.6(1 - e^{-t/0.00001}) \quad (\text{mJ})$$

Then, $w_R(\infty) = 3.6\ \text{mJ} = Q_0^2/2C$, from which $Q_0 = 120\ \mu$C.

5.4 The switch in the RL circuit shown in Fig. 5-26 is moved from position *1* to poisition *2* at $t = 0$. Obtain v_R and v_L with polarities as indicated.

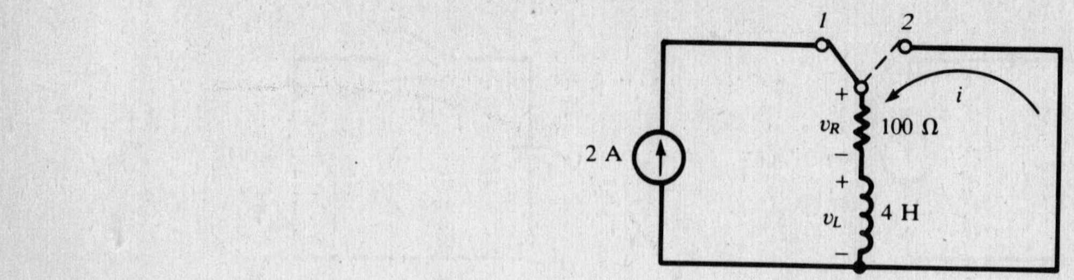

Fig. 5-26

The constant-current source drives a current through the inductance in the same direction as that of the transient current i. Then, for $t>0$,

$$i = I_0 e^{-Rt/L} = 2e^{-25t} \quad \text{(A)}$$
$$v_R = Ri = 200e^{-25t} \quad \text{(V)}$$
$$v_L = -v_R = -200e^{-25t} \quad \text{(V)}$$

5.5 For the transient of Problem 5.4 obtain p_R and p_L.

$$p_R = v_R i = 400e^{-50t} \quad \text{(W)}$$
$$p_L = v_L i = -400e^{-50t} \quad \text{(W)}$$

Negative power for the inductance is consistent with the fact that energy is leaving the element. And, since this energy is being transferred to the resistance, p_R is positive.

5.6 A series RC circuit with $R = 5\ \text{k}\Omega$ and $C = 20\ \mu\text{F}$ has a constant-voltage source of 100 V applied at $t=0$; there is no initial charge on the capacitor. Obtain i, v_R, v_C, and q, for $t>0$.

The capacitor charge, and hence v_C, must be continuous at $t=0$:

$$v_C(0^+) = v_C(0^-) = 0$$

As $t \to \infty$, $v_C \to 100$ V, the applied voltage. The time constant of the circuit is $\tau = RC = 10^{-1}$ s. Hence, from Section 5.6,

$$v_C = [v_C(0^+) - v_C(\infty)]e^{-t/\tau} + v_C(\infty) = -100e^{-10t} + 100 \quad \text{(V)}$$

The other functions follow from this. If the element voltages are both positive where the current enters, $v_R + v_C = 100$ V, and so

$$v_R = 100e^{-10t} \quad \text{(V)}$$
$$i = \frac{v_R}{R} = 20e^{-10t} \quad \text{(mA)}$$
$$q = Cv_C = 2000(1 - e^{-10t}) \quad (\mu\text{C})$$

5.7 The switch in the circuit shown in Fig. 5-27(a) is closed at $t=0$, at which moment the capacitor has charge $Q_0 = 500\ \mu\text{C}$, with the polarity indicated. Obtain i and q, for $t>0$, and sketch the graph of q.

The initial charge has a corresponding voltage $V_0 = Q_0/C = 25$ V, whence $v_C(0^+) = -25$ V. The sign is negative because the capacitor voltage, in agreement with the positive direction of the current, would be + on the top plate. Also $v_C(\infty) = +50$ V and $\tau = 0.02$ s. Thus, as in Problem 5.6,

$$v_C = -75e^{-50t} + 50 \quad \text{(V)}$$

from which

$$q = Cv_C = -1500e^{-50t} + 1000 \quad (\mu\text{C}) \qquad i = \frac{dq}{dt} = 75e^{-50t} \quad \text{(mA)}$$

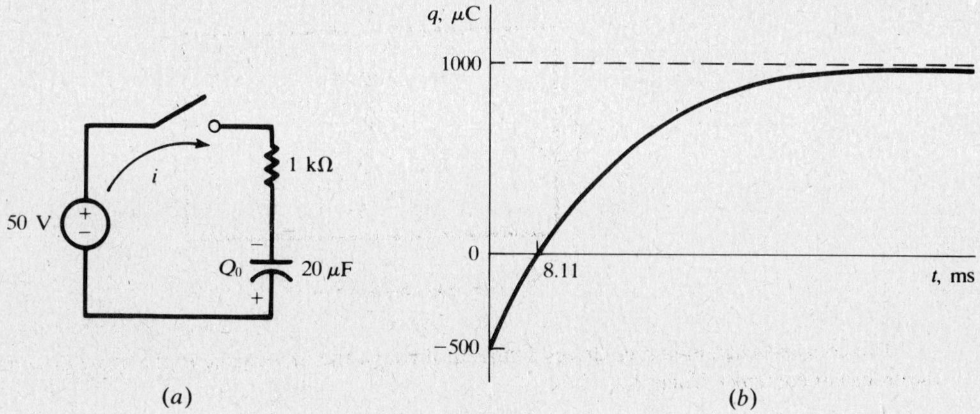

Fig. 5-27

The sketch in Fig. 5-27(*b*) shows that the charge changes from 500 μC of one polarity to 1000 μC of the opposite polarity.

5.8 The switch in the circuit of Fig. 5-28 is closed on position *1* at $t = 0$ and then moved to *2* after one time constant, at $t = \tau = 250\ \mu\text{s}$. Obtain the current for $t > 0$.

It is simplest first to find the charge on the capacitor, since it is known to be continuous (at $t = 0$ and at $t = \tau$), and then to differentiate it to obtain the current.

For $0 \le t \le \tau$, q must have the form

$$q = Ae^{-t/\tau} + B$$

From the assumption $q(0) = 0$ and the condition

$$i(0^+) = \left.\frac{dq}{dt}\right|_{0^+} = \frac{20\ \text{V}}{500\ \Omega} = 40\ \text{mA}$$

we find that $A = -B = -10\ \mu\text{C}$, or

$$q = 10(1 - e^{-4000t})\ \ (\mu\text{C}) \qquad (0 \le t \le \tau) \tag{1}$$

From (*1*), $q(\tau) = 10(1 - e^{-1})\ \mu\text{C}$; and we know that $q(\infty) = (0.5\ \mu\text{F})(-40\ \text{V}) = -20\ \mu\text{C}$. Hence, q is determined for $t \ge \tau$ as

$$q = [q(\tau) - q(\infty)]e^{-(t-\tau)/\tau} + q(\infty) = 71.55\,e^{-4000t} - 20\ \ (\mu\text{C}) \tag{2}$$

Differentiating (*1*) and (*2*).

$$i = \frac{dq}{dt} = \begin{cases} 40e^{-4000t} & (\text{mA}) \quad (0 < t < \tau) \\ -286.2\,e^{-4000t} & (\text{mA}) \quad (t > \tau) \end{cases}$$

See Fig. 5-29.

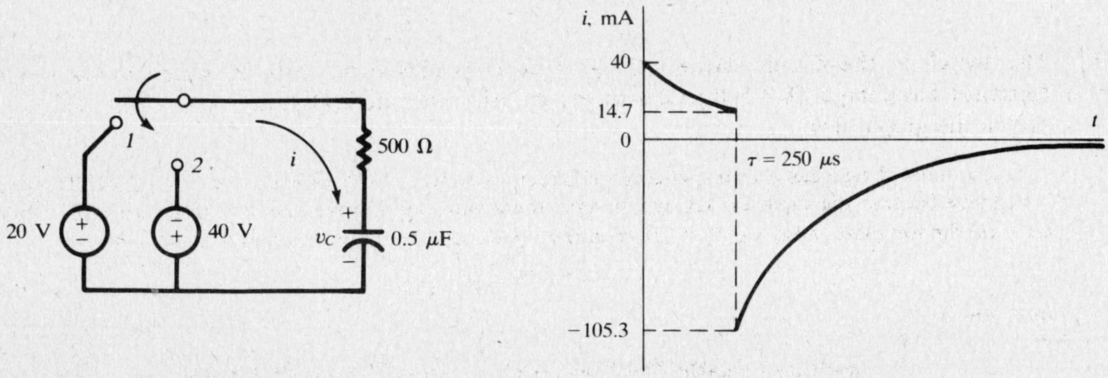

Fig. 5-28 **Fig. 5-29**

5.9 A series RL circuit has a constant voltage V applied at $t = 0$. At what time does $v_R = v_L$?

The current in an RL circuit is a continuous function, starting at zero in this case, and reaching the final value V/R. Thus, for $t > 0$,

$$i = \frac{V}{R}(1 - e^{-t/\tau}) \qquad \text{and} \qquad v_R = Ri = V(1 - e^{-t/\tau})$$

where $\tau = L/R$ is the time constant of the circuit. Since $v_R + v_L = V$, the two voltages will be equal when

$$v_R = \tfrac{1}{2}V$$
$$V(1 - e^{-t/\tau}) = \tfrac{1}{2}V$$
$$e^{-t/\tau} = \tfrac{1}{2}$$
$$\frac{t}{\tau} = \ln 2$$

that is, when $t = 0.693\,\tau$. Note that this time is independent of V.

5.10 A constant voltage is applied to a series RL circuit at $t = 0$. The voltage across the inductance is 20 V at 3.46 ms and 5 V at 25 ms. Obtain R if $L = 2$ H.

Using the two-point method of Section 5.4,

$$\tau = \frac{t_2 - t_1}{\ln v_1 - \ln v_2} = \frac{25 - 3.46}{\ln 20 - \ln 5} = 15.54 \text{ ms}$$

and so

$$R = \frac{L}{\tau} = \frac{2}{15.54 \times 10^{-3}} = 128.7\ \Omega$$

5.11 In Fig. 5-30, switch S_1 is closed at $t = 0$. Switch S_2 is opened at $t = 4$ ms. Obtain i for $t > 0$.

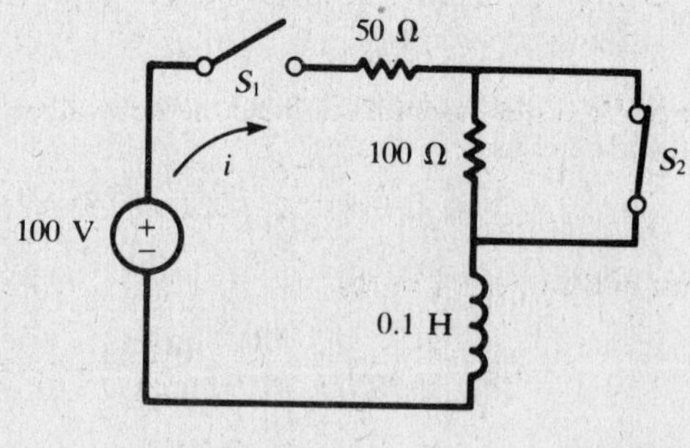

Fig. 5-30

As there is always inductance in the circuit, the current is a continuous function at all times. In the interval $0 \le t \le 4$ ms, with the 100 Ω shorted out and a time constant $\tau = (0.1\text{ H})/(50\ \Omega) = 2$ ms, i starts at zero and builds toward

$$\frac{100\text{ V}}{50\ \Omega} = 2\text{ A}$$

even though it never gets close to that value. Hence, as in Problem 5.9,

$$i = 2(1 - e^{-t/2}) \quad \text{(A)} \qquad (0 \le t \le 4) \tag{1}$$

wherein t is measured in ms. In particular,

$$i(4) = 2(1 - e^{-2}) = 1.729\text{ A}$$

In the interval $t \geq 4$ ms, i starts at 1.729 A and decays toward $100/150 = 0.667$ A, with a time constant $0.1/150 = \frac{2}{3}$ ms. Therefore, with t again in ms,

$$\begin{aligned} i &= (1.729 - 0.667)e^{-(t-4)/(2/3)} + 0.667 \\ &= 428.4e^{-3t/2} + 0.667 \quad \text{(A)} \qquad (t \geq 4) \end{aligned} \tag{2}$$

5.12 In the circuit of Fig. 5-31, the switch is closed at $t = 0$, when the 6-μF capacitor has charge $Q_0 = 300\ \mu$C. Obtain the expression for the transient voltage v_R.

The two parallel capacitors have an equivalent capacitance of 3 μF. Then this capacitance is in series with the 6 μF, so that the overall equivalent capacitance is 2 μF. Thus, $\tau = RC_{eq} = 40\ \mu$s.

At $t = 0^+$, KVL gives $v_R = 300/6 = 50$ V; and, as $t \to \infty$, $v_R \to 0$ (since $i \to 0$). Therefore,

$$v_R = 50e^{-t/\tau} = 50e^{-t/40} \quad \text{(V)}$$

in which t is measured in μs.

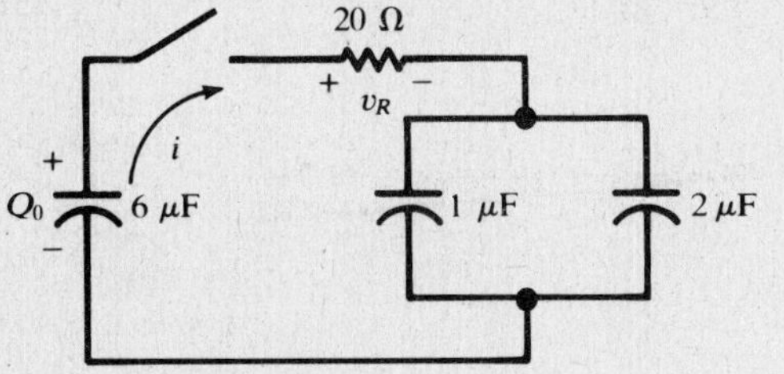

Fig. 5-31

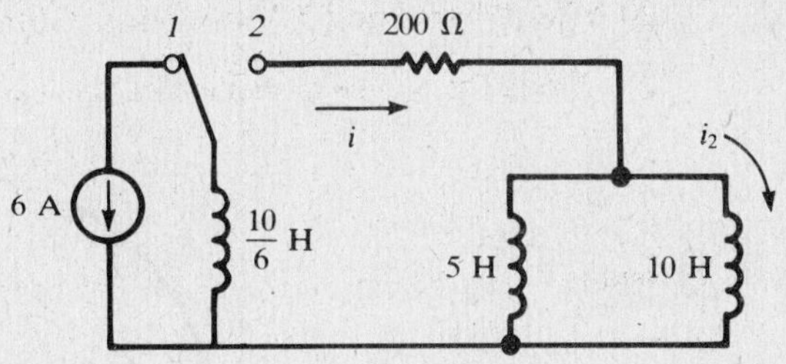

Fig. 5-32

5.13 In the circuit shown in Fig. 5-32, the switch is moved to position 2 at $t = 0$. Obtain the current i_2 at $t = 34.7$ ms.

After the switching, the three inductances have the equivalent

$$L_{eq} = \frac{10}{6} + \frac{5(10)}{15} = 5 \text{ H}$$

Then $\tau = 5/200 = 25$ ms, and so, with t in ms,

$$i = 6e^{-t/25} \quad \text{(A)} \qquad i_2 = \left(\frac{5}{15}\right)i = 2e^{-t/25} \quad \text{(A)}$$

and

$$i_2(34.7) = 2e^{-34.7/25} = 0.50 \text{ A}$$

5.14 A series RLC circuit, with $R = 3$ kΩ, $L = 10$ H, and $C = 200\ \mu$F, has a constant-voltage source, $V = 50$ V, applied at $t = 0$. (*a*) Obtain the current transient, if the capacitor has no initial charge. (*b*) Sketch the current and find the time at which it is a maximum.

(*a*)

$$\alpha = \frac{R}{2L} = 150 \text{ s}^{-1} \qquad \omega_0^2 = \frac{1}{LC} = 500 \text{ s}^{-2} \qquad \beta = \sqrt{\alpha^2 - \omega_0^2} = 148.3 \text{ s}^{-1}$$

The circuit is overdamped ($\alpha > \omega_0$).

$$s_1 = -\alpha + \beta = -1.70 \text{ s}^{-1} \qquad s_2 = -\alpha - \beta = -298.3 \text{ s}^{-1}$$

and

$$i = A_1e^{-1.70t} + A_2e^{-298.3t}$$

Since the circuit contains an inductance, $i(0^+) = i(0^-) = 0$; also, $q(0^+) = q(0^-) = 0$. Thus, at $t = 0^+$, KVL gives

$$0 + 0 + L\frac{di}{dt}\bigg|_{0^+} = V \qquad \text{or} \qquad \frac{di}{dt}\bigg|_{0^+} = \frac{V}{L} = 5 \text{ A/s}$$

Applying these initial conditions to the expression for i,

$$0 = A_1(1) + A_2(1)$$
$$5 = -1.70\,A_1(1) - 298.3\,A_2(1)$$

from which $A_1 = -A_2 = 16.9$ mA.

$$i = 16.9(e^{-1.70\,t} - e^{-298.3\,t}) \quad \text{(mA)}$$

(*b*) For the time of maximum current,

$$\frac{di}{dt} = 0 = -28.73\,e^{-1.70\,t} + 5041.3\,e^{-298.3\,t}$$

Solving by logarithms, $t = 17.4$ ms. See Fig. 5-33.

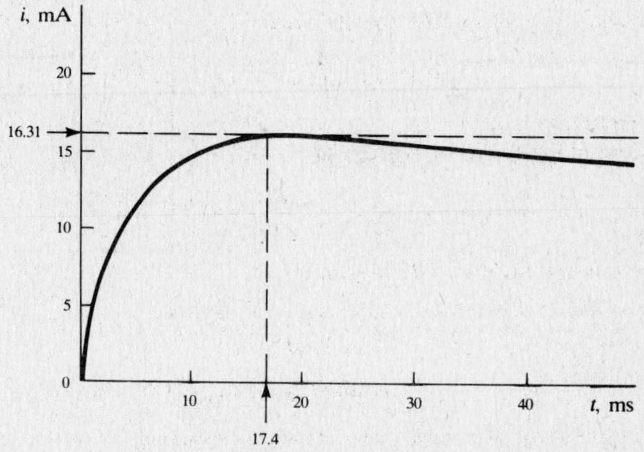

Fig. 5-33

5.15 A series *RLC* circuit, with $R = 50\ \Omega$, $L = 0.1$ H, and $C = 50\ \mu$F, has a constant voltage $V = 100$ V applied at $t = 0$. Obtain the current transient, assuming zero initial charge on the capacitor.

$$\alpha = \frac{R}{2L} = 250 \text{ s}^{-1} \qquad \omega_0^2 = \frac{1}{LC} = 2.0 \times 10^5 \text{ s}^{-2} \qquad \beta = \sqrt{\alpha^2 - \omega_0^2} = j370.8 \text{ rad/s}$$

This is an oscillatory case ($\alpha < \omega_0$), and the general current expression is

$$i = e^{-250t}(A_1 \cos 370.8\,t + A_2 \sin 370.8\,t)$$

The initial conditions, obtained as in Problem 5.14, are

$$i(0^+) = 0 \qquad \frac{di}{dt}\bigg|_{0^+} = 1000 \text{ A/s}$$

and these determine the values: $A_1 = 0$, $A_2 = 2.70$ A. Then

$$i = e^{-250t}(2.70 \sin 370.8\,t) \quad \text{(A)}$$

5.16 Rework Problem 5.15, if the capacitor has an initial charge $Q_0 = 2500\ \mu$C.

Everything remains the same as in Problem 5.15 except the second initial condition, which is now

$$0 + L\frac{di}{dt}\bigg|_{0^+} + \frac{Q_0}{C} = V \qquad \text{or} \qquad \frac{di}{dt}\bigg|_{0^+} = \frac{100 - (2500/50)}{0.1} = 500 \text{ A/s}$$

The initial values are half those in Problem 5.15, and so, by linearity,

$$i = e^{-250t}(1.35 \sin 370.8t) \quad \text{(A)}$$

5.17 In Fig. 5-34, the switch is closed at $t = 0$. Obtain the current i and capacitor voltage v_C, for $t > 0$.

As far as the natural response of the circuit is concerned, the two resistors are in parallel; hence,

$$\tau = R_{eq}C = (5\ \Omega)(2\ \mu\text{F}) = 10\ \mu\text{s}$$

By continuity, $v_C(0^+) = v_C(0^-) = 0$. Furthermore, as $t \to \infty$, the capacitor becomes an open circuit, leaving 20 Ω in series with the 50 V. That is,

$$i(\infty) = \frac{50}{20} = 2.5 \text{ A} \qquad v_C(\infty) = (2.5 \text{ A})(10\ \Omega) = 25 \text{ V}$$

Knowing the end conditions on v_C, we can write

$$v_C = [v_C(0^+) - v_C(\infty)]e^{-t/\tau} + v_C(\infty) = 25(1 - e^{-t/10}) \quad \text{(V)}$$

wherein t is measured in μs.

The current in the capacitor is given by

$$i_C = C\frac{dv_C}{dt} = 5e^{-t/10} \quad \text{(A)}$$

and the current in the parallel 10-Ω resistor is

$$i_{10\Omega} = \frac{v_C}{10\ \Omega} = 2.5(1 - e^{-t/10}) \quad \text{(A)}$$

Hence $$i = i_C + i_{10\Omega} = 2.5(1 + e^{-t/10}) \quad \text{(A)}$$

The problem might also have been solved by assigning mesh currents and solving simultaneous differential equations, as in Example 5.7.

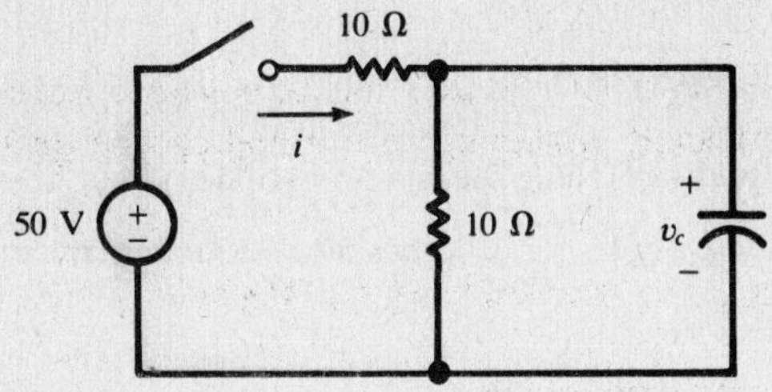

Fig. 5-34

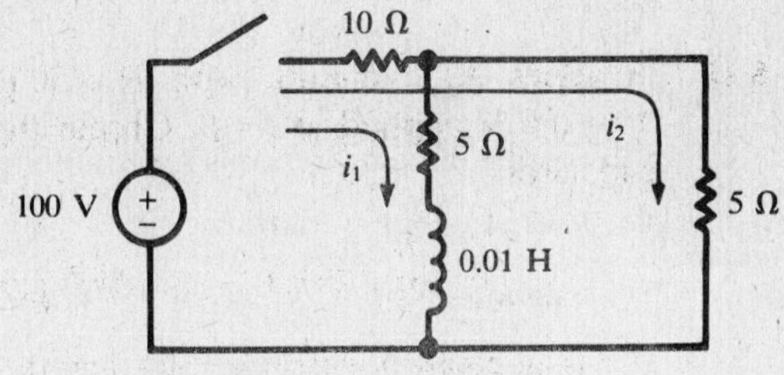

Fig. 5-35

5.18 The switch in the two-mesh circuit shown in Fig. 5-35 is closed at $t = 0$. Obtain the currents i_1 and i_2, for $t > 0$.

$$10(i_1 + i_2) + 5i_1 + 0.01\frac{di_1}{dt} = 100 \qquad (1)$$

$$10(i_1 + i_2) + 5i_2 = 100 \qquad (2)$$

From (2), $i_2 = (100 - 10i_1)/15$. Substituting in (1),

$$\frac{di_1}{dt} + 833i_1 = 3333 \qquad (3)$$

The steady-state solution (particular solution) of (3) is $i_1(\infty) = 3333/833 = 4.0$ A; hence

$$i_1 = Ae^{-833t} + 4.0 \quad \text{(A)}$$

The initial condition $i_1(0^-) = i_1(0^+) = 0$ now gives $A = -4.0$ A, so that

$$i_1 = 4.0(1 - e^{-833t}) \quad \text{(A)} \qquad \text{and} \qquad i_2 = 4.0 + 2.67\,e^{-833t} \quad \text{(A)}$$

Alternate Method

When the rest of the circuit is viewed from the terminals of the inductance, there is equivalent resistance

$$R_{\text{eq}} = 5 + \frac{5(10)}{15} = 8.33\ \Omega$$

Then $1/\tau = R_{\text{eq}}/L = 833\ \text{s}^{-1}$. At $t = \infty$, the circuit resistance is

$$R_T = 10 + \frac{5(5)}{10} = 12.5\ \Omega$$

so that the total current is $i_T = 100/12.5 = 8$ A. And, at $t = \infty$, this divides equally between the two 5-Ω resistors, yielding a final inductor current of 4 A. Consequently,

$$i_L = i_1 = 4(1 - e^{-833t}) \quad \text{(A)}$$

5.19 A series *RL* circuit, with $R = 50\ \Omega$ and $L = 0.2$ H, has a sinusoidal voltage

$$v = 150 \sin(500t + 0.785) \quad \text{(V)}$$

applied at $t = 0$. Obtain the current for $t > 0$.

The circuit equation for $t > 0$ is

$$\frac{di}{dt} + 250i = 750 \sin(500t + 0.785) \tag{1}$$

The solution is in two parts, the complementary function (i_c) and the particular solution (i_p), so that $i = i_c + i_p$. The complementary function is the general solution of (1) when the right-hand side is replaced by zero: $i_c = ke^{-250t}$. The *method of undetermined coefficients* for obtaining i_p consists in assuming that

$$i_p = A \cos 500t + B \sin 500t$$

since the right-hand side of (1) can also be expressed as a linear combination of these two functions. Then

$$\frac{di_p}{dt} = -500A \sin 500t + 500B \cos 500t$$

Substituting these expressions for i_p and di_p/dt into (1) and expanding the right-hand side,

$$-500A \sin 500t + 500B \cos 500t + 250A \cos 500t + 250B \sin 500t = 530.3 \cos 500t + 530.3 \sin 500t$$

Now equating the coefficients of like terms,

$$-500A + 250B = 530.3 \qquad \text{and} \qquad 500B + 250A = 530.3$$

Solving these simultaneous equations, $A = -0.4243$ A, $B = 1.273$ A.

$$i_p = -0.4243 \cos 500t + 1.273 \sin 500t = 1.342 \sin(500t - 0.322) \quad \text{(A)}$$

and

$$i = i_c + i_p = ke^{-250t} + 1.342 \sin(500t - 0.322) \quad \text{(A)}$$

At $t = 0$, $i = 0$. Applying this condition, $k = 0.425$ A, and, finally,

$$i = 0.425\,e^{-250t} + 1.342 \sin(500t - 0.322) \quad \text{(A)}$$

5.20 For the circuit of Fig. 5-36, obtain the current i_L, for all values of t.

For $t < 0$, the 50-V source results in inductor current $50/20 = 2.5$ A. The 5-A current source is applied for $t>0$. As $t\to\infty$, this current divides equally between the two 10-Ω resistors, whence $i_L(\infty) = -2.5$ A. The time constant of the circuit is

$$\tau = \frac{0.2\times 10^{-3}\ \text{H}}{20\ \Omega} = \frac{1}{100}\ \text{ms}$$

and so, with t in ms and using $i_L(0^+) = i_L(0^-) = 2.5$ A,

$$i_L = [i_L(0^+) - i_L(\infty)]e^{-t/\tau} + i_L(\infty) = 5.0\,e^{-100t} - 2.5 \quad \text{(A)}$$

Finally, using unit step functions to combine the expressions for $t<0$ and $t>0$,

$$i_L = 2.5\,u(-t) + (5.0\,e^{-100t} - 2.5)\,u(t) \quad \text{(A)}$$

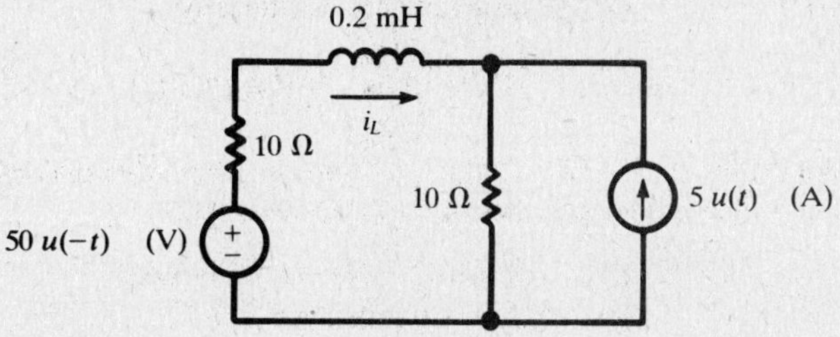

Fig. 5-36

5.21 A series RC circuit, with $R = 5$ kΩ and $C = 20\ \mu$F, has two voltage sources in series,

$$v_1 = 25\,u(-t) \quad \text{(V)} \qquad v_2 = 25\,u(t-t') \quad \text{(V)}$$

Obtain the complete expression for the voltage across the capacitor and make a sketch, if t' is positive.

The capacitor voltage is continuous. For $t \le 0$, v_1 results in a capacitor voltage of 25 V.

For $0 \le t \le t'$, both sources are zero, so that v_C decays exponentially from 25 V towards zero:

$$v_C = 25e^{-t/RC} = 25e^{-10t} \quad \text{(V)} \qquad (0 \le t \le t')$$

In particular, $v_C(t') = 25e^{-10t'}$ (V).

For $t \ge t'$, v_C builds from $v_C(t')$ towards the final value 25 V established by v_2:

$$\begin{aligned} v_C &= [v_C(t') - v_C(\infty)]e^{-(t-t')/RC} + v_C(\infty) \\ &= 25[1-(e^{10t'}-1)e^{-10t}] \quad \text{(V)} \qquad (t \ge t') \end{aligned}$$

Thus, for all t,

$$v_C = 25\,u(-t) + 25e^{-10t}[u(t) - u(t-t')] + 25[1-(e^{10t'}-1)e^{-10t}]u(t-t') \quad \text{(V)}$$

See Fig. 5-37.

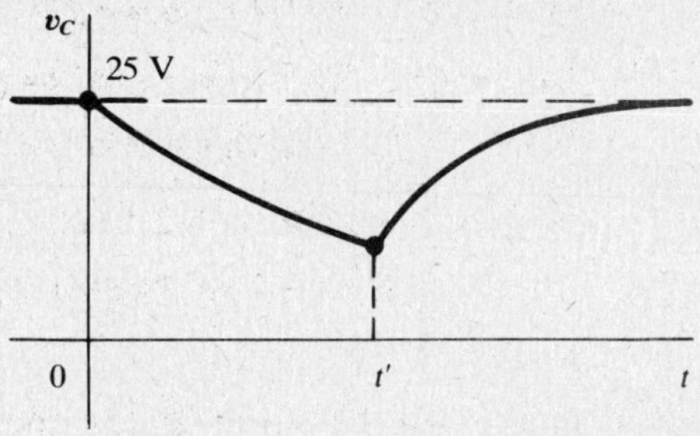

Fig. 5-37

Supplementary Problems

5.22 The capacitor in the circuit shown in Fig. 5-38 has initial charge $Q_0 = 800\ \mu C$, with polarity as indicated. If the switch is closed at $t = 0$, obtain the current and charge, for $t > 0$.
Ans. $i = -10e^{-25\,000t}$ (A), $q = 4 \times 10^{-4}(1 + e^{-25\,000t})$ (C)

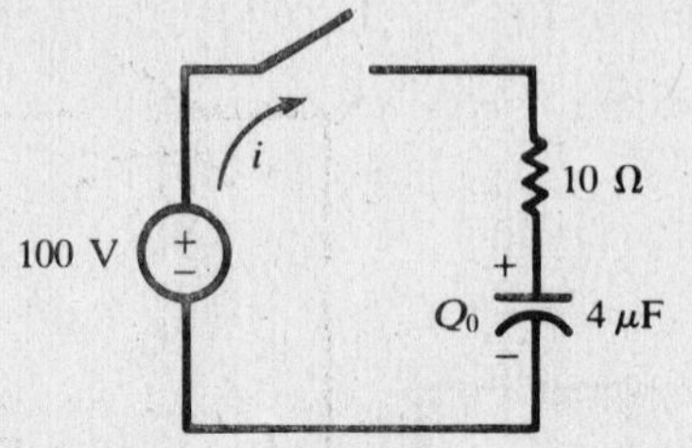

Fig. 5-38

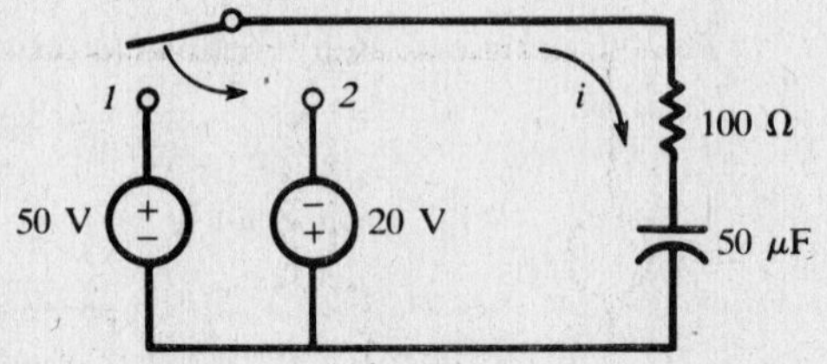

Fig. 5-39

5.23 A 2-μF capacitor, with initial charge $Q_0 = 100\ \mu C$, is connected across a 100-Ω resistor at $t = 0$. Calculate the time in which the transient voltage across the resistor drops from 40 to 10 volts.
Ans. 0.277 ms

5.24 In the *RC* circuit shown in Fig. 5-39, the switch is closed on position *1* at $t = 0$ and then moved to *2* after the passage of one time constant. Obtain the current transient for (*a*) $0 < t < \tau$, (*b*) $t > \tau$.
Ans. (*a*) $0.5e^{-200t}$ (A); (*b*) $-0.516e^{-200(t-\tau)}$ (A)

5.25 A 10-μF capacitor, with initial charge Q_0, is connected across a resistor at $t = 0$. Given that the power transient for the capacitor is $800e^{-4000t}$ (W), find R, Q_0, and the initial stored energy in the capacitor.
Ans. 50 Ω, 2000 μC, 0.20 J

5.26 A series *RL* circuit, with $R = 10\ \Omega$ and $L = 1$ H, has a 100-V source applied at $t = 0$. Find the current for $t > 0$. *Ans.* $10(1 - e^{-10t})$ (A)

5.27 In Fig. 5-40, the switch is closed on position *1* at $t = 0$, then moved to *2* at $t = 1$ ms. Find the time at which the voltage across the resistor is zero, reversing polarity. *Ans.* 1.261 ms

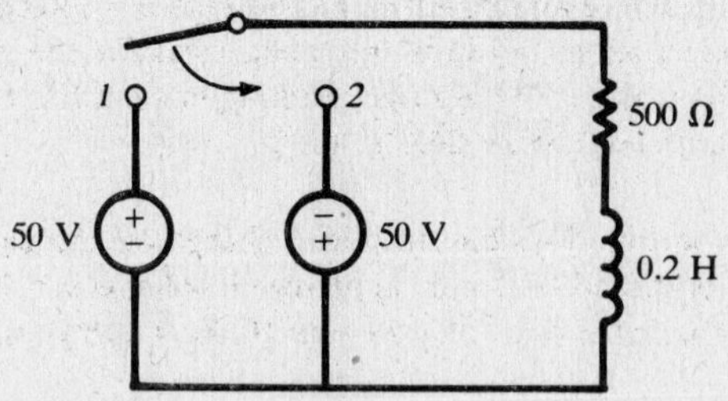

Fig. 5-40

5.28 A series *RL* circuit, with $R = 100\ \Omega$ and $L = 0.2$ H, has a 100-V source applied at $t = 0$; then a second source, of 50 V with the same polarity, is switched in at $t = t'$, replacing the first source. Find t' such that the current is constant at 0.5 A for $t > t'$. *Ans.* 1.39 ms

5.29 The circuit of Problem 5.28 has a 50-V source of *opposite* polarity switched in at $t = 0.50$ ms, replacing the first source. Obtain the current for (*a*) $0 < t < 0.50$ ms, (*b*) $t > 0.50$ ms.
Ans. (*a*) $1 - e^{-500t}$ (A); (*b*) $0.721e^{-500(t-0.0005)} - 0.50$ (A)

5.30 A voltage transient, $35e^{-500t}$ (V), has the value 25 V at $t_1 = 6.73 \times 10^{-4}$ s. Show that at $t = t_1 + \tau$ the function has a value 36.8% of that at t_1.

5.31 A transient that increases from zero toward a positive steady-state magnitude is 49.5 at $t_1 = 5.0$ ms, and 120 at $t_2 = 20.0$ ms. Obtain the time constant τ. *Ans.* 12.4 ms

5.32 The circuit shown in Fig. 5-41 is switched to position *1* at $t = 0$, then to position *2* at $t = 3\tau$. Find the transient current i for (*a*) $0 < t < 3\tau$, (*b*) $t > 3\tau$.
Ans. (*a*) $2.5e^{-50\,000t}$ (A); (*b*) $-1.58e^{-66\,700(t-0.00006)}$ (A)

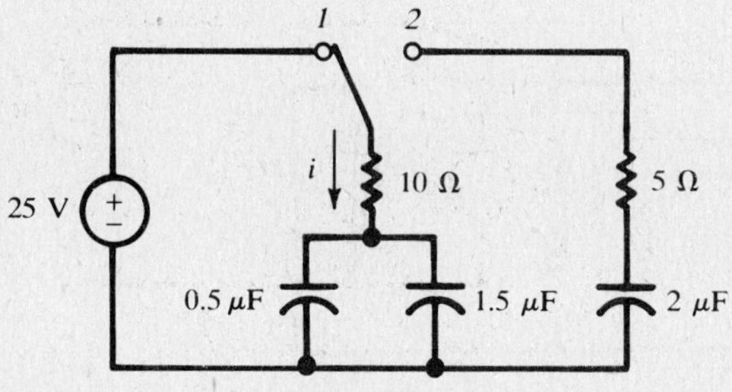

Fig. 5-41

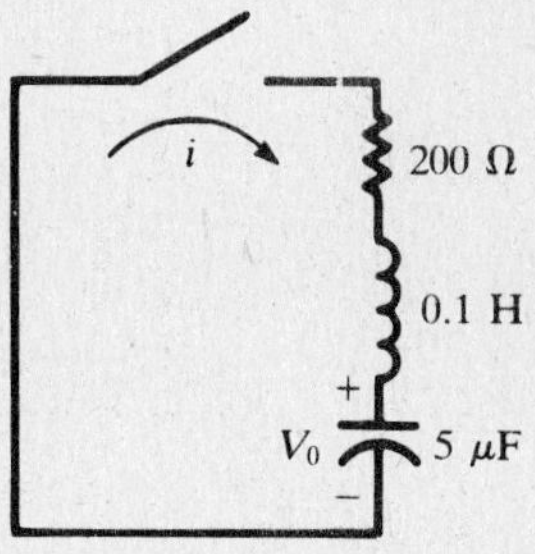

Fig. 5-42

5.33 In the *RLC* circuit of Fig. 5-42, the capacitor is initially charged to $V_0 = 200$ V. Find the current transient after the switch is closed at $t = 0$. *Ans.* $-2e^{-1000t} \sin 1000t$ (A)

5.34 A series *RLC* circuit, with $R = 200\ \Omega$, $L = 0.1$ H, and $C = 100\ \mu$F, has a voltage source of 200 V applied at $t = 0$. Find the current transient, assuming zero initial charge on the capacitor.
Ans. $1.055(e^{-52t} - e^{-1948t})$ (A)

5.35 What value of capacitance, in place of the 100 μF in Problem 5.34, results in the critically damped case? *Ans.* 10 μF

5.36 Find the natural resonant frequency, $|\beta|$, of a series *RLC* circuit with $R = 200\ \Omega$, $L = 0.1$ H, $C = 5\ \mu$F.
Ans. 1000 rad/s

5.37 A voltage of 10 V is applied at $t = 0$ to a series *RLC* circuit with $R = 5\ \Omega$, $L = 0.1$ H, $C = 500\ \mu$F. Find the transient voltage across the resistance. *Ans.* $3.60e^{-25t} \sin 139t$ (V)

5.38 An *RL* circuit, with $R = 300\ \Omega$ and $L = 1$ H, has voltage $v = 100 \cos (100t + 45°)$ (V) applied by closing a switch at $t = 0$. [A convenient notation has been used for the phase of v, which, strictly, should be indicated as $100t + (\pi/4)$ (rad).] Obtain the resulting current for $t > 0$.
Ans. $-0.282e^{-300t} + 0.316 \cos (100t + 26.6°)$ (A)

5.39 The *RC* circuit shown in Fig. 5-43 has an initial charge on the capacitor $Q_0 = 25\ \mu$C, with polarity as indicated. The switch is closed at $t = 0$, applying a voltage $v = 100 \sin (1000t + 30°)$ (V). Obtain the current for $t > 0$. *Ans.* $153.5e^{-4000t} + 48.4 \sin (1000t + 106°)$ (mA)

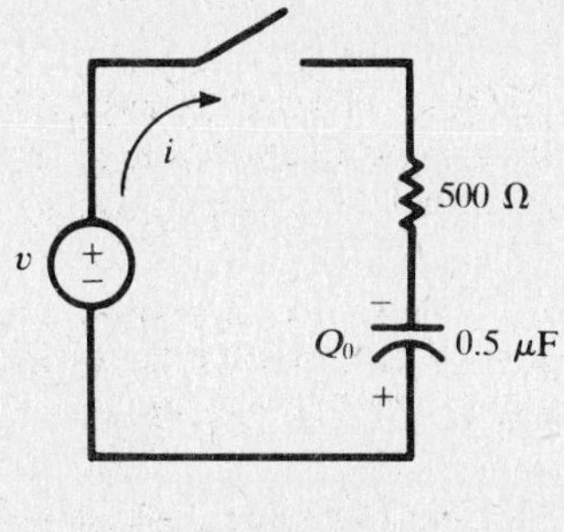

Fig. 5-43

5.40 What initial charge on the capacitor in Problem 5.39 would cause the current to go directly into the steady state without a transient? *Ans.* 13.37 μC (+ on top plate)

5.41 Write simultaneous differential equations for the circuit shown in Fig. 5-44 and solve for i_1 and i_2. The switch is closed at $t = 0$ after having been open for an extended period of time. (This problem can also be solved by applying known initial and final conditions to general solutions, as in Problem 5.17.)
Ans. $i_1 = 1.67\,e^{6.67t} + 5$ (A), $i_2 = -0.555\,e^{-6.67t} + 5$ (A)

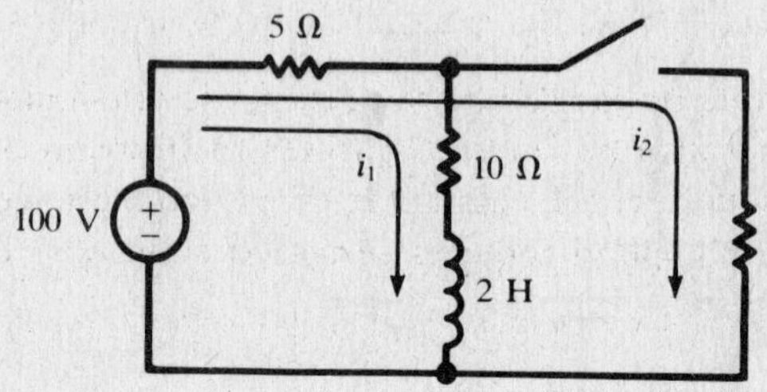

Fig. 5-44

5.42 In the two-mesh circuit shown in Fig. 5-45, the switch is closed at $t = 0$. Find i_1 and i_2, for $t > 0$.
Ans. $i_1 = 0.101\,e^{-100t} + 9.899\,e^{-9950t}$ (A), $i_2 = -5.05\,e^{-100t} + 5.00 + 0.05\,e^{-9950t}$ (A)

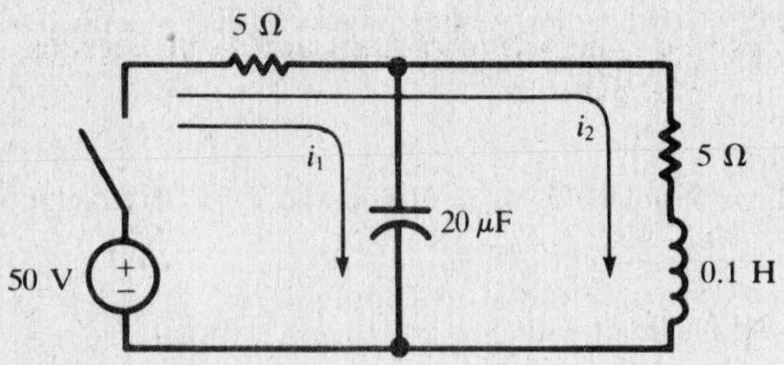

Fig. 5-45

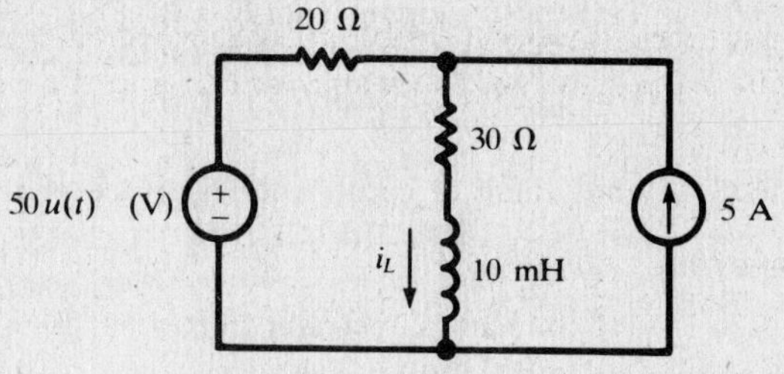

Fig. 5-46

5.43 For the *RL* circuit shown in Fig. 5-46, find the current i_L at the following times: (*a*) -1 ms, (*b*) 0^+, (*c*) 0.3 ms, (*d*) ∞. *Ans.* (*a*) 2.00 A; (*b*) 2.00 A; (*c*) 2.78 A; (*d*) 3.00 A

5.44 A series *RC* circuit, with $R = 2$ kΩ and $C = 40\ \mu$F, has two voltage sources in series with each other, $v_1 = 50$ V and $v_2 = -100\,u(t)$ (V). Find (*a*) the capacitor voltage at $t = \tau$, (*b*) the time at which the capacitor voltage is zero and reversing polarity. *Ans.* (*a*) -13.2 V; (*b*) 55.5 ms

Chapter 6

Sinusoidal Circuit Analysis

6.1 INTRODUCTION

This chapter will concentrate on the steady-state response of circuits driven by sinusoidal sources. The response in that case will also be sinusoidal (refer to Problem 5.19). For a linear circuit, the assumption of a *sinusoidal* source represents no real restriction, since *any* periodic source can be replaced by an equivalent combination (Fourier series) of sinusoids. That matter will be treated in Chapter 12.

6.2 SINUSOIDAL VOLTAGE AND CURRENT

Consider a voltage sine wave, as shown in Fig. 6-1, where the maximum value is V and where ϕ, the *phase angle*, is the phase of the wave at $t = 0$. The function may be written

$$v = V \sin(\omega t + \phi) \qquad \text{or} \qquad v = V \cos(\omega t + \phi - 90°)$$

in which, following Problem 5.38, we shall usually suppose ωt specified in radians and ϕ in degrees. The *frequency* f of the wave, in Hz, and the *period* T, in s, are given by

$$f = \frac{1}{T} = \frac{\omega}{2\pi}$$

where ω is in rad/s.

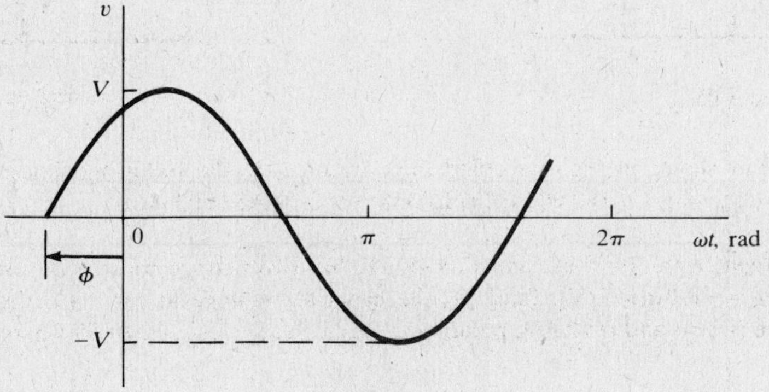

Fig. 6-1

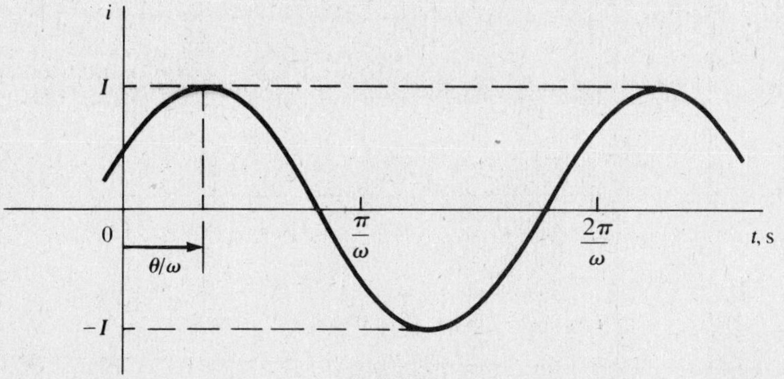

Fig. 6-2

A current cosine wave is shown in Fig. 6-2, where now the independent variable is time t.

$$i = I\cos(\omega t - \theta) \qquad \text{or} \qquad i = I\sin(\omega t - \theta + 90^\circ)$$

6.3 ELEMENT RESPONSES

Consider an inductance L carrying current $i = I\cos(\omega t + 45^\circ)$. The voltage is

$$v_L = L\frac{di}{dt} = \omega LI[-\sin(\omega t + 45^\circ)] = \omega LI\cos(\omega t + 135^\circ)$$

The current i *lags* v_L by $90^\circ = \pi/2$ rad, as shown in Fig. 6-3(b). The other circuit elements can be examined in a similar way. Table 6-1 shows responses to both sine and cosine applied functions.

Table 6-1

	$i = I\sin\omega t$	$i = I\cos\omega t$
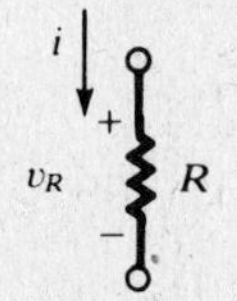	$v_R = RI\sin\omega t$	$v_R = RI\cos\omega t$
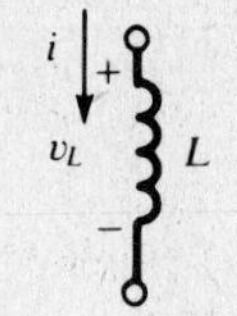	$v_L = \omega LI\sin(\omega t + 90^\circ)$	$v_L = \omega LI\cos(\omega t + 90^\circ)$
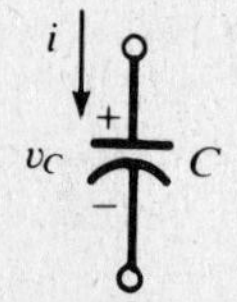	$v_C = \frac{I}{\omega C}\sin(\omega t - 90^\circ)$	$v_C = \frac{I}{\omega C}\cos(\omega t - 90^\circ)$
	$v = V\sin\omega t$	$v = V\cos\omega t$
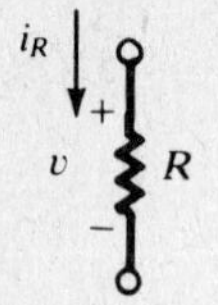	$i_R = \frac{V}{R}\sin\omega t$	$i_R = \frac{V}{R}\cos\omega t$
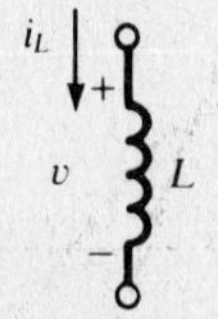	$i_L = \frac{V}{\omega L}\sin(\omega t - 90^\circ)$	$i_L = \frac{V}{\omega L}\cos(\omega t - 90^\circ)$
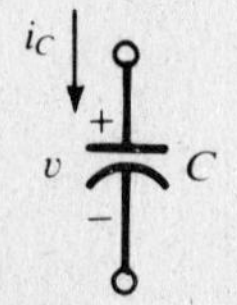	$i_C = \omega CV\sin(\omega t + 90^\circ)$	$i_C = \omega CV\cos(\omega t + 90^\circ)$

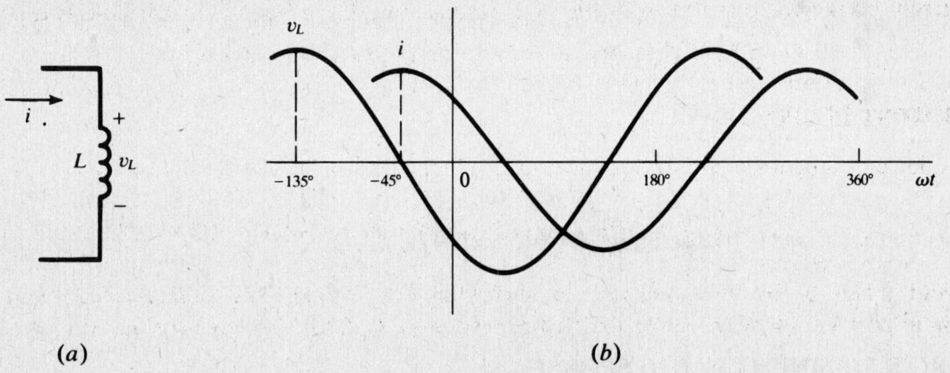

Fig. 6-3

Observe that, in a resistor, i and v are *in phase* in every case; in an inductor, i *lags* v by 90°; in a capacitor, i *leads* v by 90°.

6.4 SERIES *RL* SINUSOIDAL RESPONSE

The circuit shown in Fig. 6-4(*a*) has an applied current $i = I \sin \omega t$. Then

$$v_R = Ri = RI \sin \omega t \qquad v_L = L\frac{di}{dt} = \omega LI \sin(\omega t + 90°)$$

$$v = v_R + v_L = RI \sin \omega t + \omega LI \sin(\omega t + 90°)$$

Any number of sine and cosine terms, all of the same frequency, can be expressed as a single sine or cosine function (of that same frequency). Since the current is a sine function, assume that

$$v = V \sin(\omega t + \theta) \equiv V \sin \omega t \cos \theta + V \cos \omega t \sin \theta \qquad (1)$$

But, from the above,

$$v = RI \sin \omega t + \omega LI \sin \omega t \cos 90° + \omega LI \cos \omega t \sin 90° \qquad (2)$$

Equating coefficients of like terms in (*1*) and (*2*),

$$V \sin \theta = \omega LI \qquad \text{and} \qquad V \cos \theta = RI$$

which determine V and θ as

$$V = I\sqrt{R^2 + (\omega L)^2} \qquad \theta = \arctan \frac{\omega L}{R} \qquad (3)$$

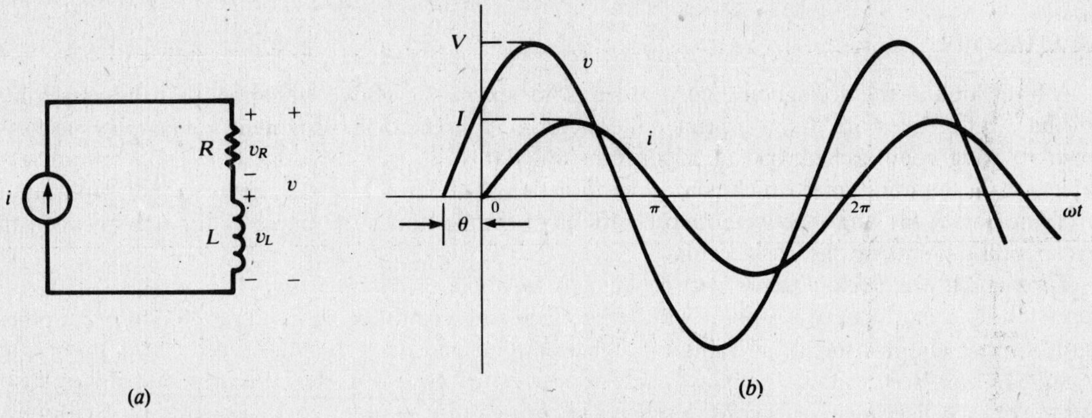

Fig. 6-4

The functions i and v are sketched in Fig. 6-4(*b*). The phase angle θ—the angle by which i lags v—lies in the range $0° \leq \theta \leq 90°$, with the limiting values attained for $\omega L \ll R$ and $\omega L \gg R$, respectively.

Conversely, if the circuit has an applied voltage $v = V \sin \omega t$, then the current response is calculated to be

$$i = \frac{V}{\sqrt{R^2 + (\omega L)^2}} \sin (\omega t - \theta)$$

where, as before, i lags v by the angle $\theta = \arctan (\omega L/R)$.

6.5 SERIES *RC* SINUSOIDAL RESPONSE

An analysis like that of Section 6.4 shows that, for an applied current $i = I \sin \omega t$, the voltage response is

$$v = I\sqrt{R^2 + (1/\omega C)^2} \sin (\omega t - \theta)$$

or, for an applied voltage $v = V \sin \omega t$, the current response is

$$i = \frac{V}{\sqrt{R^2 + (1/\omega C)^2}} \sin (\omega t + \theta)$$

Either way, i leads v by the angle $\theta = \arctan (1/\omega CR)$, where $\theta \approx 0°$ for $(1/\omega C) \ll R$ and $\theta \approx 90°$ for $(1/\omega C) \gg R$. See Fig. 6-5.

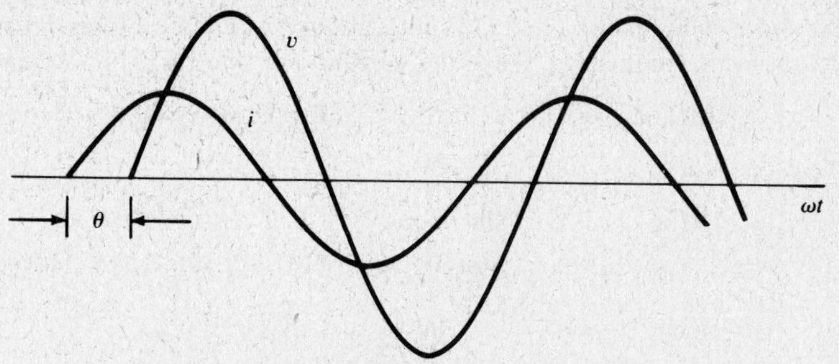

Fig. 6-5

6.6 PHASORS

A brief look at the voltage and current sinusoids examined above will show that the amplitudes and phase differences are the two principal concerns. A directed line segment, or *phasor*, such as is shown rotating counterclockwise at a constant angular velocity ω (rad/s) in Fig. 6-6, produces a projection on the horizontal which is a cosine function. The length of the phasor is the amplitude of the cosine curve; the angle between two positions of the phasor is the phase difference between the corresponding points on the cosine curve.

Throughout this book phasors will be defined from the cosine function. If a voltage or current expression is in the form of a sine, it will be changed to a cosine by subtracting 90° from the phase. Consider the examples shown in Table 6-2. Observe that phasors, which are vectorial in nature, are indicated by boldface capitals. The phase angle of the cosine function is written as the angle of the phasor. Thus the phasor diagram may be considered as a snapshot of the counterclockwise-rotating vector taken at $t = 0$.

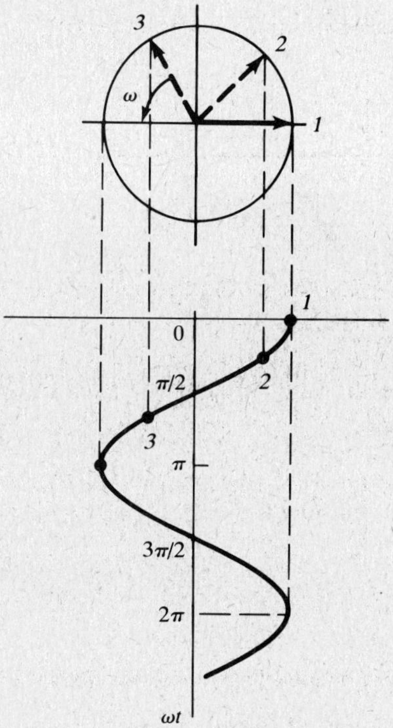

Fig. 6-6

Table 6-2

Function	Phasor Representation
$v = 150 \cos (500t + 45°)$ (V)	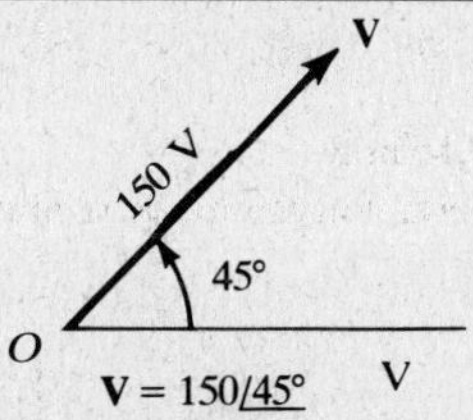 $\mathbf{V} = 150\underline{/45°}$ V
$i = (3 \times 10^{-3}) \sin (2000t + 30°)$ (A) $= (3 \times 10^{-3}) \cos (2000t - 60°)$ (A)	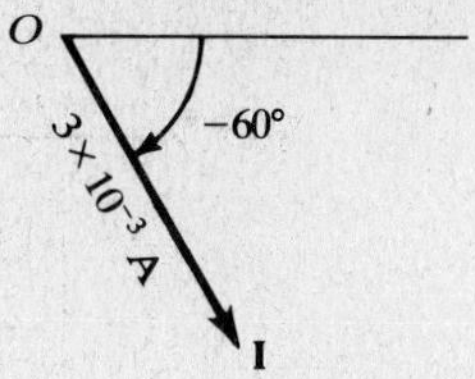 $\mathbf{I} = (3 \times 10^{-3})\underline{/-60°}$ A

EXAMPLE 6.1 A series combination of $R = 10\ \Omega$ and $L = 20$ mH has a current

$$i = 5.0 \cos (500t + 10°) \quad \text{(A)}$$

or $\mathbf{I} = 5.0\underline{/10°}$ A. Obtain the total voltage v and show the phase relationship from a plot of v and i, and also from the phasors $\mathbf{V}$ and $\mathbf{I}$.

$$v_R = Ri = 50.0 \cos (500t + 10°) \quad \text{(V)} \qquad v_L = L\frac{di}{dt} = 50.0 \cos (500t + 100°) \quad \text{(V)}$$

$$v = v_R + v_L = 70.7 \cos (500t + 55°) \quad \text{(V)}$$

or $\mathbf{V} = 70.7\underline{/55°}$ V.

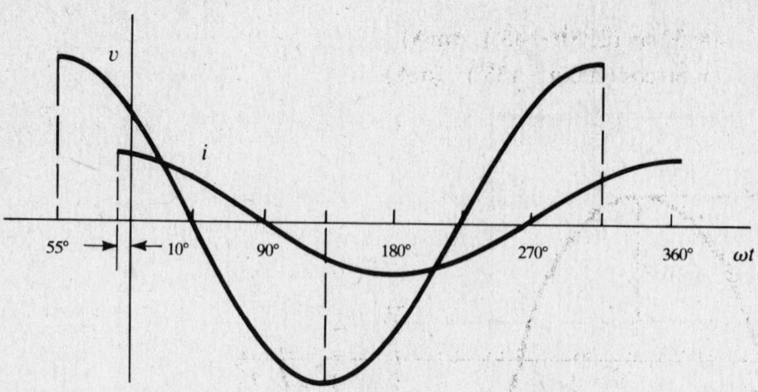

Fig. 6-7

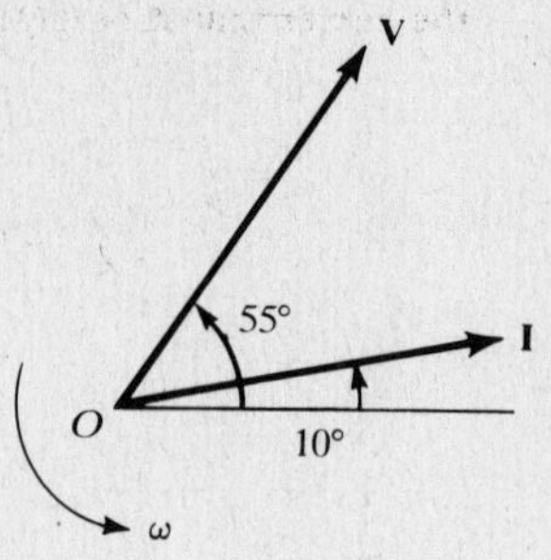

Fig. 6-8

In Fig. 6-7, the cosine functions are shown; i lags v by $55° - 10° = 45°$. Likewise in Fig. 6-8, **I** lags **V** by 45° since the two phasors rotate counterclockwise at the same radian frequency, $\omega = 500$ rad/s.

Phasors as Complex Numbers

When the horizontal projection axis is identified with the real axis of the complex plane (see Appendix B), phasors become complex numbers. Then, in view of *Euler's identity*,

$$e^{j\theta} \equiv \cos\theta + j\sin\theta$$

there are three equivalent notations for a phasor:

polar form	$\mathbf{V} = V\underline{/\theta}$
rectangular form	$\mathbf{V} = V\cos\theta + jV\sin\theta$
exponential form	$\mathbf{V} = Ve^{j\theta}$

The polar form is a convenient way to display the two parts of a phasor; however, it does not suggest how to express the product or quotient of two phasors. The exponential form suggests that

$$\mathbf{V}_1\mathbf{V}_2 = V_1V_2\underline{/\theta_1 + \theta_2} \qquad \frac{\mathbf{V}_1}{\mathbf{V}_2} = \frac{V_1}{V_2}\underline{/\theta_1 - \theta_2}$$

On the other hand, the rectangular form is useful in adding or subtracting phasors.

EXAMPLE 6.2 Add $v_1 = 25.0\cos(\omega t + 143.13°)$ (V) and $v_2 = 11.2\cos(\omega t + 26.57°)$ (V).

$$\begin{aligned} \mathbf{V}_1 &= 25.0\underline{/143.13°} = -20.0 + j15.0 \\ \mathbf{V}_2 &= 11.2\underline{/26.57°} = \underline{10.0 + j\,5.0} \\ & \qquad -10.0 + j20.0 = 22.4\underline{/116.57°} \end{aligned}$$

Hence $v_1 + v_2 = 22.4\cos(\omega t + 116.57°)$ (V).

Solved Problems

6.1 A sinusoidal current i reaches its first negative maximum, -50 mA, at $t = (7\pi/8)$ ms and has period $T = \pi$ ms. Express i as a sine and as a cosine.

For $\sin(\omega t + 0°)$, the first negative maximum occurs at $t = 3T/4$. Consequently, this waveform is shifted to the right by $T/8$ (Fig. 6-9); i.e., a phase shift of $\pi/4$ rad or 45°. Since

$$\omega = \frac{2\pi}{T} = 2000 \text{ rad/s}$$

the current must be given by

$$i = 50 \sin (2000t - 45°) \quad (\text{mA})$$
$$= 50 \cos (2000t - 135°) \quad (\text{mA})$$

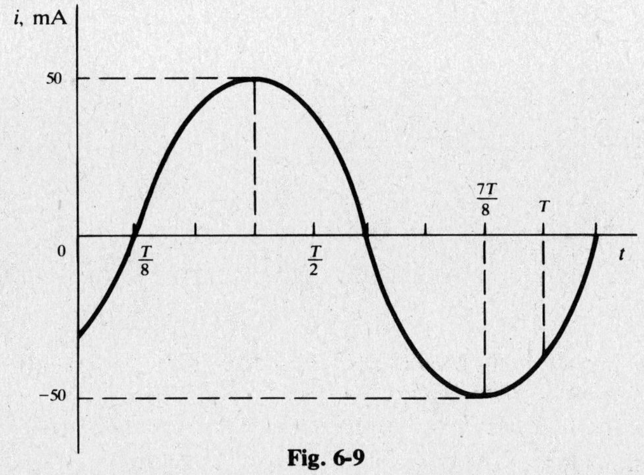

Fig. 6-9

6.2 For the voltage function $v = 150 \sin (500t - 30°)$ (V), find (*a*) the voltage at $t = 2.09$ ms, (*b*) the times at which the positive maxima occur.

(*a*) The sine can be evaluated with the phase in either radians or degrees. Using degrees,

$$V = 150 \sin [500(2.09 \times 10^{-3})(180°/\pi) - 30°] = 150 \sin 29.9° = 74.7 \text{ V}$$

(*b*) With the phase expressed in radians, positive maxima occur when

$$500t - \frac{\pi}{6} = \frac{\pi}{2} + n2\pi$$
$$t = 4.19(1 + 3n) \quad (\text{ms}) \qquad (n = 0, \pm 1, \pm 2, \ldots)$$

6.3 A 10-mH inductor has current $i = 5.0 \cos 2000t$ (A). Obtain the voltage v_L.

From Table 6-1, $v_L = \omega LI \cos (\omega t + 90°) = 100 \cos (2000t + 90°)$ (V). As a sine function,

$$v_L = 100 \sin (2000t + 180°) = -100 \sin 2000t \quad (\text{V})$$

6.4 A series circuit, with $R = 10\ \Omega$ and $L = 20$ mH, has current $i = 2.0 \sin 500t$ (A). Obtain total voltage v and the angle by which i lags v.

By the methods of Section 6.4,

$$\theta = \arctan \frac{500(20 \times 10^{-3})}{10} = 45°$$

$$v = I\sqrt{R^2 + (\omega L)^2} \sin (\omega t + \theta) = 28.3 \sin (500t + 45°) \quad (\text{V})$$

It is seen that i lags v by 45°.

6.5 Find the two elements in a series circuit, given that the current and total voltage are

$$i = 10 \cos (5000t - 23.13°) \quad (\text{A}) \qquad v = 50 \cos (5000t + 30°) \quad (\text{V})$$

Since i lags v (by 53.13°), the elements are R and L. The ratio of V_{max} to I_{max} is 50/10. Hence,

$$\frac{50}{10} = \sqrt{R^2 + (5000L)^2} \qquad \text{and} \qquad \tan 53.13° = 1.33 = \frac{5000L}{R}$$

Solving, $R = 3.0\ \Omega$, $L = 0.8$ mH.

6.6 A 17.9-μF capacitance has a voltage $v = 35.0 \sin(8000t + 25°)$ (V). Find the current i_C.

The current will lead the voltage by 90°; hence Table 6-1 gives

$$i_C = \omega CV \sin(\omega t + 25° + 90°) = 5.0 \sin(8000t + 115°) \quad \text{(A)}$$

6.7 A series circuit, with $R = 2.0\ \Omega$ and $C = 200$ pF, has a sinusoidal applied voltage with a frequency of 99.47 MHz. If the maximum voltage across the capacitance is 24 V, what is the maximum voltage across the series combination?

$$\omega = 2\pi f = 6.25 \times 10^8 \text{ rad/s}$$

From Table 6-1, $I_{max} = \omega C V_{C,max} = 3.0$ A. Then, by Section 6.5,

$$V_{max} = I_{max}\sqrt{R^2 + (1/\omega C)^2} = \sqrt{(6)^2 + (24)^2} = 24.74 \text{ V}$$

6.8 Given current $i = I \sin \omega t$ in the series *RLC* circuit (Fig. 6-10), obtain the total voltage v.

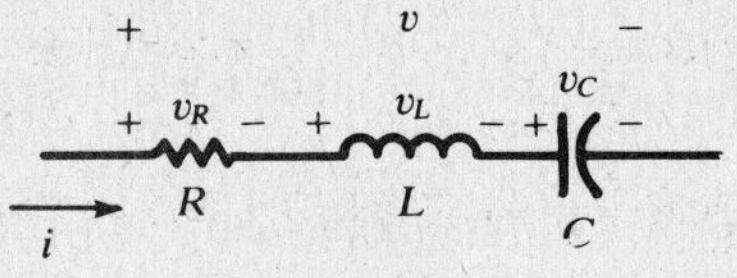

Fig. 6-10

$$v_R = Ri = RI \sin \omega t \qquad v_L = L\frac{di}{dt} = \omega LI \cos \omega t \qquad v_C = \frac{1}{C}\int i\,dt = \frac{-I}{\omega C}\cos \omega t$$

Hence
$$v = RI \sin \omega t + \left(\omega LI - \frac{I}{\omega C}\right)\cos \omega t$$

Using the methods of Section 6.4, this voltage can be expressed as a single sine function:

$$v = I\sqrt{R^2 + \left(\omega L - \frac{1}{\omega C}\right)^2}\,\sin(\omega t + \theta) \qquad \theta = \arctan\frac{\omega L - (1/\omega C)}{R}$$

It is seen that the voltage response depends critically on the relationship of the driving frequency, ω, to the *resonant frequency* of the circuit, $\omega_0 = 1/\sqrt{LC}$ (cf. Section 5.7). Thus:

If $\omega > \omega_0$, the phase angle is positive; the current *lags* the voltage, and the circuit has an overall effect which is inductive.

If $\omega < \omega_0$, the phase angle is negative; the current *leads* the voltage, and the circuit has an overall effect which is capacitive.

If $\omega = \omega_0$, the phase angle is zero; the current and voltage are *in phase*, and the circuit is said to be in *series resonance*.

6.9 An *RLC* series circuit has a current which lags the applied voltage by 30°. The inductor voltage maximum is twice the capacitor voltage maximum, and $v_L = 10.0 \sin 1000t$ (V). Determine L and C, given that $R = 20\ \Omega$.

From Problem 6.8,

$$10.0 = \omega L I_{max} \qquad \frac{1}{2}(10.0) = \frac{I_{max}}{\omega C} \qquad \tan 30° = \frac{\omega L - (1/\omega C)}{20}$$

in which $\omega = 1000$ rad/s. Solving for the three unknowns, $L = 23.1$ mH, $C = 86.5\ \mu$F (and $I_{max} = 0.433$ A).

6.10 In the parallel circuit of Fig. 6-11, the voltage is

$$v = 100.0 \sin(1000t + 50°) \quad \text{(V)}$$

Obtain the total current i.

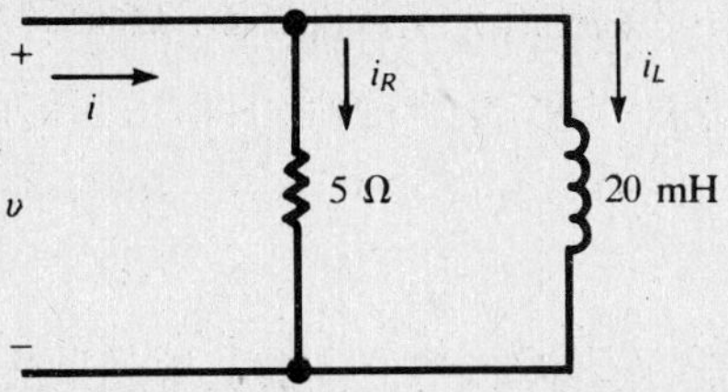

Fig. 6-11

From Table 6-1, the branch currents are

$$i_R = 20.0 \sin(1000t + 50°) \quad \text{(A)}$$
$$i_L = \frac{100.0}{\omega L} \sin(1000t + 50° - 90°) = 5.0 \sin(1000t - 40°) \quad \text{(A)}$$

Using phasors to obtain the sum $i = i_R + i_L$,

$$i_R = 20.0 \cos(1000t - 40°) \qquad \mathbf{I}_R = 20.0\underline{/-40°} = 15.3 - j12.9$$
$$i_L = 5.0 \cos(1000t - 130°) \qquad \mathbf{I}_L = 5.0\underline{/-130°} = -3.2 - j\,3.8$$
$$\mathbf{I} = \mathbf{I}_R + \mathbf{I}_L = 12.1 - j16.7 = 20.6\underline{/-54.1°}$$

Then $$i = 20.6 \cos(1000t - 54.1°) = 20.6 \sin(1000t + 35.9°) \quad \text{(A)}$$

6.11 In the parallel RC circuit of Fig. 6-12, $i_R = 15.0 \cos(5000t - 30°)$ (A). Obtain the current in the capacitance.

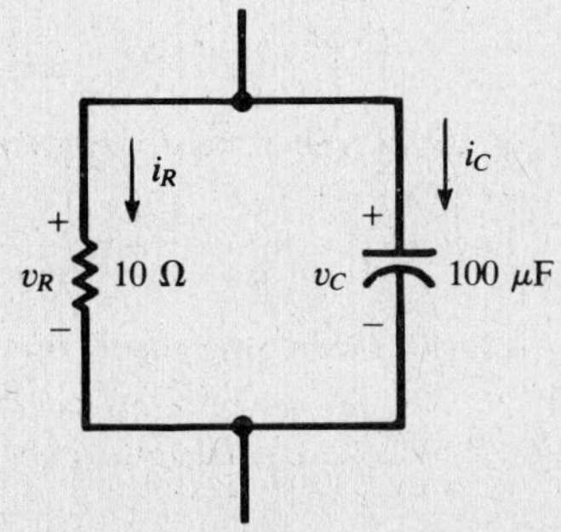

Fig. 6-12

In the parallel circuit, $v_C = v_R = Ri_R = 150.0 \cos(5000t - 30°)$ (V). Then

$$i_C = \omega CV \cos(5000t - 30° + 90°) = 75.0 \cos(5000t + 60°) \quad \text{(A)}$$

6.12 Obtain the sum of the three voltages

$$v_1 = 147.3 \cos(\omega t + 98.1°) \quad \text{(V)} \qquad v_2 = 294.6 \cos(\omega t - 45°) \quad \text{(V)}$$
$$v_3 = 88.4 \sin(\omega t + 135°) \quad \text{(V)}$$

Voltage v_3 is first changed to a cosine function. The addition is then performed from the phasor equivalents.

$$v_1 = 147.3 \cos(\omega t + 98.1°) \qquad \mathbf{V}_1 = 147.3\underline{/98.1°} = -20.8 + j145.8$$
$$v_2 = 294.6 \cos(\omega t - 45°) \qquad \mathbf{V}_2 = 294.6\underline{/-45°} = 208.3 - j208.3$$
$$v_3 = 88.4 \cos(\omega t + 45°) \qquad \mathbf{V}_3 = 88.4\underline{/45°} = \underline{62.5 + j\,62.5}$$
$$250.0 + j0$$

Thus $v_T = 250.0 \cos(\omega t + 0°) = 250.0 \sin(\omega t + 90°)$ (V).

6.13 At the junction of three branches shown in Fig. 6-13,

$$i_1 = 5.0 \cos (\omega t + 30°) \quad (A) \qquad i_3 = 3.5 \sin (\omega t + 60°) \quad (A)$$

Find i_2.

From KCL, $i_2 = i_3 - i_1$; or, in phasor form, $\mathbf{I}_2 = \mathbf{I}_3 - \mathbf{I}_1$.

$$i_3 = 3.5 \cos (\omega t - 30°) \qquad \mathbf{I}_3 = 3.5\underline{/-30°} = 3.03 - j1.75$$
$$-i_1 = -5.0 \cos (\omega t + 30°) \qquad -\mathbf{I}_1 = -5.0\underline{/30°} = -4.33 - j2.50$$
$$-1.30 - j4.25 = \mathbf{I}_2$$

whence $\mathbf{I}_2 = 4.44\underline{/-107°}$ or $i_2 = 4.44 \cos (\omega t - 107°)$ (A).

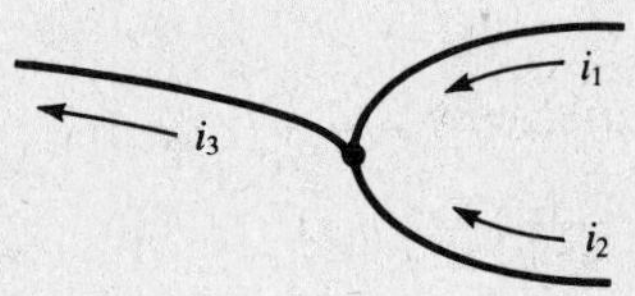

Fig. 6-13

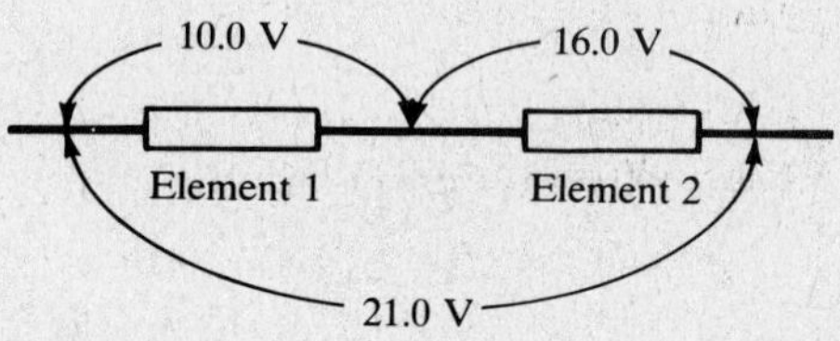

Fig. 6-14

6.14 Sinusoidal voltages across two circuit elements in series have maximum values 10.0 V and 16.0 V, while their sum has maximum value 21.0 V (see Fig. 6-14). Give complete trigonometric expressions for the three voltages.

No angles or trig functions have been specified; therefore, an angle of zero may be assigned to the sum $\mathbf{V}_T$ (Fig. 6-15). Using the law of cosines,

$$V_2^2 = V_1^2 + V_T^2 - 2V_1 V_T \cos \alpha$$

from which $\cos \alpha = 0.679$ and $\alpha = 47.27°$. Then the law of sines gives

$$\frac{V_2}{\sin \alpha} = \frac{V_T}{\sin \gamma} = \frac{V_1}{\sin \beta}$$

from which $\beta = 27.33°$ and $\gamma = 105.40°$. Assuming cosine functions, the voltages may be written:

$$v_1 = 10.0 \cos (\omega t + 47.27°) \quad (V) \qquad v_2 = 16.0 \cos (\omega t - 27.33°) \quad (V) \qquad v_T = 21.0 \cos (\omega t + 0°) \quad (V)$$

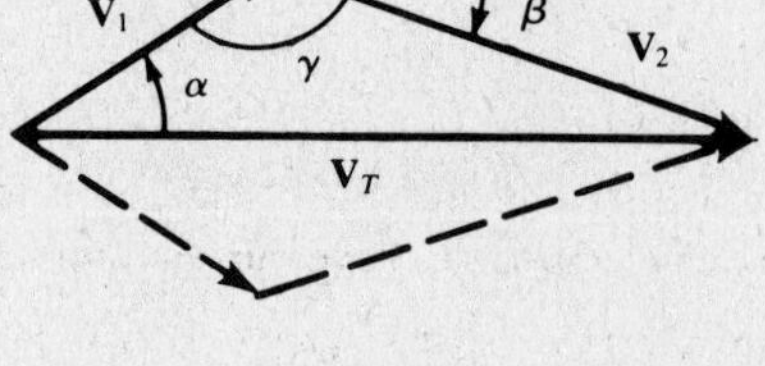

Fig. 6-15

It should be noted that the law of cosines above also has the solution $\alpha = -47.27°$, which leads to the dashed phasors in Fig. 6-15. These phasors have the same $\mathbf{V}_T$ as their sum, but give v_1 the phase angle $-47.27°$ and v_2 the phase angle $+27.33°$.

Supplementary Problems

6.15 A voltage sine wave passes through zero at $t = 0$ and each 3.93 ms thereafter. At $t = 3.12$ ms, the voltage is 30.0 V. Obtain ω, f, T and V_{max}. *Ans.* 799 rad/s, 127.2 Hz, 7.86 ms, 49.7 V

6.16 A cosine current function has phase angle $-26.0°$, period 4.19 ms, and magnitude 1.41 mA at $t = 0.826$ ms. Obtain the cosine function. *Ans.* $2.0 \cos(1500t - 26.0°)$ (mA)

6.17 Find the current in an inductance of 5.0 mH, given the voltage across the element,

$$v = 75 \sin(5000t - 45°) \quad \text{(V)}$$

Ans. $i = 3.0 \sin(5000t - 135°)$ (A)

6.18 In Fig. 6-16, $i = 2.5 \cos(1.5 \times 10^6 t + 45°)$ (A). Obtain the voltage v.
Ans. $150 \cos(1.5 \times 10^6 t - 45°)$ (V)

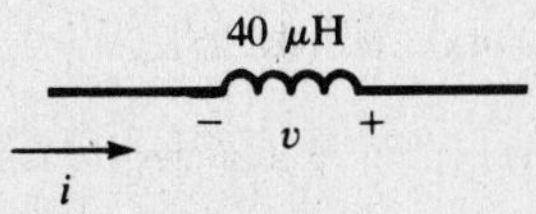

Fig. 6-16

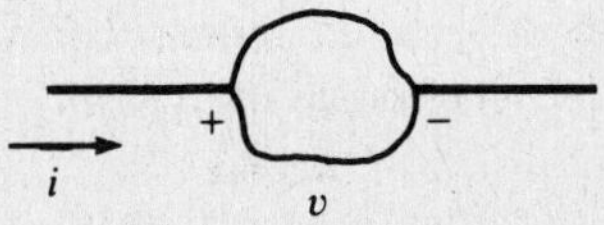

Fig. 6-17

6.19 The circuit element in Fig. 6-17 has a current $i = 2.5 \cos(2500t - 30°)$ (A) and a voltage $v = 5.0 \sin(2500t - 30°)$ (V). What is the element? *Ans.* a 200-μF capacitor

6.20 The voltage and current sine waves in Fig. 6-18 both pass through zero every 1.26 ms, and the maximum values are 120 V and 0.40 A. What single circuit element does this indicate?
Ans. a 120-mH inductor

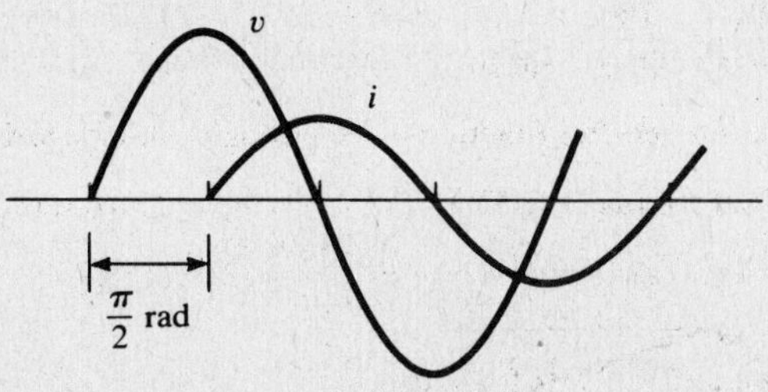

Fig. 6-18

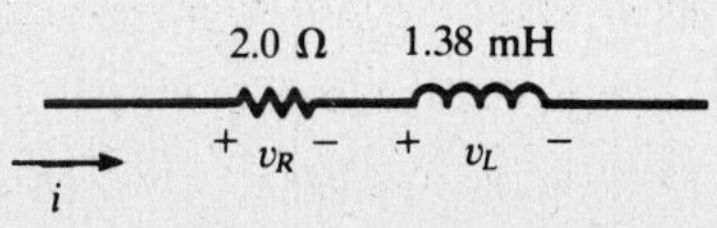

Fig. 6-19

6.21 Two circuit elements in a series connection have current and total voltage

$$i = 13.42 \sin(500t - 53.4°) \quad \text{(A)} \qquad v = 150 \sin(500t + 10°) \quad \text{(V)}$$

Identify the two elements. *Ans.* $R = 5\ \Omega$, $L = 20$ mH

6.22 Two circuit elements in a series connection have current and total voltage

$$i = 4.0 \cos(2000t + 13.2°) \quad \text{(A)} \qquad v = 200 \sin(2000t + 50.0°) \quad \text{(V)}$$

Identify the two elements. *Ans.* $R = 30\ \Omega$, $C = 12.5\ \mu$F

6.23 The circuit shown in Fig. 6-19 has a current $i = 5.0 \sin 2500t$ (mA). Find the maximum values of (*a*) v_R, (*b*) v_L, (*c*) $v_R + v_L$. *Ans.* (*a*) 10 mV; (*b*) 17.3 mV; (*c*) 20 mV

6.24 A series *RC* circuit, with $R = 27.5\ \Omega$ and $C = 66.7\ \mu$F, has sinusoidal voltages and current, with angular frequency 1500 rad/s. Find the phase angle by which the current leads the voltage. *Ans.* 20°

6.25 A series RLC circuit, with $R = 15\ \Omega$, $L = 80$ mH, and $C = 30\ \mu$F, has a sinusoidal current at angular frequency 500 rad/s. Determine the phase angle and whether the current leads or lags the total voltage.
Ans. 60.6°, leads

6.26 In Fig. 6-20, $i = 12.5 \cos (3000t - 55°)$ (A) and $v = 353.5 \cos (3000t - 10°)$ (V). Find R and C.
Ans. 20 Ω, 33.3 μF

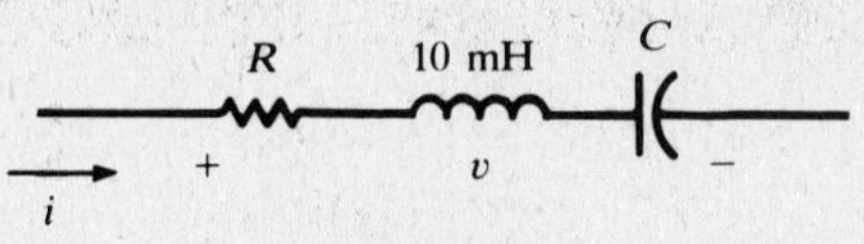

Fig. 6-20

6.27 A resistance $R = 10\ \Omega$ and an inductance $L = 5.0$ mH are in parallel. The inductive-branch current is

$$i_L = 5.0 \sin (2000t - 45°) \quad \text{(A)}$$

Obtain the total current, $i_T = i_R + i_L$, and the angle by which i_T lags the voltage v.
Ans. $7.07 \sin 2000t$ (A), 45°

6.28 A two-branch parallel circuit, with $R = 10\ \Omega$ in one branch and $C = 100\ \mu$F in the other, has a voltage $v = 150 \cos (5000t - 30°)$ (V). Find the total current, $i_T = i_R + i_C$.
Ans. $76.5 \cos (5000t + 48.7°)$ (A)

6.29 A capacitance $C = 35\ \mu$F is in parallel with a certain element. Identify the element, given that the voltage and total current are

$$v = 150 \sin 3000t \quad \text{(V)} \qquad i_T = 16.5 \sin (3000t + 72.4°) \quad \text{(A)}$$

Ans. $R = 30.1\ \Omega$

6.30 Three parallel branches respectively contain $R = 300\ \Omega$, $L = 0.50$ H, and $C = 10\ \mu$F. Given the voltage $v = 200 \sin 1000t$ (V), determine if the total current leads or lags the voltage, and by how much.
Ans. leads by 67.4°

6.31 Use phasor methods to obtain the current i in Fig. 6-21, given

$$i_1 = 14.14 \cos (800t - 45°) \text{ (A)} \qquad i_2 = 14.0 \cos (800t) \text{ (A)} \qquad i_3 = 17.0 \cos (800t + 61.9°) \text{ (A)}$$

Ans. $32.4 \cos (800t + 8.9°)$ (A)

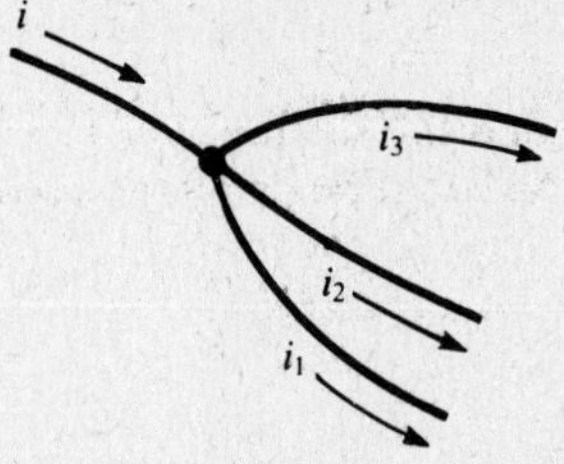

Fig. 6-21

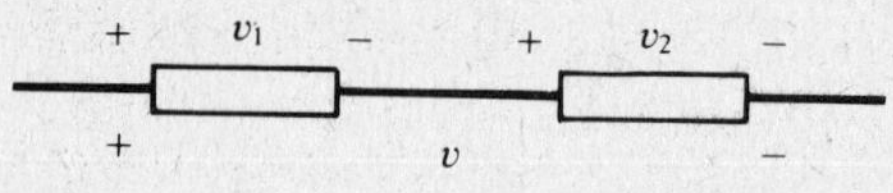

Fig. 6-22

6.32 Use phasor methods to obtain the voltage v_1 in Fig. 6-22, given

$$v_2 = 50.0 \sin (\omega t + 63.4°) \quad \text{(V)} \qquad v_T = 67.1 \cos (\omega t - 8.48°) \quad \text{(V)}$$

Ans. $25.0 \cos (\omega t + 30°)$ (V)

Chapter 7

Sinusoidal Steady State in the Frequency Domain

7.1 INTRODUCTION

In Chapter 6 it was shown how the sinusoidal steady state of a network could be described by means of complex quantities called phasors. A network so described is said to be in the *frequency domain*; for the phasors depend, generally both in magnitude and in angle, on the applied angular *frequency* ω, whereas they do not explicitly involve the time t at all. When, on the other hand, the steady state is specified in terms of sine or cosine functions such as $v(t)$ and $i(t)$, the network is said to be in the *time domain*. The present chapter is a continuation of the phasor methods of Chapter 6.

7.2 IMPEDANCE

The ratio of phasor voltage **V** to phasor current **I** is defined as the *impedance* **Z**, measured in Ω. **Z** is a complex number but not a phasor (the factor $e^{j\omega t}$, implicit in **V** and **I** and responsible for their counterclockwise rotation, cancels out when the ratio is taken). The actual circuit may contain resistance, inductance, and capacitance, and have time-domain voltages and currents, v and i, as suggested in Fig. 7-1(a). In the frequency domain, Fig. 7-1(b), the circuit elements are replaced by an equivalent impedance.

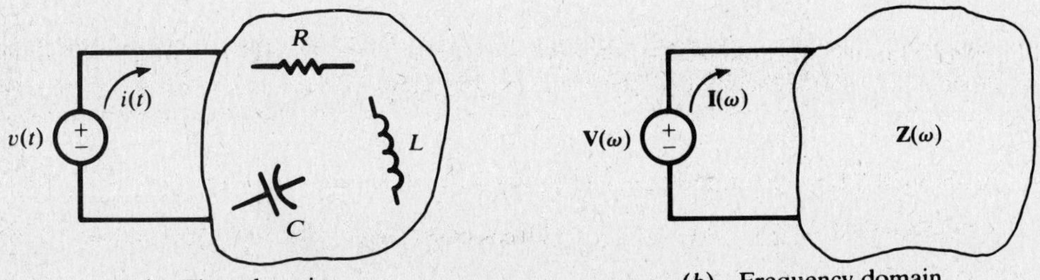

(a) Time domain (b) Frequency domain

Fig. 7-1

EXAMPLE 7.1 For the series RL circuit in the time domain, Section 6.4 showed that the applied voltage $v = V\cos\omega t$ (changing, for convenience, from the sine to the cosine function) produces the current

$$i = \frac{V}{\sqrt{R^2 + (\omega L)^2}}\cos(\omega t - \theta) \qquad \text{where} \qquad \theta = \arctan\frac{\omega L}{R}$$

In the frequency domain, these functions are represented by

$$\mathbf{V} = V\angle 0^\circ \qquad \mathbf{I} = \frac{V}{\sqrt{R^2 + (\omega L)^2}}\angle -\theta$$

and the impedance of the circuit is

$$\mathbf{Z} = \frac{\mathbf{V}}{\mathbf{I}} = \sqrt{R^2 + (\omega L)^2}\,\angle\theta = R + j\omega L$$

Figure 7-2 shows the frequency-domain circuit. The real and imaginary parts of **Z** are respectively called the resistance (as usual) and the *inductive reactance* ($X_L \equiv \omega L$).

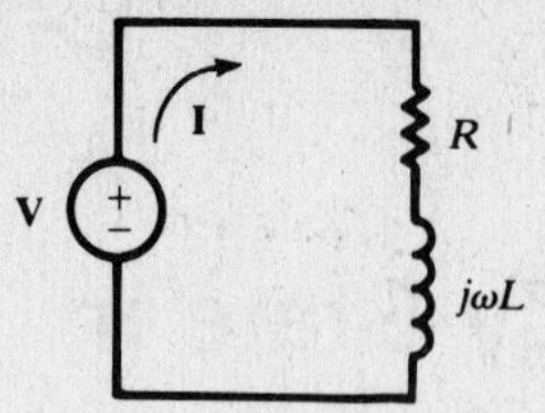

Fig. 7-2

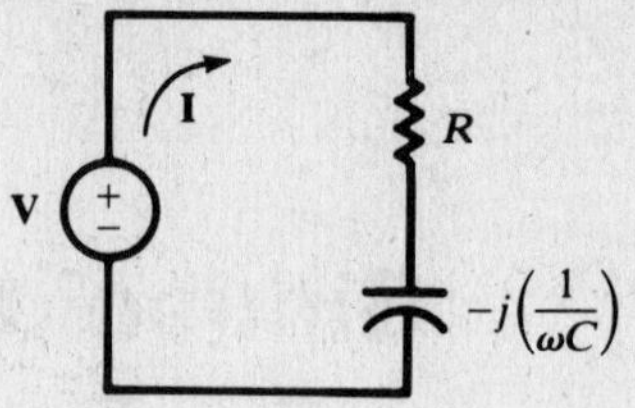

Fig. 7-3

A similar analysis of the series RC circuit leads to an impedance

$$\mathbf{Z} = \sqrt{R^2 + \left(\frac{1}{\omega C}\right)^2} \angle \theta = R - j\left(\frac{1}{\omega C}\right) \qquad \text{where} \qquad \theta = \arctan \frac{-1/\omega C}{R}$$

and a frequency-domain circuit as in Fig. 7-3. The quantity $X_C \equiv 1/\omega C$, the *negative of* the imaginary part of $\mathbf{Z}$, is called the *capacitive reactance.*

EXAMPLE 7.2 A series RL circuit, with $R = 5\ \Omega$ and $L = 2$ mH, has an applied voltage

$$v = 150 \sin 5000t \quad (\text{V})$$

Obtain $\mathbf{Z}$, calculate $\mathbf{I}$, and write a sine expression for i.

$$R = 5\ \Omega \qquad X_L = \omega L = (5000)(2 \times 10^{-3}) = 10\ \Omega$$

Then, $\mathbf{Z} = 5 + j10 = 11.18\angle 63.4^\circ\ \Omega$. Since $\mathbf{V} = 150\angle -90^\circ$ V,

$$\mathbf{I} = \frac{\mathbf{V}}{\mathbf{Z}} = \frac{150\angle -90^\circ}{11.18\angle 63.4^\circ} = 13.4\angle -153.4^\circ \ \text{A}$$

from which

$$i = 13.4 \cos (5000t - 153.4^\circ) = 13.4 \sin (5000t - 63.4^\circ) \quad (\text{A})$$

EXAMPLE 7.3 A two-element series circuit operating at $\omega = 500$ rad/s has a current which leads the voltage by 75°. One of the elements is a 20-Ω resistor; find the other element.

From the data, $\mathbf{Z} = Z\angle -75^\circ$. The negative angle means that $\mathbf{Z}$ has a negative imaginary part; i.e., capacitive reactance. Then

$$\tan -75^\circ = -\frac{1}{\omega CR}$$

$$-3.732 = -\frac{1}{(500)C(20)}$$

$$C = 26.8\ \mu\text{F}$$

Combinations of Impedances

The relation $\mathbf{V} = \mathbf{IZ}$ (in the frequency domain) is formally identical to Ohm's law, $v = iR$, for a resistive network (in the time domain). Therefore, impedances combine exactly like resistances:

impedances in series $\qquad \mathbf{Z}_{eq} = \mathbf{Z}_1 + \mathbf{Z}_2 + \cdots$

impedances in parallel $\qquad \dfrac{1}{\mathbf{Z}_{eq}} = \dfrac{1}{\mathbf{Z}_1} + \dfrac{1}{\mathbf{Z}_2} + \cdots$

In particular, for two parallel impedances, $\mathbf{Z}_{eq} = \mathbf{Z}_1\mathbf{Z}_2/(\mathbf{Z}_1 + \mathbf{Z}_2)$.

Impedance Diagram

In an *impedance diagram*, an impedance $\mathbf{Z}$ is represented by a point in the right half of the complex plane. Figure 7-4 shows two impedances; $\mathbf{Z}_1$, in the first quadrant, exhibits inductive reactance, while $\mathbf{Z}_2$, in the fourth quadrant, exhibits capacitive reactance. Their series equivalent, $\mathbf{Z}_1 + \mathbf{Z}_2$, is obtained by vector addition, as shown. Note that the "vectors" are shown without arrowheads, in order to distinguish these complex numbers from phasors.

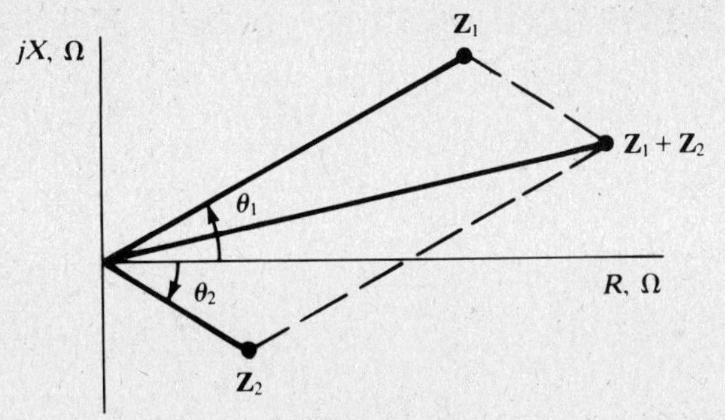

Fig. 7-4

7.3 ADMITTANCE

Admittance is defined as the reciprocal of impedance:

$$\mathbf{Y} = \frac{1}{\mathbf{Z}} \quad (\mathrm{S})$$

where (see Table 1-2) $1\ \mathrm{S} = 1\ \Omega^{-1} = 1$ mho. Like **Z**, **Y** is a complex number (not a phasor), $\mathbf{Y} \equiv G + jB$, whose real part is the *conductance* G and whose imaginary part is the *susceptance* B. Because

$$\mathbf{Y} = \frac{1}{Z\angle\theta} = \frac{1}{Z}\angle{-\theta}$$

it follows that a capacitive circuit ($-90° < \theta < 0°$) has positive susceptance, B_C, while an inductive circuit ($0° < \theta < 90°$) has negative susceptance, $-B_L$.

Combinations of Admittances

Replacing **Z** by $1/\mathbf{Y}$ in the formulas of Section 7.2 gives

admittances in series $\quad \dfrac{1}{\mathbf{Y}_{eq}} = \dfrac{1}{\mathbf{Y}_1} + \dfrac{1}{\mathbf{Y}_2} + \cdots$

admittances in parallel $\quad \mathbf{Y}_{eq} = \mathbf{Y}_1 + \mathbf{Y}_2 + \cdots$

Thus, series circuits are easiest treated in terms of impedance; parallel circuits, in terms of admittance.

Admittance Diagram

Figure 7-5, an *admittance diagram*, is analogous to Fig. 7-4 for impedance. Shown are an admittance $\mathbf{Y}_1$ having capacitive susceptance and an admittance $\mathbf{Y}_2$ having inductive susceptance, together with their vector sum, $\mathbf{Y}_1 + \mathbf{Y}_2$, which is the admittance of a parallel combination of $\mathbf{Y}_1$ and $\mathbf{Y}_2$.

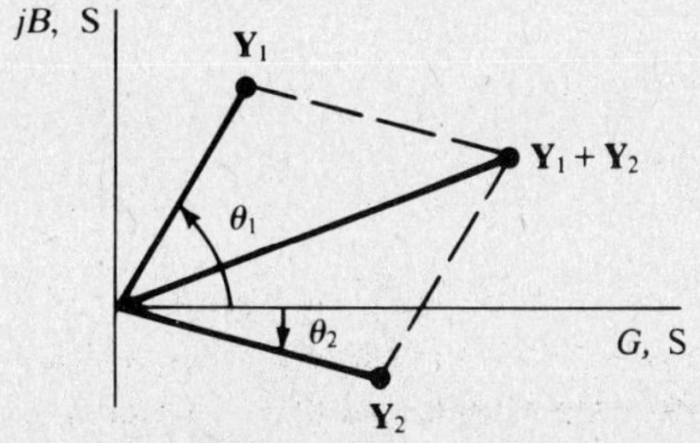

Fig. 7-5

7.4 VOLTAGE AND CURRENT DIVISION IN THE FREQUENCY DOMAIN

In view of the analogy between impedance in the frequency domain and resistance in the time domain, Section 3.3 implies the following results.

(1) Impedances in series divide the total voltage in the ratio of the impedances:

$$\frac{\mathbf{V}_r}{\mathbf{V}_s} = \frac{\mathbf{Z}_r}{\mathbf{Z}_s} \qquad \text{or} \qquad \mathbf{V}_r = \frac{\mathbf{Z}_r}{\mathbf{Z}_{\text{eq}}}\mathbf{V}_T$$

See Fig. 7-6.

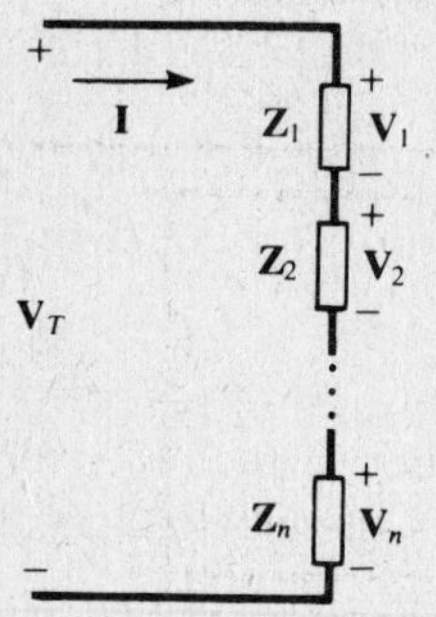

Fig. 7-6

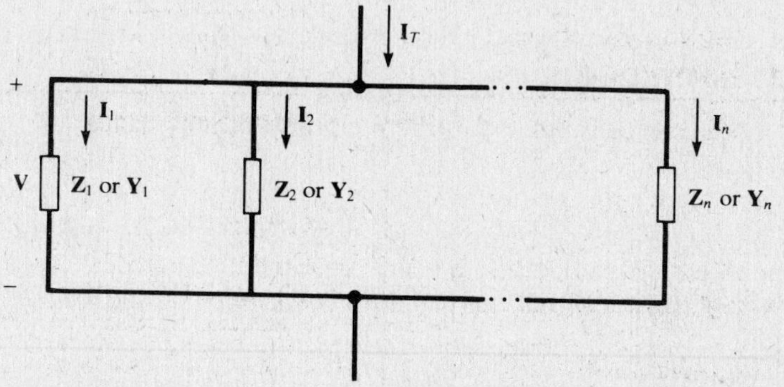

Fig. 7-7

(2) Impedances in parallel (admittances in series) divide the total current in the inverse ratio of the impedances (direct ratio of the admittances):

$$\frac{\mathbf{I}_r}{\mathbf{I}_s} = \frac{\mathbf{Z}_s}{\mathbf{Z}_r} = \frac{\mathbf{Y}_r}{\mathbf{Y}_s} \qquad \text{or} \qquad \mathbf{I}_r = \frac{\mathbf{Z}_{\text{eq}}}{\mathbf{Z}_r}\mathbf{I}_T = \frac{\mathbf{Y}_r}{\mathbf{Y}_{\text{eq}}}\mathbf{I}_T$$

See Fig. 7-7.

7.5 IMPEDANCE ANGLE

$\mathbf{Y} = \sqrt{G^2 + B_L^2}\,\angle{-\theta}$

In Example 7.1 it was seen that the polar angle of **Z** is positive for a series *RL* circuit (current lags voltage) and negative for a series *RC* circuit (current leads voltage). In general, since

$$Z\angle\theta = \frac{V\angle\phi}{I\angle\psi} = \frac{V}{I}\angle\phi - \psi$$

the impedance angle θ is *the angle by which voltage leads current.* (And the admittance angle, $-\theta$, is *the angle by which current leads voltage.*) Table 7-1 illustrates the possibilities.

Table 7-1

Phasor Diagram	Frequency-Domain Circuit	
	Impedance Form	Admittance Form
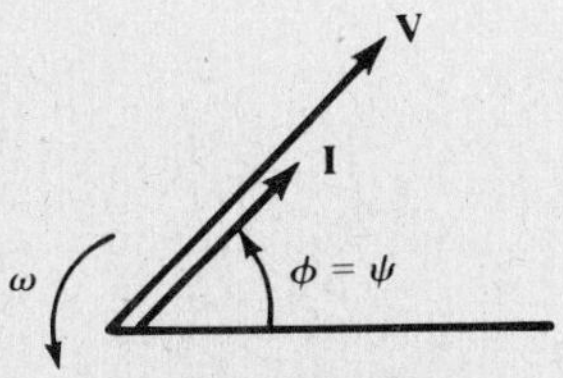 **Current and voltage in phase;** $\theta = 0°$	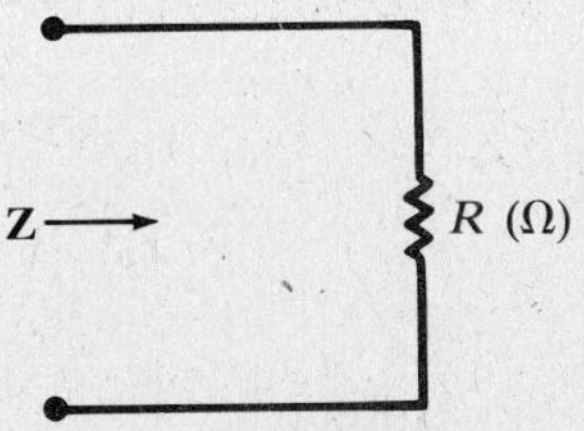 $\mathbf{Z} = R\angle 0°$	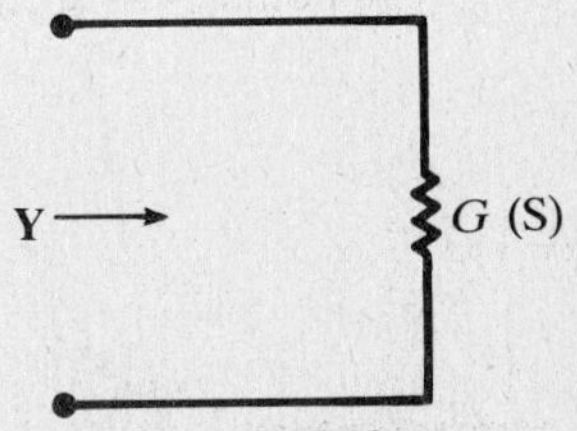 $\mathbf{Y} = G\angle 0°$
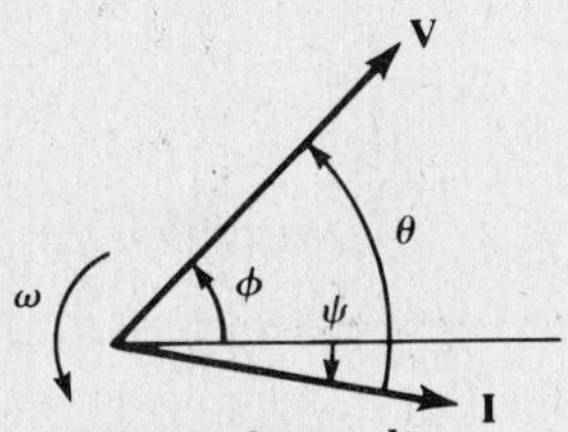 **Current lags voltage;** $0° < \theta < 90°$	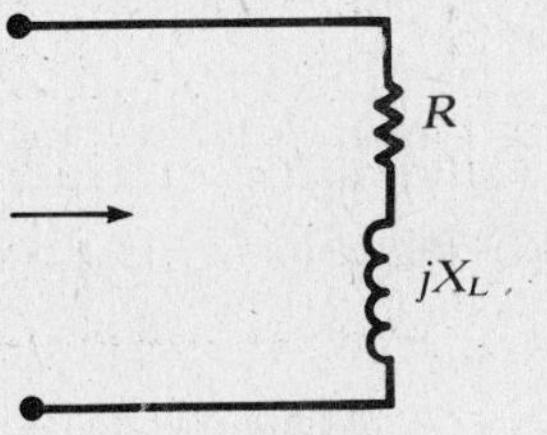 $\mathbf{Z} = \sqrt{R^2 + X_L^2}\angle\theta$	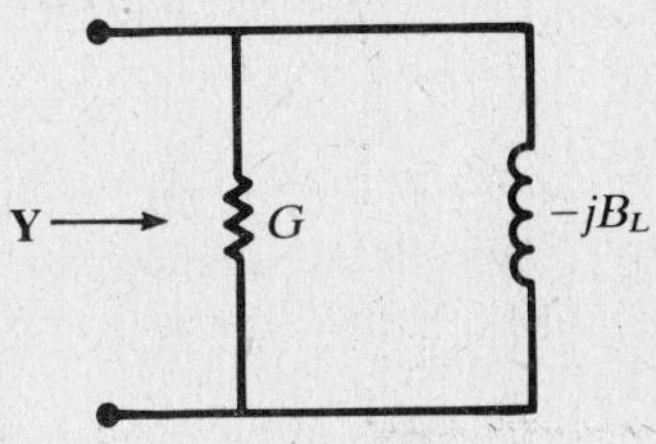 $\mathbf{Y} = \sqrt{G^2 + B_L^2}\angle -\theta$
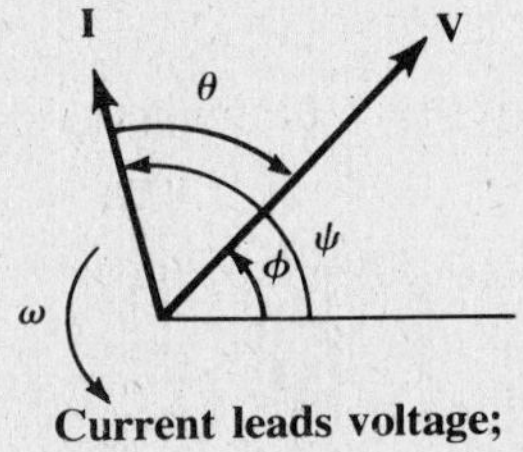 **Current leads voltage;** $-90° < \theta < 0°$	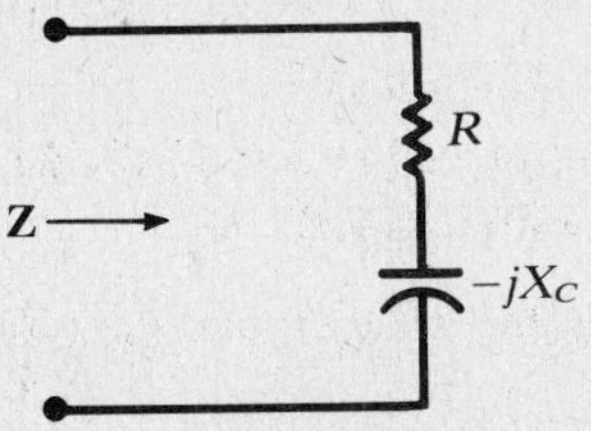 $\mathbf{Z} = \sqrt{R^2 + X_C^2}\angle\theta$	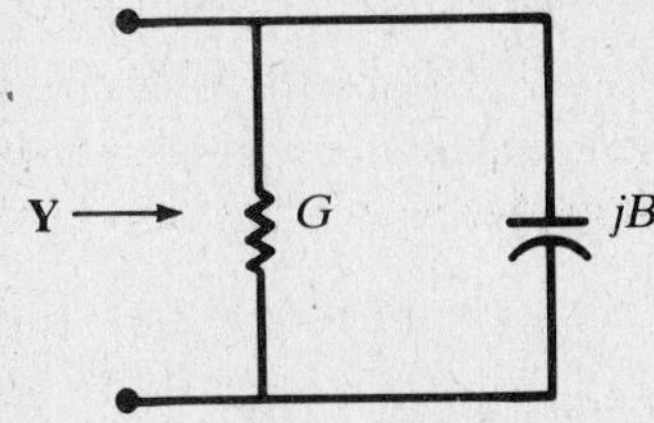 $\mathbf{Y} = \sqrt{G^2 + B_C^2}\angle -\theta$

Solved Problems

7.1 The current in a series circuit of $R = 5\ \Omega$ and $L = 30$ mH lags the applied voltage by 80°. Determine the source frequency and the impedance $\mathbf{Z}$.

From the impedance diagram, Fig. 7-8,

$$5 + jX_L = Z\angle 80° \qquad X_L = 5 \tan 80° = 28.4\ \Omega$$

Then $28.4 = \omega(30 \times 10^{-3})$, whence $\omega = 945.2$ rad/s and $f = 150.4$ Hz.

$$\mathbf{Z} = 5 + j28.4\quad \Omega$$

7.2 At what frequency will the current lead the voltage by 30° in a series circuit with $R = 8\ \Omega$ and $C = 30\ \mu$F?

From the impedance diagram, Fig. 7-9,

$$8 - jX_C = Z\angle -30° \qquad -X_C = 8 \tan(-30°) = -4.62\ \Omega$$

Then
$$4.62 = \frac{1}{2\pi f(30 \times 10^{-6})} \qquad \text{or} \qquad f = 1149 \text{ Hz}$$

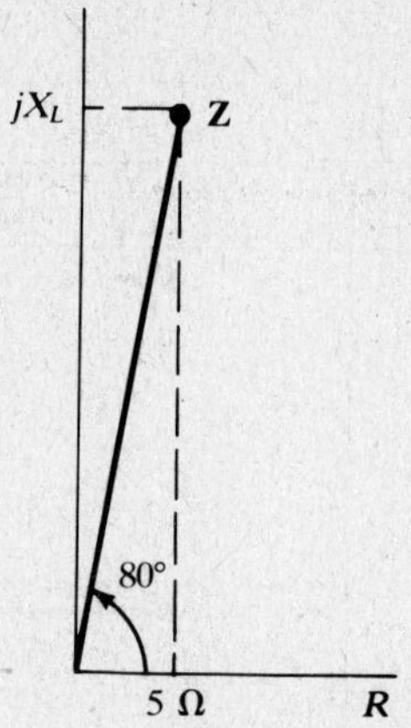

Fig. 7-8

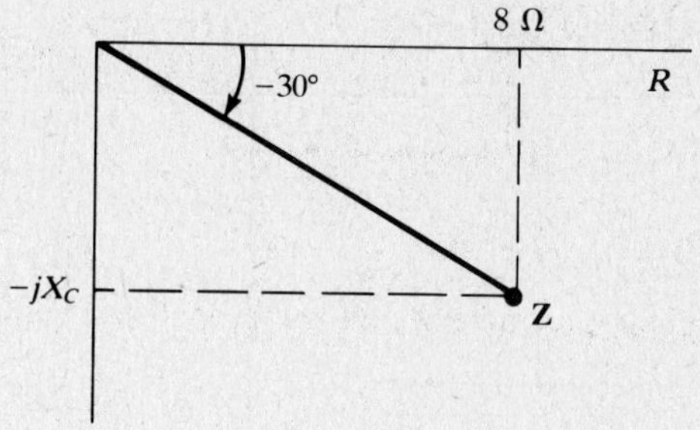

Fig. 7-9

7.3 A series RC circuit, with $R = 10\ \Omega$, has an impedance with an angle of $-45°$ at $f_1 = 500$ Hz. Find the frequency for which the magnitude of the impedance is (*a*) twice that at f_1, (*b*) one-half that at f_1.

From $10 - jX_C = Z_1\underline{/-45°}$, $X_C = 10\ \Omega$ and $Z_1 = 14.14\ \Omega$.

(*a*) For twice the magnitude,

$$10 - jX_C = 28.28\underline{/\theta_2} \qquad \text{or} \qquad X_C = \sqrt{(28.28)^2 - (10)^2} = 26.45\ \Omega$$

Then, since X_C is inversely proportional to f,

$$\frac{10}{26.45} = \frac{f_2}{500} \qquad \text{or} \qquad f_2 = 189\ \text{Hz}$$

(*b*) A magnitude $Z_3 = 7.07\ \Omega$ is impossible; the smallest magnitude possible is $Z = R = 10\ \Omega$.

7.4 A two-element series circuit has voltage $\mathbf{V} = 240\underline{/0°}$ V and current $\mathbf{I} = 50\underline{/-60°}$ A. Determine the current which results when the resistance is reduced to (*a*) 30%, (*b*) 60%, of its former value.

$$\mathbf{Z} = \frac{\mathbf{V}}{\mathbf{I}} = \frac{240\underline{/0°}}{50\underline{/-60°}} = 4.8\underline{/60°} = 2.40 + j4.16\ \ \Omega$$

(*a*) $30\% \times 2.40 = 0.72 \qquad \mathbf{Z}_1 = 0.72 + j4.16 = 4.22\underline{/80.2°}\ \ \Omega$

$$\mathbf{I}_1 = \frac{240\underline{/0°}}{4.22\underline{/80.2°}} = 56.8\underline{/-80.2°}\ \ \text{A}$$

(*b*) $60\% \times 2.40 = 1.44 \qquad \mathbf{Z}_2 = 1.44 + j4.16 = 4.40\underline{/70.9°}\ \ \Omega$

$$\mathbf{I}_2 = \frac{240\underline{/0°}}{4.40\underline{/70.9°}} = 54.5\underline{/-70.9°}\ \ \text{A}$$

7.5 For the circuit shown in Fig. 7-10, obtain $\mathbf{Z}_{eq}$ and compute $\mathbf{I}$.

For series impedances,

$$\mathbf{Z}_{eq} = 10\underline{/0°} + 4.47\underline{/63.4°} = 12.0 + j4.0 = 12.65\underline{/18.43°}\ \ \Omega$$

Then

$$\mathbf{I} = \frac{\mathbf{V}}{\mathbf{Z}_{eq}} = \frac{100\underline{/0°}}{12.65\underline{/18.43°}} = 7.91\underline{/-18.43°}\ \ \text{A}$$

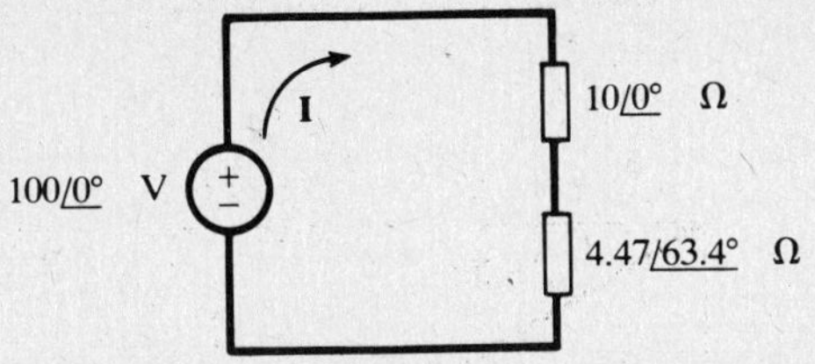

Fig. 7-10

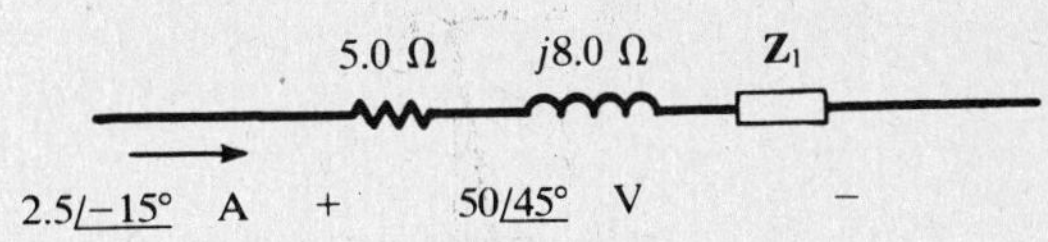

Fig. 7-11

7.6 Evaluate the impedance $\mathbf{Z}_1$ in the circuit of Fig. 7-11.

$$\mathbf{Z} = \frac{\mathbf{V}}{\mathbf{I}} = 20\angle 60^\circ = 10.0 + j17.3 \quad \Omega$$

Then, since impedances in series add,

$$5.0 + j8.0 + \mathbf{Z}_1 = 10.0 + j17.3 \qquad \text{or} \qquad \mathbf{Z}_1 = 5.0 + j9.3 \quad \Omega$$

7.7 Compute the equivalent impedance $\mathbf{Z}_{eq}$ and admittance $\mathbf{Y}_{eq}$ for the four-branch circuit of Fig. 7-12.

Using admittances,

$$\mathbf{Y}_1 = \frac{1}{j5} = -j0.20 \text{ S} \qquad \mathbf{Y}_3 = \frac{1}{15} = 0.067 \text{ S}$$

$$\mathbf{Y}_2 = \frac{1}{5 + j8.66} = 0.05 - j0.087 \quad \text{S} \qquad \mathbf{Y}_4 = \frac{1}{-j10} = j0.10 \text{ S}$$

Then
$$\mathbf{Y}_{eq} = \mathbf{Y}_1 + \mathbf{Y}_2 + \mathbf{Y}_3 + \mathbf{Y}_4 = 0.117 - j0.187 = 0.221\angle -58.0^\circ \quad \text{S}$$

and
$$\mathbf{Z}_{eq} = \frac{1}{\mathbf{Y}_{eq}} = 4.53\angle 58.0^\circ \quad \Omega$$

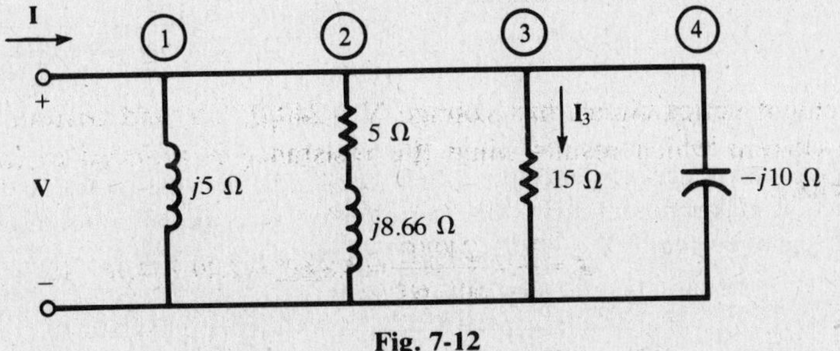

Fig. 7-12

7.8 The total current $\mathbf{I}$ entering the circuit shown in Fig. 7-12 is $33.0\angle -13.0^\circ$ A. Obtain the branch current $\mathbf{I}_3$ and the voltage $\mathbf{V}$.

$$\mathbf{V} = \mathbf{I}\mathbf{Z}_{eq} = (33.0\angle -13.0^\circ)(4.53\angle 58.0^\circ) = 149.5\angle 45.0^\circ \quad \text{V}$$

$$\mathbf{I}_3 = \mathbf{V}\mathbf{Y}_3 = (149.5\angle 45.0^\circ)\left(\frac{1}{15}\angle 0^\circ\right) = 9.97\angle 45.0^\circ \quad \text{A}$$

7.9 Obtain $\mathbf{Z}_{eq}$ and $\mathbf{Y}_{eq}$ for the circuit of Fig. 7-13.

$$\mathbf{Z}_1 = 10 + j20 = 22.4\angle 63.43^\circ \quad \Omega \qquad \mathbf{Z}_2 = 15 - j15 = 21.2\angle -45^\circ \quad \Omega$$

$$\mathbf{Z}_{eq} = \frac{\mathbf{Z}_1\mathbf{Z}_2}{\mathbf{Z}_1 + \mathbf{Z}_2} = \frac{(22.4\angle 63.43^\circ)(21.2\angle -45.0^\circ)}{(10 + j20) + (15 - j15)} = 18.63\angle 7.12^\circ \quad \Omega$$

$$\mathbf{Y}_{eq} = \frac{1}{\mathbf{Z}} = 0.0537\angle -7.12^\circ \quad \text{S}$$

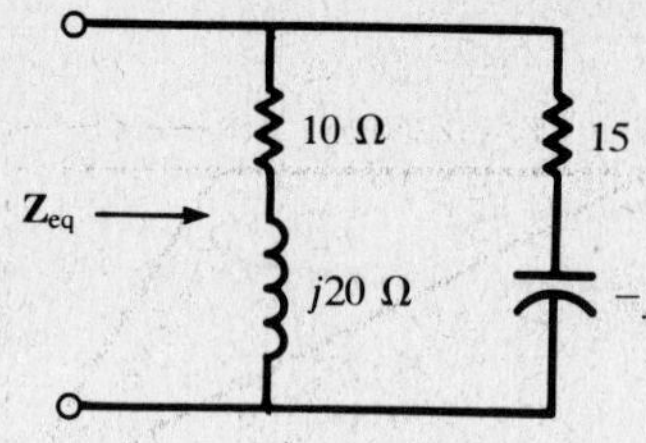

Fig. 7-13

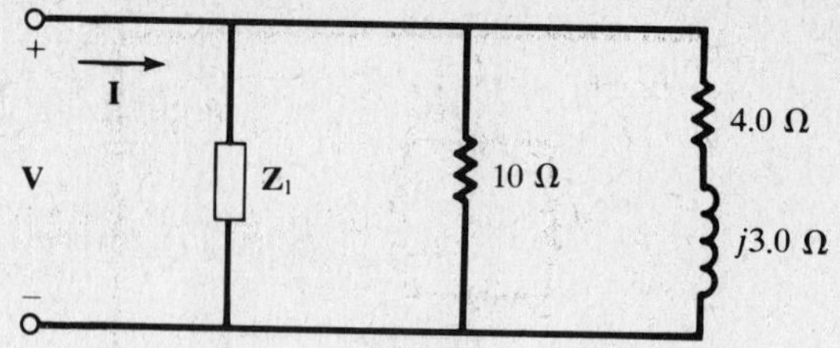

Fig. 7-14

7.10 Find $\mathbf{Z}_1$ in the three-branch network of Fig. 7-14, if $\mathbf{I} = 31.5\underline{/24.0^\circ}$ A for an applied voltage $\mathbf{V} = 50.0\underline{/60.0^\circ}$ V.

$$\mathbf{Y} = \frac{\mathbf{I}}{\mathbf{V}} = 0.630\underline{/-36.0^\circ} = 0.510 - j0.370 \quad \text{S}$$

Then

$$0.510 - j0.370 = \mathbf{Y}_1 + \frac{1}{10} + \frac{1}{4.0 + j3.0}$$

whence $\mathbf{Y}_1 = 0.354\underline{/-45^\circ}$ S and $\mathbf{Z}_1 = 2.0 + j2.0$ Ω.

7.11 The constants R and L of a coil can be obtained by connecting the coil in series with a known resistance and measuring the coil voltage V_x, the resistor voltage V_1, and the total voltage V_T (Fig. 7-15). The frequency must also be known, but the phase angles of the voltages are not known. Given that $f = 60$ Hz, $V_1 = 20$ V, $V_x = 22.4$ V, and $V_T = 36.0$ V, find R and L.

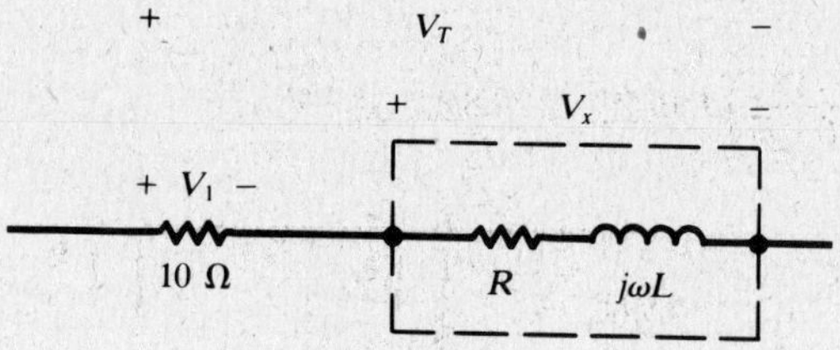

Fig. 7-15

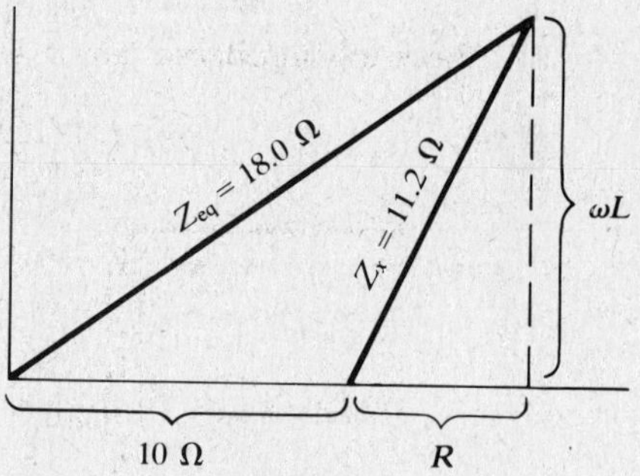

Fig. 7-16

The measured voltages are effective values (see Appendix A); but, as far as impedance calculations are concerned, it makes no difference whether effective or peak values are used.

The (effective) current is $I = V_1/10 = 2.0$ A. Then

$$Z_x = \frac{22.4}{2.0} = 11.2 \ \Omega \qquad Z_{eq} = \frac{36.0}{2.0} = 18.0 \ \Omega$$

From the impedance diagram, Fig. 7-16,

$$(18.0)^2 = (10 + R)^2 + (\omega L)^2$$
$$(11.2)^2 = R^2 + (\omega L)^2$$

where $\omega = 2\pi 60 = 377$ rad/s. Solving simultaneously,

$$R = 4.92 \ \Omega \qquad L = 26.7 \text{ mH}$$

7.12 In the parallel circuit shown in Fig. 7-17, the effective values of the currents are: $I_x = 18.0$ A, $I_1 = 15.0$ A, $I_T = 30.0$ A. Determine R and X_L.

The problem can be solved in a manner similar to that used in Problem 7.11, but with the admittance diagram.

The (effective) voltage is $V = I_1(4.0) = 60.0$ V. Then

$$Y_x = \frac{I_x}{V} = 0.300 \text{ S} \qquad Y_{eq} = \frac{I_T}{V} = 0.500 \text{ S} \qquad Y_1 = \frac{1}{4.0} = 0.250 \text{ S}$$

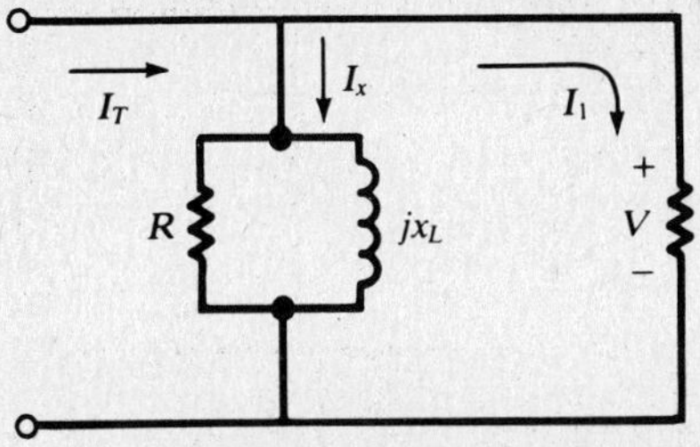

Fig. 7-17

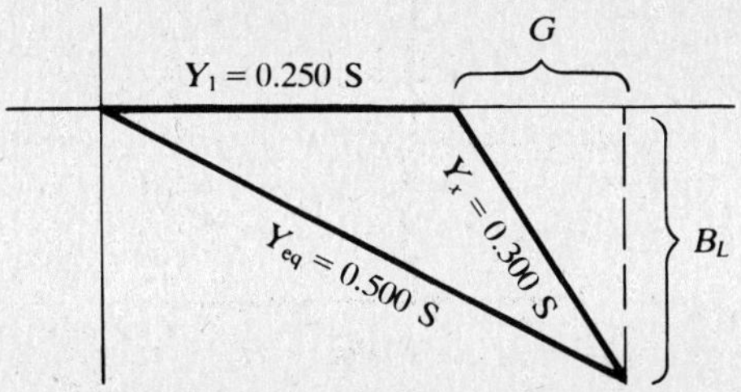

Fig. 7-18

From the admittance diagram, Fig. 7-18,

$$(0.500)^2 = (0.250 + G)^2 + B_L^2$$
$$(0.300)^2 = G^2 + B_L^2$$

which yield $G = 0.195$ S, $B_L = 0.228$ S. Then

$$R = \frac{1}{G} = 5.13\ \Omega \qquad \text{and} \qquad jX_L = \frac{1}{-jB_L} = j4.39\ \Omega$$

i.e. $X_L = 4.39\ \Omega$.

7.13 Evaluate $\mathbf{V}_3$ in Fig. 7-19.

Since the impedances are in series,

$$\mathbf{V}_3 = \left(\frac{\mathbf{Z}_3}{\mathbf{Z}_1 + \mathbf{Z}_2 + \mathbf{Z}_3}\right)\mathbf{V}_T = \left(\frac{j15}{16 + j7}\right) 150\underline{/45^\circ} = 128.8\underline{/111.4^\circ}\ \ \text{V}$$

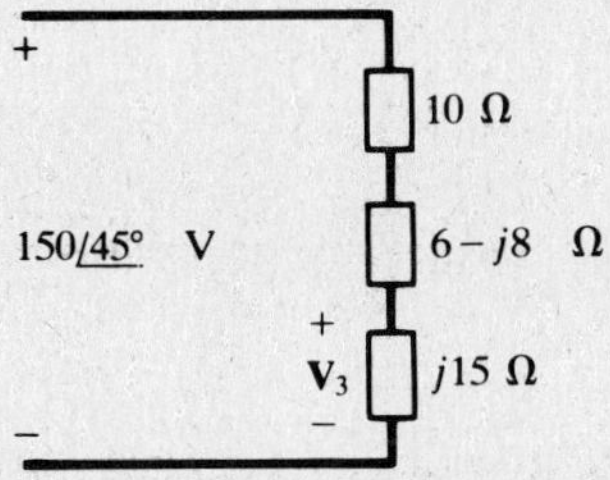

Fig. 7-19

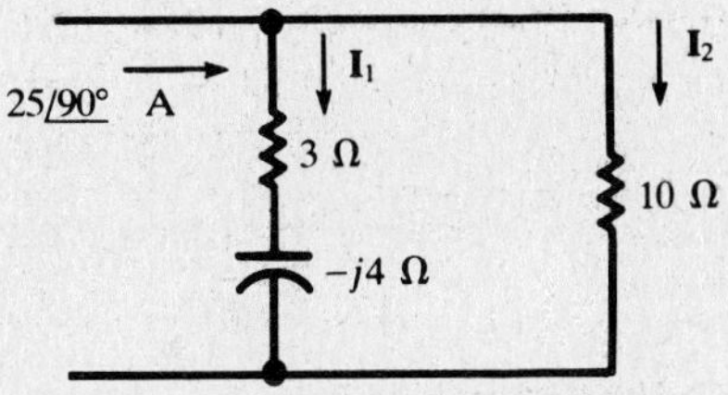

Fig. 7-20

7.14 Find $\mathbf{I}_1$ and $\mathbf{I}_2$ in the parallel circuit of Fig. 7-20.

In this case, $\mathbf{Z}_{eq} = \mathbf{Z}_1\mathbf{Z}_2/(\mathbf{Z}_1 + \mathbf{Z}_2)$, and the current-division formula becomes

$$\mathbf{I}_1 = \left(\frac{\mathbf{Z}_2}{\mathbf{Z}_1 + \mathbf{Z}_2}\right)\mathbf{I}_T = \left(\frac{10}{13 - j4}\right) 25\underline{/90^\circ} = 18.4\underline{/107.1^\circ}\ \ \text{A}$$

$$\mathbf{I}_2 = \left(\frac{\mathbf{Z}}{\mathbf{Z}_1 + \mathbf{Z}_2}\right)\mathbf{I}_T = \left(\frac{3 - j4}{13 - j4}\right) 25\underline{/90^\circ} = 9.19\underline{/54.0^\circ}\ \ \text{A}$$

(or, better, $\mathbf{I}_2 = \mathbf{I}_T - \mathbf{I}_1$).

7.15 Find the phasor voltage $\mathbf{V}_{AB}$ (the potential at point A minus the potential at point B) in the circuit shown in Fig. 7-21.

The two loops contain currents $\mathbf{I}_1$ and $\mathbf{I}_2$ as shown; hence, no current passes through the $j20$ reactance.

$$\mathbf{I}_1 = \frac{20\underline{/30^\circ}}{10 + j10} = 1.41\underline{/-15.0^\circ}\ \ \text{A} \qquad \mathbf{I}_2 = \frac{50\underline{/-45^\circ}}{5 - j5} = 7.07\underline{/0^\circ}\ \ \text{A}$$

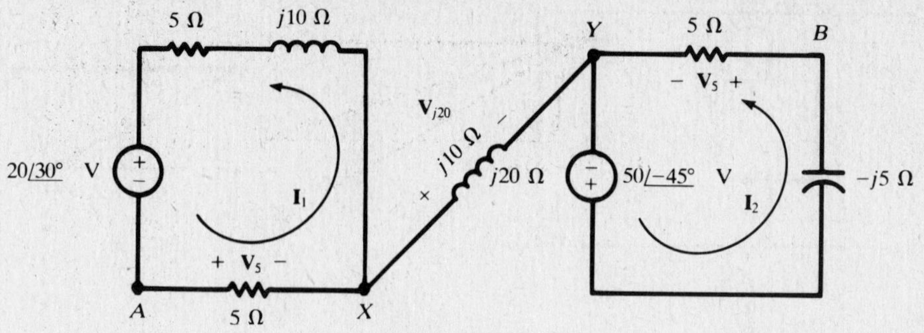

Fig. 7-21

Now, $\mathbf{V}_{AB} = \mathbf{V}_{AX} + \mathbf{V}_{XY} + \mathbf{V}_{YB}$. In the first loop, $\mathbf{V}_{AX} = \mathbf{V}_5 = \mathbf{I}_1(5) = 7.07\underline{/-15.0^\circ}$ V. In the connecting branch, $\mathbf{V}_{XY} = \mathbf{I}(j20) = 0$. And in the second loop, $\mathbf{V}_{YB} = -\mathbf{V}_5 = -\mathbf{I}_2(5) = -35.4\underline{/0^\circ}$ V. Then,

$$\mathbf{V}_{AB} = 7.07\underline{/-15^\circ} - 35.4\underline{/0^\circ} = -28.6 - j1.83 = 28.7\underline{/183.7^\circ}\ \text{V}$$

7.16 Obtain the phasor voltage $\mathbf{V}_{AB}$ in the two-branch parallel circuit of Fig. 7-22.

By current-division methods, $\mathbf{I}_1 = 4.64\underline{/120.1^\circ}$ A and $\mathbf{I}_2 = 17.4\underline{/30.1^\circ}$ A. Either path AXB or path AYB may be considered. Choosing the former,

$$\mathbf{V}_{AB} = \mathbf{V}_{AX} + \mathbf{V}_{XB} = \mathbf{I}_1(20) - \mathbf{I}_2(j6) = 92.8\underline{/120.1^\circ} + 104.4\underline{/-59.9^\circ} = 11.6\underline{/-59.9^\circ}\ \text{V}$$

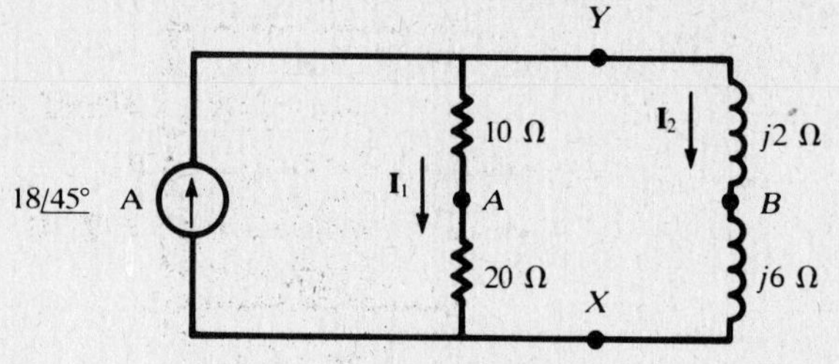

Fig. 7-22

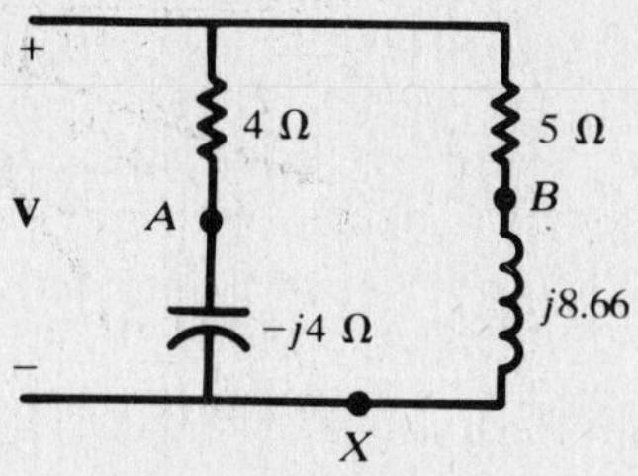

Fig. 7-23

7.17 In the parallel circuit shown in Fig. 7-23, $\mathbf{V}_{AB} = 48.3\underline{/30^\circ}$ V. Find the applied voltage $\mathbf{V}$.

By voltage division in the two branches:

$$\mathbf{V}_{AX} = \frac{-j4}{4-j4}\mathbf{V} = \frac{1}{1+j}\mathbf{V} \qquad \mathbf{V}_{BX} = \frac{j8.66}{5+j8.66}\mathbf{V}$$

and so

$$\mathbf{V}_{AB} = \mathbf{V}_{AX} - \mathbf{V}_{BX} = \left(\frac{1}{1+j} - \frac{j8.66}{5+j8.66}\right)\mathbf{V} = \frac{1}{-0.268+j1}\mathbf{V}$$

or

$$\mathbf{V} = (-0.268 + j1)\mathbf{V}_{AB} = (1.035\underline{/105^\circ})(48.3\underline{/30^\circ}) = 50.0\underline{/135^\circ}\ \text{V}$$

7.18 The voltage-current phasor diagram shown in Fig. 7-24 is for a two-branch parallel circuit. Find the branch impedances and the equivalent impedance.

$$\mathbf{Z}_1 = \frac{\mathbf{V}}{\mathbf{I}_1} = 5.39\underline{/21.8^\circ} = 5.0 + j2.0\ \Omega$$

$$\mathbf{Z}_2 = \frac{\mathbf{V}}{\mathbf{I}_2} = 4.24\underline{/-45.0^\circ} = 3.0 - j3.0\ \Omega$$

$$\mathbf{Z}_{eq} = \frac{\mathbf{Z}_1\mathbf{Z}_2}{\mathbf{Z}_1 + \mathbf{Z}_2} = \frac{22.85\underline{/-23.2^\circ}}{8.0 - j1.0} = 2.83\underline{/-16.2^\circ}\ \Omega$$

or $\mathbf{Z}_{eq} = \mathbf{V}/(\mathbf{I}_1 + \mathbf{I}_2)$.

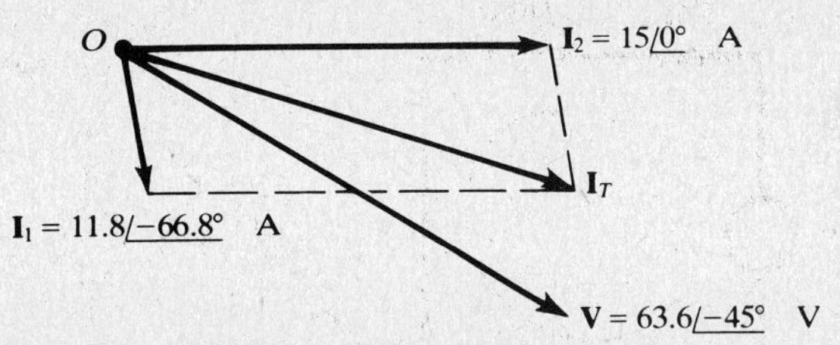

Fig. 7-24

7.19 Obtain the elements of the series circuit corresponding to the voltage-current phasor diagram Fig. 7-25, if the frequency is 21.2 kHz.

Here the current leads the voltage by 14°:

$$\mathbf{Z}_1 = \frac{85\angle -155^\circ}{41.2\angle -141^\circ} = 2.06\angle -14^\circ = 2.0 - j0.50 \quad \Omega$$

The elements are a resistance, $R = 2.0\ \Omega$, and a capacitance such that

$$X_C = 0.50 = \frac{1}{2\pi(21.2 \times 10^3)C} \qquad \text{or} \qquad C = 15.0\ \mu\text{F}$$

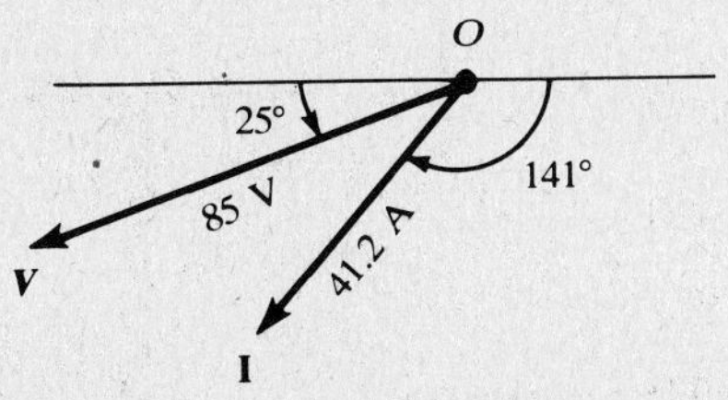

Fig. 7-25

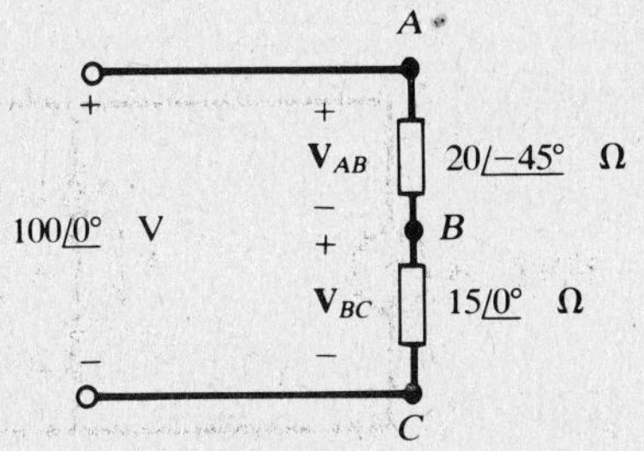

Fig. 7-26

7.20 Obtain the phasor voltages $\mathbf{V}_{AB}$ and $\mathbf{V}_{BC}$ for the circuit shown in Fig. 7-26.

Using voltage division,

$$\mathbf{V}_{AB} = \frac{20\angle -45^\circ}{32.4\angle -25.9^\circ}(100\angle 0^\circ) = 61.8\angle -19.1^\circ \quad \text{V}$$

$$\mathbf{V}_{BC} = \frac{15\angle 0^\circ}{32.4\angle -25.9^\circ}(100\angle 0^\circ) = 46.4\angle 25.9^\circ \quad \text{V}$$

Supplementary Problems

7.21 A two-element series circuit, with $R = 20\ \Omega$ and $L = 20$ mH, has an impedance $40.0\angle\theta\ \ \Omega$. Determine the angle θ and the frequency. *Ans.* 60°, 276 Hz

7.22 Determine the impedance of the series RL circuit, with $R = 25\ \Omega$ and $L = 10$ mH, at (*a*) 100 Hz, (*b*) 500 Hz, (*c*) 1000 Hz. *Ans.* (*a*) $25.8\angle 14.1^\circ\ \Omega$; (*b*) $40.1\angle 51.5^\circ\ \Omega$; (*c*) $67.6\angle 68.3^\circ\ \Omega$

7.23 A resistance of 25 Ω is in series with a second circuit element; the circuit frequency is 500 Hz. Find the element if the current (*a*) lags the applied voltage by 20°; (*b*) leads by 20°.
Ans. (*a*) 2.9 mH; (*b*) 35 μF

7.24 Determine the circuit constants of a two-element series circuit if the applied voltage

$$v = 150 \sin (5000t + 45°) \quad \text{(V)}$$

results in a current $i = 3.0 \sin (5000t - 15°)$ (A). *Ans.* 25 Ω, 8.66 mH

7.25 A series circuit of $R = 10\ \Omega$ and $C = 40\ \mu\text{F}$ has an applied voltage $v = 500 \cos (2500t - 20°)$ (V). Find the resulting current i. *Ans.* $25\sqrt{2} \cos (2500t + 25°)$ (A)

7.26 Figure 7-27 is the voltage-current phasor diagram for a two-element series circuit at angular frequency 300 rad/s. Find the elements if the phasor magnitudes are 200 V and 20 A. *Ans.* 8 Ω, 20 mH

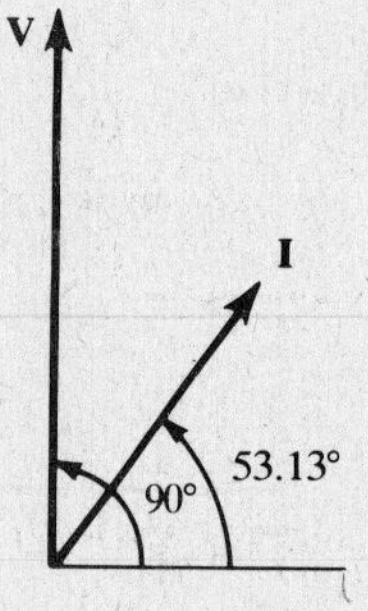

Fig. 7-27

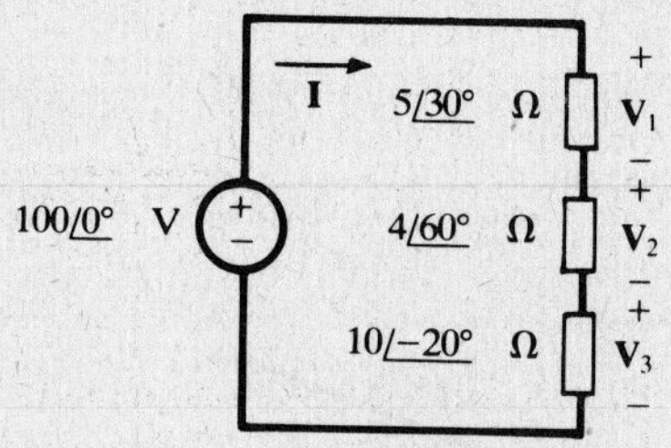

Fig. 7-28

7.27 Three impedances are in series: $\mathbf{Z}_1 = 3.0\angle 45°\ \Omega$, $\mathbf{Z}_2 = 10\sqrt{2}\angle 45°\ \Omega$, $\mathbf{Z}_3 = 5.0\angle -90°\ \Omega$. Find the applied voltage **V**, if the voltage across $\mathbf{Z}_1$ is $27.0\angle -10°$ V. *Ans.* $126.5\angle -24.6°$ V

7.28 For the three-element series circuit in Fig. 7-28, (*a*) find the current **I**; (*b*) find the voltage across each impedance and construct the voltage phasor diagram which shows that $\mathbf{V}_1 + \mathbf{V}_2 + \mathbf{V}_3 = 100\angle 0°$ V.
Ans. (*a*) $6.28\angle -9.17°$ A; (*b*) See Fig. 7-29.

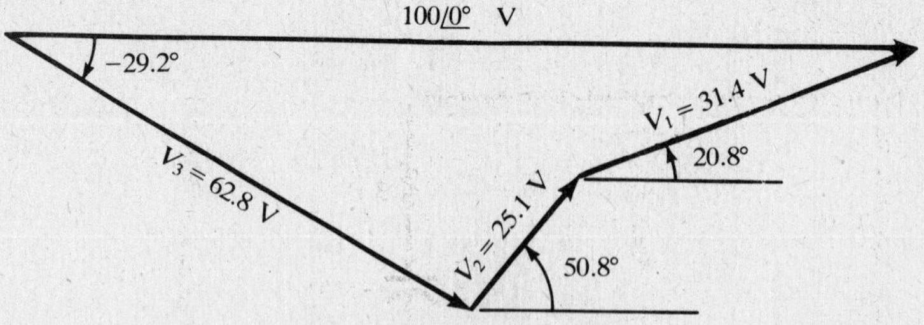

Fig. 7-29

7.29 Find **Z** in the series circuit shown in Fig. 7-30, if $\mathbf{V} = 13.05\angle 15.0°$ V. *Ans.* $4.0 - j15.0$ Ω

7.30 Determine the equivalent impedance and admittance of the four-branch parallel circuit shown in Fig. 7-31. *Ans.* $4.54\angle 57.9°$ Ω, $0.220\angle -57.9°$ S

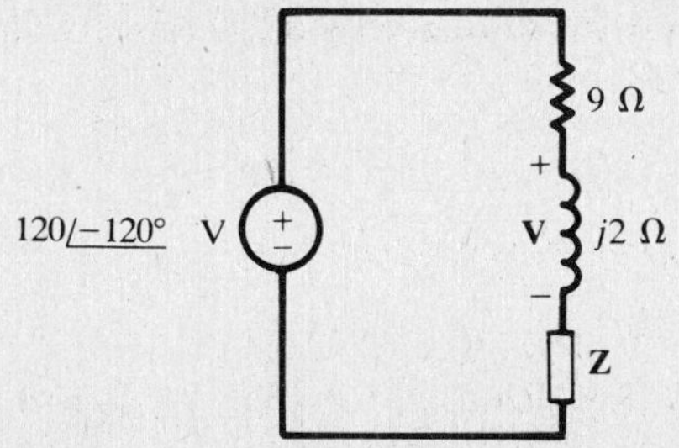

Fig. 7-30

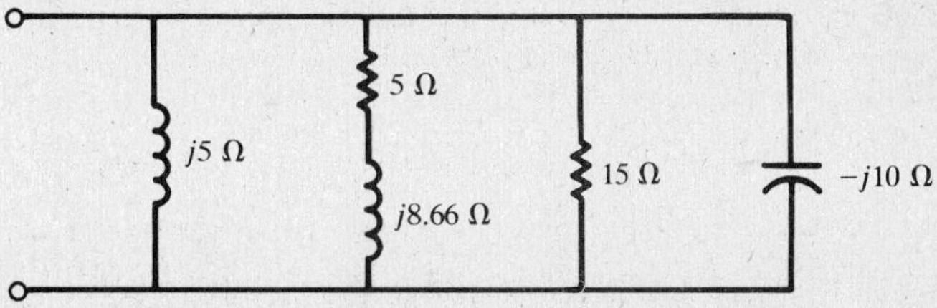

Fig. 7-31

7.31 Find **Z** in the parallel circuit of Fig. 7-32, if $\mathbf{V} = 50.0\angle{30.0^\circ}$ V and $\mathbf{I} = 27.9\angle{57.8^\circ}$ A.
Ans. $5.0\angle{-30^\circ}$ Ω

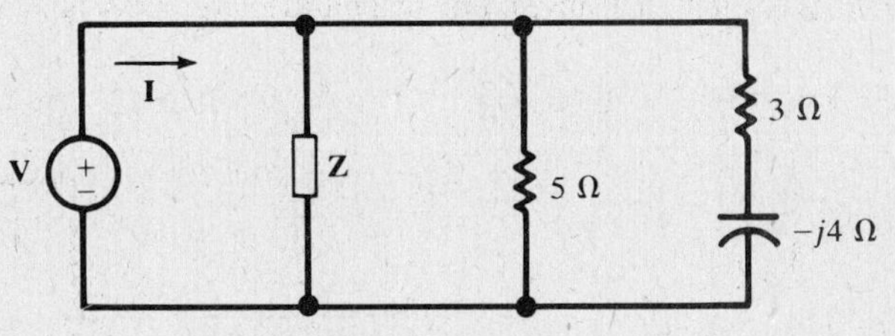

Fig. 7-32

7.32 Obtain the conductance and susceptance corresponding to a voltage $\mathbf{V} = 85.0\angle{205^\circ}$ V and a resulting current $\mathbf{I} = 41.2\angle{-141.0^\circ}$ A. *Ans.* 0.471 S, 0.117 S (capacitive)

7.33 A two-branch parallel circuit, with branch impedances

$$\mathbf{Z}_1 = 15.0\angle{20.0^\circ}\ \Omega \qquad \mathbf{Z}_2 = 20.0\angle{45.0^\circ}\ \Omega$$

has an applied voltage $\mathbf{V} = 240\angle{45.0^\circ}$ V. Determine the total current by first obtaining the equivalent admittance. *Ans.* $27.4\angle{14.3^\circ}$ A

7.34 An admittance consists of 1.77 mS conductance and 1.77 mS inductive susceptance. Find the resistance and reactance of the corresponding impedance. *Ans.* 282 Ω, 282 Ω (inductive)

7.35 For the series-parallel circuit shown in Fig. 7-33, obtain the equivalent admittance.
Ans. $0.135\angle{-62.13^\circ}$ S

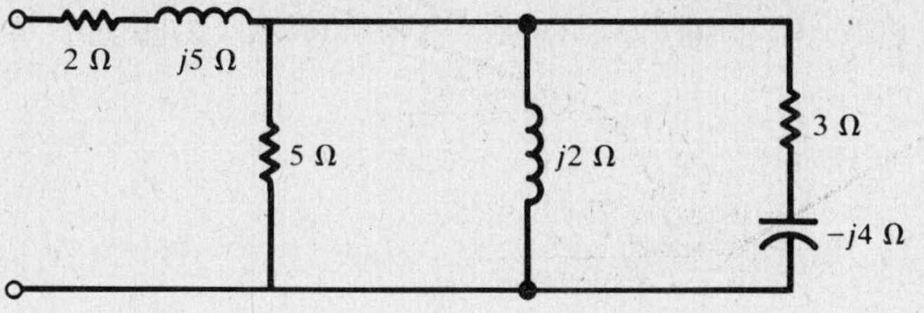

Fig. 7-33

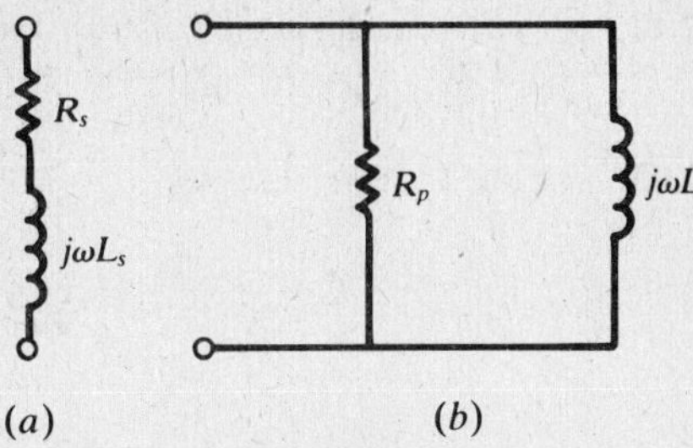

Fig. 7-34

7.36 A practical coil contains resistance as well as inductance and can be represented by either a series or parallel circuit, as suggested in Fig. 7-34. Obtain R_p and L_p in terms of R_s and L_s.

Ans. $R_p = R_s + \dfrac{(\omega L_s)^2}{R_s},\ L_p = L_s + \dfrac{R_s^2}{\omega^2 L_s}$

7.37 Obtain $\mathbf{V}_1$ in Fig. 7-35, given an applied voltage $\mathbf{V} = 150\angle{-45^\circ}$ V. *Ans.* $120.0\angle{-8.13^\circ}$ V

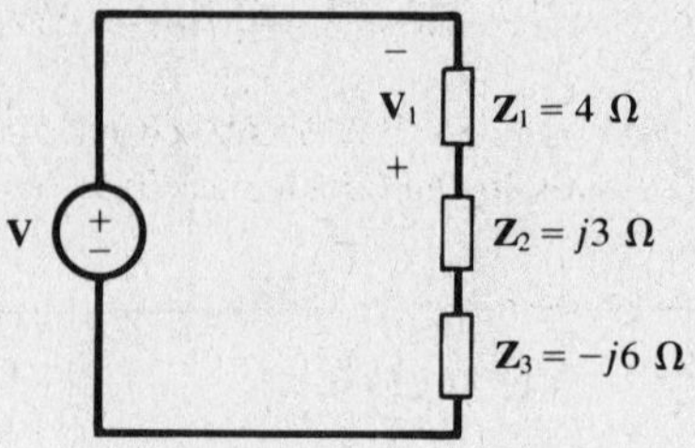

Fig. 7-35

7.38 Five impedances are series connected. Impedance $\mathbf{Z}_1 = 15.0\angle{30^\circ}$ Ω has a voltage $\mathbf{V}_1 = 2.5\angle{-15^\circ}$ V. Find $\mathbf{V}_4$ for $\mathbf{Z}_4 = 6.5\angle{0^\circ}$ Ω. *Ans.* $1.08\angle{-45^\circ}$ V

7.39 Obtain the currents $\mathbf{I}_1$ and $\mathbf{I}_2$ in Fig. 7-36. *Ans.* $3.80\angle{-64.3^\circ}$ A, $2.85\angle{-89.3^\circ}$ A

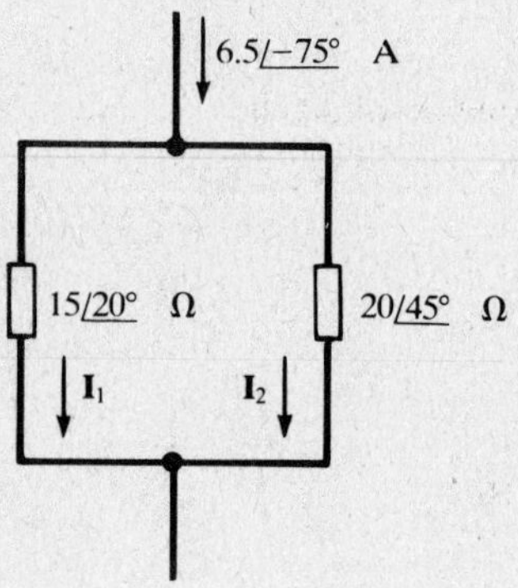

Fig. 7-36

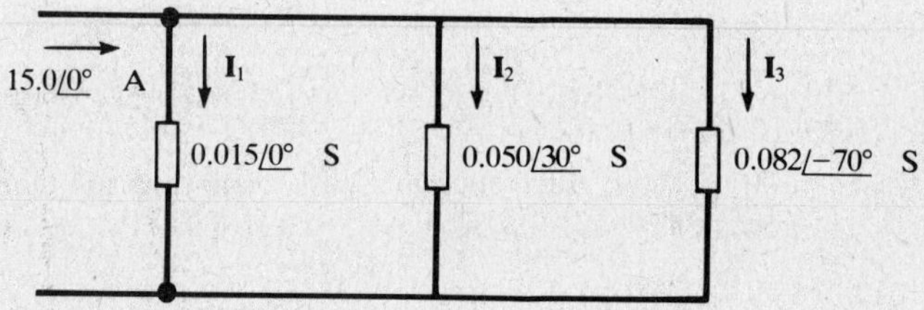

Fig. 7-37

7.40 Referring to Fig. 7-13, obtain the total current if the current in the $15 - j15$ branch is $2.5\angle{-90^\circ}$ mA. *Ans.* $2.85\angle{-142.1^\circ}$ mA

7.41 Three parallel admittances are shown in Fig. 7-37. Obtain the three branch currents. *Ans.* $2.24\angle{31.16^\circ}$ A, $7.46\angle{61.16^\circ}$ A, $12.2\angle{-38.84^\circ}$ A

7.42 A two-branch parallel circuit, with $\mathbf{Z}_1 = 9\angle{90^\circ}$ Ω, divides $\mathbf{I}_T = 5.0\angle{45^\circ}$ A such that $\mathbf{I}_2 = 3.0\angle{98.13^\circ}$ A. Find $\mathbf{Z}_2$. *Ans.* $12.0 + j0$ Ω

7.43 In the network shown in Fig. 7-38, find the voltage $\mathbf{V}_{AB}$. *Ans.* $5.96\angle{105^\circ}$ V

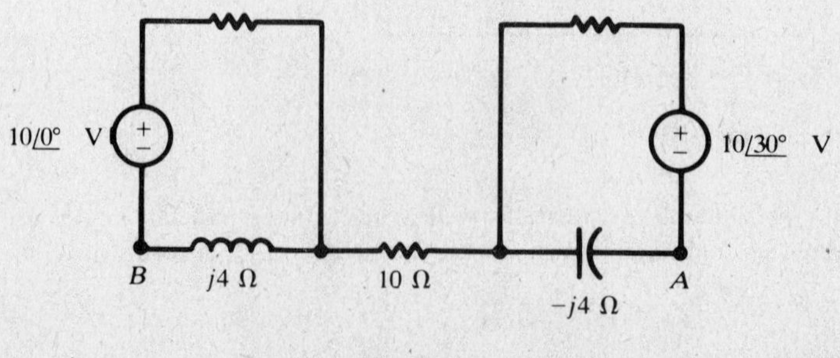

Fig. 7-38

7.44 A series combination of R and C is in parallel with a resistance of 20.0 Ω. At a source frequency of 60 Hz, the total current of 7.02 A (amplitude) divides so that the 20.0 Ω resistor takes 6.0 A and the RC branch 2.3 A (amplitudes). Evaluate R and C. *Ans.* 15.0 Ω, 53.2 μF

7.45 In the network shown in Fig. 7-39 the 60-Hz current magnitudes are known to be: $I_T = 29.9$ A, $I_1 = 22.3$ A, and $I_2 = 8.0$ A. Obtain the circuit constants R and L. *Ans.* 5.8 Ω, 38.5 mH

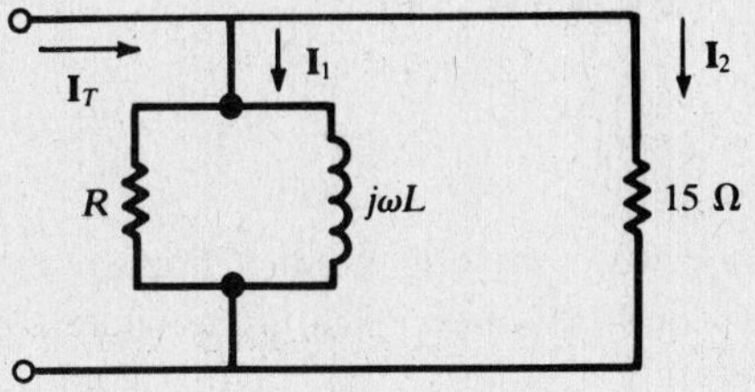

Fig. 7-39

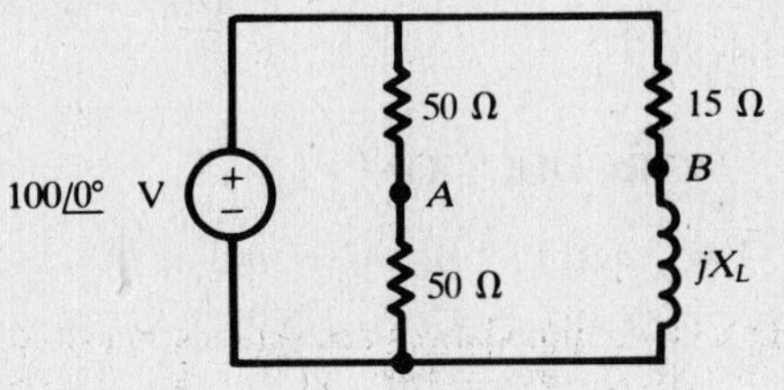

Fig. 7-40

7.46 Obtain the magnitude of the voltage $\mathbf{V}_{AB}$ in the two-branch parallel network of Fig. 7-40, if X_L is (*a*) 5 Ω, (*b*) 15 Ω, (*c*) 0 Ω. *Ans.* 50 V, whatever X_L.

7.47 In the network shown in Fig. 7-41, $\mathbf{V}_{AB} = 36.1\angle 3.18^\circ$ V. Find the source voltage $\mathbf{V}$.
Ans. $75\angle -90^\circ$ V

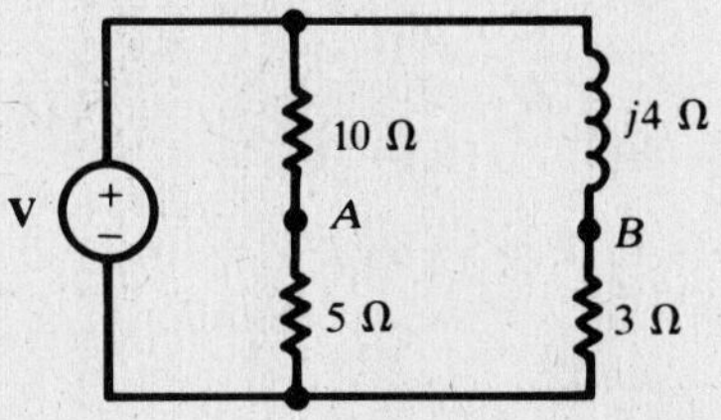

Fig. 7-41

Chapter 8

Network Analysis in the Frequency Domain

8.1 INTRODUCTION

Two methods of analyzing a dc resistive network were introduced in Chapter 4. Along with Thévenin's and Norton's theorems (Section 3.6), these methods apply, with no change in form, to the sinusoidal steady state of an ac network. It is merely a matter of having complex numbers (impedances) instead of real numbers (resistances) as the coefficients in the equations, and having phasors instead of real currents or voltages as the unknowns.

8.2 MESH CURRENT METHOD

Consider the frequency-domain network of Fig. 8-1. Applying KVL, as in Section 4.2, or simply by inspection, we find the matrix equation

$$\begin{bmatrix} \mathbf{Z}_{11} & \mathbf{Z}_{12} & \mathbf{Z}_{13} \\ \mathbf{Z}_{21} & \mathbf{Z}_{22} & \mathbf{Z}_{23} \\ \mathbf{Z}_{31} & \mathbf{Z}_{32} & \mathbf{Z}_{33} \end{bmatrix} \begin{bmatrix} \mathbf{I}_1 \\ \mathbf{I}_2 \\ \mathbf{I}_3 \end{bmatrix} = \begin{bmatrix} \mathbf{V}_1 \\ \mathbf{V}_2 \\ \mathbf{V}_3 \end{bmatrix}$$

for the unknown mesh currents $\mathbf{I}_1$, $\mathbf{I}_2$, $\mathbf{I}_3$. Here, $\mathbf{Z}_{11} \equiv \mathbf{Z}_A + \mathbf{Z}_B$, the *self-impedance* of mesh 1, is the sum of all impedances through which $\mathbf{I}_1$ passes. Similarly, $\mathbf{Z}_{22} \equiv \mathbf{Z}_B + \mathbf{Z}_C + \mathbf{Z}_D$ and $\mathbf{Z}_{33} \equiv \mathbf{Z}_D + \mathbf{Z}_E$ are the self-impedances of meshes 2 and 3.

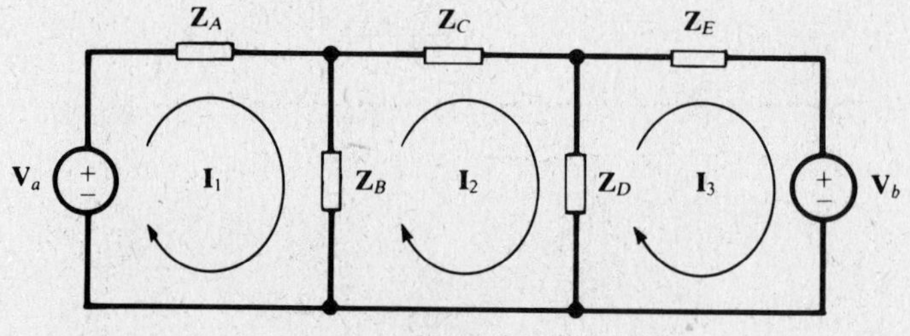

Fig. 8-1

The 1,2-element of the **Z**-matrix is defined as:

$$\mathbf{Z}_{12} \equiv \sum \pm (\text{impedance common to } \mathbf{I}_1 \text{ and } \mathbf{I}_2)$$

where a summand takes the plus sign if the two currents pass through the impedance in the same direction, and takes the minus sign in the opposite case. It follows that, invariably, $\mathbf{Z}_{12} = \mathbf{Z}_{21}$. In Fig. 8-1, $\mathbf{I}_1$ and $\mathbf{I}_2$ thread $\mathbf{Z}_B$ in opposite directions, whence

$$\mathbf{Z}_{12} = \mathbf{Z}_{21} = -\mathbf{Z}_B$$

Similarly,

$$\mathbf{Z}_{13} = \mathbf{Z}_{31} \equiv \sum \pm (\text{impedance common to } \mathbf{I}_1 \text{ and } \mathbf{I}_3) = 0$$

$$\mathbf{Z}_{23} = \mathbf{Z}_{32} \equiv \sum \pm (\text{impedance common to } \mathbf{I}_2 \text{ and } \mathbf{I}_3) = -\mathbf{Z}_D$$

The **Z**-matrix is symmetric.

In the **V**-column on the right-hand side of the equation, the entries $\mathbf{V}_k$ ($k = 1, 2, 3$) are defined exactly as in Section 4.2:

$$\mathbf{V}_k \equiv \sum \pm(\text{driving voltage in mesh } k)$$

where a summand takes the plus sign if the voltage drives in the direction of $\mathbf{I}_k$, and takes the minus sign in the opposite case. For the network of Fig. 8-1,

$$\mathbf{V}_1 = +\mathbf{V}_a \qquad \mathbf{V}_2 = 0 \qquad \mathbf{V}_3 = -\mathbf{V}_b$$

Instead of using the meshes, or "windows" of the (planar) network, it is sometimes expedient to choose an appropriate set of *loops*, each containing one or more meshes in its interior. It is easy to see that two loop currents might have the same direction in one impedance and opposite directions in another. Nevertheless, the above rules for writing the **Z**-matrix and the **V**-column have been formulated in such a way as to apply either to meshes or to loops. These rules are, of course, identical to those used in Section 4.2 to write the **R**-matrix and **V**-column.

EXAMPLE 8.1 Suppose that the phasor voltage across $\mathbf{Z}_B$, with polarity as indicated in Fig. 8-2, is sought. Choosing meshes as in Fig. 8-1 would entail solving for both $\mathbf{I}_1$ and $\mathbf{I}_2$, then obtaining the voltage as $\mathbf{V}_B = (\mathbf{I}_2 - \mathbf{I}_1)\mathbf{Z}_B$. In Fig. 8-2, three loops (two of which are meshes) are chosen so as to make $\mathbf{I}_1$ the only current in $\mathbf{Z}_B$. Furthermore, the direction of $\mathbf{I}_1$ is chosen such that $\mathbf{V}_B = \mathbf{I}_1\mathbf{Z}_B$. Setting up the matrix equation:

$$\begin{bmatrix} \mathbf{Z}_A + \mathbf{Z}_B & -\mathbf{Z}_A & 0 \\ -\mathbf{Z}_A & \mathbf{Z}_A + \mathbf{Z}_C + \mathbf{Z}_D & \mathbf{Z}_D \\ 0 & \mathbf{Z}_D & \mathbf{Z}_D + \mathbf{Z}_E \end{bmatrix} \begin{bmatrix} \mathbf{I}_1 \\ \mathbf{I}_2 \\ \mathbf{I}_3 \end{bmatrix} = \begin{bmatrix} -\mathbf{V}_a \\ \mathbf{V}_a \\ \mathbf{V}_b \end{bmatrix}$$

from which

$$\mathbf{V}_B = \mathbf{Z}_B\mathbf{I}_1 = \frac{\mathbf{Z}_B}{\Delta_\mathbf{Z}} \begin{vmatrix} -\mathbf{V}_a & -\mathbf{Z}_A & 0 \\ \mathbf{V}_a & \mathbf{Z}_A + \mathbf{Z}_B + \mathbf{Z}_C & \mathbf{Z}_D \\ \mathbf{V}_b & \mathbf{Z}_D & \mathbf{Z}_D + \mathbf{Z}_E \end{vmatrix}$$

where $\Delta_\mathbf{Z}$ is the determinant of the **Z**-matrix.

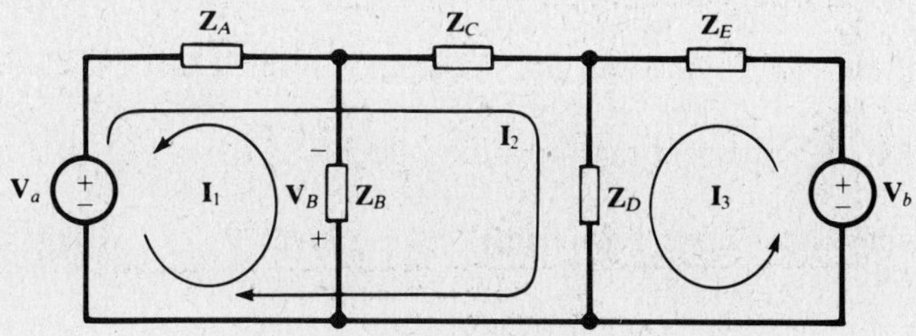

Fig. 8-2

Input and Transfer Impedances

The notions of input resistance (Section 4.4) and transfer resistance (Section 4.5) have their exact counterparts in the frequency domain. Thus, for the single-source network of Fig. 8-3, the *input impedance* is

$$\mathbf{Z}_{\text{input},r} \equiv \frac{\mathbf{V}_r}{\mathbf{I}_r} = \frac{\Delta_\mathbf{Z}}{\Delta_{rr}}$$

where Δ_{rr} is the cofactor of $\mathbf{Z}_{rr}$ in $\Delta_\mathbf{Z}$; and the *transfer impedance* between mesh (or loop) r and mesh (loop) s is

$$\mathbf{Z}_{\text{transfer},rs} \equiv \frac{\mathbf{V}_r}{\mathbf{I}_s} = \frac{\Delta_\mathbf{Z}}{\Delta_{rs}}$$

where Δ_{rs} is the cofactor of $\mathbf{Z}_{rs}$ in $\Delta_\mathbf{Z}$.

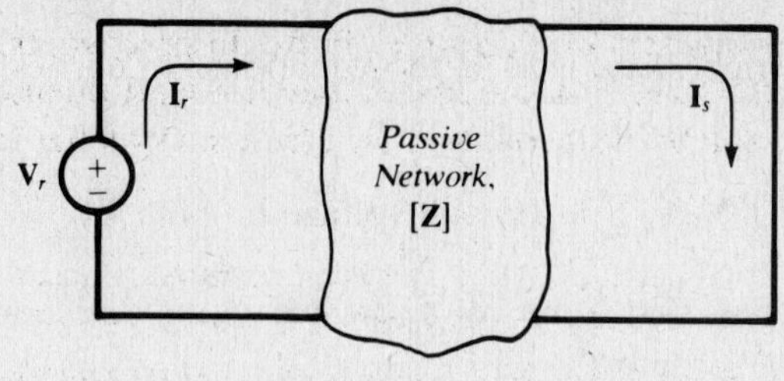

Fig. 8-3

As before, the superposition principle for an arbitrary n-mesh or n-loop network may be expressed as

$$\mathbf{I}_k = \frac{\mathbf{V}_1}{\mathbf{Z}_{\text{transfer},1k}} + \cdots + \frac{\mathbf{V}_{k-1}}{\mathbf{Z}_{\text{transfer},(k-1)k}} + \frac{\mathbf{V}_k}{\mathbf{Z}_{\text{input},k}} + \frac{\mathbf{V}_{k+1}}{\mathbf{Z}_{\text{transfer},(k+1)k}} + \cdots + \frac{\mathbf{V}_n}{\mathbf{Z}_{\text{transfer},nk}}$$

8.3 NODE VOLTAGE METHOD

The procedure is exactly as in Section 4.6, with admittances replacing reciprocal resistances. A frequency-domain network with n principal nodes, one of them designated as the reference node, requires $n - 1$ node voltage equations. Thus, for $n = 4$, the matrix equation would be

$$\begin{bmatrix} \mathbf{Y}_{11} & \mathbf{Y}_{12} & \mathbf{Y}_{13} \\ \mathbf{Y}_{21} & \mathbf{Y}_{22} & \mathbf{Y}_{23} \\ \mathbf{Y}_{31} & \mathbf{Y}_{32} & \mathbf{Y}_{33} \end{bmatrix} \begin{bmatrix} \mathbf{V}_1 \\ \mathbf{V}_2 \\ \mathbf{V}_3 \end{bmatrix} = \begin{bmatrix} \mathbf{I}_1 \\ \mathbf{I}_2 \\ \mathbf{I}_3 \end{bmatrix}$$

in which the unknowns, $\mathbf{V}_1$, $\mathbf{V}_2$, and $\mathbf{V}_3$, are the voltages of principal nodes 1, 2, and 3 with respect to principal node 4, the reference node.

$\mathbf{Y}_{11}$ is the *self-admittance* of node 1, given by the sum of all admittances connected to node 1. Similarly, $\mathbf{Y}_{22}$ and $\mathbf{Y}_{33}$ are the self-admittances of nodes 2 and 3.

$\mathbf{Y}_{12}$, the *coupling admittance* between nodes 1 and 2, is given by *minus* the sum of all admittances connecting nodes 1 and 2. It follows that $\mathbf{Y}_{12} = \mathbf{Y}_{21}$. Similarly, for the other coupling admittances: $\mathbf{Y}_{13} = \mathbf{Y}_{31}$, $\mathbf{Y}_{23} = \mathbf{Y}_{32}$. The $\mathbf{Y}$-matrix is therefore symmetric.

On the right-hand side of the equation, the **I**-column is formed just as in Section 4.6; i.e.

$$\mathbf{I}_k \equiv \sum \text{(current driving into node } k) \qquad (k = 1, 2, 3)$$

in which a current driving *out of* node k is counted as negative.

Input and Transfer Admittances

The matrix equation of the node voltage method,

$$[\mathbf{Y}][\mathbf{V}] = [\mathbf{I}]$$

is identical in form to the matrix equation of the mesh current method,

$$[\mathbf{Z}][\mathbf{I}] = [\mathbf{V}]$$

Therefore, in theory at least, *input* and *transfer admittances* can be defined by analogy with input and transfer impedances:

$$\mathbf{Y}_{\text{input},r} \equiv \frac{\mathbf{I}_r}{\mathbf{V}_r} = \frac{\Delta_{\mathbf{Y}}}{\Delta_{rr}}$$

$$\mathbf{Y}_{\text{transfer},rs} \equiv \frac{\mathbf{I}_r}{\mathbf{V}_s} = \frac{\Delta_{\mathbf{Y}}}{\Delta_{rs}}$$

where now Δ_{rr} and Δ_{rs} are the cofactors of $\mathbf{Y}_{rr}$ and $\mathbf{Y}_{rs}$ in $\Delta_{\mathbf{Y}}$. In practice, these definitions are often of limited use, since, as suggested by Fig. 8-4, the nodes may not lend themselves to a simple system of labeling. However, they are valuable in providing an expression of the superposition principle (for voltages):

$$\mathbf{V}_k = \frac{\mathbf{I}_1}{\mathbf{Y}_{\text{transfer},1k}} + \cdots + \frac{\mathbf{I}_{k-1}}{\mathbf{Y}_{\text{transfer},(k-1)k}} + \frac{\mathbf{I}_k}{\mathbf{Y}_{\text{input},k}} + \frac{\mathbf{I}_{k+1}}{\mathbf{Y}_{\text{transfer},(k+1)k}} + \cdots + \frac{\mathbf{I}_{n-1}}{\mathbf{Y}_{\text{transfer},(n-1)k}}$$

for $k = 1, 2, \ldots, n-1$. In words: the voltage at any principal node (relative to the reference node) is obtained by adding the voltages produced at that node by the various driving currents, these currents acting one at a time.

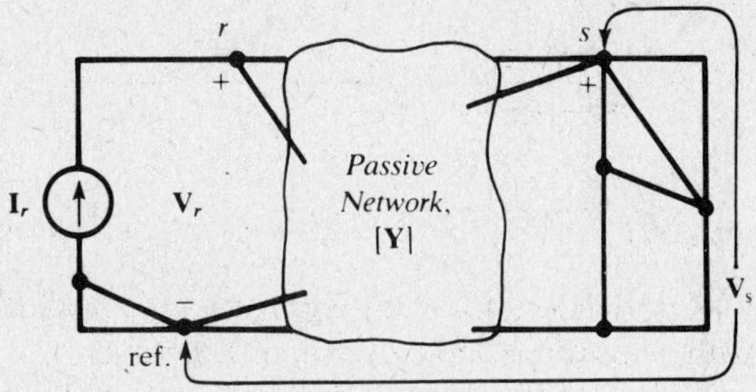

Fig. 8-4

8.4 THEVENIN'S AND NORTON'S THEOREMS

These are exactly as given in Section 3.6, with the open-circuit voltage V', short-circuit current I', and representative resistance R' replaced by the open-circuit phasor voltage $\mathbf{V}'$, short-circuit phasor current $\mathbf{I}'$, and representative impedance $\mathbf{Z}'$. See Fig. 8-5.

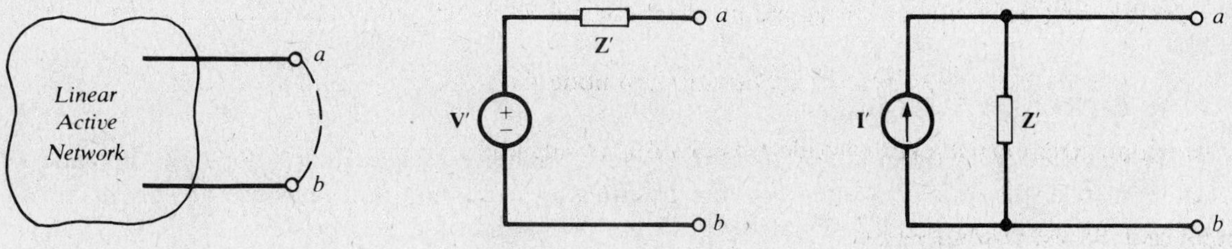

(*a*) Frequency-domain network (*b*) Thévenin equivalent (*c*) Norton equivalent

Fig. 8-5

8.5 EQUIVALENT Y- AND Δ-CONNECTIONS

Figure 8-6 shows three impedances connected in a Δ (*delta*) configuration, and three impedances connected in a Y (*wye*) configuration. Let the terminals of the two connections be identified in pairs as indicated by the labels α, β, γ. Then $\mathbf{Z}_1$ is the impedance "adjoining" terminal α in the Y-connection, and $\mathbf{Z}_C$ is the impedance "opposite" terminal α in the Δ-connection; etc. Looking into any two terminals, the two connections will be equivalent if corresponding input, output, and transfer impedances are equal. The criteria for equivalence are as follows:

Y-to-Δ Transformation

$$\mathbf{Z}_A = \frac{\mathbf{Z}_1\mathbf{Z}_2 + \mathbf{Z}_1\mathbf{Z}_3 + \mathbf{Z}_2\mathbf{Z}_3}{\mathbf{Z}_3}$$

$$\mathbf{Z}_B = \frac{\mathbf{Z}_1\mathbf{Z}_2 + \mathbf{Z}_1\mathbf{Z}_3 + \mathbf{Z}_2\mathbf{Z}_3}{\mathbf{Z}_2}$$

$$\mathbf{Z}_C = \frac{\mathbf{Z}_1\mathbf{Z}_2 + \mathbf{Z}_1\mathbf{Z}_3 + \mathbf{Z}_2\mathbf{Z}_3}{\mathbf{Z}_1}$$

Δ-to-Y Transformation

$$\mathbf{Z}_1 = \frac{\mathbf{Z}_A\mathbf{Z}_B}{\mathbf{Z}_A + \mathbf{Z}_B + \mathbf{Z}_C}$$

$$\mathbf{Z}_2 = \frac{\mathbf{Z}_A\mathbf{Z}_C}{\mathbf{Z}_A + \mathbf{Z}_B + \mathbf{Z}_C}$$

$$\mathbf{Z}_3 = \frac{\mathbf{Z}_B\mathbf{Z}_C}{\mathbf{Z}_A + \mathbf{Z}_B + \mathbf{Z}_C}$$

It should be noted that if the three impedances of one connection are equal, so are those of the equivalent connection, with $\mathbf{Z}_\Delta/\mathbf{Z}_Y = 3$. This fact will be applied in Polyphase Circuits, Chapter 10.

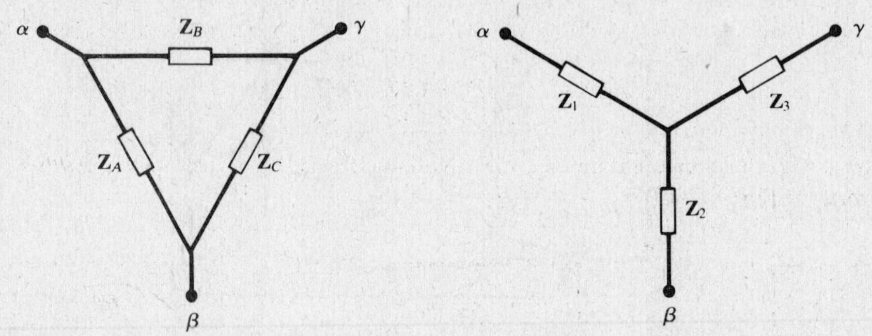

(*a*) Δ-connection (*b*) Y-connection

Fig. 8-6

8.6 SUPERPOSITION THEOREM

The superposition theorem states that the response in any element of a linear, bilateral (all elements two-way) network containing more than one source is the sum of the responses produced by the sources, each acting alone. By "response" is meant the current in the element or the voltage across the element. Superposition does not apply to power, which is proportional to the square of the current and, therefore, not a linear function. For mesh currents, an explicit expression of the theorem was developed in Section 8.2; for node voltages, in Section 8.3.

8.7 RECIPROCITY THEOREM

The reciprocity theorem states that in a linear, bilateral, *single-source* network the ratio of excitation to response is constant when the positions of excitation and response are interchanged. The basis of the theorem is the symmetry of **Z**- and **Y**-matrices (cf. Section 4.5).

For networks with a single voltage source, the theorem implies that the current produced in mesh *r* when the voltage source is in mesh *s*, is the same as the current in mesh *s* when the voltage source is moved to mesh *r*. It must be noted that *currents in other parts of the network will not remain the same.*

For networks containing a single current source, the theorem implies that the voltage which results at a pair of terminals *mn* due to the action of the current source at terminals *ab*, is the same as the voltage at terminals *ab* when the current source is moved to terminals *mn*. It should be noted that *voltages at other points in the network will not remain the same.*

8.8 COMPENSATION THEOREM

According to the compensation theorem, if an impedance **Z** which carries current **I** is replaced by a *compensation voltage source* $\mathbf{V}_c = \mathbf{IZ}$, then all other currents and voltages in the network will remain unchanged. The polarity of $\mathbf{V}_c$ must agree with the phasor voltage across **Z**.

When this theorem is combined with the superposition theorem for linear, bilateral networks, the following theorem (also called "the compensation theorem") is obtained: If the impedance $\mathbf{Z}$ of any branch is changed by $\delta\mathbf{Z}$, the *incremental* current $\delta\mathbf{I}$ in that branch is that which would be produced by a voltage source $\mathbf{V}_c = \mathbf{I}\,\delta\mathbf{Z}$ introduced in that same branch, with polarity opposing the original current $\mathbf{I}$.

EXAMPLE 8.2 The compensation theorem is useful in bridge and potentiometer circuits, where a slight change in one impedance results in a shift from a null condition. In Fig. 8-7(*a*), current $\mathbf{I}$ results from a driving voltage $\mathbf{V}$ and a circuit impedance $\mathbf{Z}$. When a change in impedance $\delta\mathbf{Z}$ is introduced, in Fig. 8-7(*b*), the current becomes $\mathbf{I}+\delta\mathbf{I}$; i.e., the change in the current is $\delta\mathbf{I}$. From

$$\mathbf{V} = (\mathbf{I}+\delta\mathbf{I})(\mathbf{Z}+\delta\mathbf{Z}) = \mathbf{V} + \mathbf{I}\,\delta\mathbf{Z} + \delta\mathbf{I}\,(\mathbf{Z}+\delta\mathbf{Z})$$

it follows that

$$\delta\mathbf{I} = \frac{-\mathbf{I}\,\delta\mathbf{Z}}{\mathbf{Z}+\delta\mathbf{Z}}$$

Thus, in accordance with the compensation theorem, a voltage source $-\mathbf{I}\,\delta\mathbf{Z}$, *with the original source* $\mathbf{V}$ *removed*, will produce current $\delta\mathbf{I}$ [Fig. 8-7(*c*)].

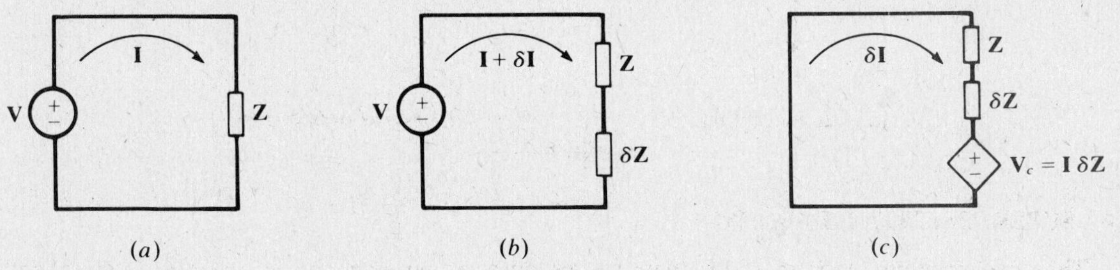

Fig. 8-7

Solved Problems

8.1 Obtain the voltage $\mathbf{V}_x$ in the network of Fig. 8-8, using the mesh current method.

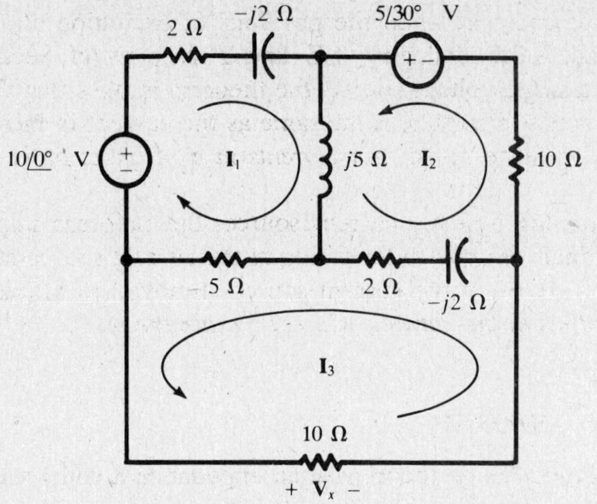

Fig. 8-8

One choice of mesh currents is shown on the circuit diagram, with $\mathbf{I}_3$ passing through the 10-Ω resistor in a direction such that $\mathbf{V}_x = \mathbf{I}_3(10)$ (V). The matrix equation can be written by inspection:

$$\begin{bmatrix} 7+j3 & j5 & 5 \\ j5 & 12+j3 & -(2-j2) \\ 5 & -(2-j2) & 17-j2 \end{bmatrix} \begin{bmatrix} \mathbf{I}_1 \\ \mathbf{I}_2 \\ \mathbf{I}_3 \end{bmatrix} = \begin{bmatrix} 10\angle 0^\circ \\ 5\angle 30^\circ \\ 0 \end{bmatrix}$$

Solving by determinants,

$$\mathbf{I}_3 = \frac{\begin{vmatrix} 7+j3 & j5 & 10\angle 0^\circ \\ j5 & 12+j3 & 5\angle 30^\circ \\ 5 & -2+j2 & 0 \end{vmatrix}}{\begin{vmatrix} 7+j3 & j5 & 5 \\ j5 & 12+j3 & -2+j2 \\ 5 & -2+j2 & 17-j2 \end{vmatrix}} = \frac{667.96\angle -169.09^\circ}{1534.5\angle 25.06^\circ} = 0.435\angle -194.15^\circ \quad \text{A}$$

and $\mathbf{V}_x = \mathbf{I}_3(10) = 4.35\angle -194.15^\circ$ V.

8.2 In the network of Fig. 8-9, determine the voltage **V** which results in a zero current through the $2+j3$ Ω impedance.

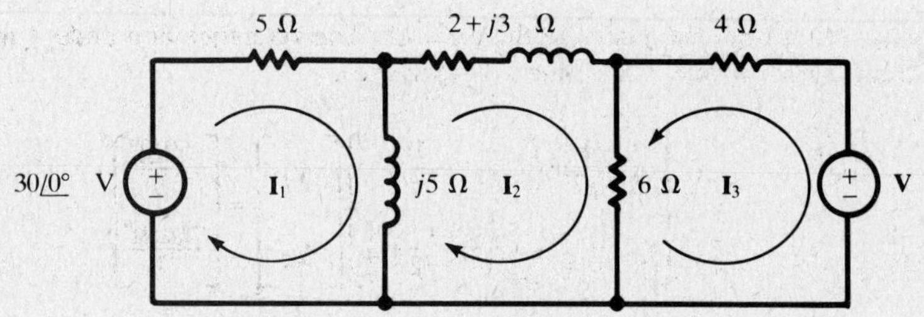

Fig. 8-9

Choosing mesh currents as shown on the circuit diagram,

$$\mathbf{I}_2 = \frac{1}{\Delta_\mathbf{z}} \begin{vmatrix} 5+j5 & 30\angle 0^\circ & 0 \\ -j5 & 0 & 6 \\ 0 & \mathbf{V} & 10 \end{vmatrix} = 0$$

Expanding the numerator determinant by cofactors of the second column,

$$-(30\angle 0^\circ)\begin{vmatrix} -j5 & 6 \\ 0 & 10 \end{vmatrix} - \mathbf{V}\begin{vmatrix} 5+j5 & 0 \\ -j5 & 6 \end{vmatrix} = 0 \qquad \text{whence} \qquad \mathbf{V} = 35.4\angle 45.0^\circ \quad \text{V}$$

8.3 Solve Problem 8.2 by the node voltage method.

The network is redrawn in Fig. 8-10, with one end of the $2+j3$ Ω impedance as the reference node. By the rules of Section 8.3, the matrix equation is

$$\begin{bmatrix} \frac{1}{5}+\frac{1}{j5}+\frac{1}{2+j3} & -\left(\frac{1}{5}+\frac{1}{j5}\right) \\ -\left(\frac{1}{5}+\frac{1}{j5}\right) & \frac{1}{5}+\frac{1}{j5}+\frac{1}{4}+\frac{1}{6} \end{bmatrix} \begin{bmatrix} \mathbf{V}_1 \\ \mathbf{V}_2 \end{bmatrix} = \begin{bmatrix} \frac{30\angle 0^\circ}{5} \\ \frac{-30\angle 0^\circ}{5} - \frac{\mathbf{V}}{4} \end{bmatrix}$$

For node voltage $\mathbf{V}_1$ to be zero, it is necessary that the numerator determinant in the solution for $\mathbf{V}_1$ vanish.

$$N_1 = \begin{vmatrix} \dfrac{30\angle 0^\circ}{5} & -0.200 + j0.200 \\ \dfrac{-30\angle 0^\circ}{5} - \dfrac{\mathbf{V}}{4} & 0.617 - j0.200 \end{vmatrix} = 0 \qquad \text{from which} \qquad \mathbf{V} = 35.4\angle 45^\circ \ \text{V}$$

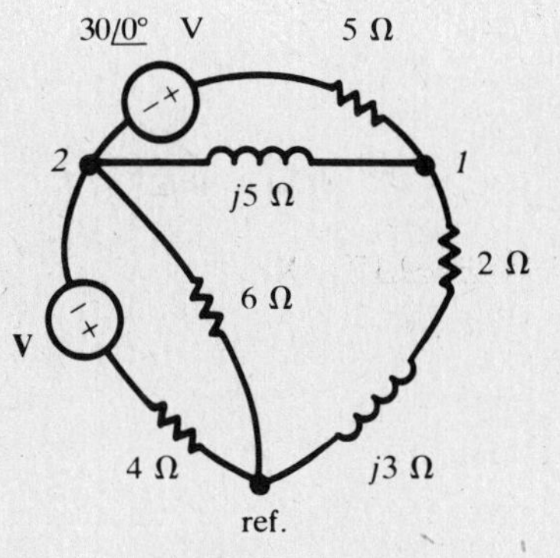

Fig. 8-10

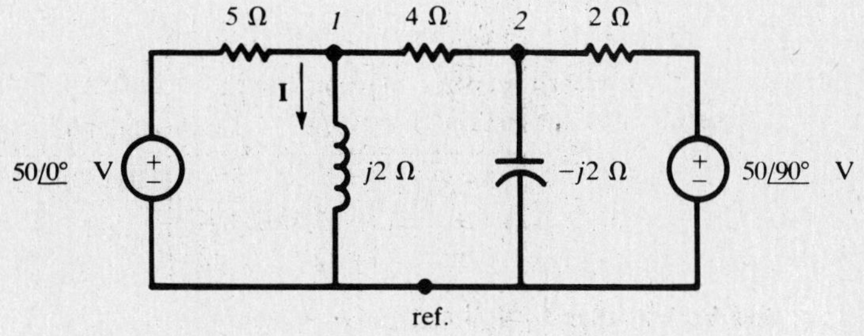

Fig. 8-11

8.4 Use the node voltage method to obtain the current **I** in the network of Fig. 8-11.

There are three principal nodes in the network. The reference and node *1* are selected so that the node *1* voltage is the voltage across the $j2$-Ω reactance.

$$\begin{bmatrix} \dfrac{1}{5} + \dfrac{1}{j2} + \dfrac{1}{4} & -\dfrac{1}{4} \\ -\dfrac{1}{4} & \dfrac{1}{4} + \dfrac{1}{-j2} + \dfrac{1}{2} \end{bmatrix} \begin{bmatrix} \mathbf{V}_1 \\ \mathbf{V}_2 \end{bmatrix} = \begin{bmatrix} \dfrac{50\angle 0^\circ}{5} \\ \dfrac{50\angle 90^\circ}{2} \end{bmatrix}$$

from which

$$\mathbf{V}_1 = \frac{\begin{vmatrix} 10 & -0.250 \\ j25 & 0.750 + j0.500 \end{vmatrix}}{\begin{vmatrix} 0.450 - j0.500 & -0.250 \\ -0.250 & 0.750 + j0.500 \end{vmatrix}} = \frac{13.52\angle 56.31^\circ}{0.546\angle -15.94^\circ} = 24.76\angle 72.25^\circ \ \text{V}$$

and

$$\mathbf{I} = \frac{24.76\angle 72.25^\circ}{2\angle 90^\circ} = 12.38\angle -17.75^\circ \ \text{A}$$

8.5 Find the input impedance at terminals ab for the network of Fig. 8-12.

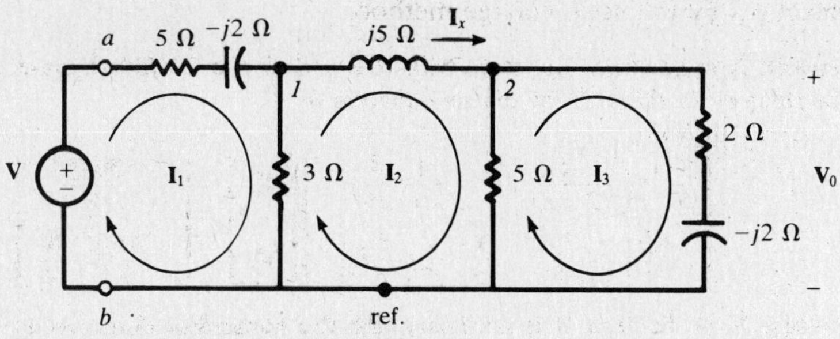

Fig. 8-12

With mesh current $\mathbf{I}_1$ selected as shown on the diagram,

$$\mathbf{Z}_{\text{input},1}=\frac{\Delta_{\mathbf{z}}}{\Delta_{11}}=\frac{\begin{vmatrix}8-j2 & -3 & 0\\ -3 & 8+j5 & -5\\ 0 & -5 & 7-j2\end{vmatrix}}{\begin{vmatrix}8+j5 & -5\\ -5 & 7-j2\end{vmatrix}}=\frac{315.5\underline{/16.19^\circ}}{45.2\underline{/24.86^\circ}}=6.98\underline{/-8.67^\circ}\ \ \Omega$$

8.6 For the network in Fig. 8-12, obtain the current in the inductor, $\mathbf{I}_x$, by first obtaining the transfer impedance. Let $\mathbf{V}=10\underline{/30^\circ}$ V.

$$\mathbf{Z}_{\text{transfer},12}=\frac{\Delta_{\mathbf{z}}}{\Delta_{12}}=\frac{315.5\underline{/16.19^\circ}}{-\begin{vmatrix}-3 & -5\\ 0 & 7-j2\end{vmatrix}}=14.45\underline{/32.14^\circ}\ \ \Omega$$

Then $$\mathbf{I}_x=\mathbf{I}_2=\frac{\mathbf{V}}{\mathbf{Z}_{\text{transfer},12}}=\frac{10\underline{/30^\circ}}{14.45\underline{/32.14^\circ}}=0.692\underline{/-2.14^\circ}\ \ \text{A}$$

8.7 For the network in Fig. 8-12, find the value of the source voltage $\mathbf{V}$ which results in $\mathbf{V}_0=5.0\underline{/0^\circ}$ V.

The transfer impedance can be used to compute the current in the $2-j2\ \ \Omega$ impedance, from which $\mathbf{V}_0$ is readily obtained.

$$\mathbf{Z}_{\text{transfer},13}=\frac{\Delta_{\mathbf{z}}}{\Delta_{13}}=\frac{315.5\underline{/16.19^\circ}}{15\underline{/0^\circ}}=21.0\underline{/16.19^\circ}\ \ \Omega$$

$$\mathbf{V}_0=\mathbf{I}_3(2-j2)=\frac{\mathbf{V}}{\mathbf{Z}_{\text{transfer},13}}(2-j2)=\mathbf{V}(0.135\underline{/-61.19^\circ})$$

Thus, if $\mathbf{V}_0=5.0\underline{/0^\circ}$ V,

$$\mathbf{V}=\frac{5.0\underline{/0^\circ}}{0.135\underline{/-61.19^\circ}}=37.0\underline{/61.19^\circ}\ \ \text{V}$$

Alternate Method

The node voltage method may be used. $\mathbf{V}_0$ is the node voltage $\mathbf{V}_2$ for the selection of nodes indicated in Fig. 8-12.

$$\mathbf{V}_0=\mathbf{V}_2=\frac{\begin{vmatrix}\dfrac{1}{5-j2}+\dfrac{1}{3}+\dfrac{1}{j5} & \dfrac{\mathbf{V}}{5-j2}\\ -\dfrac{1}{j5} & 0\end{vmatrix}}{\begin{vmatrix}\dfrac{1}{5-j2}+\dfrac{1}{3}+\dfrac{1}{j5} & -\dfrac{1}{j5}\\ -\dfrac{1}{j5} & \dfrac{1}{j5}+\dfrac{1}{5}+\dfrac{1}{2-j2}\end{vmatrix}}=\mathbf{V}(0.134\underline{/-61.15^\circ})$$

For $\mathbf{V}_0=5.0\underline{/0^\circ}$ V, $\mathbf{V}=37.3\underline{/61.15^\circ}$ V, which agrees with the previous answer to within roundoff errors.

8.8 For the network shown in Fig. 8-13, obtain the input admittance and use it to compute node voltage $\mathbf{V}_1$.

$$\mathbf{Y}_{\text{input},1}=\frac{\Delta_{\mathbf{Y}}}{\Delta_{11}}=\frac{\begin{vmatrix}\dfrac{1}{10}+\dfrac{1}{j5}+\dfrac{1}{2} & -\dfrac{1}{2}\\ -\dfrac{1}{2} & \dfrac{1}{2}+\dfrac{1}{3+j4}+\dfrac{1}{-j10}\end{vmatrix}}{\dfrac{1}{2}+\dfrac{1}{3+j4}+\dfrac{1}{-j10}}=0.311\underline{/-49.97^\circ}\ \ \text{S}$$

$$\mathbf{V}_1 = \frac{\mathbf{I}_1}{\mathbf{Y}_{\text{input},1}} = \frac{5.0\underline{/0^\circ}}{0.311\underline{/-49.97^\circ}} = 16.1\underline{/49.97^\circ} \text{ V}$$

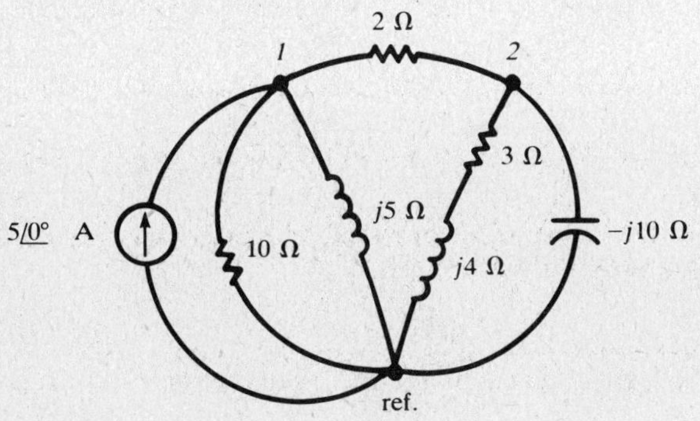

Fig. 8-13

8.9 For the network of Problem 8.8, compute the transfer admittance $\mathbf{Y}_{\text{transfer},12}$ and use it to obtain node voltage $\mathbf{V}_2$.

$$\mathbf{Y}_{\text{transfer},12} = \frac{\Delta_\mathbf{Y}}{\Delta_{12}} = \frac{0.194\underline{/-55.49^\circ}}{-(-0.50)} = 0.388\underline{/-55.49^\circ} \text{ S}$$

$$\mathbf{V}_2 = \frac{\mathbf{I}_1}{\mathbf{Y}_{\text{transfer},12}} = 12.9\underline{/55.49^\circ} \text{ V}$$

8.10 Replace the active network in Fig. 8-14(a) at terminals ab with a Thévenin equivalent.

$$\mathbf{Z}' = j5 + \frac{5(3+j4)}{5+3+j4} = 2.50 + j6.25 \ \Omega$$

The open-circuit voltage $\mathbf{V}'$ at terminals ab is the voltage across the $3 + j4 \ \Omega$ impedance:

$$\mathbf{V}' = \left(\frac{10\underline{/0^\circ}}{8+j4}\right)(3+j4) = 5.59\underline{/26.56^\circ} \text{ V}$$

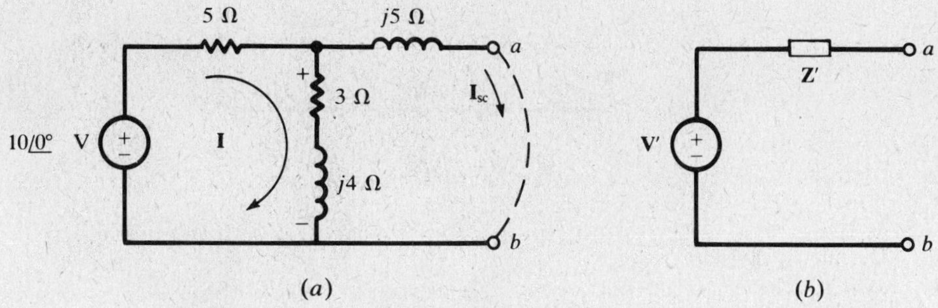

Fig. 8-14

8.11 For the network of Problem 8.10, obtain a Norton equivalent circuit (Fig. 8-15).

At terminals ab, $\mathbf{I}_{sc}$ is the Norton current $\mathbf{I}'$. By current division,

$$\mathbf{I}' = \frac{10\underline{/0^\circ}}{5 + \dfrac{j5(3+j4)}{3+j9}}\left(\frac{3+j4}{3+j9}\right) = 0.830\underline{/-41.63^\circ} \text{ A}$$

The shunt impedance $\mathbf{Z}'$ is as found in Problem 8.10.

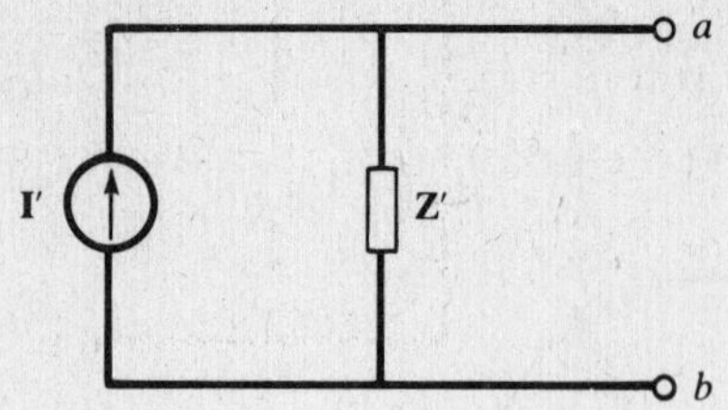

Fig. 8-15

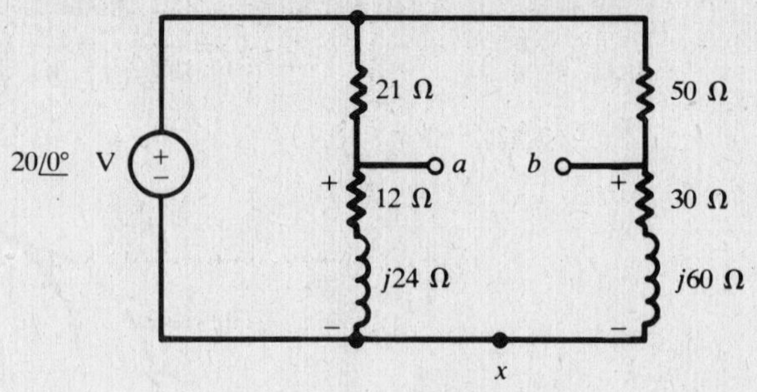

Fig. 8-16

8.12 Obtain the Thévenin equivalent for the bridge circuit of Fig. 8-16. Make $\mathbf{V}'$ the voltage of a with respect to b.

By voltage division in either branch,

$$\mathbf{V}_{ax} = \frac{12+j24}{33+j24}(20\underline{/0^\circ}) \qquad \mathbf{V}_{bx} = \frac{30+j60}{80+j60}(20\underline{/0^\circ})$$

Hence
$$\mathbf{V}_{ab} = \mathbf{V}_{ax} - \mathbf{V}_{bx} = (20\underline{/0^\circ})\left(\frac{12+j24}{33+j24} - \frac{30+j60}{80+j60}\right) = 0.326\underline{/169.4^\circ} \quad \text{V} = \mathbf{V}'$$

Viewed from ab with the voltage source shorted out, the circuit is two parallel combinations in series, and so

$$\mathbf{Z}' = \frac{21(12+j24)}{33+j24} + \frac{50(30+j60)}{80+j60} = 47.35\underline{/26.81^\circ} \quad \Omega$$

8.13 Replace the network of Fig. 8-17 at terminals ab with a Norton equivalent and with a Thévenin equivalent.

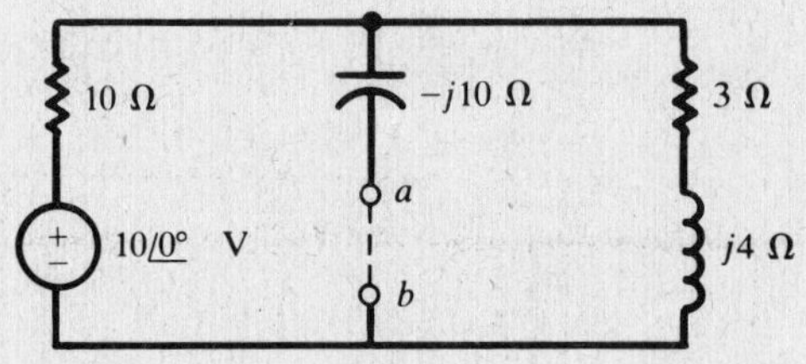

Fig. 8-17

By current division,

$$\mathbf{I}_{sc} = \mathbf{I}' = \left[\frac{10\underline{/0^\circ}}{10 + \dfrac{(-j10)(3+j4)}{3-j6}}\right]\left(\frac{3+j4}{3-j6}\right) = 0.439\underline{/105.26^\circ} \quad \text{A}$$

and by voltage division in the open circuit,

$$\mathbf{V}_{ab} = \mathbf{V}' = \frac{3+j4}{13+j4}(10\underline{/0^\circ}) = 3.68\underline{/36.03^\circ} \quad \text{V}$$

Then
$$\mathbf{Z}' = \frac{\mathbf{V}'}{\mathbf{I}'} = \frac{3.68\underline{/36.03^\circ}}{0.439\underline{/105.26^\circ}} = 8.37\underline{/-69.23^\circ} \quad \Omega$$

See Fig. 8-18.

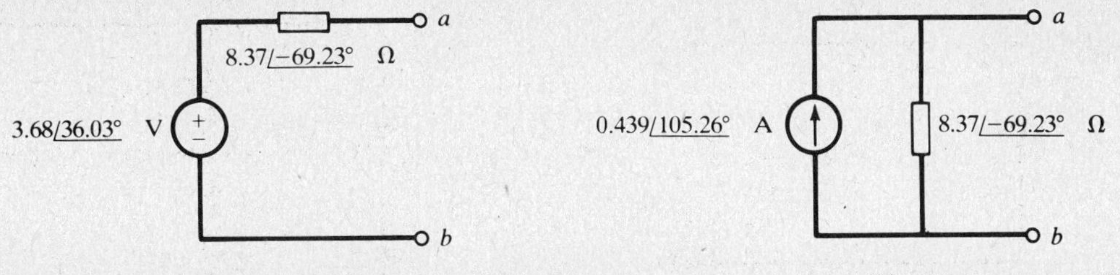

(*a*) Thévenin (*b*) Norton

Fig. 8-18

8.14 Convert the Y-circuit of Fig. 8-19(*a*) into an equivalent Δ.

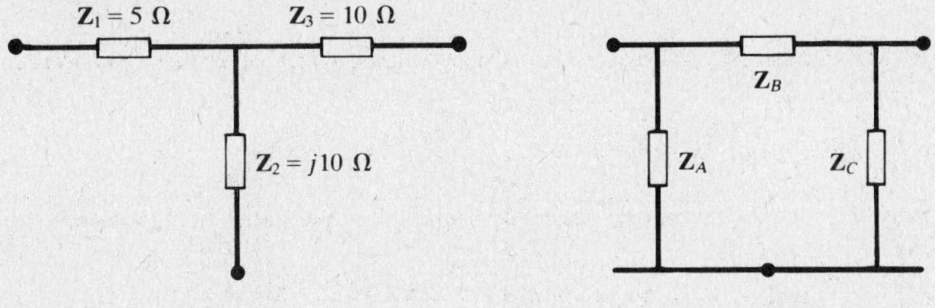

(*a*) (*b*)

Fig. 8-19

The Δ-impedances are obtained using the formulas from Section 8.5.

$$\mathbf{Z}_A = \frac{\mathbf{Z}_1\mathbf{Z}_2 + \mathbf{Z}_1\mathbf{Z}_3 + \mathbf{Z}_2\mathbf{Z}_3}{\mathbf{Z}_3} = \frac{5(j10) + 5(10) + (j10)(10)}{10} = 5 + j15 \quad \Omega$$

Similarly, $\mathbf{Z}_B = 15 - j5 \quad \Omega$, $\mathbf{Z}_C = 10 + j30 \quad \Omega$. As a check, these Δ-impedances should be converted back to a Y and compared with Fig. 8-19(*a*).

8.15 Show that an n-mesh, passive, three-terminal network (Fig. 8-20) may be replaced with a Δ-connection of three impedances.

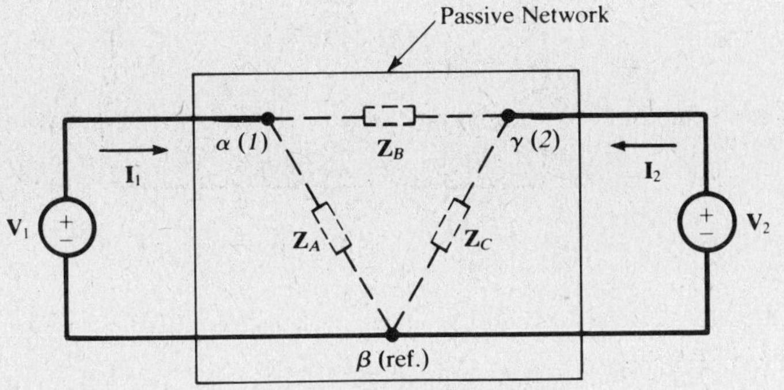

Fig. 8-20

Apply a voltage $\mathbf{V}_1$ to terminals $\alpha\beta$ and label the current which enters the network $\mathbf{I}_1$. Also, apply $\mathbf{V}_2$ to terminals $\gamma\beta$ and label the current $\mathbf{I}_2$. Since the network is passive, all other driving voltages are zero.

The mesh current equations in matrix form are

$$\begin{bmatrix} \mathbf{Z}_{11} & \mathbf{Z}_{12} & \cdots & \mathbf{Z}_{1n} \\ \mathbf{Z}_{21} & \mathbf{Z}_{22} & \cdots & \mathbf{Z}_{2n} \\ \cdots & \cdots & \cdots & \cdots \\ \mathbf{Z}_{n1} & \mathbf{Z}_{n2} & \cdots & \mathbf{Z}_{nn} \end{bmatrix} \begin{bmatrix} \mathbf{I}_1 \\ \mathbf{I}_2 \\ \cdots \\ \mathbf{I}_n \end{bmatrix} = \begin{bmatrix} \mathbf{V}_1 \\ \mathbf{V}_2 \\ 0 \\ 0 \\ 0 \end{bmatrix}$$

from which $$\mathbf{I}_1 = \frac{\Delta_{11}}{\Delta_\mathbf{z}}\mathbf{V}_1 + \frac{\Delta_{21}}{\Delta_\mathbf{z}}\mathbf{V}_2 \qquad \mathbf{I}_2 = \frac{\Delta_{12}}{\Delta_\mathbf{z}}\mathbf{V}_1 + \frac{\Delta_{22}}{\Delta_\mathbf{z}}\mathbf{V}_2 \tag{1}$$

Now, the given networks will be equivalent to the Δ-configuration that is shown dashed in Fig. 8-20, provided the Δ-configuration has (*1*) as its node voltage equations under the driving currents $\mathbf{I}_1$ and $\mathbf{I}_2$. But Section 8.3 gives these node voltage equations as

$$\begin{bmatrix} \frac{1}{\mathbf{Z}_A}+\frac{1}{\mathbf{Z}_B} & -\frac{1}{\mathbf{Z}_B} \\ -\frac{1}{\mathbf{Z}_B} & \frac{1}{\mathbf{Z}_B}+\frac{1}{\mathbf{Z}_C} \end{bmatrix}\begin{bmatrix}\mathbf{V}_1\\ \mathbf{V}_2\end{bmatrix} = \begin{bmatrix}\mathbf{I}_1\\ \mathbf{I}_2\end{bmatrix}$$

that is, $$\mathbf{I}_1 = \left(\frac{1}{\mathbf{Z}_A}+\frac{1}{\mathbf{Z}_B}\right)\mathbf{V}_1 - \frac{1}{\mathbf{Z}_B}\mathbf{V}_2 \qquad \mathbf{I}_2 = -\frac{1}{\mathbf{Z}_B}\mathbf{V}_1 + \left(\frac{1}{\mathbf{Z}_B}+\frac{1}{\mathbf{Z}_C}\right)\mathbf{V}_2 \tag{2}$$

Equating corresponding coefficients in systems (*1*) and (*2*) gives for the three impedances of the equivalent Δ-connection (recall that $\Delta_{21} = \Delta_{12}$):

$$\mathbf{Z}_A = \frac{\Delta_\mathbf{z}}{\Delta_{11}+\Delta_{21}} \qquad \mathbf{Z}_B = -\frac{\Delta_\mathbf{z}}{\Delta_{21}} \qquad \mathbf{Z}_C = \frac{\Delta_\mathbf{z}}{\Delta_{22}+\Delta_{21}}$$

It may be that one or more of these impedances has a negative real part and is therefore not physically realizable.

8.16 Apply the results of Problem 8.15 to obtain a Δ-circuit equivalent to the network of Fig. 8-21.

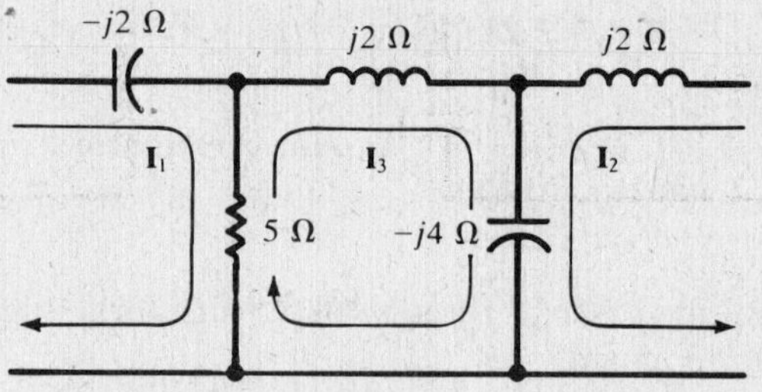

Fig. 8-21

Assign mesh currents as shown in the diagram (compare with Fig. 8-20). Then

$$\Delta_\mathbf{z} = \begin{vmatrix} 5-j2 & 0 & -5 \\ 0 & -j2 & -j4 \\ -5 & -j4 & 5-j2 \end{vmatrix} = 40 - j24$$

$$\Delta_{11} = \begin{vmatrix} -j2 & -j4 \\ -j4 & 5-j2 \end{vmatrix} = 12 - j10 \qquad \Delta_{22} = \begin{vmatrix} 5-j2 & -5 \\ -5 & 5-j2 \end{vmatrix} = -4 - j20 \qquad \Delta_{21} = -\begin{vmatrix} 0 & -5 \\ -j4 & 5-j2 \end{vmatrix} = j20$$

Using the expressions from Problem 8.15,

$$\mathbf{Z}_A = \frac{\Delta_\mathbf{z}}{\Delta_{11}+\Delta_{21}} = \frac{46.6\underline{/-31^\circ}}{12-j10+j20} = 2.98\underline{/-70.8^\circ}\ \Omega$$

$$\mathbf{Z}_B = -\frac{\Delta_\mathbf{z}}{\Delta_{21}} = -\frac{46.6\underline{/-31^\circ}}{j20} = 2.33\underline{/59^\circ}\ \Omega$$

$$\mathbf{Z}_C = \frac{\Delta_\mathbf{z}}{\Delta_{22}+\Delta_{21}} = \frac{46.6\underline{/-31^\circ}}{-4-j20+j20} = 11.65\underline{/149^\circ}\ \Omega$$

Note that the impedance $\mathbf{Z}_A$ can be realized as a resistance and capacitance in series, and $\mathbf{Z}_B$ as a resistance and inductance in series. However, the impedance $\mathbf{Z}_C$ would require a negative resistance. Hence a circuit with the three calculated impedances cannot be constructed.

8.17 Convert the Δ-circuit of Fig. 8-22(*a*) into an equivalent Y.

$$\mathbf{Z}_1 = \frac{\mathbf{Z}_A\mathbf{Z}_B}{\mathbf{Z}_A + \mathbf{Z}_B + \mathbf{Z}_C} = \frac{(3-j2)(2+j3)}{7+j17} = 0.50 - j0.50 \quad \Omega$$

Similarly, $\mathbf{Z}_2 = 3.0 - j1.0 \quad \Omega$ and $\mathbf{Z}_3 = 1.0 + j3.0 \quad \Omega$.

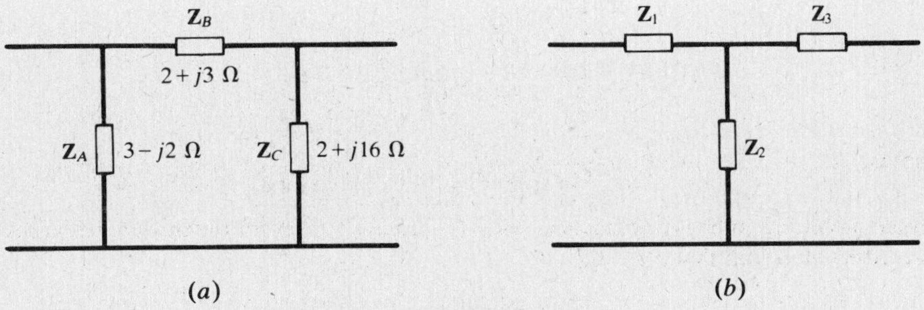

Fig. 8-22

8.18 Demonstrate the reciprocity theorem by computing voltage $\mathbf{V}_x$ in Fig. 8-23(*a*), then transposing the current source and finding $\mathbf{V}_x$ in the circuit of Fig. 8-23(*b*).

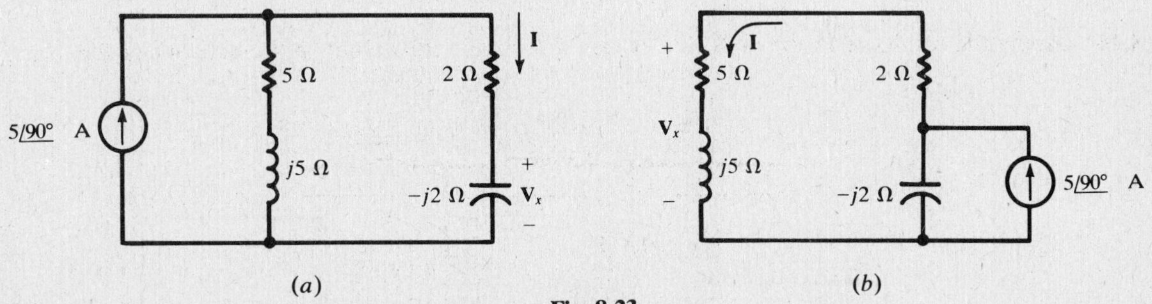

Fig. 8-23

In Fig. 8-23(*a*),

$$\mathbf{I} = \frac{5+j5}{7+j3}(5\angle 90^\circ) \qquad \text{and} \qquad \mathbf{V}_x = \mathbf{I}(-j2) = \frac{5+j5}{7+j3}(5\angle 90^\circ)(-j2) \quad \text{V}$$

In Fig. 8-23(*b*),

$$\mathbf{V}_x = \mathbf{I}(5+j5) = \frac{-j2}{7+j3}(5\angle 90^\circ)(5+j5) \quad \text{V}$$

8.19 Demonstrate the reciprocity theorem by computing $\mathbf{I}_x$ in Fig. 8-24(*a*), then inserting the voltage source in the branch where $\mathbf{I}_x$ flowed and finding the current in the branch which formerly contained the source.

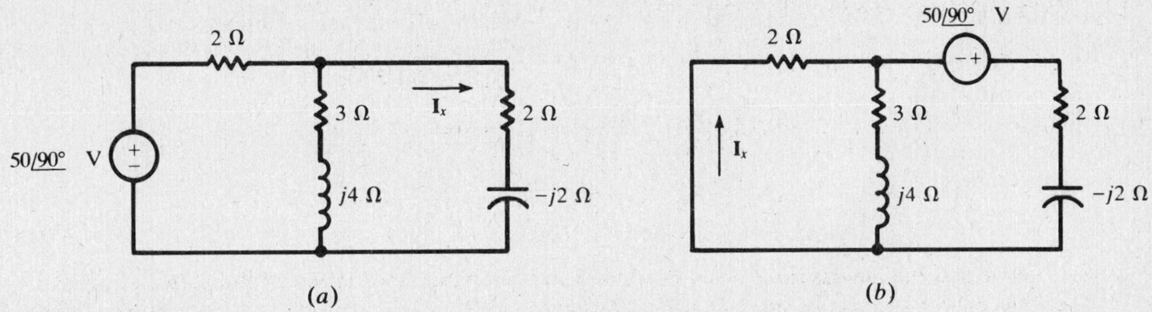

Fig. 8-24

In Fig. 8-24(a) the circuit impedance as seen by the source is

$$\mathbf{Z}_T = 2 + \frac{(3+j4)(2-j2)}{(5+j2)} = \frac{24+j6}{5+j2} \quad \Omega$$

Then, by current division,

$$\mathbf{I}_x = \frac{50\angle 90^\circ}{\mathbf{Z}_T}\left(\frac{3+j4}{5+j2}\right) = \frac{(50\angle 90^\circ)(3+j4)}{24+j6} \quad \text{A}$$

In Fig. 8-24(b) the circuit impedance as seen by the source is

$$\mathbf{Z}_T = (2-j2) + \frac{2(3+j4)}{5+j4} = \frac{24+j6}{5+j4} \quad \Omega$$

Then,
$$\mathbf{I}_x = \frac{50\angle 90^\circ}{\mathbf{Z}_T}\left(\frac{3+j4}{5+j4}\right) = \frac{(50\angle 90^\circ)(3+j4)}{24+j6} \quad \text{A}$$

8.20 Using the compensation theorem, calculate the change in the current in the circuit shown in Fig. 8-25 when the reactance is changed to $j35\ \Omega$.

$$\mathbf{I} = \frac{100\angle 45^\circ}{30+j40} = 2.00\angle -8.13^\circ \ \text{A} \qquad \mathbf{I} + \delta\mathbf{I} = \frac{100\angle 45^\circ}{30+j35} = 2.17\angle -4.40^\circ \ \text{A}$$

$$\delta\mathbf{I} = (2.17\angle -4.40^\circ) - (2.00\angle -8.13^\circ) = 0.218\angle 32.45^\circ \ \text{A}$$

Using the compensation theorem, the circuit is as shown in Fig. 8-26.

$$\mathbf{V}_c = \mathbf{I}\,\delta\mathbf{Z} = (2.00\angle -8.13^\circ)(-j5) = 10.0\angle -98.13^\circ \ \text{V}$$

$$\delta\mathbf{I} = \frac{-\mathbf{V}_c}{\mathbf{Z}+\delta\mathbf{Z}} = \frac{-10\angle -98.13^\circ}{30+j35} = 0.217\angle 32.47^\circ \ \text{A}$$

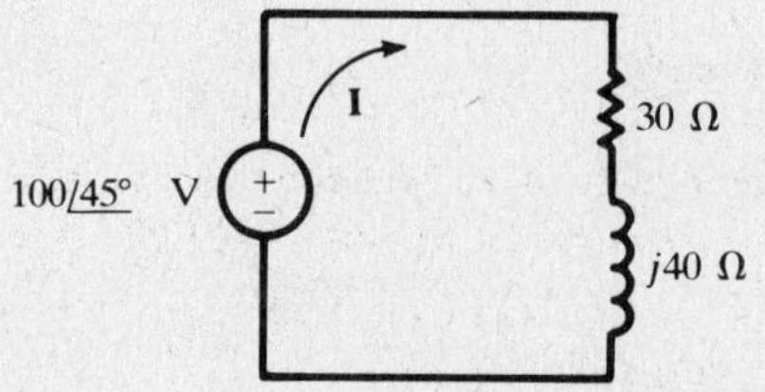

Fig. 8-25

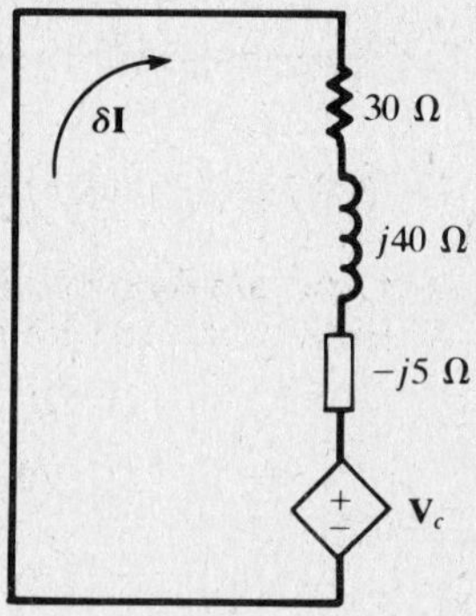

Fig. 8-26

8.21 In the circuit shown in Fig. 8-27(a), the 5-Ω resistor is changed to 8 Ω. Find the compensation source $\mathbf{V}_c$ and use it to calculate the change in current through the 3-Ω resistor.

$$\mathbf{I} = \frac{10\angle 0^\circ}{5 + \dfrac{(j5)(3+j4)}{3+j9}} = 1.58\angle -23.23^\circ \ \text{A} \qquad \mathbf{V}_c = \mathbf{I}(8-5) = 4.74\angle -23.23^\circ \ \text{V}$$

In Fig. 8-27(b),
$$\mathbf{Z}_T = 8 + \frac{j5(3+j4)}{3+j9} = 9.18\angle 15.82^\circ \ \Omega$$

Then, by current division,

$$\delta\mathbf{I} = \frac{-4.73\angle -23.13^\circ}{9.18\angle 15.82^\circ}\left(\frac{j5}{3+j9}\right) = 0.271\angle 159.5^\circ \ \text{A}$$

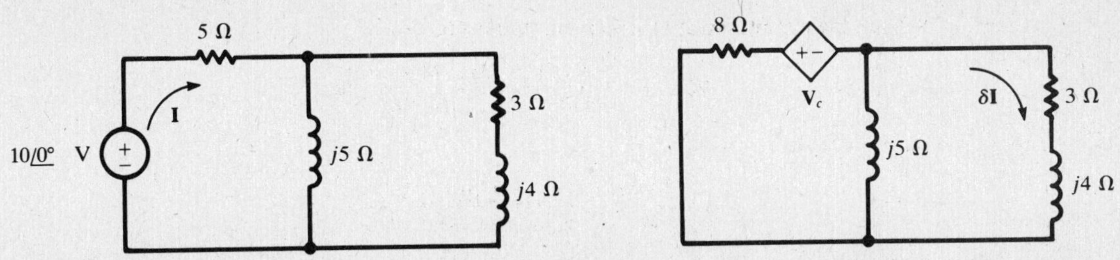

(a) (b)

Fig. 8-27

Supplementary Problems

8.22 For the network of Fig. 8-28, assign two different sets of mesh currents and show that for each, $\Delta_Z = 55.9\underline{/-26.57^\circ}$ Ω^2. For each choice, calculate the phasor voltage **V**. Obtain the phasor voltage across the $3+j4$ Ω impedance and compare with **V**. *Ans.* $\mathbf{V} = \mathbf{V}_{3+j4} = 22.36\underline{/-10.30^\circ}$ V

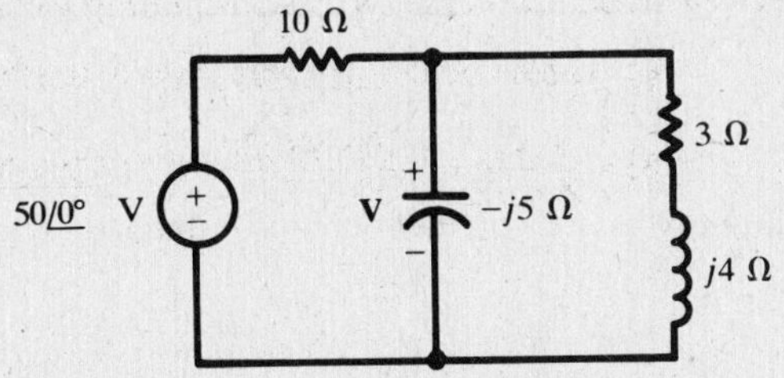

Fig. 8-28

8.23 For the network of Fig. 8-29, use the mesh current method to find the current in the $2+j3$ Ω impedance due to each of the sources $\mathbf{V}_1$ and $\mathbf{V}_2$. *Ans.* $2.41\underline{/6.45^\circ}$ A, $1.36\underline{/141.45^\circ}$ A

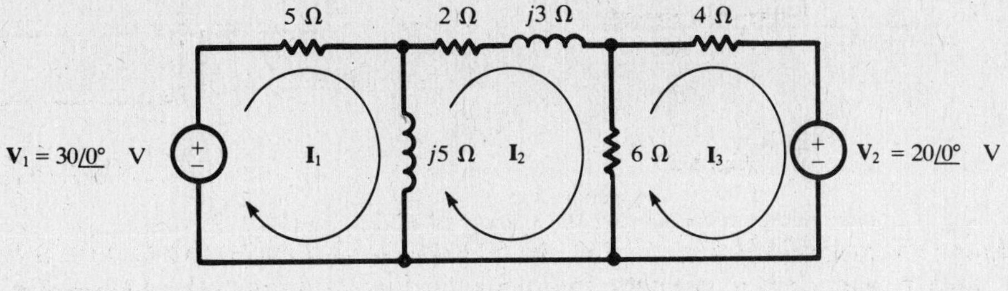

Fig. 8-29

8.24 In the network shown in Fig. 8-30, the two equal capacitances C and the shunting resistance R are adjusted until the detector current $\mathbf{I}_D$ is zero. Assuming a source angular frequency ω, determine the values of R_x and L_x. *Ans.* $R_x = 1/(\omega^2 C^2 R)$, $L_x = 1/(2\omega C)$

8.25 For the network of Fig. 8-31, obtain the current ratio $\mathbf{I}_1/\mathbf{I}_3$. *Ans.* $3.3\underline{/-90^\circ}$

8.26 For the network of Fig. 8-31, obtain $\mathbf{Z}_{\text{input},1}$ and $\mathbf{Z}_{\text{transfer},13}$. Show that $\mathbf{Z}_{\text{transfer},31} = \mathbf{Z}_{\text{transfer},13}$. *Ans.* $1.31\underline{/21.8^\circ}$ Ω, $4.31\underline{/-68.2^\circ}$ Ω

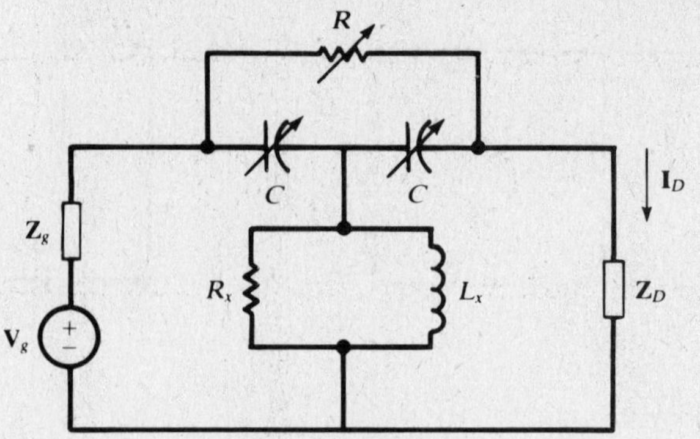

Fig. 8-30

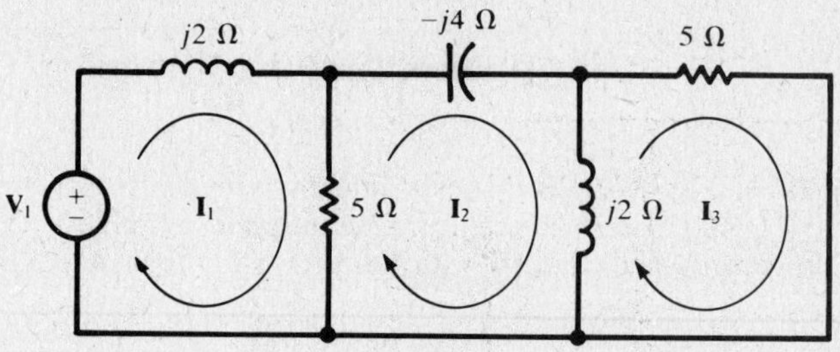

Fig. 8-31

8.27 Obtain the input impedance as seen by the source **V** in the network of Fig. 8-32.
Ans. $20.2\angle{-36.1^\circ}$ Ω

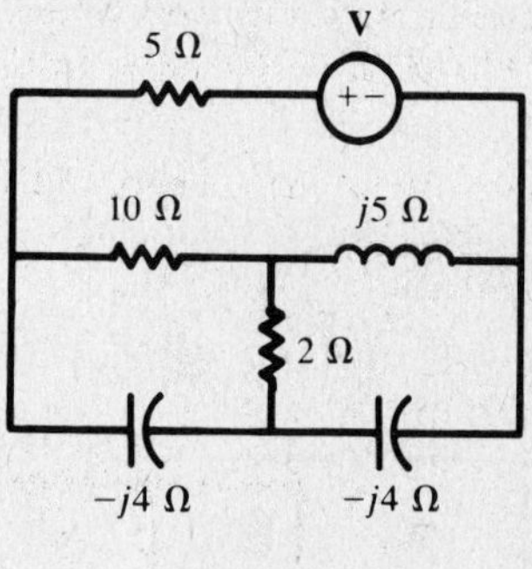

Fig. 8-32

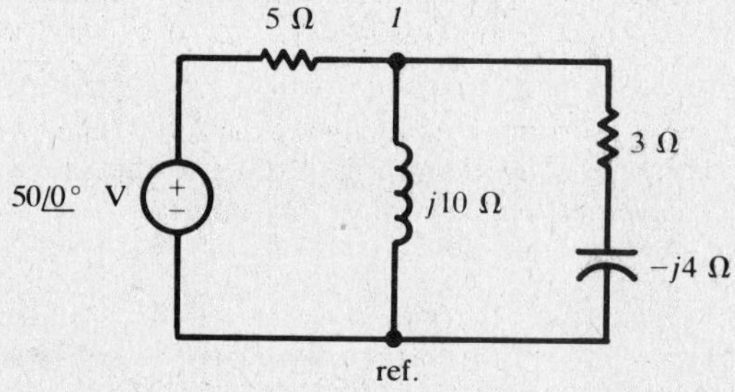

Fig. 8-33

8.28 Using the node voltage method, find the voltage of node *1* in Fig. 8-33 with respect to the reference.
Ans. $30.7\angle{-10.6^\circ}$ V

8.29 The circuit of Fig. 8-34 contains no principal nodes. Even so, obtain the voltage of node *1* with respect to the reference, by use of the node voltage method. *Ans.* $11.8\angle{54.68^\circ}$ V

8.30 In the network of Fig. 8-35, obtain the voltage ratio $\mathbf{V}_1/\mathbf{V}_2$ by application of the node voltage method.
Ans. $\dfrac{\Delta_{11}}{\Delta_{12}} = 1.61\angle{-29.8^\circ}$

8.31 For the network of Fig. 8-35, obtain the driving-point impedance $\mathbf{Z}_{\text{input},1}$. *Ans.* $5.59\angle{17.35^\circ}$ Ω

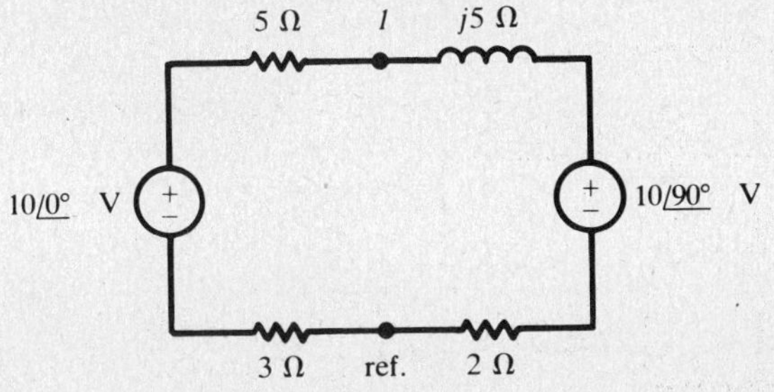

Fig. 8-34

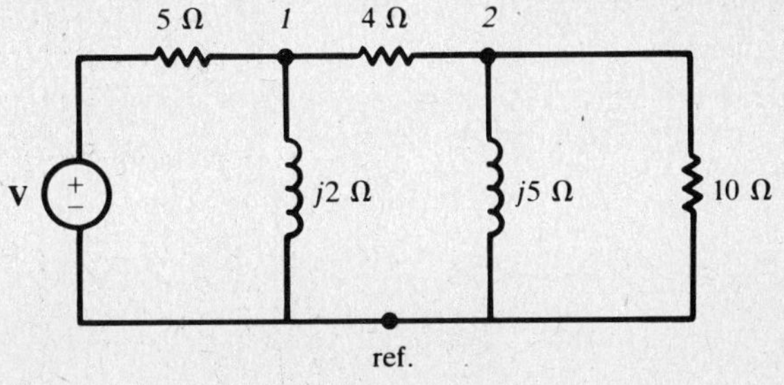

Fig. 8-35

8.32 Obtain the node voltage $\mathbf{V}_1$ in the network of Fig. 8-36. *Ans.* $74.9\angle 63.0^\circ$ V

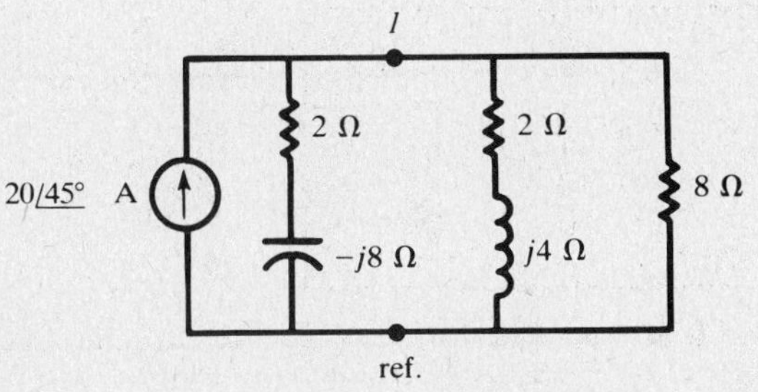

Fig. 8-36

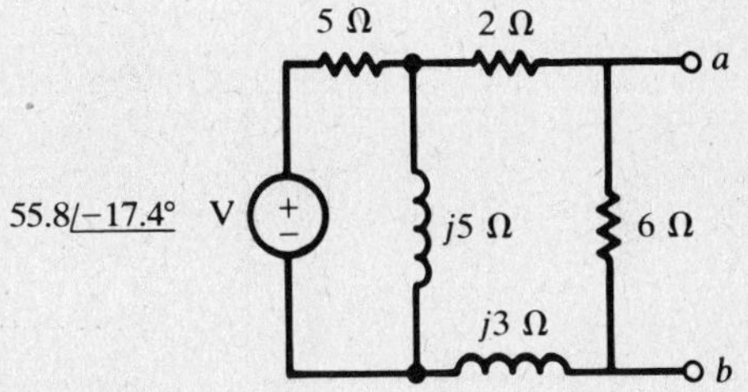

Fig. 8-37

8.33 Obtain the Thévenin and Norton equivalent circuits at terminals *ab* for the network of Fig. 8-37. Choose the polarity such that $\mathbf{V}' = \mathbf{V}_{ab}$. *Ans.* $\mathbf{V}' = 20.0\angle 0^\circ$ V, $\mathbf{I}' = 5.56\angle -23.06^\circ$ A, $\mathbf{Z}' = 3.60\angle 23.06^\circ$ Ω

8.34 Obtain the Thévenin and Norton equivalent circuits at terminals *ab* for the network of Fig. 8-38. *Ans.* $\mathbf{V}' = 11.5\angle -95.8^\circ$ V, $\mathbf{I}' = 1.39\angle -80.6^\circ$ A, $\mathbf{Z}' = 8.26\angle -15.2^\circ$ Ω

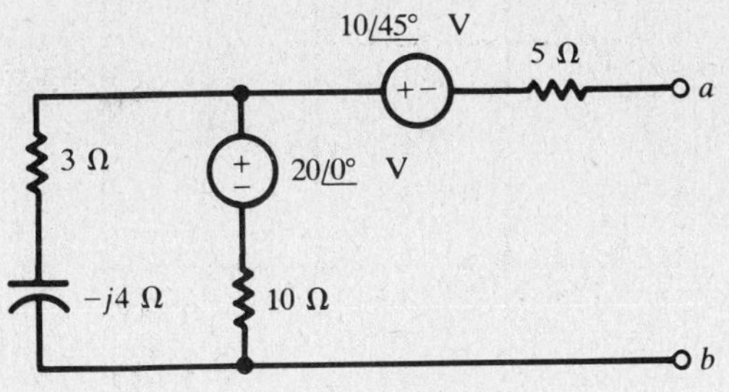

Fig. 8-38

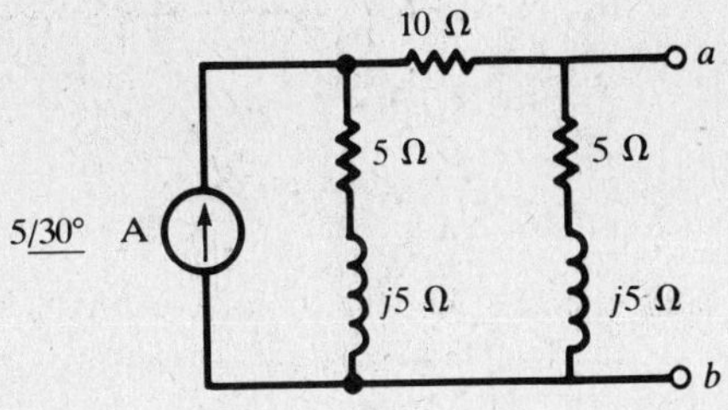

Fig. 8-39

8.35 Obtain the Thévenin and Norton equivalent circuits at terminals *ab* for the network of Fig. 8-39. *Ans.* $\mathbf{V}' = 11.18\angle 93.43^\circ$ V, $\mathbf{I}' = 2.24\angle 56.56^\circ$ A, $\mathbf{Z}' = 5.0\angle 36.87^\circ$ Ω

8.36 Use the node voltage method, instead of the mesh current method as in Problem 8.15, to show that an *n*-mesh, three-terminal, passive network can be replaced by a Y-equivalent.
Ans. See Fig. 8-40: $\mathbf{Z}_1 = (\Delta_{11} - \Delta_{12})/\Delta_\mathbf{Y}$, $\mathbf{Z}_2 = \Delta_{12}/\Delta_\mathbf{Y}$, $\mathbf{Z}_3 = (\Delta_{22} - \Delta_{12})/\Delta_\mathbf{Y}$.

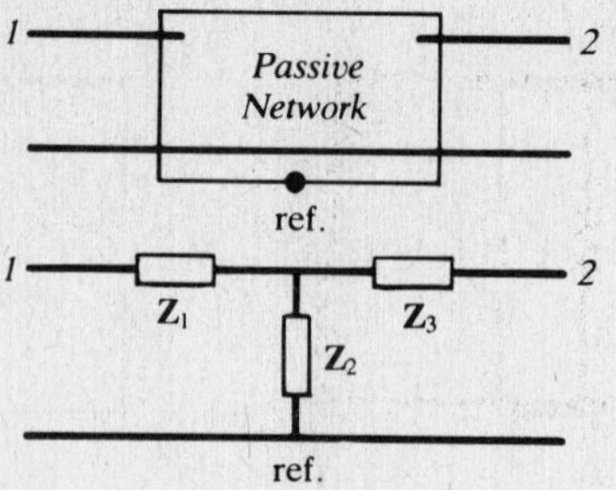

Fig. 8-40

8.37 Using the expressions developed in Problem 8.36, obtain the Y-equivalent of the network shown in Fig. 8-41. *Ans.* 6.25 Ω, 2.5 Ω, 10.5 Ω

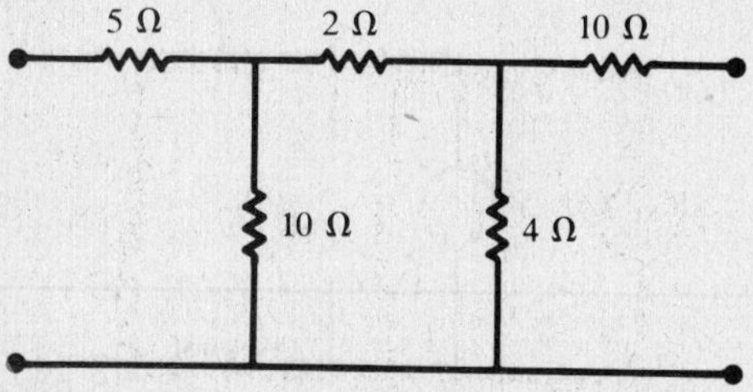

Fig. 8-41

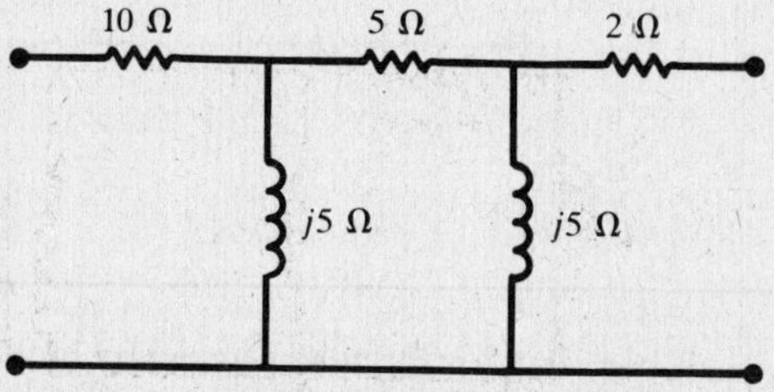

Fig. 8-42

8.38 Convert the three impedances forming a Δ in Fig. 8-42 into an equivalent Y, using the transformation given in Section 8.5. Combine the result with the 10-Ω and 2-Ω resistors for an overall Y-equivalent. *Ans.* $12+j1$ Ω, $-1+j2$ Ω, $4+j1$ Ω

8.39 The three-terminal network shown in Fig. 8-43 contains a balanced Δ in parallel with a balanced Y. Obtain the Y-connected equivalent. *Ans.* $2.28\angle -3.54°$ Ω (balanced)

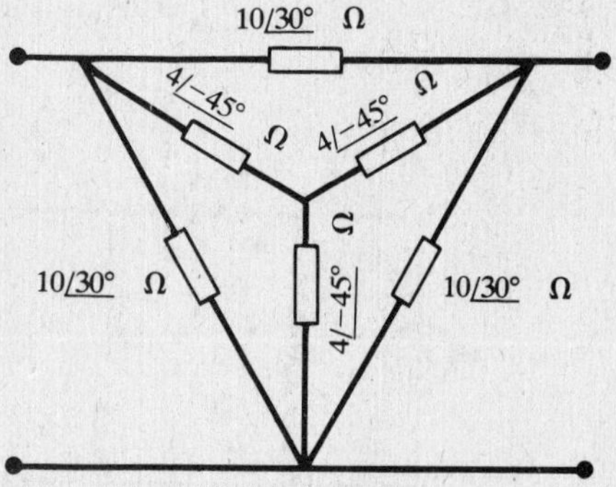

Fig. 8-43

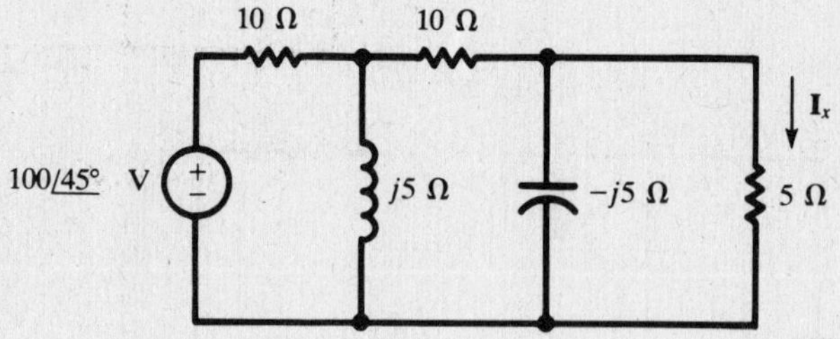

Fig. 8-44

8.40 Compute $\mathbf{I}_x$ in the single-source network of Fig. 8-44. Demonstrate the reciprocity theorem by interchanging the positions of the source and $\mathbf{I}_x$ and again computing $\mathbf{I}_x$. *Ans.* $2.16\angle 57.5°$ A

8.41 The network of Fig. 8-45 has a single current source. Obtain the phasor voltage $\mathbf{V}_{ab}$, and demonstrate the reciprocity theorem by transposing the current source to branch *ab* and computing the voltage across the $3+j4$ Ω impedance. *Ans.* $7.45\angle 99.6°$ V

8.42 In the circuit of Fig. 8-46, replace the parallel combination of impedances with a compensation source $\mathbf{V}_c$. Show that the current in the 5-Ω resistor is the same. *Ans.* $\mathbf{V}_c = 9.77\angle 39.1°$ V

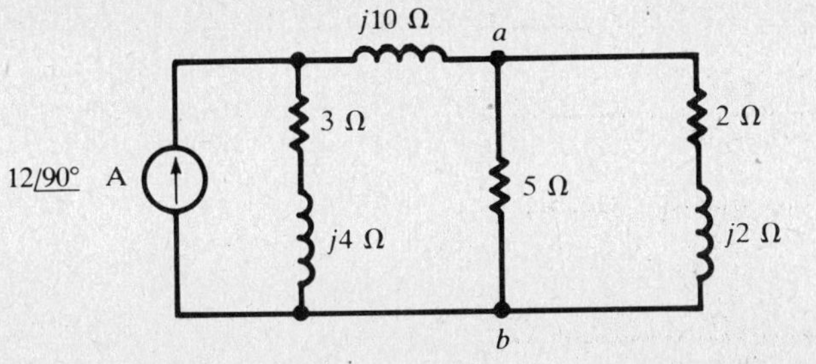

Fig. 8-45

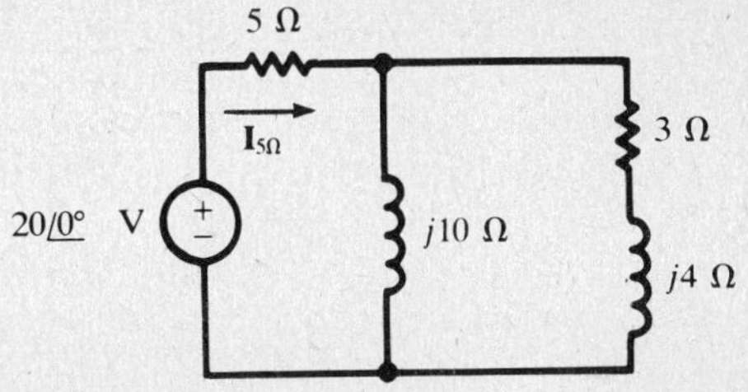

Fig. 8-46

8.43 In Fig. 8-47(a), the 3-Ω resistor is to be changed to 4 Ω, as shown in Fig. 8-47(b). Use the compensation theorem to obtain the change in current $\delta \mathbf{I}_T$.
Ans. See Fig. 8-47(c): $\mathbf{V}_c = 2.37\underline{/5.44^\circ}$ V, $\delta \mathbf{I}_T = 0.106\underline{/195.7^\circ}$ A.

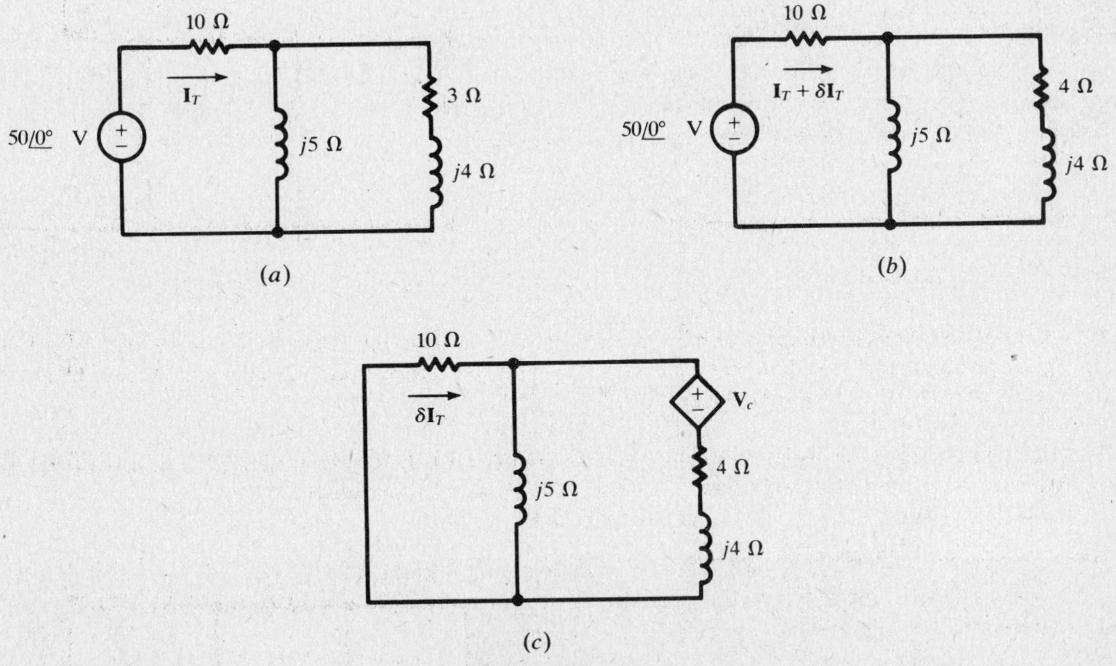

Fig. 8-47

Chapter 9

Power and Power Factor

9.1 POWER IN THE TIME DOMAIN

A generalized passive network, with a voltage $v(t)$ and resulting current $i(t)$, is shown in Fig. 9-1. The *time-variable* or *instantaneous power* into the network is the product of the voltage and current:

$$p(t) = v(t)\, i(t) \quad \text{(W)}$$

If $v(t)$ is sinusoidal, then, after the transient period has passed, the periodic voltage and periodic current result in a periodic power. By the sign convention of Section 2.4, a positive power corresponds to a transfer of energy from the source to the network, and negative power to a return of energy from the network to the source. Of course, for a passive network, the net energy transfer must be from the source to the network; hence the time-average power must be either positive or zero. Zero average power would result from a purely reactive network, where energy storage could take place periodically but without energy dissipation.

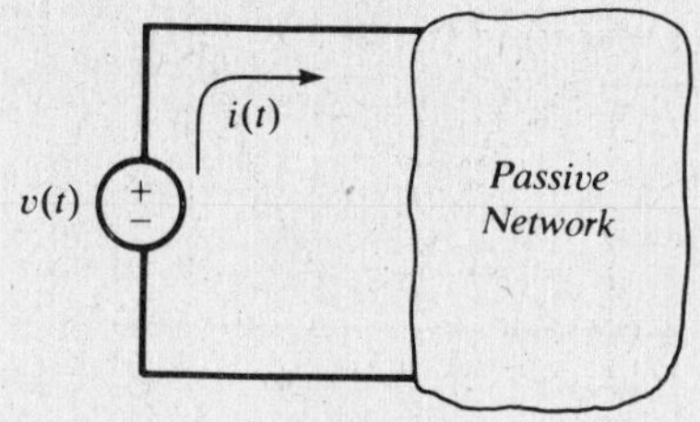

Fig. 9-1

9.2 POWER IN THE SINUSOIDAL STEADY STATE

For a passive network containing a single inductive circuit element, the applied sinusoidal voltage $v = V_m \cos \omega t$ results in a sinusoidal current which lags by 90°, $i = I_m \cos(\omega t - 90°)$. Then the instantaneous power is given by

$$p = vi = V_m I_m \cos \omega t \cos(\omega t - 90°) = \frac{1}{2} V_m I_m \sin 2\omega t$$

This result is illustrated in Fig. 9-2. In the intervals where v and i are of the same sign, e.g. $0 < \omega t < \pi/2$, p is positive. Energy will be transferred from the source to the inductive circuit element during such times. In other intervals, such as $\pi/2 < \omega t < \pi$, p is negative and energy is returned from the circuit element to the source. Over a cycle, the average value of p is zero. Note that the power frequency is twice that of the voltage and current.

A more general case is where a voltage such as $v = V_m \cos \omega t$ results in a current $i = I_m \cos(\omega t - \theta)$, where θ could be positive or negative, corresponding to inductive or capacitive equivalent impedance, respectively. The case of leading current, $\theta < 0$, is shown in Fig. 9-3. Then

$$p = V_m I_m \cos \omega t \cos(\omega t - \theta) = \frac{1}{2} V_m I_m [\cos \theta + \cos(2\omega t - \theta)]$$

The term $\cos(2\omega t - \theta)$ has zero average value, so that

$$P_{\text{avg}} = \frac{1}{2} V_m I_m \cos \theta$$

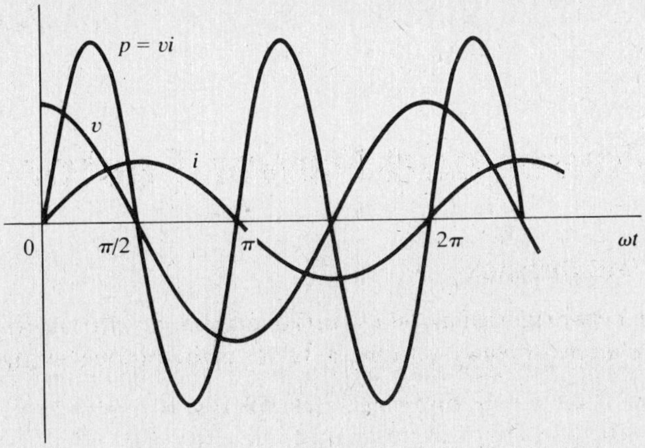

Fig. 9-2

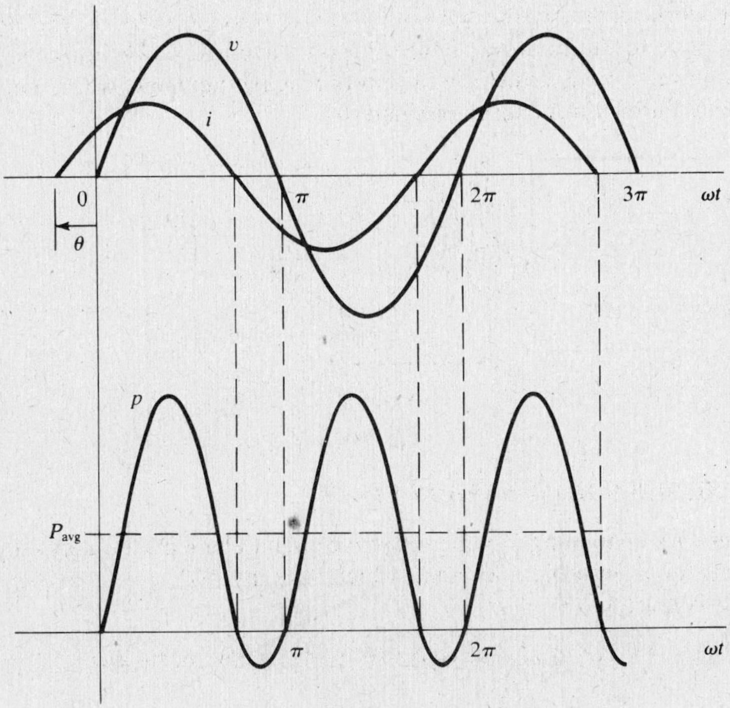

Fig. 9-3

(For the remainder of this chapter, the subscript avg will be omitted and P will denote average power.) *Effective* or *rms* values of sine or cosine functions, illustrated in Appendix A, are $V_{eff} = V_m/\sqrt{2}$, $I_{eff} = I_m/\sqrt{2}$. Therefore:

$$\textbf{average power} \qquad P = V_{eff} I_{eff} \cos\theta \quad \text{(W)}$$

The product $V_{eff}I_{eff}$ is called the *apparent power*, which is given the symbol S and is measured in *voltamperes*, VA, where $1\ \text{VA} = 1\ \text{V}\cdot\text{A} = 1\ \text{W}$. The factor by which the apparent power must be multiplied to obtain the average power is called the *power factor*,

$$\text{pf} \equiv \cos\theta$$

When citing the pf, it is usual to include information about the sign of θ by speaking of a *lagging* pf if $\theta > 0$ (current *lags* voltage), or a *leading* pf if $\theta < 0$ (current *leads* voltage). In any case, $0 \le \text{pf} \le 1$.

9.3 POWER TRIANGLE, COMPLEX POWER

Representing the passive network in the frequency domain by the equivalent impedance $\mathbf{Z} = Z\angle\theta$, one has

$$\text{pf} = \cos\theta = \frac{R}{Z}$$

and, since $V_{\text{eff}} = I_{\text{eff}}Z$,

$$\textbf{apparent power} \qquad S = V_{\text{eff}}I_{\text{eff}} = I_{\text{eff}}^2 Z \quad \text{(VA)}$$
$$\textbf{average power} \qquad P = V_{\text{eff}}I_{\text{eff}}\cos\theta = I_{\text{eff}}^2 R \quad \text{(W)}$$

It follows that S and P may be represented geometrically as the hypotenuse and horizontal leg of a right triangle, this triangle being simply the impedance diagram (Chapter 7) scaled by the factor I_{eff}^2. See Figs. 9-4 and 9-5. The vertical leg of the triangle is the

$$\textbf{quadrature power} \qquad Q \equiv V_{\text{eff}}I_{\text{eff}}\sin\theta = I_{\text{eff}}^2 X \quad \text{(var)}$$

The unit of Q is the *voltampere reactive*, var, where, again, 1 var = 1 W. It is customary always to take Q as nonnegative. Thus, when $\theta < 0$ (as in Fig. 9-5), we cite "$Q = 13.2$ var (capacitive)" instead of "$Q = -13.2$ var." When $\theta > 0$, Q is quoted in var (inductive).

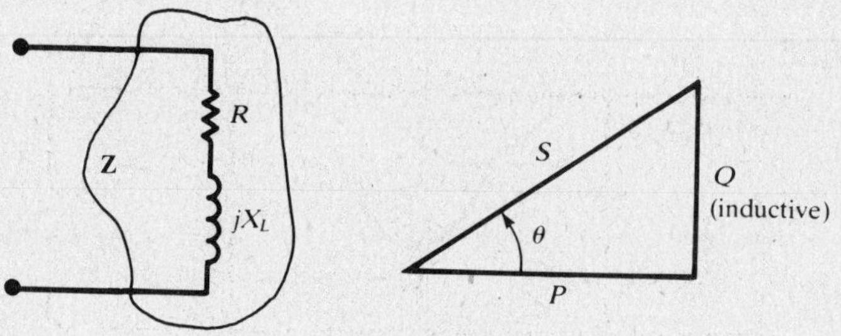

Fig. 9-4

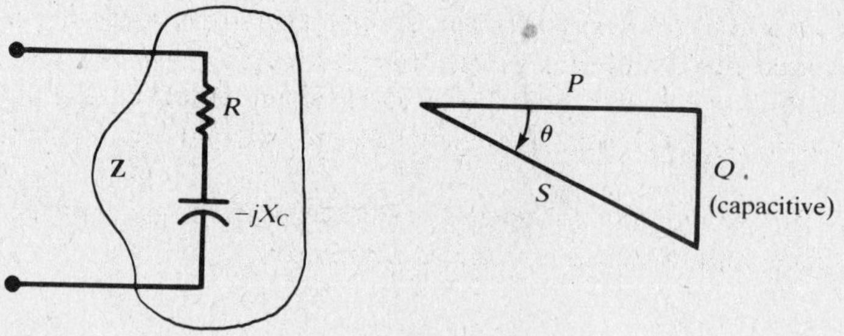

Fig. 9-5

The various power quantities may all be conveniently derived from the

$$\textbf{complex power} \qquad \mathbf{S} \equiv \mathbf{V}_{\text{eff}}\mathbf{I}_{\text{eff}}^* = S\angle\theta = P + jQ$$

in which $\mathbf{V}_{\text{eff}}$ is the phasor effective voltage and $\mathbf{I}_{\text{eff}}^*$ is the complex conjugate (Appendix B) of the phasor effective current. An equivalent formula is $\mathbf{S} = I_{\text{eff}}^2\mathbf{Z}$.

EXAMPLE 9.1 A certain passive network has equivalent impedance $\mathbf{Z} = 3 + j4\ \Omega$ and an applied voltage

$$v = 42.5\cos(1000t + 30°) \quad \text{(V)}$$

Give complete power information.

$$\mathbf{V}_{\text{eff}} = \frac{42.5}{\sqrt{2}} \angle 30^\circ \text{ V} \qquad \mathbf{I}_{\text{eff}} = \frac{\mathbf{V}_{\text{eff}}}{\mathbf{Z}_{\text{eq}}} = \frac{(42.5/\sqrt{2})\angle 30^\circ}{5\angle 53.13^\circ} = \frac{8.5}{\sqrt{2}} \angle -23.13^\circ \text{ A}$$

$$\mathbf{S} = \mathbf{V}_{\text{eff}}\mathbf{I}^*_{\text{eff}} = 180.6\angle 53.13^\circ = 108.4 + j144.5$$

Hence $P = 108.4$ W, $Q = 144.5$ var (inductive), $S = 180.6$ VA, and pf $= \cos 53.13^\circ = 0.6$ lagging.

Parallel-connected Networks

The complex power **S** is also useful in analyzing practical networks, e.g. the collection of households drawing on the same power lines. Referring to Fig. 9-6,

$$\mathbf{S}_T = \mathbf{V}_{\text{eff}}\mathbf{I}^*_{\text{eff}} = \mathbf{V}_{\text{eff}}(\mathbf{I}^*_{1\text{eff}} + \mathbf{I}^*_{2\text{eff}} + \cdots + \mathbf{I}^*_{n\text{eff}}) = \mathbf{S}_1 + \mathbf{S}_2 + \cdots + \mathbf{S}_n$$

from which

$$P_T = P_1 + P_2 + \cdots + P_n$$
$$Q_T = Q_1 + Q_2 + \cdots + Q_n$$
$$S_T = \sqrt{P_T^2 + Q_T^2}$$
$$\text{pf}_T = \frac{P_T}{S_T}$$

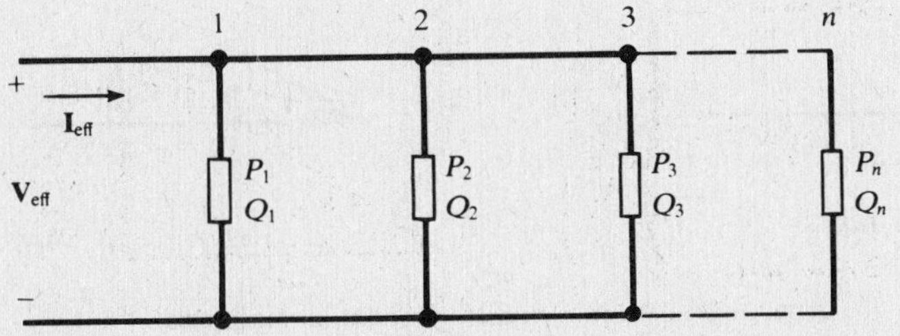

Fig. 9-6

These results (which also hold for series-connected networks) imply that the power triangle for the network may be obtained by joining the power triangles for the branches vertex to vertex. In the example shown in Fig. 9-7, $n = 3$, with branches 1 and 3 assumed inductive and branch 2 capacitive. In such diagrams, some of the triangles may degenerate into straight-line segments if the corresponding R or X is zero.

If the power data for the individual branches are not important, the network may be replaced by its equivalent admittance, and this used directly to compute $\mathbf{S}_T$.

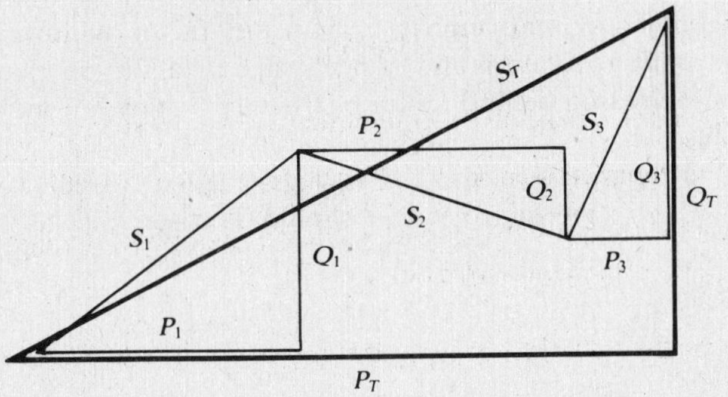

Fig. 9-7

9.4 POWER FACTOR IMPROVEMENT

Electrical service to industrial customers is three-phase, as opposed to the single-phase power supplied to residential and small commercial customers. While metering and billing practices vary among the utilities, the large consumers will always find it advantageous to reduce the quadrature component of their power triangle; this is called "improving the power factor." Industrial systems generally have an overall inductive component because of the large number of motors. Each individual load tends to be either pure resistance, with unity power factor, or resistance and inductive reactance, with a lagging power factor. All of the loads are parallel-connected, and the equivalent impedance results in a lagging current and a corresponding inductive quadrature power Q. To improve the power factor, capacitors, in three-phase banks, are connected to the system either on the primary or secondary side of the main transformer, such that the combination of the plant load and the capacitor banks presents a load to the serving utility which is nearer to unity power factor.

EXAMPLE 9.2 How much capacitive Q must be provided by the capacitor bank in Fig. 9-8 to improve the power factor to 0.95 lagging?

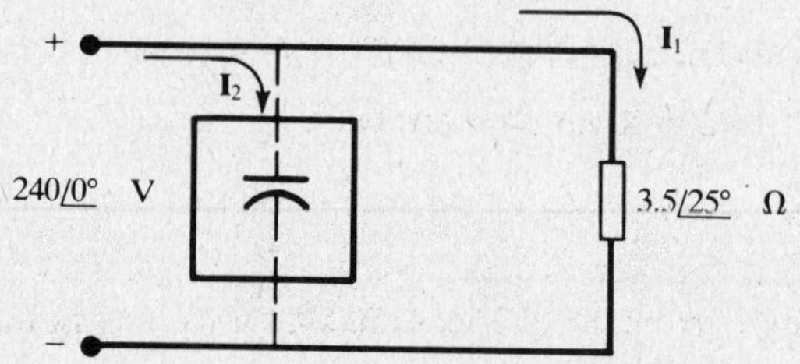

Fig. 9-8

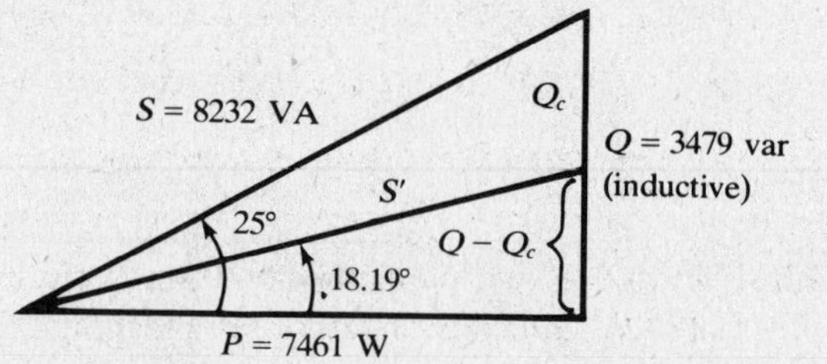

Fig. 9-9

Before addition of the capacitor bank, pf = cos 25° = 0.906 lagging, and

$$\mathbf{I}_1 = \frac{240\angle 0^\circ}{3.5\angle 25^\circ} = 68.6\angle -25^\circ \ \ \text{A}$$

$$\mathbf{S} = \mathbf{V}_{\text{eff}}\mathbf{I}^*_{1\text{eff}} = \left(\frac{240}{\sqrt{2}}\angle 0^\circ\right)\left(\frac{68.6}{\sqrt{2}}\angle +25^\circ\right) = 8232\angle 25^\circ = 7461 + j3479$$

After the improvement, the triangle has the same P, but its angle is $\cos^{-1} 0.95 = 18.19^\circ$. Then (see Fig. 9-9),

$$\frac{3479 - Q_c}{7461} = \tan 18.19^\circ \qquad \text{or} \qquad Q_c = 1027 \text{ var (capacitive)}$$

The new value of apparent power is $S' = 7854$ VA, as compared to the original $S = 8232$ VA. The decrease, 378 VA, amounts to 4.6%.

The transformers, the distribution systems, and the utility company alternators are all rated in kVA or MVA. Consequently, an improvement in the power factor, with its corresponding reduction in kVA, releases some of this generation and transmission capability so that it can be used to serve other customers. This is the reason behind the rate structures which, in one way or another, make it more costly for an industrial customer to operate with a low power factor. An economic study comparing the cost of the capacitor bank to the savings realized is frequently made. The results of such a study will show whether the improvement should be made and also what final power factor should be attained.

Solved Problems

9.1 Obtain the complete power information for a passive circuit with an applied voltage $v = 150\cos(\omega t + 10°)$ V and a resulting current $i = 5.0\cos(\omega t - 50°)$ A.

Using the complex power,

$$\mathbf{S} = \mathbf{V}_{\text{eff}}\mathbf{I}^*_{\text{eff}} = \left(\frac{150}{\sqrt{2}}\angle 10°\right)\left(\frac{5.0}{\sqrt{2}}\angle 50°\right) = 375\angle 60° = 187.5 + j324.8$$

Thus $P = 187.5$ W, $Q = 324.8$ var (inductive), $S = 375$ VA, and pf $= \cos 60° = 0.50$ lagging.

9.2 A two-element series circuit has average power 940 W and power factor 0.707 leading. Determine the circuit elements if the applied voltage is $v = 99.0\cos(6000t + 30°)$ V.

The effective applied voltage is $99.0/\sqrt{2} = 70.0$ V. Substituting in $P = V_{\text{eff}} I_{\text{eff}} \cos\theta$

$$940 = (70.0)I_{\text{eff}}(0.707) \qquad \text{or} \qquad I_{\text{eff}} = 19.0 \text{ A}$$

Then, $(19.0)^2R = 940$, from which $R = 2.60\ \Omega$. For a leading pf, $\theta = \cos^{-1} 0.707 = -45°$, and so

$$\mathbf{Z} = R - jX_C \qquad \text{where} \qquad X_C = R\tan 45° = 2.60\ \Omega$$

Finally, from $2.60 = 1/\omega C$, $C = 64.1\ \mu$F.

9.3 Find the two elements of a series circuit having current $i = 4.24\cos(5000t + 45°)$ A, power 180 W, and power factor 0.80 lagging.

The effective value of the current is $I_{\text{eff}} = 4.24/\sqrt{2} = 3.0$ A. Then,

$$180 = (3.0)^2R \qquad \text{or} \qquad R = 20.0\ \Omega$$

The impedance angle is $\theta = \cos^{-1} 0.80 = +36.87°$, wherefore the second element must be an inductor. From the power triangle,

$$\frac{Q}{P} = \frac{I^2_{\text{eff}}X_L}{180} = \tan 36.87° \qquad \text{or} \qquad X_L = 15.0\ \Omega$$

Finally, from $15.0 = 5000L$, $L = 3.0$ mH.

9.4 Obtain the power information for each element in Fig. 9-10 and construct the power triangle.

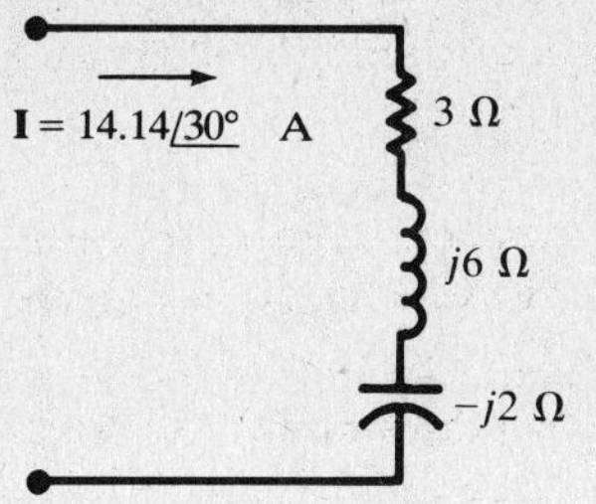

Fig. 9-10

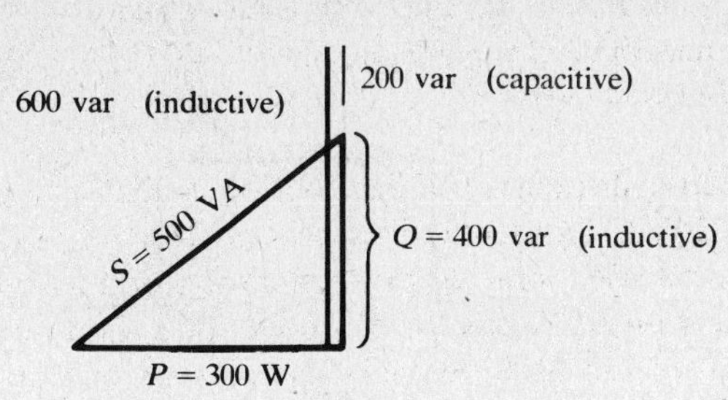

Fig. 9-11

The effective current is $14.14/\sqrt{2} = 10$ A.

$$P = (10)^2 3 = 300 \text{ W} \qquad Q_{j6\Omega} = (10)^2 6 = 600 \text{ var (inductive)} \qquad Q_{-j2\Omega} = (10)^2 2 = 200 \text{ var (capacitive)}$$

$$S = \sqrt{(300)^2 + (600 - 200)^2} = 500 \text{ VA} \qquad \text{pf} = P/S = 0.6 \text{ lagging}$$

The power triangle is shown in Fig. 9-11.

9.5 A series circuit of $R = 10\ \Omega$ and $X_C = 5\ \Omega$ has an effective applied voltage of 120 V. Determine the complete power information.

$$Z = \sqrt{10^2 + 5^2} = 11.18\ \Omega \qquad I_{\text{eff}} = \frac{120}{11.18} = 10.73\ \text{A}$$

Then:

$$P = I_{\text{eff}}^2 R = 1152\ \text{W} \qquad Q = I_{\text{eff}}^2 X_C = 576\ \text{var (capacitive)} \qquad S = \sqrt{(1152)^2 + (576)^2} = 1288\ \text{VA}$$

and pf = 1152/1288 = 0.894 leading.

9.6 Impedances $\mathbf{Z}_1 = 5.83\angle{-59.0^\circ}\ \Omega$ and $\mathbf{Z}_2 = 8.94\angle{63.43^\circ}\ \Omega$ are in series and carry an effective current of 5.0 A. Determine the complete power information.

$$\mathbf{Z}_T = \mathbf{Z}_1 + \mathbf{Z}_2 = 7.0 + j3.0\ \Omega$$

Hence,

$$P_T = (5.0)^2(7.0) = 175\ \text{W} \qquad Q_T = (5.0)^2(3.0) = 75\ \text{var (inductive)}$$

$$S_T = \sqrt{(175)^2 + (75)^2} = 190.4\ \text{VA} \qquad \text{pf} = \frac{175}{190.4} = 0.919\ \text{lagging}$$

9.7 Obtain the total power information for the parallel circuit shown in Fig. 9-12.

By current division,

$$\mathbf{I}_5 = 17.88\angle{18.43^\circ}\ \text{A} \qquad \mathbf{I}_4 = 26.05\angle{-12.53^\circ}\ \text{A}$$

Then,

$$P_T = \left(\frac{17.88}{\sqrt{2}}\right)^2(5) + \left(\frac{26.05}{\sqrt{2}}\right)^2(4) = 2156\ \text{W}$$

$$Q_T = \left(\frac{17.88}{\sqrt{2}}\right)^2(3) = 480\ \text{var (capacitive)}$$

$$S_T = \sqrt{(2156)^2 + (480)^2} = 2209\ \text{VA}$$

$$\text{pf} = \frac{2156}{2209} = 0.976\ \text{leading}$$

Alternate Method

$$\mathbf{Z}_{\text{eq}} = \frac{4(5 - j3)}{9 - j3} = 2.40 - j0.53\ \Omega$$

Then, $P = (42.4/\sqrt{2})^2(2.40) = 2157$ W and $Q = (42.4/\sqrt{2})^2(0.53) = 476$ var (capacitive).

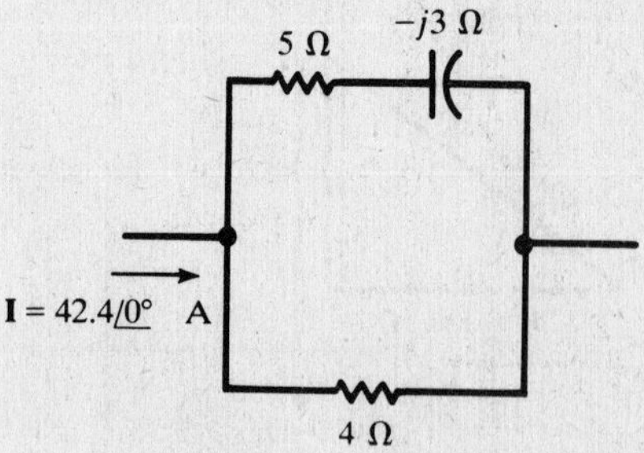

Fig. 9-12

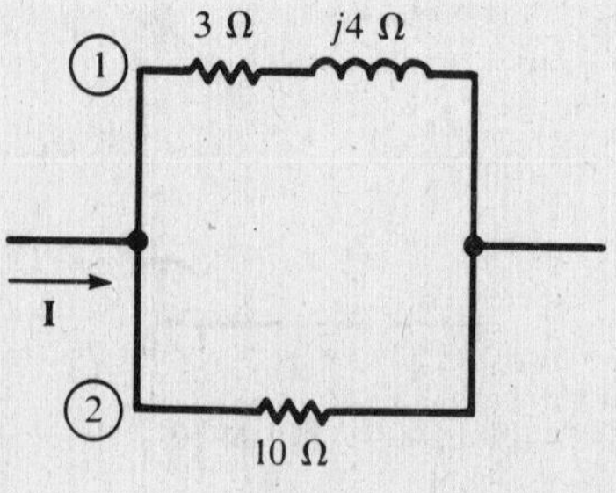

Fig. 9-13

9.8 Find the power factor for the circuit shown in Fig. 9-13.

With no voltage or current specified, P, Q, and S cannot be calculated. However, the power factor is the cosine of the angle on the equivalent impedance.

$$\mathbf{Z}_{eq} = \frac{(3+j4)(10)}{13+j4} = 3.68\underline{/36.03^\circ} \quad \Omega$$

$$\text{pf} = \cos 36.03^\circ = 0.809 \text{ lagging}$$

9.9 If the total power in the circuit of Fig. 9-13 is 1100 W, what are the powers in the two resistors?

By current division,

$$\frac{I_{1\text{eff}}}{I_{2\text{eff}}} = \frac{Z_2}{Z_1} = \frac{10}{\sqrt{3^2+4^2}} = 2$$

and so

$$\frac{P_{3\Omega}}{P_{10\Omega}} = \frac{I_{1\text{eff}}^2(3)}{I_{2\text{eff}}^2(10)} = \frac{6}{5}$$

Solving simultaneously with $P_{3\Omega} + P_{10\Omega} = 1100$ W gives $P_{3\Omega} = 600$ W, $P_{10\Omega} = 500$ W.

9.10 Obtain the power factor of a two-branch parallel circuit where the first branch has $\mathbf{Z}_1 = 2 + j4 \quad \Omega$, and the second $\mathbf{Z}_2 = 6 + j0 \quad \Omega$. To what value must the 6-Ω resistor be changed to result in the overall power factor 0.90 lagging?

Since the angle on the equivalent admittance is the negative of the angle on the equivalent impedance, its cosine also gives the power factor.

$$\mathbf{Y}_{eq} = \frac{1}{2+j4} + \frac{1}{6} = 0.334\underline{/-36.84^\circ} \quad \text{S}$$

$$\text{pf} = \cos(-36.84^\circ) = 0.80 \text{ lagging}$$

The pf is lagging, because the impedance angle is positive.

Now, for a change in power factor to 0.90, the admittance angle must become $\cos^{-1} 0.90 = -25.84^\circ$. Then,

$$\mathbf{Y}'_{eq} = \frac{1}{2+j4} + \frac{1}{R} = \left(\frac{1}{10} + \frac{1}{R}\right) - j\frac{1}{5}$$

requires

$$\frac{1/5}{\dfrac{1}{10} + \dfrac{1}{R}} = \tan 25.84^\circ \qquad \text{or} \qquad R = 3.20 \ \Omega$$

9.11 A voltage, $28.28\underline{/60^\circ}$ V, is applied to a two-branch parallel circuit in which $\mathbf{Z}_1 = 4\underline{/30^\circ} \quad \Omega$ and $\mathbf{Z}_2 = 5\underline{/60^\circ} \quad \Omega$. Obtain the power triangles for the branches and combine them into the total power triangle.

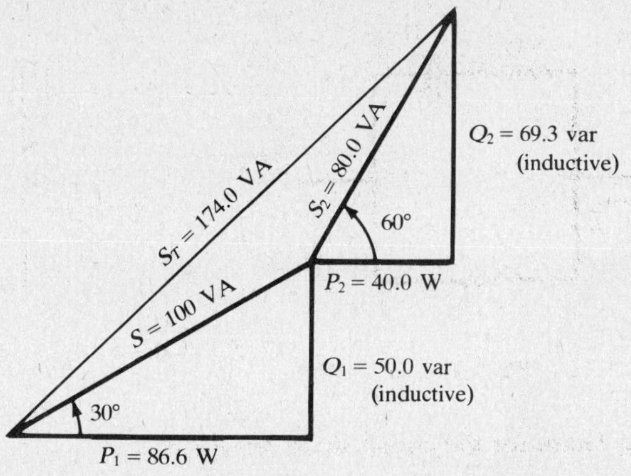

Fig. 9-14

$$\mathbf{I}_1 = \frac{\mathbf{V}}{\mathbf{Z}_1} = 7.07\underline{/30^\circ} \quad \text{A} \qquad \mathbf{I}_2 = \frac{\mathbf{V}}{\mathbf{Z}_2} = 5.66\underline{/0^\circ} \quad \text{A}$$

$$\mathbf{S}_1 = \left(\frac{28.28}{\sqrt{2}}\underline{/60^\circ}\right)\left(\frac{7.07}{\sqrt{2}}\underline{/-30^\circ}\right) = 100\underline{/30^\circ} = 86.6 + j50.0$$

$$\mathbf{S}_2 = \left(\frac{28.28}{\sqrt{2}}\underline{/60^\circ}\right)\left(\frac{5.66}{\sqrt{2}}\underline{/0^\circ}\right) = 80.0\underline{/60^\circ} = 40.0 + j69.3$$

$$\mathbf{S}_T = \mathbf{S}_1 + \mathbf{S}_2 = 126.6 + j119.3 = 174.0\underline{/43.3^\circ} \quad \text{VA}$$

The power triangles and their summation are shown in Fig. 9-14.

9.12 Determine the total power information for three parallel-connected loads: load #1, 250 VA, pf = 0.50 lagging; load #2, 180 W, pf = 0.80 leading; load #3, 300 VA, 100 var (inductive).

Calculate the average power P and the reactive power Q for each load.

Load #1 Given $S = 250$ VA, $\cos\theta = 0.50$ lagging. Then,

$$P = 250(0.50) = 125 \text{ W} \qquad Q = \sqrt{(250)^2 - (125)^2} = 216.5 \text{ var (inductive)}$$

Load #2 Given $P = 180$ W, $\cos\theta = 0.80$ leading. Then, $\theta = \cos^{-1} 0.80 = -36.87^\circ$ and

$$Q = 180 \tan(-36.87^\circ) = 135 \text{ var (capacitive)}$$

Load #3 Given $S = 300$ VA, $Q = 100$ var (inductive). Then,

$$P = \sqrt{(300)^2 - (100)^2} = 282.8 \text{ W}$$

Combining componentwise:

$$P_T = 125 + 180 + 282.8 = 587.8 \text{ W}$$
$$Q_T = 216.5 - 135 + 100 = 181.5 \text{ var (inductive)}$$
$$\mathbf{S}_T = 587.8 + j181.5 = 615.2\underline{/17.16^\circ}$$

Therefore, $S_T = 615.2$ VA and pf = cos 17.16° = 0.955 lagging.

9.13 Obtain the complete power triangle and the total current for the parallel circuit shown in Fig. 9-15, if for branch 2, $S_2 = 1490$ VA.

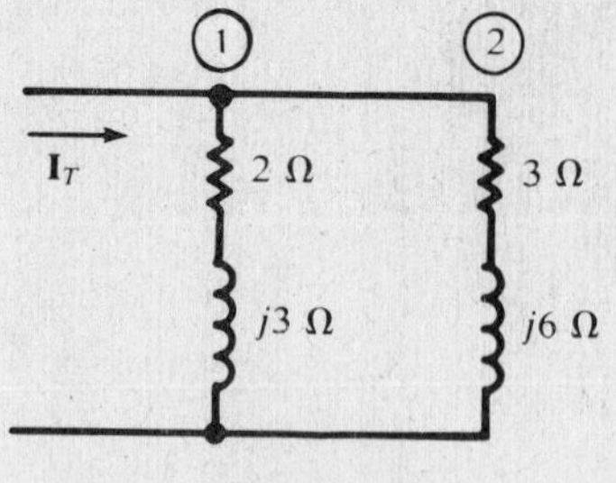

Fig. 9-15

From $S_2 = I_{2\text{eff}}^2 Z_2$,

$$I_{2\text{eff}}^2 = \frac{1490}{\sqrt{3^2 + 6^2}} = 222 \text{ A}^2$$

and, by current division,

$$\frac{\mathbf{I}_1}{\mathbf{I}_2} = \frac{3 + j6}{2 + j3} \qquad \text{whence} \qquad I_{1\text{eff}}^2 = \frac{3^2 + 6^2}{2^2 + 3^2} I_{2\text{eff}}^2 = \frac{45}{13}(222) = 768 \text{ A}^2$$

Then,

$$\mathbf{S}_1 = I_{1\text{eff}}^2\mathbf{Z}_1 = 768(2 + j3) = 1536 + j2304$$
$$\mathbf{S}_2 = I_{2\text{eff}}^2\mathbf{Z}_2 = 222(3 + j6) = 666 + j1332$$
$$\mathbf{S}_T = \mathbf{S}_1 + \mathbf{S}_2 = 2202 + j3636$$

that is, $P_T = 2202$ W, $Q_T = 3636$ var (inductive),

$$S_T = \sqrt{(2202)^2 + (3636)^2} = 4251 \text{ VA} \qquad \text{and} \qquad \text{pf} = \frac{2202}{4251} = 0.518 \text{ lagging}$$

Since the phase angle of the voltage is unknown, only the magnitude of $\mathbf{I}_T$ can be given. By current division,

$$\mathbf{I}_2 = \frac{2 + j3}{5 + j9}\mathbf{I}_T \qquad \text{or} \qquad I_{2\text{eff}}^2 = \frac{2^2 + 3^2}{5^2 + 9^2} I_{T\text{eff}}^2 = \frac{13}{106} I_{T\text{eff}}^2$$

and so

$$I_{T\text{eff}}^2 = \frac{106}{13}(222) = 1811 \text{ A}^2 \qquad \text{or} \qquad I_{T\text{eff}} = 42.6 \text{ A}$$

9.14 Obtain the complete power triangle for the circuit shown in Fig. 9-16, if the total reactive power is 2500 var (inductive). Find the branch powers P_1 and P_2.

The equivalent admittance allows the calculation of the total power triangle.

$$\mathbf{Y}_{eq} = \mathbf{Y}_1 + \mathbf{Y}_2 = 0.2488\underline{/-39.57^\circ} \text{ S}$$

Then,

$$P_T = 2500 \cot 39.57^\circ = 3025 \text{ W} \qquad \mathbf{S}_T = 3025 + j2500 = 3924\underline{/39.57^\circ} \text{ VA}$$

and pf $= P_T/S_T = 0.771$ lagging.

The current ratio is $I_1/I_2 = Y_1/Y_2 = 0.177/0.0745$.

$$\frac{P_1}{P_2} = \frac{I_1^2(4)}{I_2^2(12)} = 1.88 \qquad \text{and} \qquad P_1 + P_2 = 3025 \text{ W}$$

from which $P_1 = 1975$ W and $P_2 = 1050$ W.

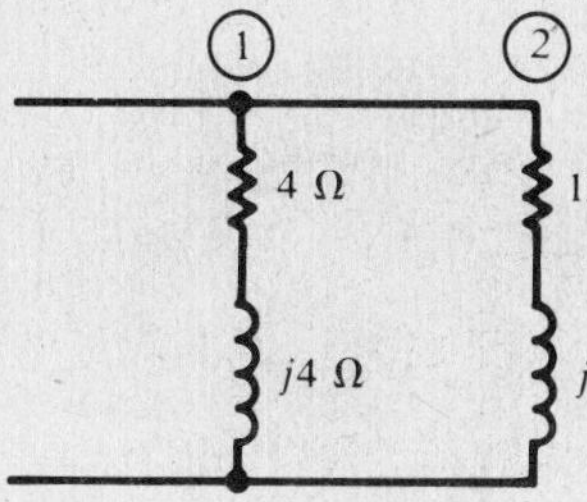

Fig. 9-16

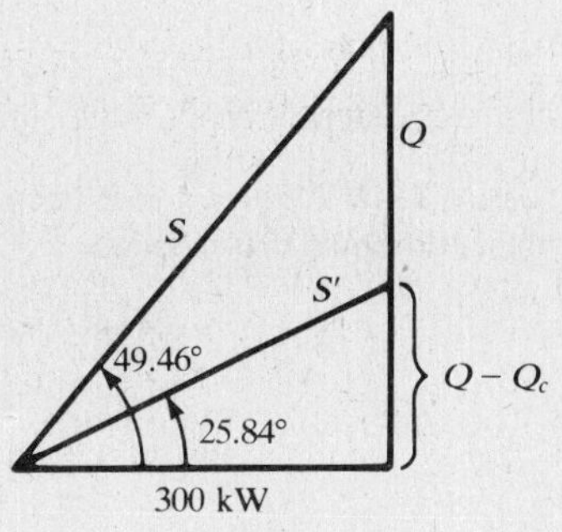

Fig. 9-17

9.15 A load of 300 kW, with power factor 0.65 lagging, has the power factor improved to 0.90 lagging by parallel capacitors. How many kvar must these capacitors furnish, and what is the resulting percent reduction in apparent power?

The angles corresponding to the power factors are first obtained:

$$\cos^{-1} 0.65 = 49.46^\circ \qquad \cos^{-1} 0.90 = 25.84^\circ$$

Then (see Fig. 9-17),

$$Q = 300 \tan 49.46^\circ = 350.7 \text{ kvar (inductive)}$$
$$Q - Q_c = 300 \tan 25.84^\circ = 145.3 \text{ kvar (inductive)}$$

whence $Q_c = 205.4$ kvar (capacitive). Since

$$S = \frac{300}{0.65} = 461.5 \text{ kVA} \qquad S' = \frac{300}{0.90} = 333.3 \text{ kVA}$$

the reduction is

$$\frac{461.5 - 333.3}{461.5}(100\%) = 27.8\%$$

9.16 Find the capacitance C necessary to improve the power factor to 0.95 lagging in the circuit shown in Fig. 9-18, if the effective voltage of 120 V has a frequency of 60 Hz.

Admittance provides a good approach.

$$Y_{eq} = j\omega C + \frac{1}{20\underline{/30^\circ}} = 0.0433 - j(0.0250 - \omega C) \quad \text{(S)}$$

The admittance diagram, Fig. 9-19, illustrates the next step.

$$\begin{aligned} \theta &= \cos^{-1} 0.95 = 18.19^\circ \\ 0.0250 - \omega C &= (0.0433)(\tan 18.19^\circ) \\ \omega C &= 0.0108 \\ C &= 28.6\ \mu\text{F} \end{aligned}$$

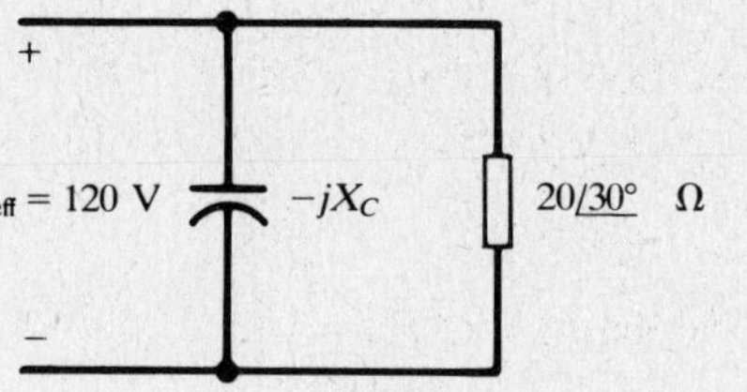

Fig. 9-18

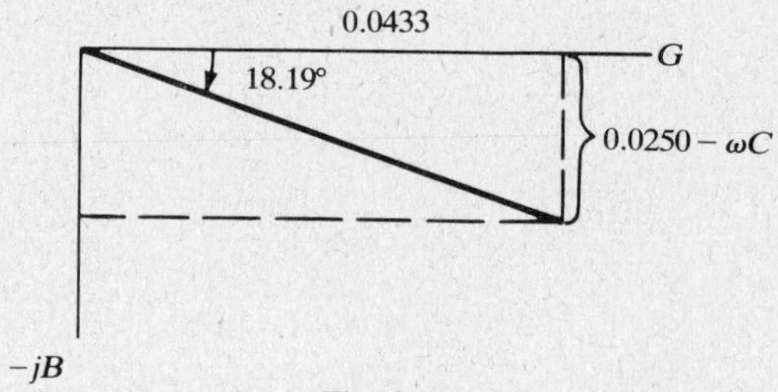

Fig. 9-19

9.17 A circuit with impedance $\mathbf{Z} = 10.0\underline{/60^\circ}\ \Omega$ has the power factor improved by a parallel capacitive reactance of 20 Ω. What percent reduction in current results?

Since $\mathbf{I} = \mathbf{VY}$, the current reduction can be obtained from the ratio of the admittances after and before addition of the capacitors.

$$\begin{aligned} \mathbf{Y}_{\text{before}} &= 0.100\underline{/-60^\circ} \quad \text{S} \\ \mathbf{Y}_{\text{after}} &= 0.050\underline{/90^\circ} + 0.100\underline{/-60^\circ} = 0.062\underline{/-36.20^\circ} \quad \text{S} \\ \frac{I_{\text{after}}}{I_{\text{before}}} &= \frac{0.062}{0.100} = 0.620 \end{aligned}$$

so the reduction is 38%.

9.18 A transformer rated at a maximum of 25 kVA supplies a 12-kW load at power factor 0.60 lagging. What percent of the transformer rating does this load represent? How many kW in additional load may be added at unity power factor before the transformer exceeds its rated kVA?

For the 12-kW load, $S = 12/0.60 = 20$ kVA. The transformer is at $(20/25)(100\%) = 80\%$ of full rating.

The additional load at unity power factor does not change the reactive power,

$$Q = \sqrt{(20)^2 - (12)^2} = 16 \text{ kvar (inductive)}$$

Then, at full capacity,

$$\theta' = \sin^{-1}(16/25) = 39.79^\circ$$
$$P' = 25 \cos 39.79^\circ = 19.2 \text{ kW}$$
$$P_{add} = 19.2 - 12.0 = 7.2 \text{ kW}$$

Note that the full-rated kVA is shown by an arc in Fig. 9-20, of radius 25.

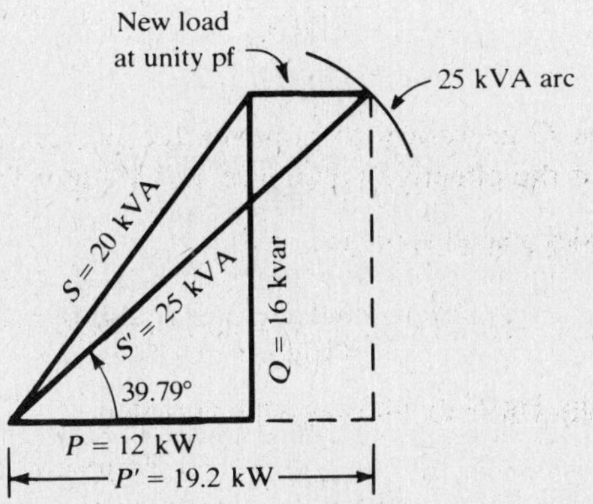

Fig. 9-20

9.19 Referring to Problem 9.18, if the additional load has power factor 0.866 leading, how many kVA may be added without exceeding the transformer rating?

The original load is $\mathbf{S} = 12 + j16$ kVA and the added load is

$$\mathbf{S}_2 = S_2\underline{/-30^\circ} = S_2(0.866) - jS_2(0.500) \quad \text{(kVA)}$$

The total is $\mathbf{S}_T = (12 + 0.866\,S_2) + j(16 - 0.500\,S_2)$ (kVA). Then,

$$S_T^2 = (12 + 0.866\,S_2)^2 + (16 - 0.500\,S_2)^2 = (25)^2$$

gives $S_2 = 12.8$ kVA.

9.20 An induction motor with a shaft power output of 1.5 kW has an efficiency of 85%. At this load, the power factor is 0.80 lagging. Give complete input power information.

$$\frac{P_{out}}{P_{in}} = 0.85 \qquad \text{or} \qquad P_{in} = \frac{1.5}{0.85} = 1.765 \text{ kW}$$

Then, from the power triangle,

$$S_{in} = \frac{1.765}{0.80} = 2.206 \text{ kVA} \qquad Q_{in} = \sqrt{(2.206)^2 - (1.765)^2} = 1.324 \text{ kvar (inductive)}$$

The equivalent circuit of an induction motor contains a variable resistance which is a function of the shaft load. The power factor is therefore variable, ranging from values near 0.30 at starting to 0.85 at full load.

Supplementary Problems

9.21 Given a circuit with an applied voltage $v = 14.14 \cos \omega t$ (V) and a resulting current $i = 17.1 \cos(\omega t - 14.05^\circ)$ (mA), determine the complete power triangle.
Ans. $P = 117$ mW, $Q = 29.3$ mvar (inductive), pf = 0.970 lagging

9.22 Given a circuit with an applied voltage $v = 340 \sin(\omega t - 60^\circ)$ (V) and a resulting current $i = 13.3 \sin(\omega t - 48.7^\circ)$ (A), determine the complete power triangle.
Ans. $P = 2217$ W, $Q = 443$ var (capacitive), pf = 0.981 leading

9.23 A two-element series circuit with $R = 5.0\ \Omega$ and $X_L = 15.0\ \Omega$, has an effective voltage 31.6 V across the resistance. Find the complex power and the power factor. *Ans.* $200 + j600$ VA, 0.316 lagging

9.24 A circuit with impedance $\mathbf{Z} = 8.0 - j6.0\ \Omega$ has an applied phasor voltage $70.7\underline{/-90.0^\circ}$ V. Obtain the complete power triangle. *Ans.* $P = 200$ W, $Q = 150$ var (capacitive), pf = 0.80 leading

9.25 Determine the circuit impedance which has a complex power $\mathbf{S} = 5031\underline{/-26.57^\circ}$ VA for an applied phasor voltage $212.1\underline{/0^\circ}$ V. *Ans.* $4.0 - j2.0\ \Omega$

9.26 Determine the impedance corresponding to apparent power 3500 VA, power factor 0.76 lagging, and effective current 18.0 A. *Ans.* $10.8\underline{/40.54^\circ}\ \Omega$

9.27 A two-branch parallel circuit, with $\mathbf{Z}_1 = 10\underline{/0^\circ}\ \Omega$ and $\mathbf{Z}_2 = 8.0\underline{/-30.0^\circ}\ \Omega$, has a total current $i = 7.07 \cos(\omega t - 90^\circ)$ (A). Obtain the complete power triangle.
Ans. $P = 110$ W, $Q = 32.9$ var (capacitive), pf = 0.958 leading

9.28 A two-branch parallel circuit has branch impedances $\mathbf{Z}_1 = 2.0 - j5.0\ \Omega$ and $\mathbf{Z}_2 = 1.0 + j1.0\ \Omega$. Obtain the complete power triangle for the circuit if the 2.0-Ω resistor consumes 20 W.
Ans. $P = 165$ W, $Q = 95$ var (inductive), pf = 0.867 lagging

9.29 A two-branch parallel circuit, with impedances $\mathbf{Z}_1 = 4.0\underline{/-30^\circ}\ \Omega$ and $\mathbf{Z}_2 = 5.0\underline{/60^\circ}\ \Omega$, has an applied effective voltage of 20 V. Obtain the power triangles for the branches and combine them to obtain the total power triangle. *Ans.* $S_T = 128.1$ VA, pf = 0.989 lagging

9.30 Obtain the complex power for the complete circuit of Fig. 9-21 if branch 1 takes 8.0 kvar.
Ans. $\mathbf{S} = 8 + j12$ kVA, pf = 0.555 lagging

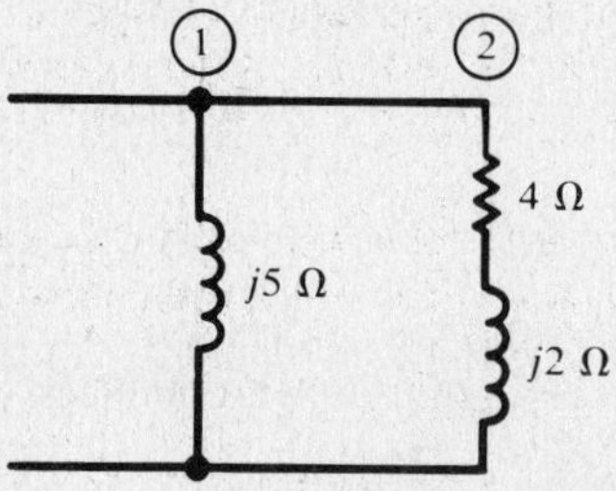

Fig. 9-21

9.31 In the circuit of Fig. 9-22, find **Z** if $S_T = 3373$ VA, pf = 0.938 leading, and the 3-Ω resistor has an average power of 666 W. *Ans.* $2 - j2\ \Omega$

9.32 The parallel circuit in Fig. 9-23 has a total average power of 1500 W. Obtain the total power-triangle information. *Ans.* $\mathbf{S} = 1500 + j2471$ VA, pf = 0.519 lagging

9.33 Determine the average power in the 15-Ω and 8-Ω resistances in Fig. 9-24, if the total average power in the circuit is 2000 W. *Ans.* 723 W, 1277 W

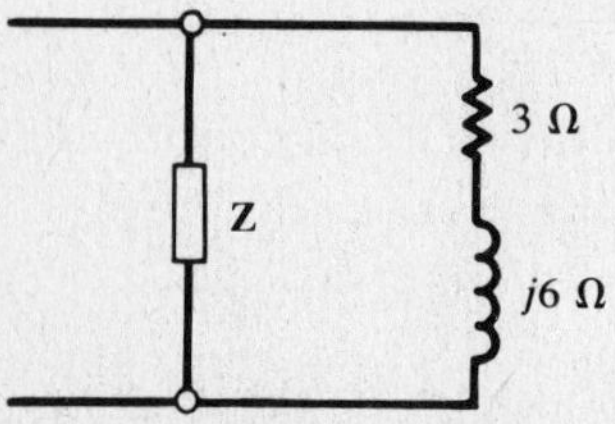

Fig. 9-22

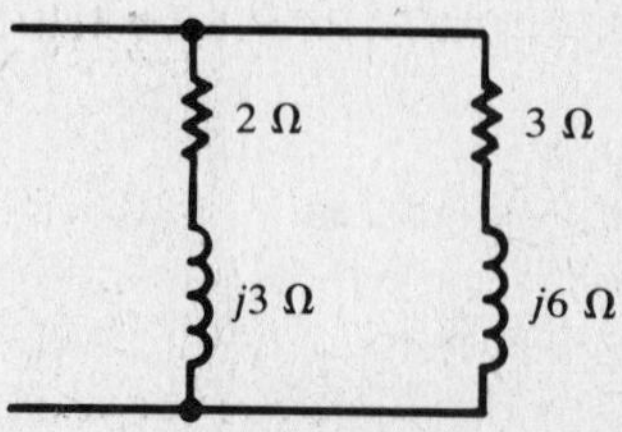

Fig. 9-23

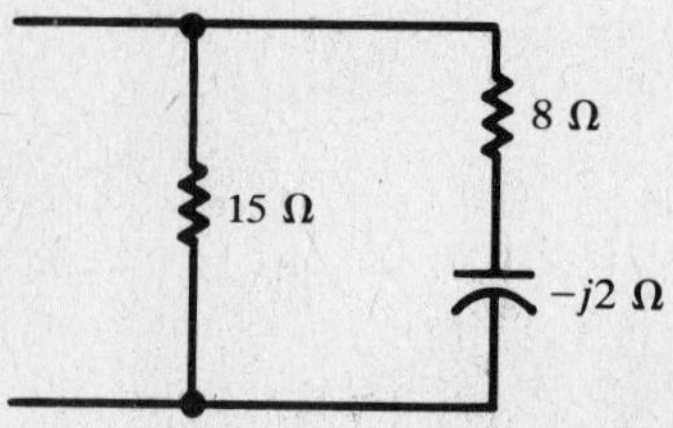

Fig. 9-24

9.34 A three-branch parallel circuit, with $\mathbf{Z}_1 = 25\angle 15^\circ\ \Omega$, $\mathbf{Z}_2 = 15\angle -60^\circ\ \Omega$, and $\mathbf{Z}_3 = 15\angle 90^\circ\ \Omega$, has an applied voltage $\mathbf{V} = 339.4\angle -30^\circ$ V. Obtain the total apparent power and the overall power factor.
Ans. 4291 VA, 0.966 lagging

9.35 Obtain the complete power triangle for the following parallel-connected loads: load #1, 5 kW, pf = 0.80 lagging; load #2, 4 kVA, 2 kvar (capacitive); load #3, 6 kVA, pf = 0.90 lagging.
Ans. 14.535 kVA, pf = 0.954 lagging

9.36 Obtain the complete power triangle for the following parallel-connected loads: load #1, 200 VA, pf = 0.70 lagging; load #2, 350 VA, pf = 0.50 lagging; load #3, 275 VA, pf = 1.00.
Ans. $\mathbf{S} = 590 + j444$ VA, pf = 0.799 lagging

9.37 A 4500-VA load at power factor 0.75 lagging is supplied by a 60-Hz source at effective voltage 240 V. Determine the parallel capacitance in microfarads necessary to improve the power factor to (*a*) 0.90 lagging, (*b*) 0.90 leading. *Ans.* (*a*) 61.8 μF; (*b*) 212 μF

9.38 In Problem 9.37, what percent reduction in line current and total voltamperes was achieved in part (*a*)? What further reduction was achieved in part (*b*)? *Ans.* 16.1%, none

9.39 The addition of a 20-kvar capacitor bank improved the power factor of a certain load to 0.90 lagging. Determine the complex power before the addition of the capacitors, if the final apparent power is 185 kVA. *Ans.* $\mathbf{S} = 166.5 + j100.6$ kVA

9.40 A 25-kVA load with power factor 0.80 lagging has a group of resistive heating units added at unity power factor. How many kW do these units take, if the new overall power factor is 0.85 lagging?
Ans. 4.2 kW

9.41 A 500-kVA transformer is at full load and 0.60 lagging power factor. A capacitor bank is added, improving the power factor to 0.90 lagging. After improvement, what percent of rated kVA is the transformer carrying? *Ans.* 66.7%

9.42 A 100-kVA transformer is at 80% of rated load at power factor 0.85 lagging. How many kVA in additional load at 0.60 lagging power factor will bring the transformer to full rated load?
Ans. 21.2 kVA

9.43 A 250-kVA transformer is at full load with power factor 0.80 lagging. (*a*) How many kvar of capacitors must be added to improve this power factor to 0.90 lagging? (*b*) After improvement of the power factor, a new load is to be added at 0.50 lagging power factor. How many kVA of this new load will bring the transformer back to rated kVA, and what is the final power factor?
Ans. (*a*) 53.1 kvar (capacitive); (*b*) 33.35 kVA, 0.867 lagging

9.44 A 65-kVA load with a lagging power factor is combined with a 25-kVA synchronous motor load which operates at pf = 0.60 leading. Find the power factor of the 65-kVA load, if the overall power factor is 0.85 lagging. *Ans.* 0.585 lagging

9.45 An induction motor load of 2000 kVA has power factor 0.80 lagging. Synchronous motors totaling 500 kVA are added and operated at a leading power factor. If the overall power factor is then 0.90 lagging, what is the power factor of the synchronous motors? *Ans.* 0.92 leading

Chapter 10

Polyphase Circuits

10.1 THREE-PHASE VOLTAGES

The nature of a three-phase system of voltages is best illustrated by considering a method by which the voltages could be generated. In Fig. 10-1, three coils are equally distributed about the circumference of the rotor; i.e. the coils are displaced from one another by 120 mechanical degrees. Coil ends and slip rings are not shown; however, it is evident that counterclockwise rotation results in the coil sides A, B, and C passing under the pole pieces in the order . . . A-B-C-A-B-C. . . . Voltage polarities reverse for each change of pole. Assuming that the pole shape and corresponding magnetic flux density are such that the induced voltages are sinusoidal, the result for the three coils is as shown in Fig. 10-2. Voltage B is 120 electrical degrees later than A, and C is 240° later. This is referred to as the *ABC sequence*. Changing the direction of rotation would result in . . . A-C-B-A-C-B . . . , which is called the *CBA sequence*.

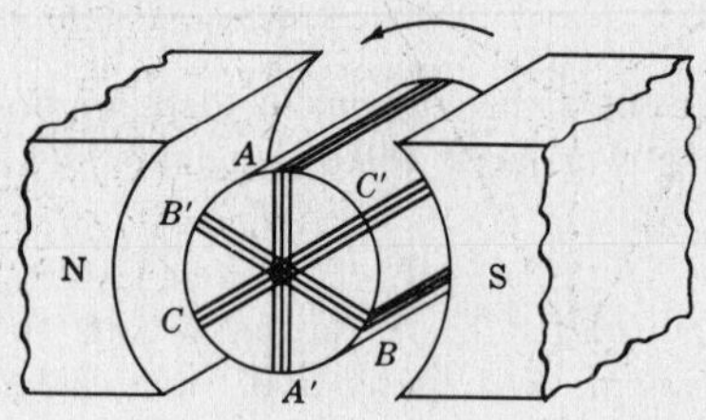

Fig. 10-1

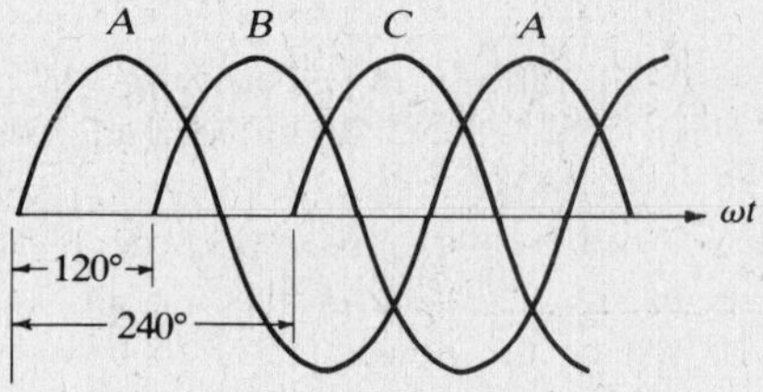

Fig. 10-2

10.2 WYE AND DELTA SYSTEMS

The ends of the coils can be connected in wye (also designated Y; see Section 8.5), with ends A', B', and C' joined at a common point designated the *neutral*, N; and with ends A, B, and C brought out to become the *lines* A, B, and C of the three-phase system. If the neutral point is carried along with the lines, it is a *three-phase, four-wire* system. In Fig. 10-3, the lines are designated by lowercase a, b, and c at the supply, which could either be a transformer bank or a three-phase alternator, and by uppercase A, B, and C at the load. If line impedances must be considered, then the current direction through, e.g., line aA would be $\mathbf{I}_{aA}$, and the phasor line voltage drop $\mathbf{V}_{aA}$.

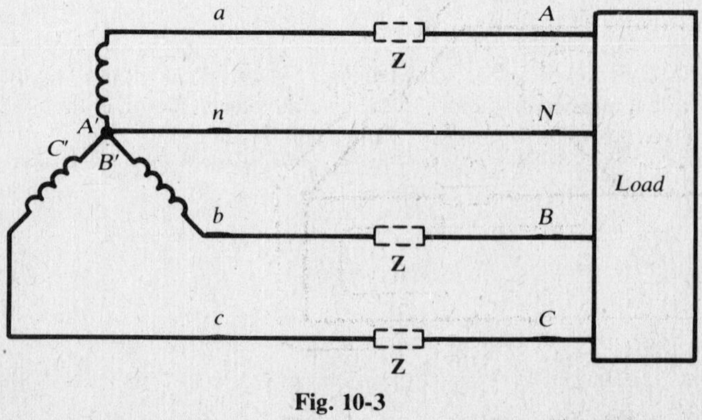

Fig. 10-3

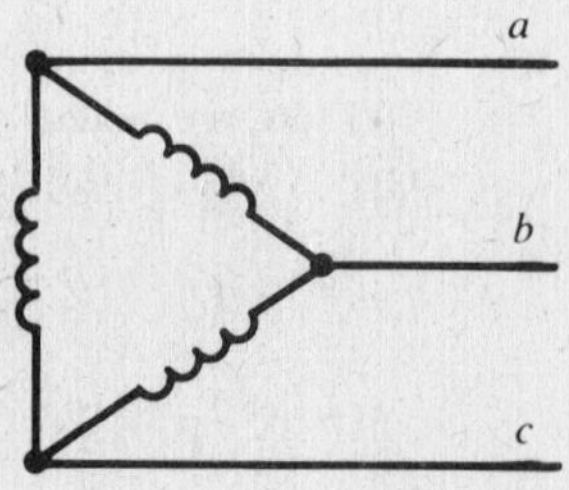

Fig. 10-4

The coil ends can be connected as shown in Fig. 10-4, making a delta-connected (or Δ-connected), three-phase system with lines *a*, *b*, and *c*. A delta-connected set of coils has no neutral point to produce a four-wire system, except through the use of special transformers.

10.3 PHASOR VOLTAGES

The selection of a phase angle for one voltage in a three-phase system fixes the angles of all other voltages. This is tantamount to fixing the $t = 0$ point on the horizontal axis of Fig. 10-2, which can be done quite arbitrarily. In this chapter, an angle of zero will always be associated with the phasor voltage of line B with respect to line C: $\mathbf{V}_{BC} \equiv V_L\underline{/0°}$.

It is shown in Problem 10.1 that the line-to-line voltage V_L is $\sqrt{3}$ times the line-to-neutral voltage. All ABC-sequence voltages are shown in Fig. 10-5(*a*), and CBA voltages in Fig. 10-5(*b*). These phasor voltages, in keeping with the previous chapters, reflect maximum values. In the three-phase, four-wire, 480-volt system, widely used for industrial loads, and the 208-volt system, common in commercial buildings, effective values are specified. In this chapter, a line-to-line voltage in the former system would be $\mathbf{V}_{BC} = 678.8\underline{/0°}$ V, making $V_{BC\text{eff}} = 678.8/\sqrt{2} = 480$ V. People who regularly work in this field use effective-valued phasors, and would write $\mathbf{V}_{BC} = 480\underline{/0°}$ V.

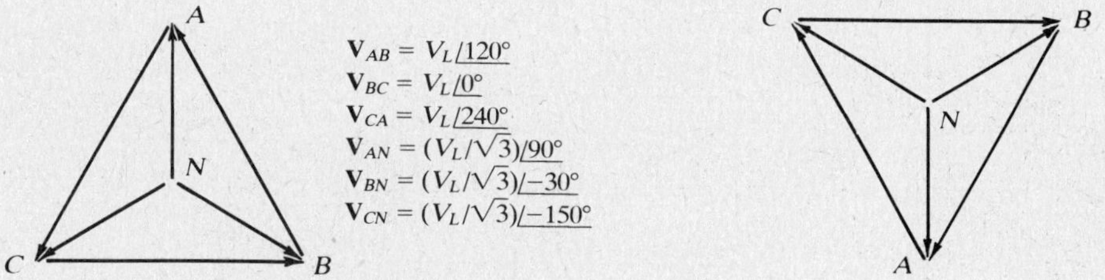

$$\mathbf{V}_{AB} = V_L\underline{/120°} \quad \mathbf{V}_{BC} = V_L\underline{/0°} \quad \mathbf{V}_{CA} = V_L\underline{/240°}$$
$$\mathbf{V}_{AN} = (V_L/\sqrt{3})\underline{/90°} \quad \mathbf{V}_{BN} = (V_L/\sqrt{3})\underline{/-30°} \quad \mathbf{V}_{CN} = (V_L/\sqrt{3})\underline{/-150°}$$

(*a*) Sequence *ABC*

$$\mathbf{V}_{AB} = V_L\underline{/240°} \quad \mathbf{V}_{BC} = V_L\underline{/0°} \quad \mathbf{V}_{CA} = V_L\underline{/120°}$$
$$\mathbf{V}_{AN} = (V_L/\sqrt{3})\underline{/-90°} \quad \mathbf{V}_{BN} = (V_L/\sqrt{3})\underline{/30°} \quad \mathbf{V}_{CN} = (V_L/\sqrt{3})\underline{/150°}$$

(*b*) Sequence *CBA*

Fig. 10-5

10.4 BALANCED DELTA-CONNECTED LOAD

Three identical impedances connected as shown in Fig. 10-6 make up a balanced Δ-connected load. The currents in the impedances are referred to either as *phase currents* or *load currents*, and the three will be equal in magnitude and mutually displaced in phase by 120°. The line currents will also be equal in magnitude and displaced from one another by 120°; by convention, they are given a direction from the source to the load.

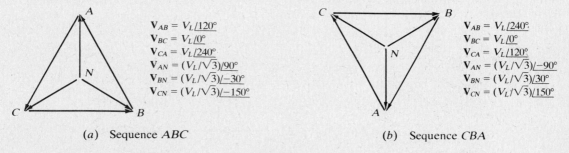

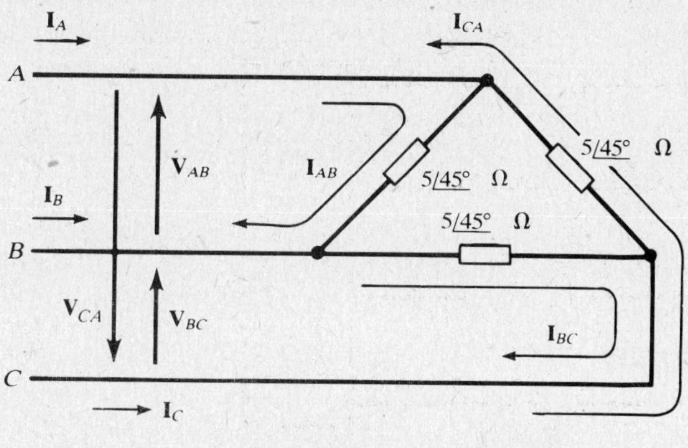

Fig. 10-6

EXAMPLE 10.1 A three-phase, three-wire, *ABC* system, with an effective line voltage of 120 V, has three impedances of $5.0\underline{/45^\circ}$ Ω in a Δ-connection. Determine the line currents and draw the voltage-current phasor diagram.

The maximum line voltage is $120\sqrt{2} = 169.7$ V. Referring to Fig. 10-5(*a*), the voltages are:

$$\mathbf{V}_{AB} = 169.7\underline{/120^\circ}\ \text{V} \qquad \mathbf{V}_{BC} = 169.7\underline{/0^\circ}\ \text{V} \qquad \mathbf{V}_{CA} = 169.7\underline{/240^\circ}\ \text{V}$$

Double subscripts give the phase-current directions; for example, $\mathbf{I}_{AB}$ passes through the impedance from line *A* to line *B*. All current directions are shown in Fig. 10-6. Then the phase currents are:

$$\mathbf{I}_{AB} = \frac{\mathbf{V}_{AB}}{\mathbf{Z}} = \frac{169.7\underline{/120^\circ}}{5\underline{/45^\circ}} = 33.9\underline{/75^\circ}\ \text{A}$$

$$\mathbf{I}_{BC} = \frac{\mathbf{V}_{BC}}{\mathbf{Z}} = \frac{169.7\underline{/0^\circ}}{5\underline{/45^\circ}} = 33.9\underline{/-45^\circ}\ \text{A}$$

$$\mathbf{I}_{CA} = \frac{\mathbf{V}_{CA}}{\mathbf{Z}} = \frac{169.7\underline{/240^\circ}}{5\underline{/45^\circ}} = 33.9\underline{/195^\circ}\ \text{A}$$

By KCL, line current $\mathbf{I}_A$ is given by

$$\mathbf{I}_A = \mathbf{I}_{AB} + \mathbf{I}_{AC} = 33.9\underline{/75^\circ} - 33.9\underline{/195^\circ} = 58.7\underline{/45^\circ}\ \text{A}$$

Similarly, $\mathbf{I}_B = 58.7\underline{/-75^\circ}$ A and $\mathbf{I}_C = 58.7\underline{/165^\circ}$ A.

The line-to-line voltages and all currents are shown on the phasor diagram, Fig. 10-7. Note particularly the balanced currents. After one phase current has been computed, all other currents may be obtained through the symmetry of the phasor diagram. Note also that $33.9 \times \sqrt{3} = 58.7$; that is, $I_L = \sqrt{3}\, I_{Ph}$ for a balanced delta load.

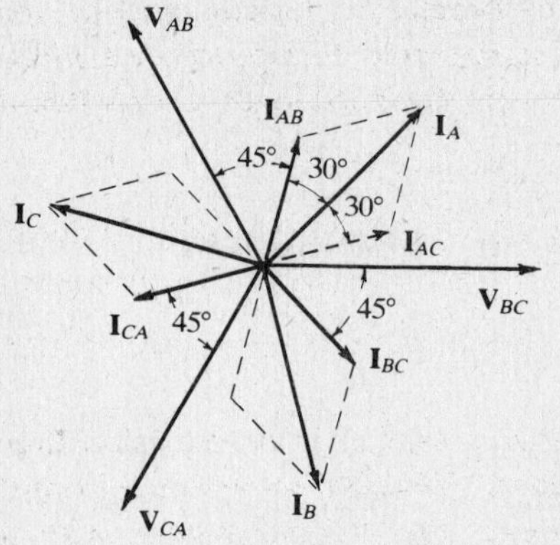

Fig. 10-7

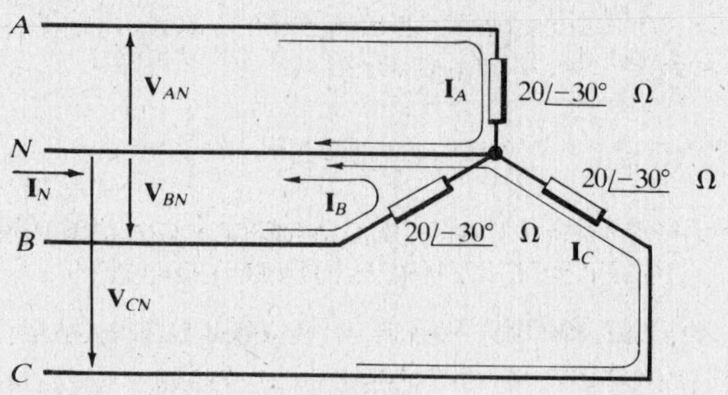

Fig. 10-8

10.5 BALANCED FOUR-WIRE, WYE-CONNECTED LOAD

Three identical impedances connected as shown in Fig. 10-8 make up a balanced Y-connected load. The currents in the impedances are also the line currents; so the directions are chosen from the source to the load, as before.

EXAMPLE 10.2 A three-phase, four-wire, *CBA* system, with an effective line voltage of 120 V, has three impedances of $20\underline{/-30^\circ}$ Ω in a Y-connection (Fig. 10-8). Determine the line currents and draw the voltage-current phasor diagram.

The maximum line voltage is 169.7 V, and the line-to-neutral magnitude, $169.7/\sqrt{3} = 98.0$ V. From Fig. 10-5(*b*),

$$\mathbf{V}_{AN} = 98.0\underline{/-90^\circ}\ \text{V} \qquad \mathbf{V}_{BN} = 98.0\underline{/30^\circ}\ \text{V} \qquad \mathbf{V}_{CN} = 98.0\underline{/150^\circ}\ \text{V}$$

Then

$$\mathbf{I}_A = \frac{\mathbf{V}_{AN}}{\mathbf{Z}} = \frac{98.0\underline{/-90^\circ}}{20\underline{/-30^\circ}} = 4.90\underline{/-60^\circ}\ \text{A}$$

and, similarly, $\mathbf{I}_B = 4.90\underline{/60^\circ}$ A, $\mathbf{I}_C = 4.90\underline{/180^\circ}$ A.

The voltage-current phasor diagram is shown in Fig. 10-9. Note that with one line current calculated, the other two can be obtained through the symmetry of the phasor diagram. All three line currents return through the neutral. Therefore, the neutral current is the negative sum of the line currents:

$$\mathbf{I}_N = -(\mathbf{I}_A + \mathbf{I}_B + \mathbf{I}_C) = 0$$

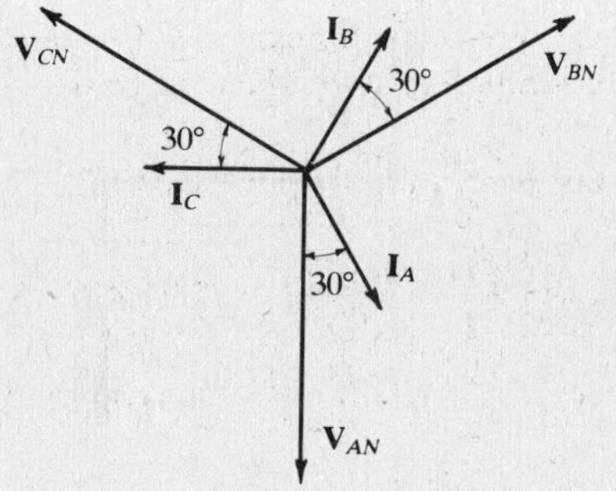

Fig. 10-9

Since the neutral current of a balanced, Y-connected, three-phase load is always zero, the neutral conductor may, for computational purposes, be removed, with no change in the results. In actual power circuits, it must not be physically removed, since it carries the (small) unbalance of the currents, carries short-circuit or fault currents for operation of protective devices, and prevents overvoltages on the phases of the load. Since the computation in Example 10.2 proceeded without difficulty, the neutral will be included when calculating line currents in balanced loads, even when the system is actually three-wire.

10.6 SINGLE-LINE EQUIVALENT CIRCUIT FOR BALANCED THREE-PHASE LOADS

Figure 10-10(*a*) shows a balanced Y-connected load. In many cases, e.g. in power calculations, only the common magnitude, I_L, of the three line currents is needed. This may be obtained from the *single-line equivalent*, Fig. 10-10(*b*), which represents one phase of the original system, with the line-to-neutral voltage arbitrarily given a zero phase angle. This makes $\mathbf{I}_L = I_L\underline{/-\theta}$, where θ is the impedance angle. If the actual line currents $\mathbf{I}_A$, $\mathbf{I}_B$, and $\mathbf{I}_C$ are desired, their phase angles may be found by adding $-\theta$ to the phase angles of $\mathbf{V}_{AN}$, $\mathbf{V}_{BN}$, and $\mathbf{V}_{CN}$ as given in Fig. 10-5. Observe that the angle on $\mathbf{I}_L$ gives the power factor for each phase, pf = cos θ.

The method may be applied to a balanced Δ-connected load if the load is replaced by its Y-equivalent, where $\mathbf{Z}_Y = \frac{1}{3}\mathbf{Z}_\Delta$ (Section 8.5).

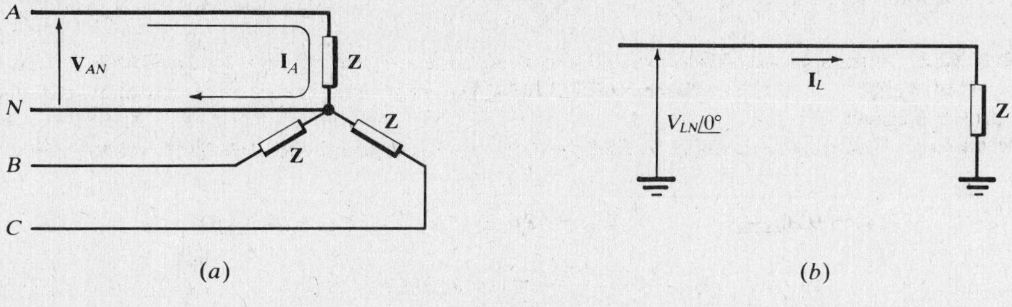

Fig. 10-10

EXAMPLE 10.3 Rework Example 10.2 by the single-line equivalent method.

Referring to Fig. 10-11 (in which the symbol Y indicates the type of connection of the original load),

$$\mathbf{I}_L = \frac{\mathbf{V}_{LN}}{\mathbf{Z}} = \frac{98.0\angle 0^\circ}{20\angle -30^\circ} = 4.90\angle 30^\circ \quad \text{A}$$

From Fig. 10-5(*b*), the phase angles of $\mathbf{V}_{AN}$, $\mathbf{V}_{BN}$, and $\mathbf{V}_{CN}$ are −90°, 30°, and 150°. Hence,

$$\mathbf{I}_A = 4.90\angle -60^\circ \quad \text{A} \qquad \mathbf{I}_B = 4.90\angle 60^\circ \quad \text{A} \qquad \mathbf{I}_C = 4.90\angle 180^\circ \quad \text{A}$$

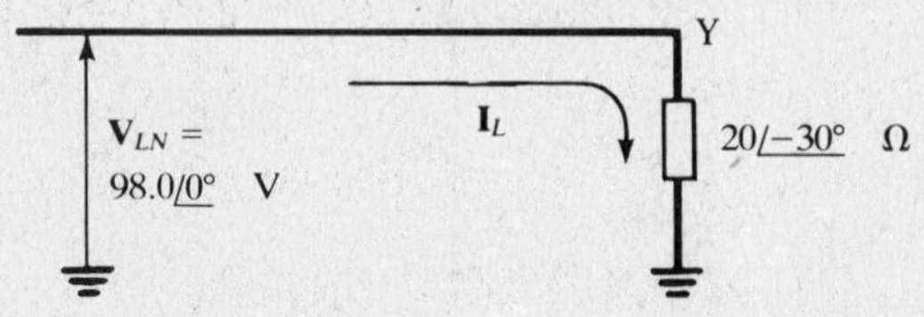

Fig. 10-11

10.7 UNBALANCED DELTA-CONNECTED LOAD

The solution of the unbalanced delta-connected load consists in computing the phase currents and then applying KCL to obtain the line currents. The currents will be unequal and will not have the symmetry of the balanced case.

EXAMPLE 10.4 A three-phase, 339.4-V, *ABC* system [Fig. 10-12(*a*)] has a Δ-connected load, with

$$\mathbf{Z}_{AB} = 10\angle 0^\circ \quad \Omega \qquad \mathbf{Z}_{BC} = 10\angle 30^\circ \quad \Omega \qquad \mathbf{Z}_{CA} = 15\angle -30^\circ \quad \Omega$$

Obtain phase and line currents and draw the phasor diagram.

$$\mathbf{I}_{AB} = \frac{\mathbf{V}_{AB}}{\mathbf{Z}_{AB}} = \frac{339.4\angle 120^\circ}{10\angle 0^\circ} = 33.94\angle 120^\circ \quad \text{A}$$

Similarly, $\mathbf{I}_{BC} = 33.94\angle -30^\circ$ A and $\mathbf{I}_{CA} = 22.63\angle 270^\circ$ A. Then,

$$\mathbf{I}_A = \mathbf{I}_{AB} + \mathbf{I}_{AC} = 33.94\angle 120^\circ - 22.63\angle 270^\circ = 54.72\angle 108.1^\circ \quad \text{A}$$

Also, $\mathbf{I}_B = 65.56\angle -45^\circ$ A and $\mathbf{I}_C = 29.93\angle -169.1^\circ$ A.

The voltage-current phasor diagram is shown in Fig. 10-12(*b*), with magnitudes and angles to scale.

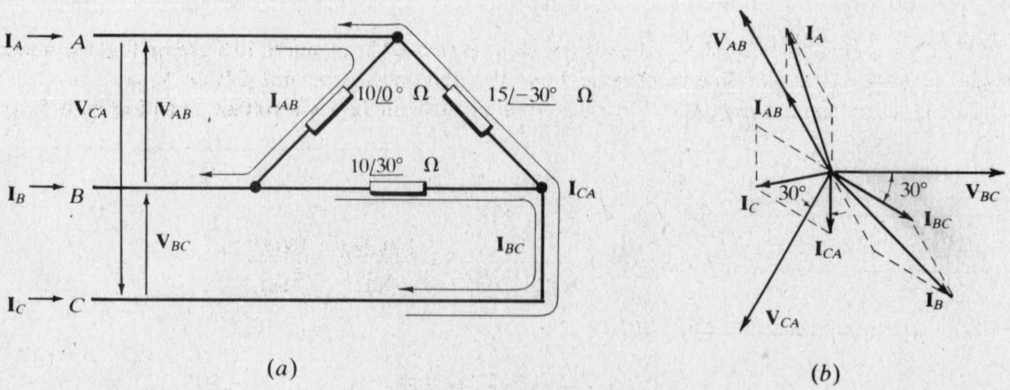

Fig. 10-12

10.8 UNBALANCED WYE-CONNECTED LOAD

Four-Wire

The neutral conductor carries the unbalanced current of a wye-connected load and maintains the line-to-neutral voltage magnitude across each phase of the load. The line currents are unequal and the currents on the phasor diagram have no symmetry.

EXAMPLE 10.5 A three-phase, four-wire, 150-V, *CBA* system has a Y-connected load, with

$$\mathbf{Z}_A = 6\underline{/0^\circ}\ \Omega \qquad \mathbf{Z}_B = 6\underline{/30^\circ}\ \Omega \qquad \mathbf{Z}_C = 5\underline{/45^\circ}\ \Omega$$

Obtain all line currents and draw the phasor diagram.

$$\mathbf{I}_A = \frac{\mathbf{V}_{AN}}{\mathbf{Z}_A} = \frac{86.6\underline{/-90^\circ}}{6\underline{/0^\circ}} = 14.43\underline{/-90^\circ}\ \text{A}$$

$$\mathbf{I}_B = \frac{\mathbf{V}_{BN}}{\mathbf{Z}_B} = \frac{86.6\underline{/30^\circ}}{6\underline{/30^\circ}} = 14.43\underline{/0^\circ}\ \text{A}$$

$$\mathbf{I}_C = \frac{\mathbf{V}_{CN}}{\mathbf{Z}_C} = \frac{86.6\underline{/150^\circ}}{5\underline{/45^\circ}} = 17.32\underline{/105^\circ}\ \text{A}$$

$$\mathbf{I}_N = -(14.43\underline{/-90^\circ} + 14.43\underline{/0^\circ} + 17.32\underline{/105^\circ}) = 10.21\underline{/-167.0^\circ}\ \text{A}$$

Figure 10-13(*b*) gives the phasor diagram.

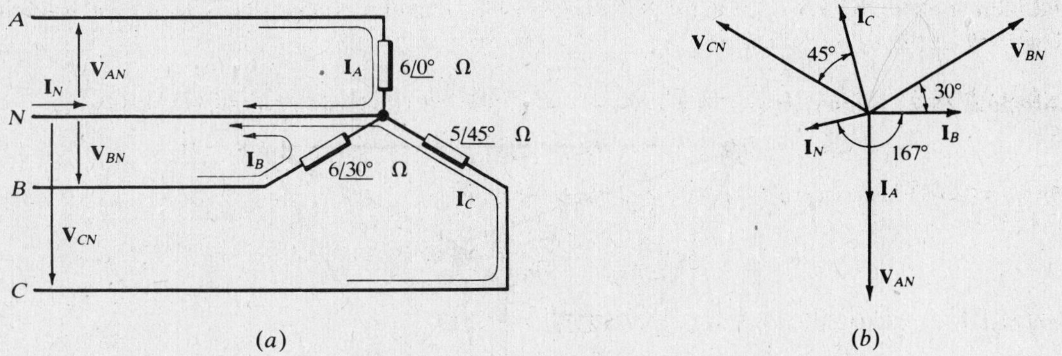

Fig. 10-13

Three-Wire

Without the neutral conductor, the Y-connected impedances will have voltages which vary considerably from the line-to-neutral magnitude.

EXAMPLE 10.6 Figure 10-14(*a*) shows the same system as treated in Example 10.5 except that the neutral wire is no longer present. Obtain the line currents and find the *displacement neutral voltage*, $\mathbf{V}_{ON}$.

The circuit is redrawn in Fig. 10-14(*b*) so as to suggest a single node-voltage equation with $\mathbf{V}_{OB}$ as the unknown.

$$\frac{\mathbf{V}_{OB} - \mathbf{V}_{AB}}{\mathbf{Z}_A} + \frac{\mathbf{V}_{OB}}{\mathbf{Z}_B} + \frac{\mathbf{V}_{OB} + \mathbf{V}_{BC}}{\mathbf{Z}_C} = 0$$

$$\mathbf{V}_{OB}\left(\frac{1}{6\underline{/0^\circ}} + \frac{1}{6\underline{/30^\circ}} + \frac{1}{5\underline{/45^\circ}}\right) = \frac{150\underline{/240^\circ}}{6\underline{/0^\circ}} - \frac{150\underline{/0^\circ}}{5\underline{/45^\circ}}$$

from which $\mathbf{V}_{OB} = 66.76\underline{/-152.85^\circ}$ V. Then,

$$\mathbf{I}_B = -\frac{\mathbf{V}_{OB}}{\mathbf{Z}_B} = 11.13\underline{/-2.85^\circ}\ \text{A}$$

From $\mathbf{V}_{OA} + \mathbf{V}_{AB} = \mathbf{V}_{OB}$, $\mathbf{V}_{OA} = 100.7\underline{/81.08^\circ}$ V, and

$$\mathbf{I}_A = -\frac{\mathbf{V}_{OA}}{\mathbf{Z}_A} = 16.78\underline{/-98.92^\circ} \quad \text{A}$$

Similarly, $\mathbf{V}_{OC} = \mathbf{V}_{OB} - \mathbf{V}_{CB} = 95.58\underline{/-18.58^\circ}$ V, and

$$\mathbf{I}_C = 19.12\underline{/116.4^\circ} \quad \text{A}$$

Point O is displaced from the neutral N by a phasor voltage $\mathbf{V}_{ON}$, given by

$$\mathbf{V}_{ON} = \mathbf{V}_{OA} + \mathbf{V}_{AN} = 100.7\underline{/81.08^\circ} + \frac{150}{\sqrt{3}}\underline{/-90^\circ} = 20.24\underline{/39.53^\circ} \quad \text{V}$$

The phasor diagram, Fig. 10-15, shows the shift of point O from the centroid of the equilateral triangle. See Problem 10.10 for an alternate method.

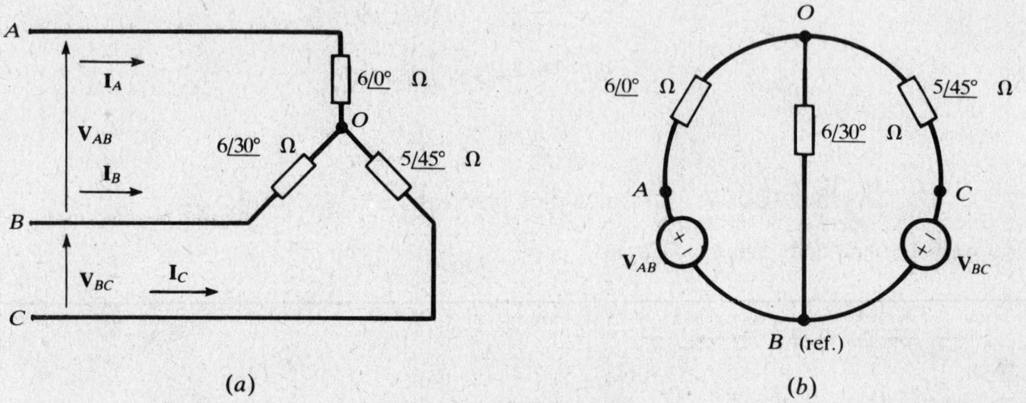

Fig. 10-14

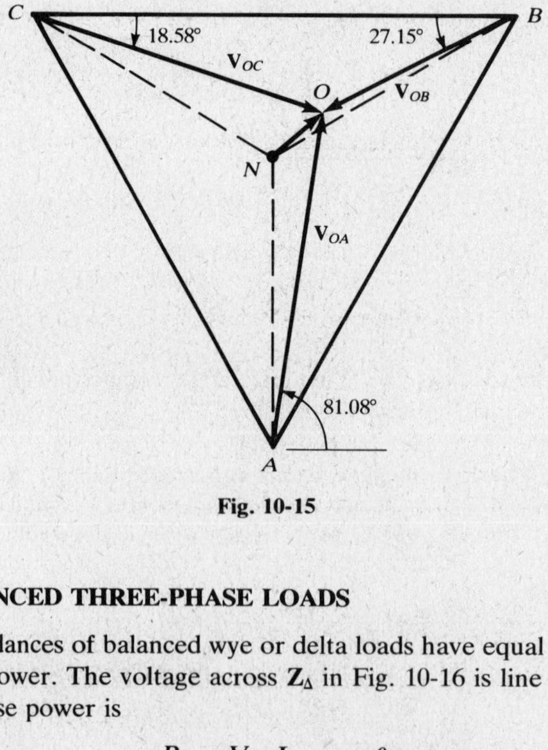

Fig. 10-15

10.9 POWER IN BALANCED THREE-PHASE LOADS

Since the phase impedances of balanced wye or delta loads have equal currents, the phase power is one-third of the total power. The voltage across $\mathbf{Z}_\Delta$ in Fig. 10-16 is line voltage and the current is phase current; so the phase power is

$$P_{Ph} = V_{L\text{eff}} I_{Ph\text{eff}} \cos\theta$$

where θ is the angle on the impedance. But, for balanced loads, $I_{L\text{eff}} = \sqrt{3} I_{Ph_\text{eff}}$; hence

$$P_T = 3P_{Ph} = \sqrt{3} V_{L_\text{eff}} I_{L\text{eff}} \cos\theta$$

which formula is also valid for balanced loads in wye.

The effective line voltage for an industrial or commercial load is always known. If the system is balanced, the total power can be computed if the line current and power factor are known. The corresponding formulas for the total quadrature power and the total apparent power are

$$Q_T = \sqrt{3}V_{L\text{eff}}I_{L\text{eff}} \sin\theta \qquad S_T = \sqrt{3}V_{L\text{eff}}I_{L\text{eff}}$$

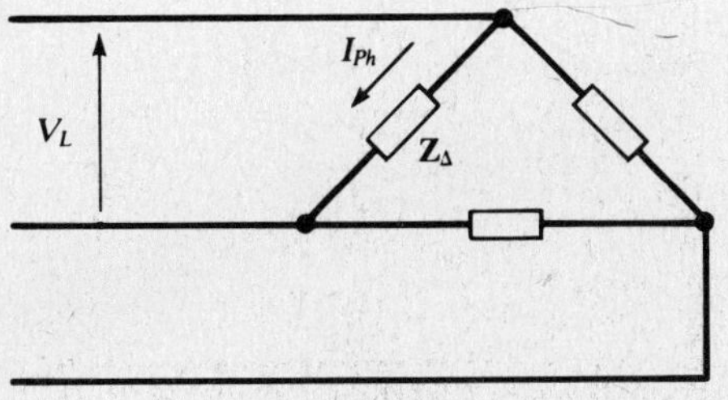

Fig. 10-16

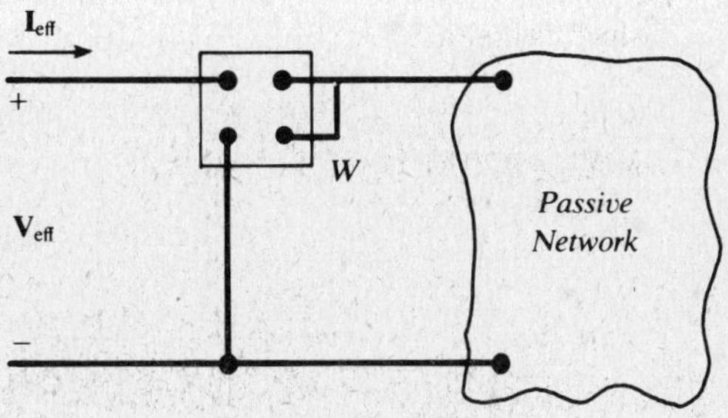

Fig. 10-17

10.10 THREE-PHASE LOADS AND THE TWO-WATTMETER METHOD

An ac wattmeter has a potential coil and a current coil and responds to the product of the effective voltage, the effective current, and the cosine of the phase angle between them. Thus, in Fig. 10-17, the wattmeter will indicate the average power supplied to the passive network,

$$P = V_{\text{eff}}I_{\text{eff}} \cos\theta = \text{Re}\,(\mathbf{V}_{\text{eff}}\mathbf{I}^*_{\text{eff}})$$

(see Section 9.3).

Two wattmeters connected in any two lines of a three-phase, three-wire system will correctly indicate the total three-phase power by the sum of the two meter readings. A meter will attempt to go downscale if the phase angle between the voltage and current exceeds 90°. In this event, the current-coil connections can be reversed and the upscale meter reading treated as negative in the sum. In Fig. 10-18 the meters are inserted in lines *A* and *C*, with the potential-coil reference connections in line *B*. Their readings will be:

$$W_A = \text{Re}\,(\mathbf{V}_{AB\text{eff}}\mathbf{I}^*_{A\text{eff}}) = \text{Re}\,(\mathbf{V}_{AB\text{eff}}\mathbf{I}^*_{AB\text{eff}}) + \text{Re}\,(\mathbf{V}_{AB\text{eff}}\mathbf{I}^*_{AC\text{eff}})$$
$$W_C = \text{Re}\,(\mathbf{V}_{CB\text{eff}}\mathbf{I}^*_{C\text{eff}}) = \text{Re}\,(\mathbf{V}_{CB\text{eff}}\mathbf{I}^*_{CA\text{eff}}) + \text{Re}\,(\mathbf{V}_{CB\text{eff}}\mathbf{I}^*_{CB\text{eff}})$$

in which the KCL expressions $\mathbf{I}_A = \mathbf{I}_{AB} + \mathbf{I}_{AC}$ and $\mathbf{I}_C = \mathbf{I}_{CA} + \mathbf{I}_{CB}$ have been used to replace line currents by phase currents. The first term in W_A is recognized as P_{AB}, the average power in phase *AB* of the delta load; likewise, the second term in W_C is P_{CB}. Adding the two equations and recombining the middle terms then yields

$$W_A + W_C = P_{AB} + \text{Re}\,[(\mathbf{V}_{AB\text{eff}} - \mathbf{V}_{CB\text{eff}})\mathbf{I}^*_{AC\text{eff}}] + P_{CB} = P_{AB} + P_{AC} + P_{CB}$$

since, by KVL, $\mathbf{V}_{AB} - \mathbf{V}_{CB} = \mathbf{V}_{AC}$.

The same reasoning establishes the analogous result for a Y-connected load.

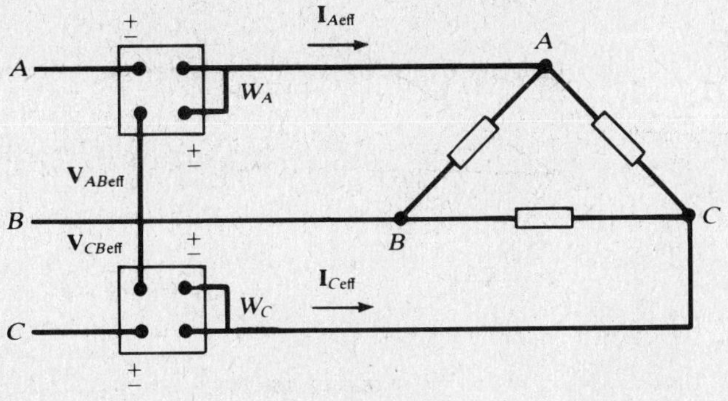

Fig. 10-18

Balanced Loads

When three equal impedances $Z\underline{/\theta}$ are connected in delta, the phase currents make 30° angles with their resultant line currents. Figure 10-19 corresponds to Fig. 10-18 under the assumption of *ABC* sequencing. It is seen that $\mathbf{V}_{AB}$ leads $\mathbf{I}_A$ by $\theta+30°$, while $\mathbf{V}_{CB}$ leads $\mathbf{I}_C$ by $\theta-30°$. Consequently, the two wattmeters will read

$$W_A = V_{AB\text{eff}} I_{A\text{eff}} \cos(\theta+30°) \qquad W_C = V_{CB\text{eff}} I_{C\text{eff}} \cos(\theta-30°)$$

or, since in general we do not know the relative order in the voltage sequence of the two lines chosen for the wattmeters,

$$W_1 = V_{L\text{eff}} I_{L\text{eff}} \cos(\theta+30°)$$

$$W_2 = V_{L\text{eff}} I_{L\text{eff}} \cos(\theta-30°)$$

These expressions also hold for a balanced Y-connection.

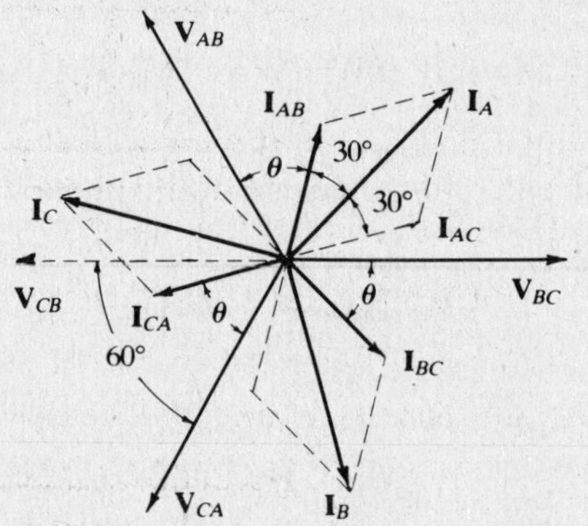

Fig. 10-19

Elimination of $V_{L\text{eff}} I_{L\text{eff}}$ between the two readings leads to

$$\tan\theta = \sqrt{3}\left(\frac{W_2 - W_1}{W_2 + W_1}\right)$$

Thus, from the two wattmeter readings, the *magnitude* of the impedance angle θ can be inferred. The sign on $\tan\theta$ suggested by the above formula is meaningless, since the arbitrary subscripts 1 and 2 might just as well be interchanged. However, in the practical case, the balanced load is usually known to be inductive ($\theta > 0$).

Solved Problems

10.1 Show that the line-to-line voltage V_L in a three-phase system is $\sqrt{3}$ times the line-to-neutral voltage V_{Ph}.

See the voltage phasor diagram (for the *ABC* sequence), Fig. 10-20.

10.2 A three-phase, *ABC* system, with an effective voltage 70.7 V, has a balanced Δ-connected load with impedances $20\underline{/45°}\ \Omega$. Obtain the line currents and draw the voltage-current phasor diagram.

The circuit is shown in Fig. 10-21. The phasor voltages have magnitudes $V_{\max} = \sqrt{2}V_{\text{eff}} = 100$ V. Phase angles are obtained from Fig. 10-5(*a*). Then,

$$\mathbf{I}_{AB} = \frac{\mathbf{V}_{AB}}{\mathbf{Z}} = \frac{100\angle 120^\circ}{20\angle 45^\circ} = 5.0\angle 75^\circ \text{ A}$$

Similarly, $\mathbf{I}_{BC} = 5.0\angle -45^\circ$ A and $\mathbf{I}_{CA} = 5.0\angle 195^\circ$ A. The line currents are:

$$\mathbf{I}_A = \mathbf{I}_{AB} + \mathbf{I}_{AC} = 5\angle 75^\circ - 5\angle 195^\circ = 8.65\angle 45^\circ \text{ A}$$

Similarly, $\mathbf{I}_B = 8.65\angle -75^\circ$ A, $\mathbf{I}_C = 8.65\angle 165^\circ$ A.

The voltage-current phasor diagram is shown in Fig. 10-22.

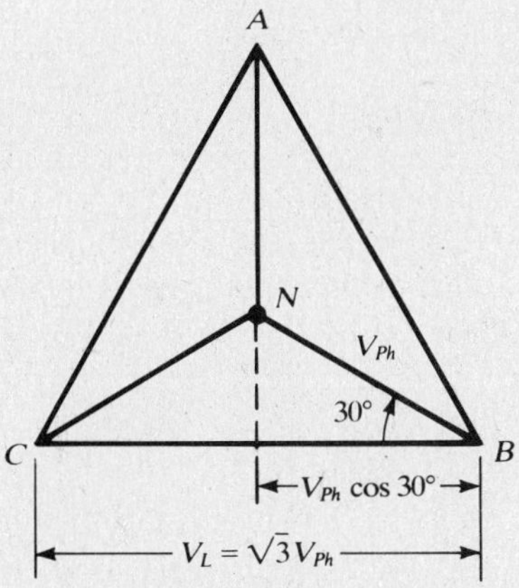

Fig. 10-20

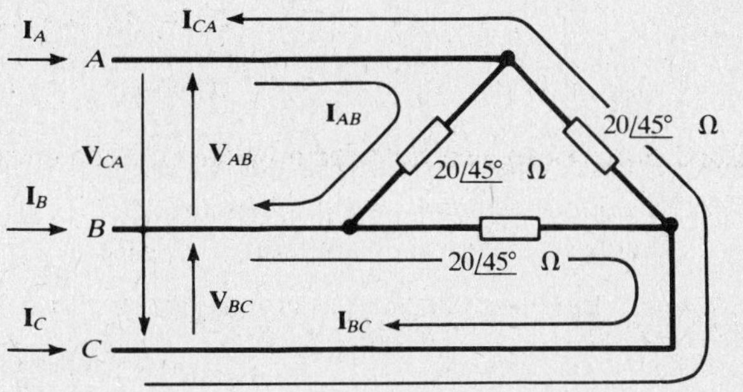

Fig. 10-21

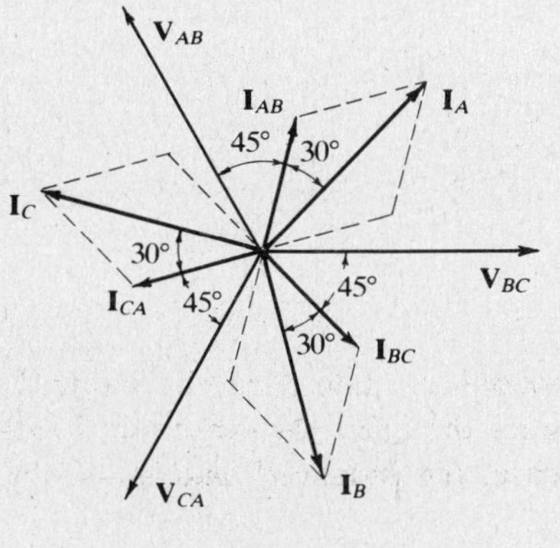

Fig. 10-22

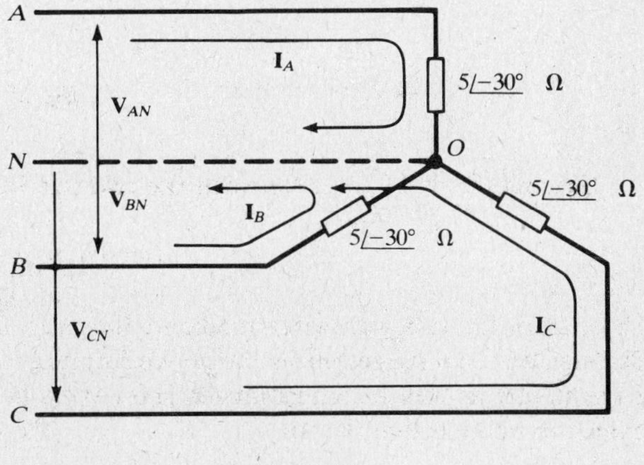

Fig. 10-23

10.3 A three-phase, three-wire *CBA* system, with an effective line voltage 106.1 V, has a balanced Y-connected load with impedances $5\angle -30^\circ\ \Omega$ (Fig. 10-23). Obtain the currents and draw the voltage-current phasor diagram.

With balanced Y-loads, the neutral conductor carries no current. Even though this system is three-wire, the neutral may be added to simplify computation of the line currents. The magnitude of the line voltage is $V_L = \sqrt{2}(106.1) = 150$ V. Then the line-to-neutral magnitude is $V_{LN} = 150/\sqrt{3} = 86.6$ V.

$$\mathbf{I}_A = \frac{\mathbf{V}_{AN}}{\mathbf{Z}} = \frac{86.6\angle -90^\circ}{5\angle -30^\circ} = 17.32\angle -60^\circ \text{ A}$$

Similarly, $\mathbf{I}_B = 17.32\angle 60^\circ$ A, $\mathbf{I}_C = 17.32\angle 180^\circ$ A. See the phasor diagram, Fig. 10-24, in which the balanced set of line currents leads the set of line-to-neutral voltages by 30°, the (negative of the) angle of the impedances.

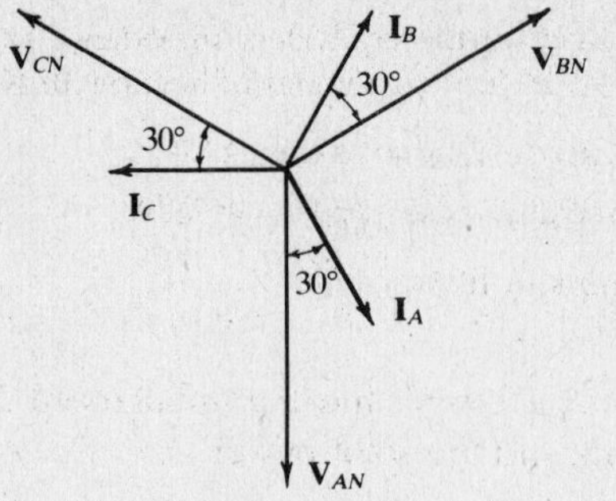

Fig. 10-24

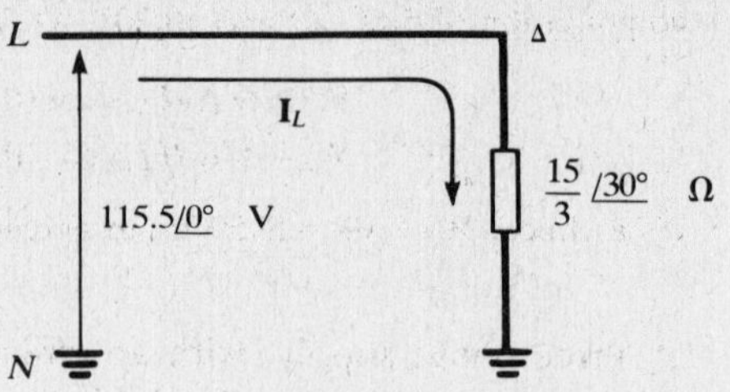

Fig. 10-25

10.4 A three-phase, three-wire *CBA* system, with an effective line voltage 106.1 V, has a balanced Δ-connected load with impedances $\mathbf{Z} = 15\angle 30^\circ\ \Omega$. Obtain the line and phase currents by the single-line equivalent method.

Referring to Fig. 10-25, $V_{LN} = (141.4\sqrt{2})/\sqrt{3} = 115.5$ V, and so

$$\mathbf{I}_L = \frac{115.5\angle 0^\circ}{(15/3)\angle 30^\circ} = 23.1\angle -30^\circ \text{ A}$$

The line currents lag the *ABC*-sequence, line-to-neutral voltages by 30°:

$$\mathbf{I}_A = 23.1\angle 60^\circ \text{ A} \qquad \mathbf{I}_B = 23.1\angle -60^\circ \text{ A} \qquad \mathbf{I}_C = 23.1\angle 180^\circ \text{ A}$$

The phase currents, of magnitude $I_{Ph} = I_L/\sqrt{3} = 13.3$ A, lag the corresponding line-to-line voltages by 30°:

$$\mathbf{I}_{AB} = 13.3\angle 90^\circ \text{ A} \qquad \mathbf{I}_{BC} = 13.3\angle -30^\circ \text{ A} \qquad \mathbf{I}_{CA} = 13.3\angle 210^\circ \text{ A}$$

A sketch of the phasor diagram will make all of the foregoing angles evident.

10.5 A three-phase, three-wire system, with an effective line voltage 176.8 V, supplies two balanced loads, one in delta with $\mathbf{Z}_\Delta = 15\angle 0^\circ\ \Omega$ and the other in wye with $\mathbf{Z}_Y = 10\angle 30^\circ\ \Omega$. Obtain the total power.

First convert the Δ-load to Y, and then use the single-line equivalent circuit, Fig. 10-26, to obtain the line current.

$$\mathbf{I}_L = \frac{144.3\angle 0^\circ}{5\angle 0^\circ} + \frac{144.3\angle 0^\circ}{10\angle 30^\circ} = 42.0\angle -9.9^\circ \text{ A}$$

Then $$P = \sqrt{3}V_{L\text{eff}}I_{L\text{eff}} \cos\theta = \sqrt{3}(176.8)(29.7)\cos 9.9^\circ = 8959 \text{ W}$$

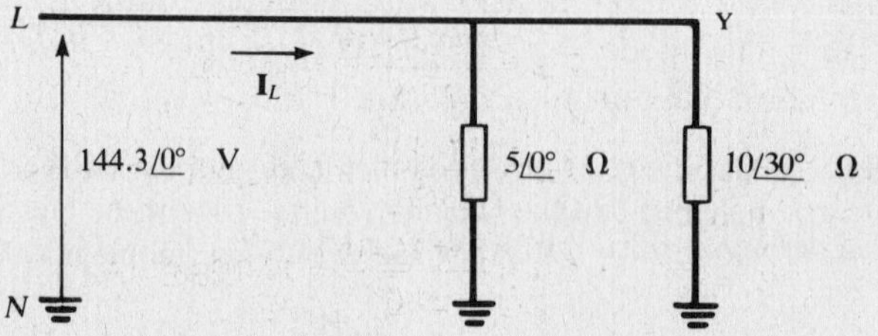

Fig. 10-26

10.6 Obtain the readings when the two-wattmeter method is applied to the circuit of Problem 10.5.

The angle on $\mathbf{I}_L$, $-9.9°$, is the negative of the angle on the equivalent impedance of the parallel combination of $5\underline{/0°}$ Ω and $10\underline{/30°}$ Ω. Therefore, $\theta = 9.9°$ in the formulas of Section 10.10.

$$W_1 = V_{L\text{eff}} I_{L\text{eff}} \cos(\theta + 30°) = (176.8)(29.7)\cos 39.9° = 4028 \text{ W}$$
$$W_2 = V_{L\text{eff}} I_{L\text{eff}} \cos(\theta - 30°) = (176.8)(29.7)\cos(-20.1°) = 4931 \text{ W}$$

As a check, $W_1 + W_2 = 8959$ W, in agreement with Problem 10.5.

10.7 A three-phase supply, with an effective line voltage 240 V, has the unbalanced Δ-connected load shown in Fig. 10-27. Obtain the line currents and the total power.

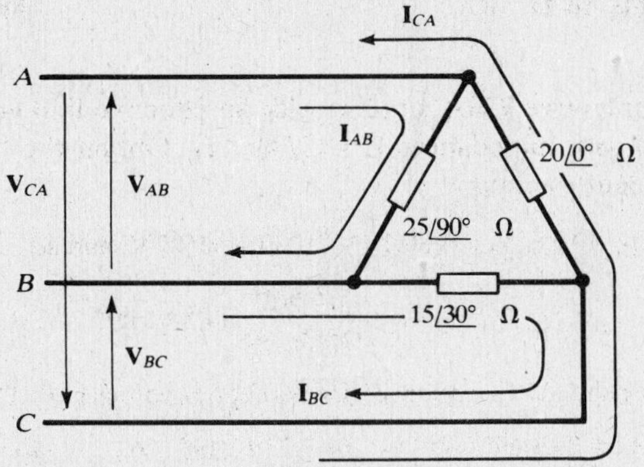

Fig. 10-27

The power calculations can be performed without knowledge of the sequence of the system. The effective values of the phase currents are

$$I_{AB\text{eff}} = \frac{240}{25} = 9.6 \text{ A} \qquad I_{BC\text{eff}} = \frac{240}{15} = 16 \text{ A} \qquad I_{CA\text{eff}} = \frac{240}{20} = 12 \text{ A}$$

Hence, the complex powers in the three phases are

$$\mathbf{S}_{AB} = (9.6)^2(25\underline{/90°}) = 2304\underline{/90°} = 0 + j2304$$
$$\mathbf{S}_{BC} = (16)^2(15\underline{/30°}) = 3840\underline{/30°} = 3325 + j1920$$
$$\mathbf{S}_{CA} = (12)^2(20\underline{/0°}) = 2880\underline{/0°} = 2880 + j0$$

and the total complex power is their sum,

$$\mathbf{S}_T = 6205 + j4224$$

that is, $P_T = 6205$ W and $Q_T = 4224$ var (inductive).

To obtain the currents a sequence must be assumed; let it be *ABC*. Then, using Fig. 10-5(*a*),

$$\mathbf{I}_{AB} = \frac{339.4\underline{/120°}}{25\underline{/90°}} = 13.6\underline{/30°} \text{ A}$$
$$\mathbf{I}_{BC} = \frac{339.4\underline{/0°}}{15\underline{/30°}} = 22.6\underline{/-30°} \text{ A}$$
$$\mathbf{I}_{CA} = \frac{339.4\underline{/240°}}{20\underline{/0°}} = 17.0\underline{/240°} \text{ A}$$

The line currents are obtained by applying KCL at the junctions.

$$\mathbf{I}_A = \mathbf{I}_{AB} + \mathbf{I}_{AC} = 13.6\underline{/30°} - 17.0\underline{/240°} = 29.6\underline{/46.7°} \text{ A}$$
$$\mathbf{I}_B = \mathbf{I}_{BC} + \mathbf{I}_{BA} = 22.6\underline{/-30°} - 13.6\underline{/30°} = 19.7\underline{/-66.7°} \text{ A}$$
$$\mathbf{I}_C = \mathbf{I}_{CA} + \mathbf{I}_{CB} = 17.0\underline{/240°} - 22.6\underline{/-30°} = 28.3\underline{/-173.1°} \text{ A}$$

10.8 Obtain the readings of wattmeters placed in lines A and B of the circuit of Problem 10.7. (Line C is the potential reference for both meters.)

$$W_A = \text{Re}\,(\mathbf{V}_{AC\text{eff}}\mathbf{I}^*_{A\text{eff}}) = \text{Re}\left[(240\angle 60^\circ)\left(\frac{29.6}{\sqrt{2}}\angle -46.7^\circ\right)\right]$$

$$= \text{Re}\,(5023\angle 13.3^\circ) = 4888\ \text{W}$$

$$W_B = \text{Re}\,(\mathbf{V}_{BC\text{eff}}\mathbf{I}^*_{B\text{eff}}) = \text{Re}\left[(240\angle 0^\circ)\left(\frac{19.7}{\sqrt{2}}\angle 66.7^\circ\right)\right]$$

$$= \text{Re}\,(3343\angle 66.7^\circ) = 1322\ \text{W}$$

Note that $W_A + W_B = 6210$ W, which agrees with P_T as found in Problem 10.7.

10.9 A three-phase, four-wire, ABC system, with line voltage $\mathbf{V}_{BC} = 294.2\angle 0^\circ$ V, has a Y-connected load of $\mathbf{Z}_A = 10\angle 0^\circ\ \Omega$, $\mathbf{Z}_B = 15\angle 30^\circ\ \Omega$, and $\mathbf{Z}_C = 10\angle -30^\circ\ \Omega$ (Fig. 10-28). Obtain the line and neutral currents.

$$\mathbf{I}_A = \frac{169.9\angle 90^\circ}{10\angle 0^\circ} = 16.99\angle 90^\circ\ \text{A}$$

$$\mathbf{I}_B = \frac{169.9\angle -30^\circ}{15\angle 30^\circ} = 11.33\angle -60^\circ\ \text{A}$$

$$\mathbf{I}_C = \frac{169.9\angle -150^\circ}{10\angle -30^\circ} = 16.99\angle -120^\circ\ \text{A}$$

$$\mathbf{I}_N = -(\mathbf{I}_A + \mathbf{I}_B + \mathbf{I}_C) = 8.04\angle 69.5^\circ\ \text{A}$$

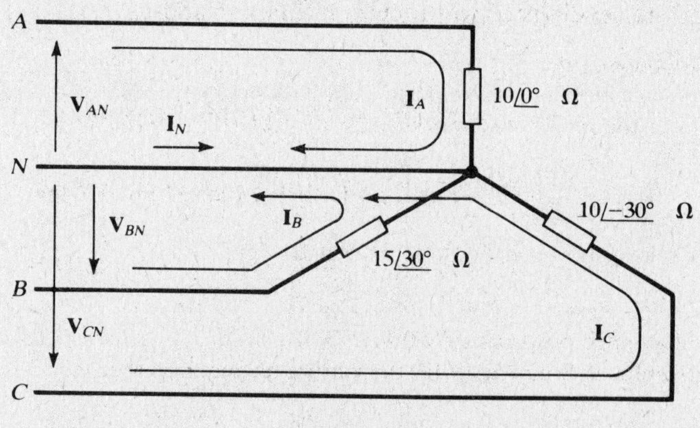

Fig. 10-28

10.10 The Y-connected load impedances $\mathbf{Z}_A = 10\angle 0^\circ\ \Omega$, $\mathbf{Z}_B = 15\angle 30^\circ\ \Omega$, and $\mathbf{Z}_C = 10\angle -30^\circ\ \Omega$, in Fig. 10-29, are supplied by a three-phase, three-wire, ABC system in which $\mathbf{V}_{BC} = 208\angle 0^\circ$ V. Obtain the voltages across the impedances and the displacement neutral voltage $\mathbf{V}_{ON}$.

The method of Example 10.6 could be applied here and one node-voltage equation solved. However, the mesh currents $\mathbf{I}_1$ and $\mathbf{I}_2$ suggested in Fig. 10-29 provide another approach.

$$\begin{bmatrix} 10\angle 0^\circ + 15\angle 30^\circ & -15\angle 30^\circ \\ -15\angle 30^\circ & 15\angle 30^\circ + 10\angle -30^\circ \end{bmatrix}\begin{bmatrix}\mathbf{I}_1 \\ \mathbf{I}_2\end{bmatrix} = \begin{bmatrix} 208\angle 120^\circ \\ 208\angle 0^\circ \end{bmatrix}$$

Solving, $\mathbf{I}_1 = 14.16\angle 86.09^\circ$ A and $\mathbf{I}_2 = 10.21\angle 52.41^\circ$ A. The line currents are then

$$\mathbf{I}_A = \mathbf{I}_1 = 14.16\angle 86.09^\circ\ \text{A}, \qquad \mathbf{I}_B = \mathbf{I}_2 - \mathbf{I}_1 = 8.01\angle -48.93^\circ\ \text{A} \qquad \mathbf{I}_C = -\mathbf{I}_2 = 10.21\angle -127.59^\circ\ \text{A}$$

Now the phasor voltages at the load may be computed.

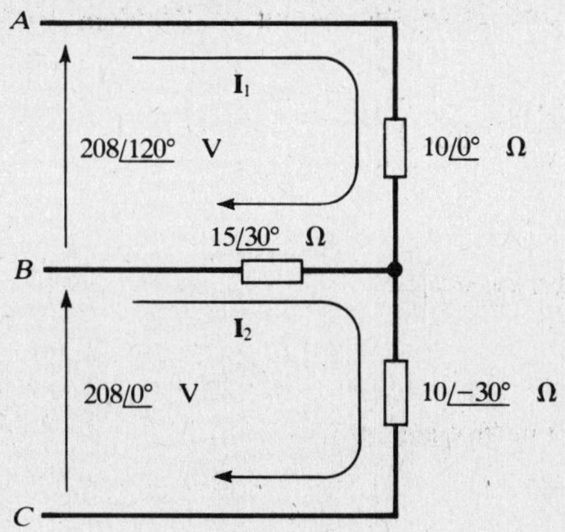

Fig. 10-29

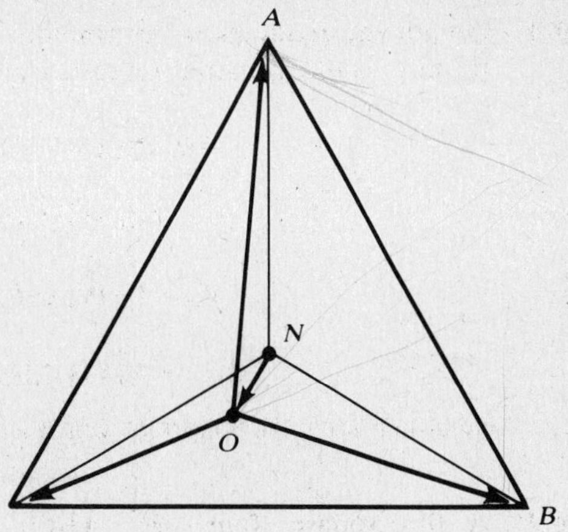

Fig. 10-30

$$\mathbf{V}_{AO} = \mathbf{I}_A\mathbf{Z}_A = 141.6\underline{/86.09°} \text{ V}$$
$$\mathbf{V}_{BO} = \mathbf{I}_B\mathbf{Z}_B = 120.2\underline{/-18.93°} \text{ V}$$
$$\mathbf{V}_{CO} = \mathbf{I}_C\mathbf{Z}_C = 102.1\underline{/-157.59°} \text{ V}$$
$$\mathbf{V}_{ON} = \mathbf{V}_{OA} + \mathbf{V}_{AN} = 141.6\underline{/-93.91°} + 120.1\underline{/90°} = 23.3\underline{/-114.53°} \text{ V}$$

The phasor diagram is given in Fig. 10-30.

10.11 Obtain the total average power for the unbalanced, Y-connected load in Problem 10.10, and compare with the readings of wattmeters in lines B and C.

The phase powers are

$$P_A = I_{A\text{eff}}^2 R_A = \left(\frac{14.16}{\sqrt{2}}\right)^2 (10) = 1002.5 \text{ W}$$

$$P_B = I_{B\text{eff}}^2 R_B = \left(\frac{8.01}{\sqrt{2}}\right)^2 (15 \cos 30°) = 417.0 \text{ W}$$

$$P_C = I_{C\text{eff}}^2 R_C = \left(\frac{10.21}{\sqrt{2}}\right)^2 (10 \cos 30°) = 451.4 \text{ W}$$

and so the total average power is 1870.9 W.

From the results of Problem 10.10, the wattmeter readings are:

$$W_B = \text{Re}\,(\mathbf{V}_{BA\text{eff}}\mathbf{I}^*_{B\text{eff}}) = \text{Re}\left[\left(\frac{208}{\sqrt{2}}\underline{/-60°}\right)\left(\frac{8.01}{\sqrt{2}}\underline{/48.93°}\right)\right] = 817.1 \text{ W}$$

$$W_C = \text{Re}\,(\mathbf{V}_{CA\text{eff}}\mathbf{I}^*_{C\text{eff}}) = \text{Re}\left[\left(\frac{208}{\sqrt{2}}\underline{/240°}\right)\left(\frac{10.21}{\sqrt{2}}\underline{/127.59°}\right)\right] = 1052.8 \text{ W}$$

The total power read by the two wattmeters is 1869.9 W.

10.12 A three-phase, three-wire, balanced, Δ-connected load yields wattmeter readings of 1154 W and 577 W. Obtain the load impedance, if the line voltage is 141.4 V.

$$\pm \tan\theta = \sqrt{3}\left(\frac{W_2 - W_1}{W_2 + W_1}\right) = \sqrt{3}\left(\frac{577}{1731}\right) = 0.577 \qquad \theta = \pm 30.0°$$

and, using $P_T = \sqrt{3}V_{L\text{eff}}I_{L\text{eff}} \cos\theta$,

$$Z_\Delta = \frac{V_{L\text{eff}}}{I_{Ph\text{eff}}} = \frac{\sqrt{3}V_{L\text{eff}}}{I_{L\text{eff}}} = \frac{3V_{L\text{eff}}^2 \cos\theta}{P_T} = \frac{3(100)^2 \cos 30.0°}{1154 + 577} = 15.0\ \Omega$$

Thus, $\mathbf{Z}_\Delta = 15.0\underline{/\pm 30.0°}\ \Omega$.

10.13 A balanced Δ-connected load, with $\mathbf{Z}_\Delta = 30\angle 30^\circ\ \Omega$, is connected to a three-phase, three-wire, 250-V system by conductors having impedances $\mathbf{Z}_c = 0.4 + j0.3\ \Omega$. Obtain the line-to-line voltage at the load.

The single-line equivalent circuit is shown in Fig. 10-31. By voltage division, the voltage across the substitute Y-load is

$$\mathbf{V}_{AN} = \left(\frac{10\angle 30^\circ}{0.4 + j0.3 + 10\angle 30^\circ}\right)\left(\frac{250}{\sqrt{3}}\angle 0^\circ\right) = 137.4\angle -0.33^\circ \ \text{V}$$

whence $V_L = (137.4)(\sqrt{3}) = 238.0$ V.

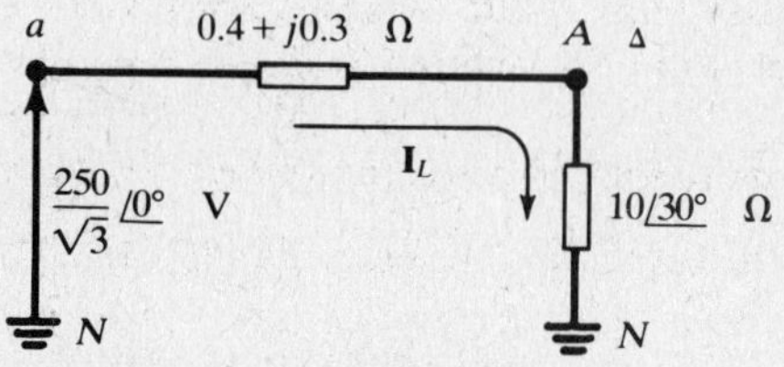

Fig. 10-31

Considering the magnitudes only, the line voltage at the load, 238.0 V, represents a drop of 12.0 V. The wire size and total length control the resistance in $\mathbf{Z}_c$, while the enclosing conduit material (e.g. steel, aluminum, or fiber), as well as the length, affects the inductive reactance.

Supplementary Problems

In the following, the voltage-current phasor diagram will not be included in the answer, even though the problem may ask specifically for one. As a general rule, a phasor diagram should be constructed for every polyphase problem.

10.14 Three impedances of $10.0\angle 53.13^\circ\ \Omega$ are connected in delta to a three-phase, *CBA* system with an effective line voltage 240 V. Obtain the line currents.
Ans. $\mathbf{I}_A = 58.8\angle -143.13^\circ$ A, $\mathbf{I}_B = 58.8\angle -23.13^\circ$ A, $\mathbf{I}_C = 58.8\angle 96.87^\circ$ A

10.15 Three impedances of $42.0\angle -35^\circ\ \Omega$ are connected in delta to a three-phase, *ABC* system having $\mathbf{V}_{BC} = 495.0\angle 0^\circ$ V. Obtain the line currents.
Ans. $\mathbf{I}_A = 20.41\angle 125^\circ$ A, $\mathbf{I}_B = 20.41\angle 5^\circ$ A, $\mathbf{I}_C = 20.41\angle -115^\circ$ A

10.16 A three-phase, three-wire system, with an effective line voltage 100 V, has currents

$$\mathbf{I}_A = 15.41\angle -160^\circ \ \text{A} \qquad \mathbf{I}_B = 15.41\angle -40^\circ \ \text{A} \qquad \mathbf{I}_C = 15.41\angle 80^\circ \ \text{A}$$

What is the sequence of the system and what are the impedances, if the connection is delta?
Ans. *CBA*, $15.9\angle 70^\circ\ \Omega$

10.17 A balanced Y-connected load, with impedances $6.0\angle 45^\circ\ \Omega$, is connected to a three-phase, four-wire *CBA* system having effective line voltage 208 V. Obtain the four line currents.
Ans. $\mathbf{I}_A = 28.31\angle -135^\circ$ A, $\mathbf{I}_B = 28.31\angle -15^\circ$ A, $\mathbf{I}_C = 28.31\angle 105^\circ$ A, $\mathbf{I}_N = 0$

10.18 A balanced Y-connected load, with impedances $65.0\angle -20^\circ\ \Omega$, is connected to a three-phase, three-wire, *CBA* system, where $\mathbf{V}_{AB} = 678.8\angle -120^\circ$ V. Obtain the three line currents.
Ans. $\mathbf{I}_A = 6.03\angle -70^\circ$ A, $\mathbf{I}_B = 6.03\angle 50^\circ$ A, $\mathbf{I}_C = 6.03\angle 170^\circ$ A

10.19 A balanced Δ-connected load, with $\mathbf{Z}_\Delta = 9.0\underline{/-30^\circ}$ Ω, and a balanced Y-connected load, with $\mathbf{Z}_Y = 5.0\underline{/45^\circ}$ Ω, are supplied by the same three-phase, *ABC* system, with effective line voltage 480 V. Obtain the line currents, using the single-line equivalent method.
Ans. $\mathbf{I}_A = 168.9\underline{/93.36^\circ}$ A, $\mathbf{I}_B = 168.9\underline{/-26.64^\circ}$ A, $\mathbf{I}_C = 168.9\underline{/-146.64^\circ}$ A

10.20 A balanced Δ-connected load having impedances $27.0\underline{/-25^\circ}$ Ω and a balanced Y-connected load having impedances $10.0\underline{/-30^\circ}$ Ω are supplied by the same three-phase, *ABC* system, with $\mathbf{V}_{CN} = 169.8\underline{/-150^\circ}$ V. Obtain the line currents.
Ans. $\mathbf{I}_A = 35.8\underline{/117.36^\circ}$ A, $\mathbf{I}_B = 35.8\underline{/-2.64^\circ}$ A, $\mathbf{I}_C = 35.8\underline{/-122.64^\circ}$ A

10.21 A balanced Δ-connected load, with impedances $10.0\underline{/-36.9^\circ}$ Ω, and a balanced Y-connected load are supplied by the same three-phase, *ABC* system having $\mathbf{V}_{CA} = 141.4\underline{/240^\circ}$ V. If $\mathbf{I}_B = 40.44\underline{/13.41^\circ}$ A, what are the impedances of the Y-connected load? *Ans.* $5.0\underline{/-53.3^\circ}$ Ω

10.22 A three-phase, *ABC* system, with effective line voltage 500 V, has a Δ-connected load for which

$$\mathbf{Z}_{AB} = 10.0\underline{/30^\circ}\ \Omega \qquad \mathbf{Z}_{BC} = 25.0\underline{/0^\circ}\ \Omega \qquad \mathbf{Z}_{CA} = 20.0\underline{/-30^\circ}\ \Omega$$

Obtain the line currents.
Ans. $\mathbf{I}_A = 106.1\underline{/90.0^\circ}$ A, $\mathbf{I}_B = 76.15\underline{/-68.20^\circ}$ A, $\mathbf{I}_C = 45.28\underline{/-128.65^\circ}$ A

10.23 A three-phase, *ABC* system, with $\mathbf{V}_{BC} = 294.2\underline{/0^\circ}$ V, has the Δ-connected load

$$\mathbf{Z}_{AB} = 5.0\underline{/0^\circ}\ \Omega \qquad \mathbf{Z}_{BC} = 4.0\underline{/30^\circ}\ \Omega \qquad \mathbf{Z}_{CA} = 6.0\underline{/-15^\circ}\ \Omega$$

Obtain the line currents.
Ans. $\mathbf{I}_A = 99.7\underline{/99.7^\circ}$ A, $\mathbf{I}_B = 127.9\underline{/-43.3^\circ}$ A, $\mathbf{I}_C = 77.1\underline{/-172.1^\circ}$ A

10.24 A three-phase, four-wire, *CBA* system, with effective line voltage 100 V, has Y-connected impedances

$$\mathbf{Z}_A = 3.0\underline{/0^\circ}\ \Omega \qquad \mathbf{Z}_B = 3.61\underline{/56.31^\circ}\ \Omega \qquad \mathbf{Z}_C = 2.24\underline{/-26.57^\circ}\ \Omega$$

Obtain the currents $\mathbf{I}_A$, $\mathbf{I}_B$, $\mathbf{I}_C$, and $\mathbf{I}_N$.
Ans. $27.2\underline{/-90^\circ}$ A, $22.6\underline{/-26.3^\circ}$ A, $36.4\underline{/176.6^\circ}$ A, $38.6\underline{/65.3^\circ}$ A

10.25 A three-phase, four-wire, *ABC* system, with $\mathbf{V}_{BC} = 294.2\underline{/0^\circ}$ V, has Y-connected impedances

$$\mathbf{Z}_A = 12.0\underline{/45^\circ}\ \Omega \qquad \mathbf{Z}_B = 10.0\underline{/30^\circ}\ \Omega \qquad \mathbf{Z}_C = 8.0\underline{/0^\circ}\ \Omega$$

Obtain the currents $\mathbf{I}_A$, $\mathbf{I}_B$, $\mathbf{I}_C$, and $\mathbf{I}_N$.
Ans. $14.16\underline{/45^\circ}$ A, $16.99\underline{/-60^\circ}$ A, $21.24\underline{/-150^\circ}$ A, $15.32\underline{/90.4^\circ}$ A

10.26 A Y-connected load, with $\mathbf{Z}_A = 10\underline{/0^\circ}$ Ω, $\mathbf{Z}_B = 10\underline{/60^\circ}$ Ω, and $\mathbf{Z}_C = 10\underline{/-60^\circ}$ Ω, is connected to a three-phase, three-wire, *ABC* system having effective line voltage 141.4 V. Find the load voltages $\mathbf{V}_{AO}$, $\mathbf{V}_{BO}$, $\mathbf{V}_{CO}$ and the displacement neutral voltage $\mathbf{V}_{ON}$. Construct a phasor diagram similar to Fig. 10-15. *Ans.* $173.2\underline{/90^\circ}$ V, $100\underline{/0^\circ}$ V, $100\underline{/180^\circ}$ V, $57.73\underline{/-90^\circ}$ V

10.27 A Y-connected load, with $\mathbf{Z}_A = 10\underline{/-60^\circ}$ Ω, $\mathbf{Z}_B = 10\underline{/0^\circ}$ Ω, and $\mathbf{Z}_C = 10\underline{/60^\circ}$ Ω, is connected to a three-phase, three-wire, *CBA* system having effective line voltage 147.1 V. Obtain the line currents $\mathbf{I}_A$, $\mathbf{I}_B$, and $\mathbf{I}_C$. *Ans.* $20.8\underline{/-60^\circ}$ A, 0, $20.8\underline{/120^\circ}$ A

10.28 A three-phase, three-wire, *ABC* system with a balanced load has effective line voltage 200 V and (maximum) line current $\mathbf{I}_A = 13.61\underline{/60^\circ}$ A. Obtain the total power. *Ans.* 2887 W

10.29 Two balanced Δ-connected loads, with impedances $20\underline{/-60^\circ}$ Ω and $18\underline{/45^\circ}$ Ω, respectively, are connected to a three-phase system for which a line voltage is $\mathbf{V}_{BC} = 212.1\underline{/0^\circ}$ V. Obtain the phase power of each load. After using the single-line equivalent method to obtain the total line current, compute the total power, and compare with the sum of the phase powers.
Ans. 562.3 W, 883.6 W, 4337.5 W = 3(562.3 W) + 3(883.6 W)

10.30 In Problem 10.2, a balanced Δ-connected load with $\mathbf{Z} = 20\underline{/45^\circ}$ Ω resulted in line currents 8.65 A for line voltages 100 V, both maximum values. Find the readings of two wattmeters used to measure the total average power. *Ans.* 111.9 W, 417.7 W

10.31 Obtain the readings of two wattmeters in a three-phase, three-wire system having effective line voltage 240 V and balanced, Δ-connected load impedances $20\angle 80^\circ$ Ω. *Ans.* −1706 W, 3206 W

10.32 A three-phase, three-wire, *ABC* system with line voltage $\mathbf{V}_{BC} = 311.1\angle 0^\circ$ V, has line currents

$$\mathbf{I}_A = 61.5\angle 116.6^\circ \text{ A} \qquad \mathbf{I}_B = 61.2\angle -48.0^\circ \text{ A} \qquad \mathbf{I}_C = 16.1\angle 218^\circ \text{ A}$$

Find the readings of wattmeters in lines (*a*) *A* and *B*, (*b*) *B* and *C*, (*c*) *A* and *C*.
Ans. (*a*) 5266 W, 6370 W; (*b*) 9312 W, 2322 W; (*c*) 9549 W, 1973 W

10.33 A three-phase, three-wire, *ABC* system has an effective line voltage 440 V. The line currents are

$$\mathbf{I}_A = 27.9\angle 90^\circ \text{ A} \qquad \mathbf{I}_B = 81.0\angle -9.9^\circ \text{ A} \qquad \mathbf{I}_C = 81.0\angle 189.9^\circ \text{ A}$$

Obtain the readings of wattmeters in lines (*a*) *A* and *B*, (*b*) *B* and *C*.
Ans. (*a*) 7.52 kW, 24.8 kW; (*b*) 16.16 kW, 16.16 kW

10.34 Two wattmeters in a three-phase, three-wire system with effective line voltage 120 V read 1500 W and 500 W. What is the impedance of the balanced Δ-connected load? *Ans.* $16.3\angle \pm 40.9^\circ$ Ω

10.35 A three-phase, three-wire, *ABC* system has effective line voltage 173.2 V. Wattmeters in lines *A* and *B* read −301 W and 1327 W, respectively. Find the impedance of the balanced Y-connected load. (Since the sequence is specified, the sign of the impedance angle can be determined.)
Ans. $10\angle -70^\circ$ Ω

10.36 A three-phase, three-wire system, with a line voltage $\mathbf{V}_{BC} = 339.4\angle 0^\circ$ V, has a balanced Y-connected load of $\mathbf{Z}_Y = 15\angle 60^\circ$ Ω. The lines between the system and the load have impedances $2.24\angle 26.57^\circ$ Ω. Find the line-voltage magnitude at the load. *Ans.* 301.1 V

10.37 Repeat Problem 10.36 with the load impedance $\mathbf{Z}_Y = 15\angle -60^\circ$ Ω. By drawing the voltage phasor diagrams for the two cases, illustrate the effect of load impedance angle on the voltage drop for a given line impedance. *Ans.* 332.9 V

Chapter 11

Frequency Response and Resonance

11.1 INTRODUCTION

By the *frequency response* of a network is meant its performance over a range of driving frequencies. *Resonance* is a specifically defined condition for a network containing R, L, and C elements; the notion has already been encountered in Problem 6.8. To display the frequency response, a plot of both the magnitude and angle of a certain complex function is made versus frequency f (Hz) or angular frequency ω (rad/s). For simplicity, the single word "frequency" shall be used in this chapter to refer to either f or ω.

11.2 ONE-PORT AND TWO-PORT NETWORKS

Figure 11-1(a) shows a passive network with two terminals where a voltage $\mathbf{V}_1$ is defined and a corresponding current $\mathbf{I}_1$. This is a *one-port* network. A *two-port* network is shown in Fig. 11-1(b), where a second voltage $\mathbf{V}_2$ and corresponding current $\mathbf{I}_2$ are defined. The indicated current directions and voltage polarities are according to convention. In the two-port network, the output condition must be defined as (i) open-circuit, $\mathbf{I}_2 = 0$; (ii) short-circuit, $\mathbf{V}_2 = 0$; or (iii) load impedance applied, $\mathbf{V}_2 = -\mathbf{I}_2\mathbf{Z}_L$.

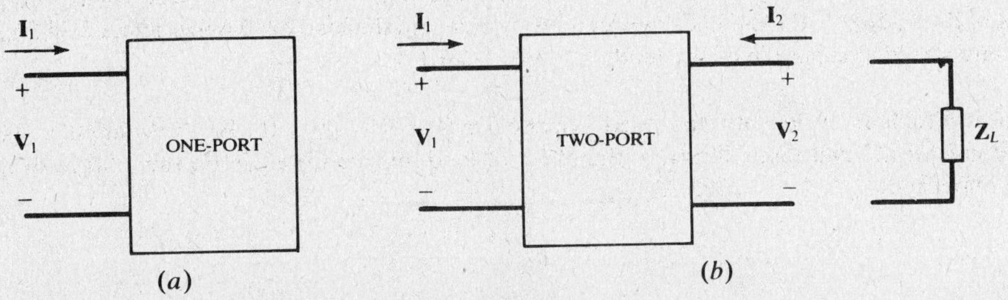

Fig. 11-1

For the one-port network, the frequency function of interest is

$$\textbf{input (driving-point) impedance or admittance} \qquad \frac{\mathbf{V}_1}{\mathbf{I}_1} \equiv \mathbf{Z}_{\text{in}}(\omega) \equiv \frac{1}{\mathbf{Y}_{\text{in}}(\omega)}$$

For the two-port network, the functions are

$$\textbf{input impedance} \qquad \frac{\mathbf{V}_1}{\mathbf{I}_1} \equiv \mathbf{Z}_{\text{in}}(\omega) \text{ or } \mathbf{H}_z(\omega)$$

$$\textbf{nondimensional voltage ratio} \qquad \frac{\mathbf{V}_2}{\mathbf{V}_1} \equiv \mathbf{H}_v(\omega)$$

$$\textbf{nondimensional current ratio} \qquad \frac{\mathbf{I}_2}{\mathbf{I}_1} \equiv \mathbf{H}_i(\omega)$$

$$\textbf{transfer impedances} \qquad \frac{\mathbf{V}_2}{\mathbf{I}_1} \text{ and } \frac{\mathbf{V}_1}{\mathbf{I}_2}$$

The transfer impedances and their reciprocals (transfer admittances), as well as the input admittance, are also given the symbol $\mathbf{H}$; collectively, these $\mathbf{H}$-functions are known as the *network* or *transfer functions.* They are all complex-valued functions (not phasors) of the real variable ω. In defining the transfer functions, a load condition of the two-port network was implicitly assumed. If an open-circuit condition ($Z_L \to \infty$) or a short-circuit condition ($Z_L \to 0$) prevails, that fact can be indicated by including the appropriate limiting value of Z_L as an extra subscript on **H**. Thus,

$$\mathbf{H}_{v\infty}(\omega) = \frac{\mathbf{V}_2}{\mathbf{V}_1}\bigg|_{\mathbf{I}_2=0}$$

is the open-circuit ratio of output voltage to input voltage as a function of frequency. Sometimes we shall drop the added subscript, to keep the notation simple.

11.3 HIGH-PASS AND LOW-PASS NETWORKS

A resistive voltage divider under a no-load condition is shown in Fig. 11-2, with the standard two-port voltages and currents. The voltage transfer function and input impedance are

$$\mathbf{H}_{v\infty}(\omega) = \frac{R_2}{R_1 + R_2} \qquad \mathbf{H}_{z\infty}(\omega) = R_1 + R_2$$

Both $\mathbf{H}_{v\infty}$ and $\mathbf{H}_{z\infty}$ are real constants, independent of frequency, since no reactive elements are present. If the network contains either an inductance or a capacitance, then $\mathbf{H}_{v\infty}$ and $\mathbf{H}_{z\infty}$ will be complex and will vary with frequency. If $|\mathbf{H}_{v\infty}|$ decreases as frequency increases, the performance is called *high-frequency roll-off* and the circuit is a *low-pass network*. On the contrary, a high-pass network will have low-frequency roll-off, with $|\mathbf{H}_{v\infty}|$ decreasing as the frequency decreases. Four, two-element circuits are shown in Fig. 11-3, two high-pass and two low-pass.

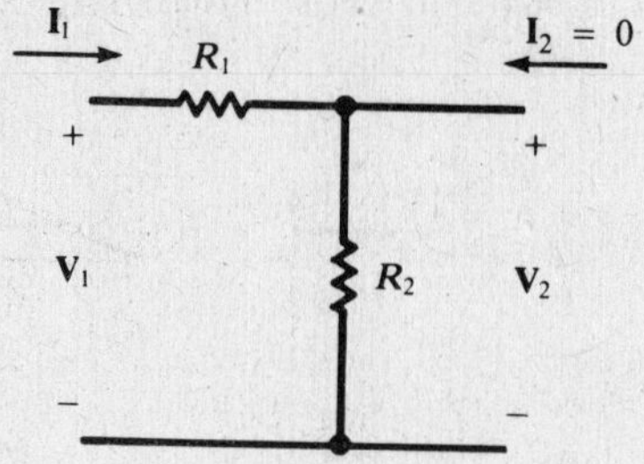

Fig. 11-2

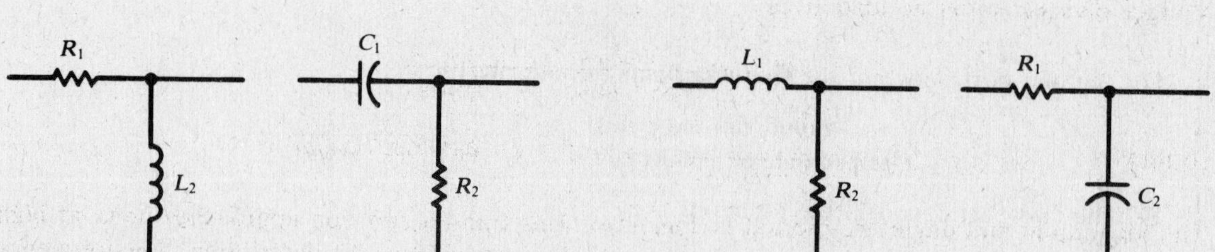

(*a*) High-Pass (*b*) Low-Pass

Fig. 11-3

The *RL* high-pass circuit shown in Fig. 11-4 is open-circuited or under no-load. The input impedance frequency response is determined by plotting the magnitude and phase angle of

$$\mathbf{H}_{z\infty}(\omega) = R_1 + j\omega L_2 \equiv |\mathbf{H}_z| \underline{/\theta_{\mathbf{H}}}$$

or, normalizing and writing $\omega_x \equiv R_1/L_2$,

$$\frac{\mathbf{H}_{z\infty}(\omega)}{R_1} = 1 + j(\omega/\omega_x) = \sqrt{1 + (\omega/\omega_x)^2} \underline{/\tan^{-1}(\omega/\omega_x)}$$

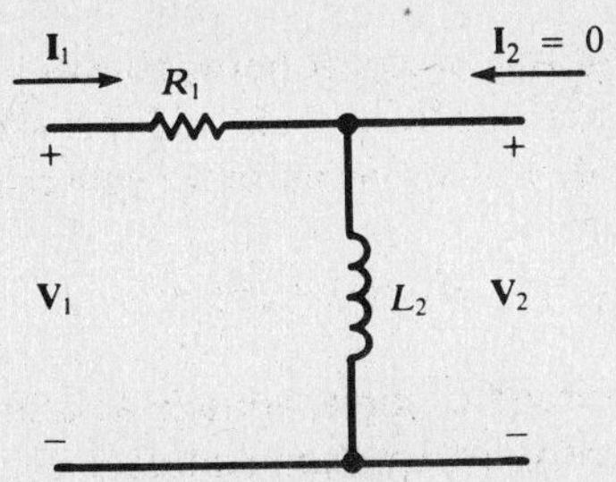

Fig. 11-4

Five values of ω provide sufficient data to plot $|\mathbf{H}_z|/R_1$ and $\theta_{\mathbf{H}}$, as shown in Fig. 11-5. The magnitude approaches infinity with increasing frequency, and so, at very high frequencies, the network current $\mathbf{I}_1$ will be zero.

ω	$\lvert\mathbf{H}_z\rvert/R_1$	$\theta_{\mathbf{H}}$
0	1	0°
$0.5\,\omega_x$	$0.5\sqrt{5}$	26.6°
ω_x	$\sqrt{2}$	45°
$2\omega_x$	$\sqrt{5}$	63.4°
∞	∞	90°

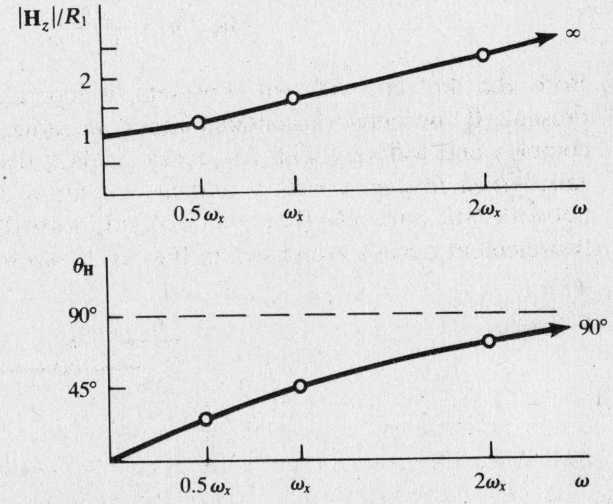

Fig. 11-5

In a similar manner, the frequency response of the output-to-input voltage ratio can be obtained. Voltage division under no-load gives

$$\mathbf{H}_{v\infty}(\omega) = \frac{j\omega L_2}{R_1 + j\omega L_2} = \frac{1}{1 - j(\omega_x/\omega)}$$

so that

$$|\mathbf{H}_v| = \frac{1}{\sqrt{1 + (\omega_x/\omega)^2}} \qquad \text{and} \qquad \theta_{\mathbf{H}} = \tan^{-1}(\omega_x/\omega)$$

The magnitude and angle are plotted in Fig. 11-6. This transfer function approaches unity at high frequency, where the output voltage is the same as the input. Hence the description "low-frequency roll-off" and the name "high-pass."

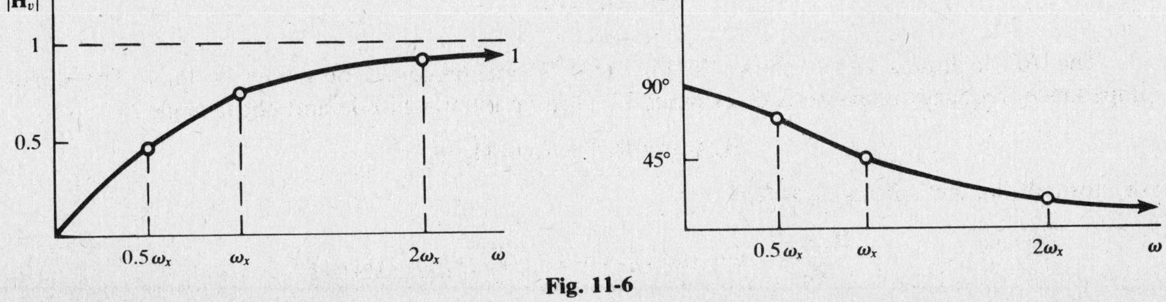

Fig. 11-6

A transfer impedance of the RL high-pass circuit under no-load is

$$\mathbf{H}_\infty(\omega) = \frac{\mathbf{V}_2}{\mathbf{I}_1} = j\omega L_2 \qquad \text{or} \qquad \frac{\mathbf{H}_\infty(\omega)}{R_1} = j\,\frac{\omega}{\omega_x}$$

The angle is constant at 90°; the graph of magnitude versus ω is a straight line, similar to a reactance plot of ωL versus ω. See Fig. 11-7.

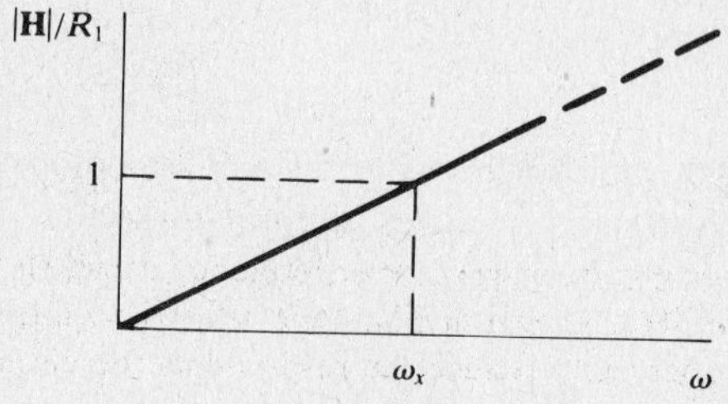

Fig. 11-7

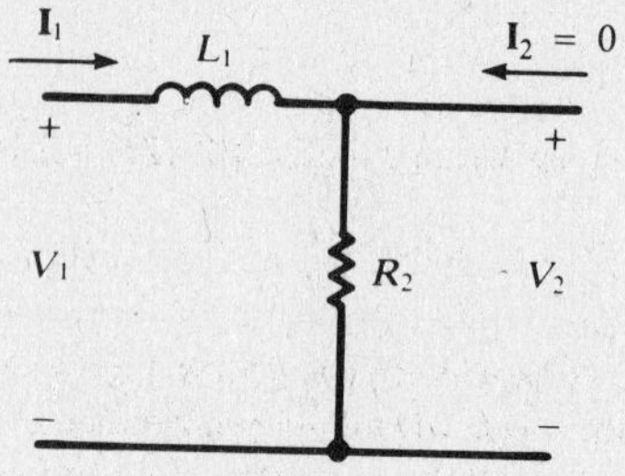

Fig. 11-8

Interchanging the positions of R and L results in a low-pass network with high-frequency roll-off (Fig. 11-8). For the open-circuit condition,

$$\mathbf{H}_{v\infty}(\omega) = \frac{R_2}{R_2 + j\omega L_1} = \frac{1}{1 + j(\omega/\omega_x)}$$

with $\omega_x \equiv R_2/L_1$; that is,

$$|\mathbf{H}_v| = \frac{1}{\sqrt{1 + (\omega/\omega_x)^2}} \qquad \text{and} \qquad \theta_{\mathbf{H}} = \tan^{-1}(-\omega/\omega_x)$$

The magnitude and angle plots are shown in Fig. 11-9. The voltage transfer function $\mathbf{H}_{v\infty}$ approaches zero at high frequencies and unity at $\omega = 0$. Hence the name "low-pass."

The other network functions of this low-pass network are obtained in the Solved Problems.

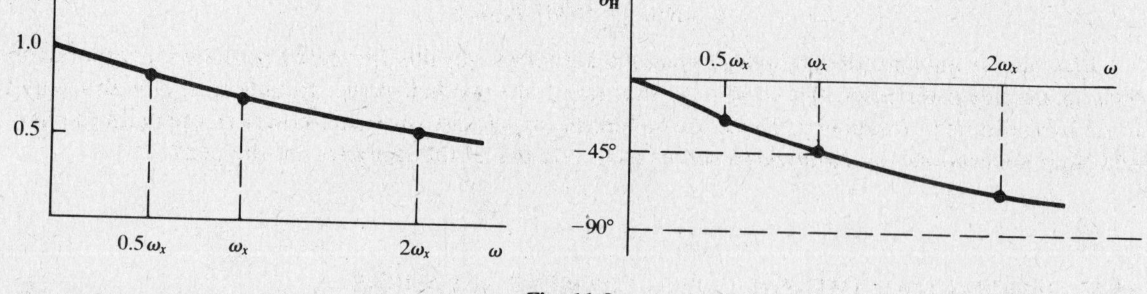

Fig. 11-9

EXAMPLE 11.1 Obtain the voltage transfer function $\mathbf{H}_{v\infty}$ for the open circuit shown in Fig. 11-10. At what frequency, in hertz, does $|\mathbf{H}_v| = 1/\sqrt{2}$ if (*a*) $C_2 = 10$ nF, (*b*) $C_2 = 1$ nF?

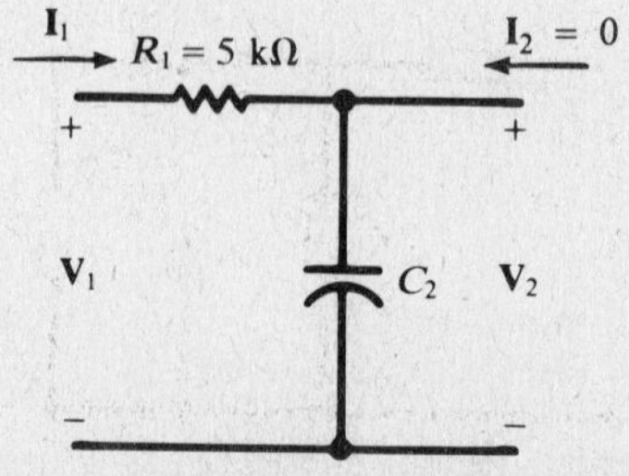

Fig. 11-10

$$\mathbf{H}_{v\infty}(\omega) = \frac{1/j\omega C_2}{R_1 + (1/j\omega C_2)} = \frac{1}{1 + j(\omega/\omega_x)} \qquad \text{where} \qquad \omega_x \equiv \frac{1}{R_1 C_2} = \frac{2\times 10^{-4}}{C_2} \quad \text{(rad/s)}$$

(*a*)
$$|\mathbf{H}_v| = \frac{1}{\sqrt{1 + (\omega/\omega_x)^2}}$$

and so $|\mathbf{H}_v| = 1/\sqrt{2}$ when

$$\omega = \omega_x = \frac{2\times 10^{-4}}{10\times 10^{-9}} = 2\times 10^4 \text{ rad/s}$$

or when $f = (2\times 10^4)/2\pi = 3.18$ kHz.

(*b*)
$$f = \frac{10}{1}(3.18) = 31.8 \text{ kHz}$$

Comparing (*a*) and (*b*), it is seen that the greater the value of C_2, the lower is the frequency at which $|\mathbf{H}_v|$ drops to 0.707 of its peak value, 1; i.e., the more is the graph of $|\mathbf{H}_v|$, shown in Fig. 11-9, shifted to the left. Consequently, any stray shunting capacitance, in parallel with C_2, serves to reduce the response of the circuit.

11.4 HALF-POWER FREQUENCIES

The frequency ω_x calculated in Example 11.1, the frequency at which

$$|\mathbf{H}_v| = 0.707\,|\mathbf{H}_v|_{\max}$$

is called the *half-power frequency*. In this case, the name is justified by Problem 11.5, which shows that the power input into the circuit of Fig. 11-10 will be half-maximum when

$$\left|\frac{1}{j\omega C_2}\right| = R_1$$

that is, when $\omega = \omega_x$.

Quite generally, any nonconstant network function $\mathbf{H}(\omega)$ will attain its greatest absolute value at some unique frequency ω_x. We shall call a frequency at which

$$|\mathbf{H}(\omega)| = 0.707\,|\mathbf{H}(\omega_x)|$$

a *half-power frequency* (or *half-power point*), whether or not this frequency actually corresponds to 50% power. In most cases, $0 < \omega_x < \infty$, so that there are two half-power frequencies, one above and one below the peak frequency. These are called the *upper* and *lower* half-power frequencies (points), and their separation, the *bandwidth*, serves as a measure of the sharpness of the peak.

11.5 GENERALIZED TWO-PORT, TWO-ELEMENT NETWORKS

The basic *RL* or *RC* network of the type examined in Section 11.3 can be generalized with $\mathbf{Z}_1$ and $\mathbf{Z}_2$, as shown in Fig. 11-11; the load impedance $\mathbf{Z}_L$ is connected at the output port.

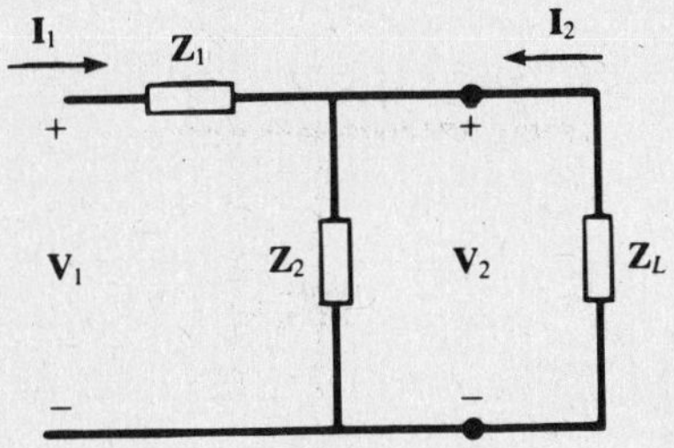

Fig. 11-11

By voltage division,

$$\mathbf{V}_2 = \frac{\mathbf{Z}'}{\mathbf{Z}_1 + \mathbf{Z}'}\mathbf{V}_1 \qquad \text{or} \qquad \mathbf{H}_v = \frac{\mathbf{Z}'}{\mathbf{Z}_1 + \mathbf{Z}'}$$

where $\mathbf{Z}' = \mathbf{Z}_2\mathbf{Z}_L/(\mathbf{Z}_2 + \mathbf{Z}_L)$, the equivalent impedance of $\mathbf{Z}_2$ and $\mathbf{Z}_L$ in parallel. The other transfer functions are calculated similarly, and are displayed in Table 11-1.

Table 11-1

Network Function / Output Condition	$\mathbf{H}_z = \dfrac{\mathbf{V}_1}{\mathbf{I}_1}$ (Ω)	$\mathbf{H}_v = \dfrac{\mathbf{V}_2}{\mathbf{V}_1}$	$\mathbf{H}_i = \dfrac{\mathbf{I}_2}{\mathbf{I}_1}$	$\mathbf{H}_v\mathbf{H}_z = \dfrac{\mathbf{V}_2}{\mathbf{I}_1}$ (Ω)	$\dfrac{\mathbf{H}_i}{\mathbf{H}_z} = \dfrac{\mathbf{I}_2}{\mathbf{V}_1}$ (S)
Short-circuit, $\mathbf{Z}_L = 0$	$\mathbf{Z}_1$	0	-1	0	$-\dfrac{1}{\mathbf{Z}_1}$
Open-circuit, $\mathbf{Z}_L = \infty$	$\mathbf{Z}_1 + \mathbf{Z}_2$	$\dfrac{\mathbf{Z}_2}{\mathbf{Z}_1 + \mathbf{Z}_2}$	0	$\mathbf{Z}_2$	0
Load, $\mathbf{Z}_L$	$\mathbf{Z}_1 + \mathbf{Z}'$	$\dfrac{\mathbf{Z}'}{\mathbf{Z}_1 + \mathbf{Z}'}$	$\dfrac{-\mathbf{Z}_2}{\mathbf{Z}_2 + \mathbf{Z}_L}$	$\mathbf{Z}'$	$\dfrac{-\mathbf{Z}'}{\mathbf{Z}_L(\mathbf{Z}_1 + \mathbf{Z}')}$

11.6 *RLC* SERIES CIRCUIT; SERIES RESONANCE

The *RLC* circuit shown in Fig. 11-12 has, under open-circuit condition, an input or driving-point impedance

$$\mathbf{Z}_{in}(\omega) = R + j\left(\omega L - \frac{1}{\omega C}\right)$$

The circuit is said to be in *series resonance* (or *low-impedance resonance*) when $\mathbf{Z}_{in}(\omega)$ is real (and so $|\mathbf{Z}_{in}(\omega)|$ is a minimum); that is, when

$$\omega L - \frac{1}{\omega C} = 0 \qquad \text{or} \qquad \omega = \omega_0 \equiv \frac{1}{\sqrt{LC}}$$

Figure 11-13 shows the frequency response. The capacitive reactance, inversely proportional to ω, is higher at low frequencies, while the inductive reactance, directly proportional to ω, is greater at the higher frequencies. Consequently, the net reactance at frequencies below ω_0 is capacitive, and the angle on $\mathbf{Z}_{in}$ is negative. At frequencies above ω_0, the circuit appears inductive, and the angle on $\mathbf{Z}_{in}$ is positive.

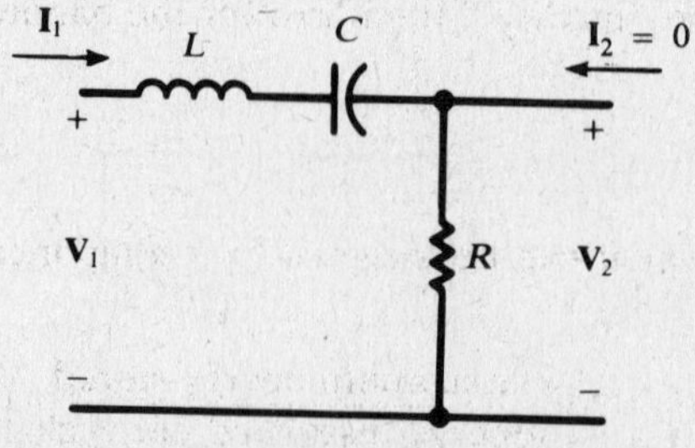

Fig. 11-12

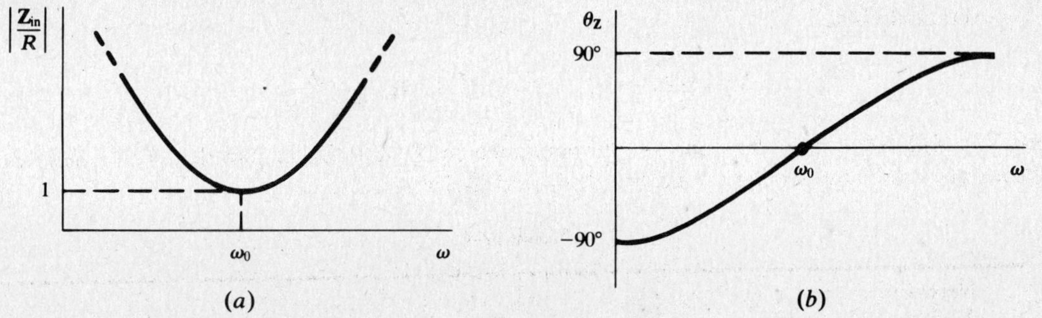

Fig. 11-13

By voltage division, the voltage transfer function for Fig. 11-12 is

$$\mathbf{H}_{v\infty}(\omega) = \frac{R}{\mathbf{Z}_{\text{in}}(\omega)} = R\mathbf{Y}_{\text{in}}(\omega)$$

The frequency response (magnitude only) is plotted in Fig. 11-14; the curve is just the reciprocal of that in Fig. 11-13(*a*). Note that roll-off occurs both below and above the series resonant frequency ω_0. The points where the response is 0.707, the half-power points (Section 11.4), are at frequencies ω_l and ω_h. The bandwidth is the width between these two frequencies: $\beta = \omega_h - \omega_l$.

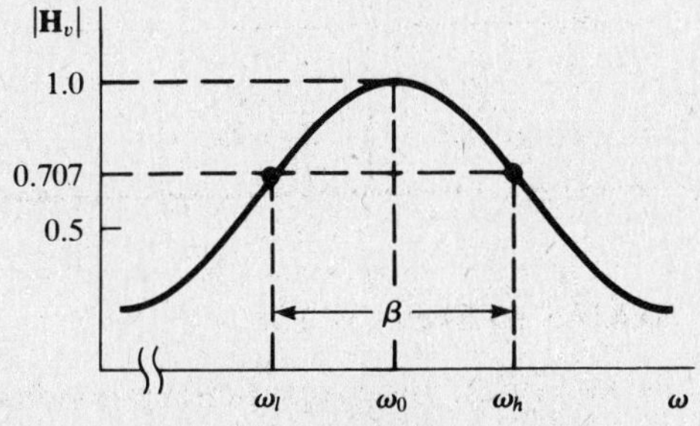

Fig. 11-14

A *quality factor*, $Q_0 = \omega_0 L/R$, may be defined for the series *RLC* circuit at resonance. (See Section 11.7 for the general development of *Q*.) The half-power frequencies can be expressed in terms of the circuit elements, or in terms of ω_0 and Q_0, as follows:

$$\omega_h = \frac{R}{2L} + \sqrt{\left(\frac{R}{2L}\right)^2 + \frac{1}{LC}} = \omega_0\left(\sqrt{1 + \frac{1}{4Q_0^2}} + \frac{1}{2Q_0}\right)$$

$$\omega_l = -\frac{R}{2L} + \sqrt{\left(\frac{R}{2L}\right)^2 + \frac{1}{LC}} = \omega_0\left(\sqrt{1 + \frac{1}{4Q_0^2}} - \frac{1}{2Q_0}\right)$$

See Problem 11.5. Subtraction of the expressions gives

$$\beta = \frac{R}{L} = \frac{\omega_0}{Q_0}$$

which suggests that the higher the "quality," the narrower the bandwidth.

11.7 QUALITY FACTOR

A *quality factor* or *figure of merit* can be assigned to a component or to a complete circuit. It is defined as

$$Q \equiv 2\pi\left(\frac{\text{maximum energy stored}}{\text{energy dissipated per cycle}}\right)$$

a dimensionless number.

A practical inductor, in which both resistance and inductance are present, is modeled in Fig. 11-15. The maximum stored energy is $\frac{1}{2}LI_{max}^2$, while the energy dissipated per cycle is

$$(I_{eff}^2 R)\left(\frac{2\pi}{\omega}\right) = \frac{I_{max}^2 R\pi}{\omega}$$

Hence

$$Q_{ind} = \frac{\omega L}{R}$$

A practical capacitor can be modeled by a parallel combination of R and C, as shown in Fig. 11-16. The maximum stored energy is $\frac{1}{2}CV_{max}^2$ and the energy dissipated per cycle is $V_{max}^2\pi/R\omega$. Thus, $Q_{cap} = \omega CR$.

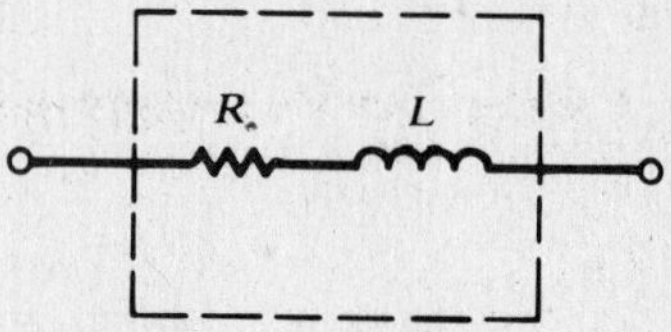

Fig. 11-15

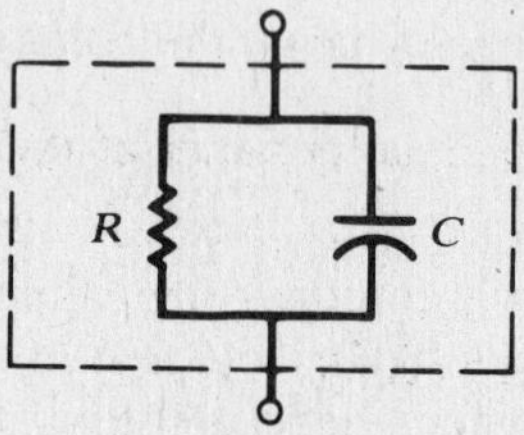

Fig. 11-16

The Q of the series RLC circuit is derived in Problem 11.6(a). It is usually applied at resonance, in which case it has the equivalent forms

$$Q_0 = \frac{\omega_0 L}{R} = \frac{1}{\omega_0 CR} = \frac{1}{R}\sqrt{\frac{L}{C}}$$

11.8 *RLC* PARALLEL CIRCUIT; PARALLEL RESONANCE

A parallel RLC network is shown in Fig. 11-17. Observe that $\mathbf{V}_2 = \mathbf{V}_1$. Under the open-circuit condition, the input admittance is

$$\mathbf{Y}_{in}(\omega) = \frac{1}{R} + \frac{1}{j\omega L} + j\omega C = \frac{1}{\mathbf{Z}_{in}(\omega)}$$

The network will be in *parallel resonance* (or *high-impedance resonance*) when $\mathbf{Y}_{in}(\omega)$, and thus $\mathbf{Z}_{in}(\omega)$, is real (and so $|\mathbf{Y}_{in}(\omega)|$ is a minimum and $|\mathbf{Z}_{in}(\omega)|$ is a maximum); that is, when

$$-\frac{1}{\omega L} + \omega C = 0 \qquad \text{or} \qquad \omega = \omega_a \equiv \frac{1}{\sqrt{LC}}$$

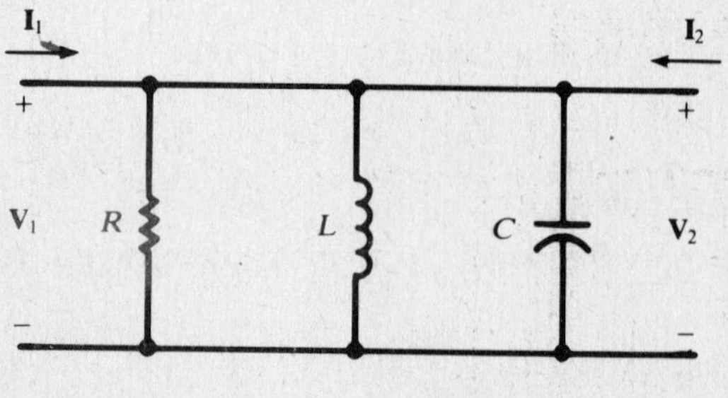

Fig. 11-17

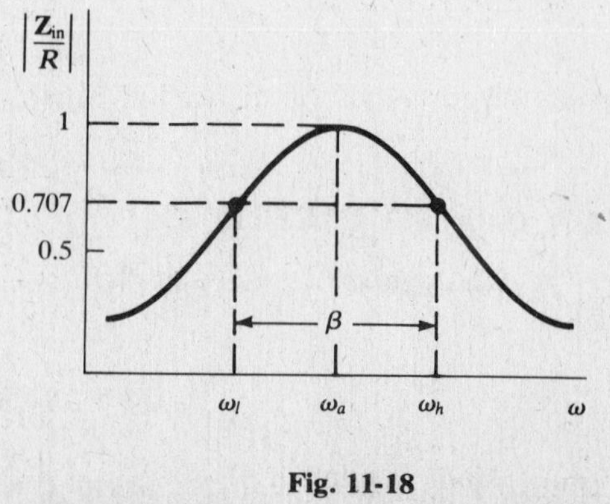

Fig. 11-18

The symbol ω_a is now used to denote the quantity $1/\sqrt{LC}$ in order to distinguish the resonance from a low-impedance resonance. Complex series-parallel networks may have several high-impedance resonant frequencies ω_a and several low-impedance resonant frequencies ω_0.

The normalized input impedance

$$\frac{\mathbf{Z}_{\text{in}}(\omega)}{R} = \frac{1}{1 + jR\left(\omega C - \frac{1}{\omega L}\right)}$$

is plotted (magnitude only) in Fig. 11-18. Half-power frequencies ω_l and ω_h are indicated on the plot. Analogous to series resonance, the bandwidth is given by

$$\beta = \frac{\omega_a}{Q_a}$$

where Q_a, the quality factor of the parallel circuit at $\omega = \omega_a$, has the equivalent expressions

$$Q_a = \frac{R}{\omega_a L} = \omega_a RC = R\sqrt{\frac{C}{L}}$$

See Problem 11.6(*b*).

11.9 PRACTICAL *LC* PARALLEL CIRCUIT

A parallel *LC* "tank" circuit has frequent applications in electronics as a tuning or frequency selection device. While the capacitor may often be treated as "pure *C*," the losses in the inductor should be included. A reasonable model for the practical tank is shown in Fig. 11-19. The input admittance is

$$\mathbf{Y}_{\text{in}}(\omega) = j\omega C + \frac{1}{R + j\omega L} = \frac{R}{R^2 + (\omega L)^2} + j\left[\omega C - \frac{\omega L}{R^2 + (\omega L)^2}\right]$$

For resonance,

$$\omega_a C = \frac{\omega_a L}{R^2 + (\omega_a L)^2} \qquad \text{or} \qquad \omega_a = \frac{1}{\sqrt{LC}}\sqrt{1 - \frac{R^2 C}{L}}$$

At the resonant frequency, $\mathbf{Y}_{\text{in}}(\omega_a) = RC/L$ and, from Section 11.6, the Q of the inductance is

$$Q_{\text{ind}} = \frac{\omega_a L}{R} = \sqrt{\frac{L}{CR^2} - 1}$$

If $Q_{\text{ind}} \geq 10$, then $\omega_a \approx 1/\sqrt{LC}$ and

$$\left|\frac{\mathbf{Z}_{\text{in}}(\omega_a)}{R}\right| \approx Q_{\text{ind}}^2$$

The frequency response is similar to that of the parallel *RLC* circuit, except that the high-impedance resonance occurs at a lower frequency for low Q_{ind}. This becomes evident when the expression for ω_a above is rewritten as

$$\omega_a = \left(\frac{1}{\sqrt{LC}}\right)\frac{1}{\sqrt{1 + (1/Q_{\text{ind}}^2)}}$$

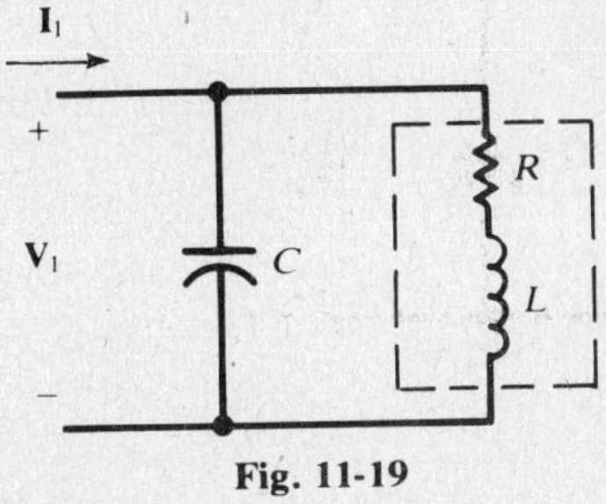

Fig. 11-19

11.10 SERIES-PARALLEL CONVERSIONS

It is often convenient in the analysis of circuits to convert the series RL to the parallel form (see Fig. 11-20). Given R_s, L_s, and the operating frequency ω, the elements R_p, L_p of the equivalent parallel circuit are determined by equating the admittances

$$\mathbf{Y}_s = \frac{R_s - j\omega L_s}{R_s^2 + (\omega L_s)^2} \quad \text{and} \quad \mathbf{Y}_p = \frac{1}{R_p} + \frac{1}{j\omega L_p}$$

The results are:

$$R_p = R_s\left[1 + \left(\frac{\omega L_s}{R_s}\right)^2\right] = R_s(1 + Q_s^2)$$

$$L_p = L_s\left[1 + \left(\frac{R_s}{\omega L_s}\right)^2\right] = L_s\left(1 + \frac{1}{Q_s^2}\right)$$

If $Q_s \geq 10$, $R_p \approx R_s Q_s^2$ and $L_p \approx L_s$.

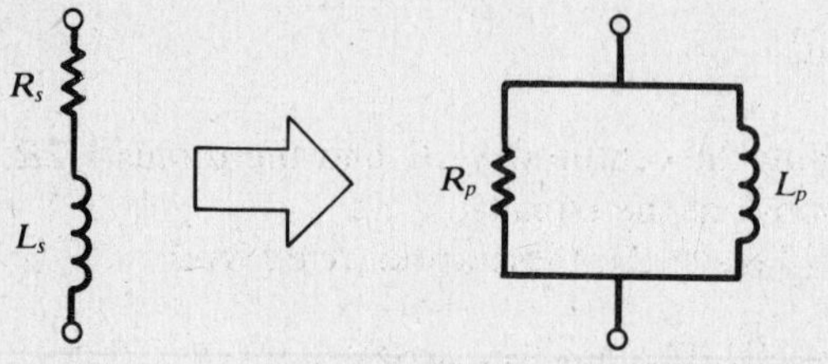

Fig. 11-20

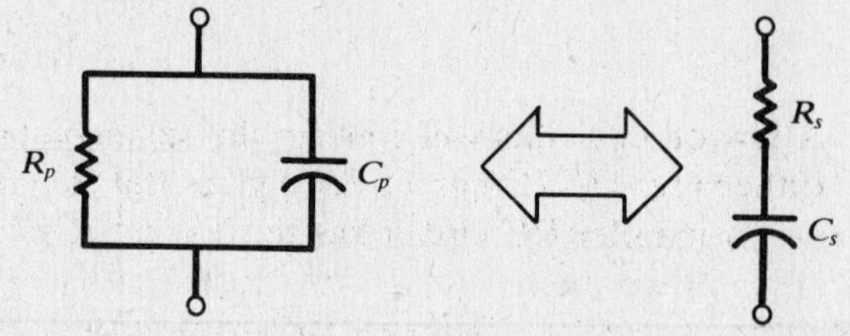

Fig. 11-21

There are times when the RC circuit in either form should be converted to the other form (see Fig. 11-21). Equating either the impedances or the admittances, one finds

$$R_s = \frac{R_p}{1 + (\omega C_p R_p)^2} = \frac{R_p}{1 + Q_p^2}$$

$$C_s = C_p\left[1 + \frac{1}{(\omega C_p R_p)^2}\right] = C_p\left(1 + \frac{1}{Q_p^2}\right)$$

as the parallel-to-series transformation, and

$$R_p = R_s\left[1 + \frac{1}{(\omega C_s R_s)^2}\right] = R_s(1 + Q_s^2)$$

$$C_p = \frac{C_s}{1 + (\omega C_s R_s)^2} = \frac{C_s}{1 + (1/Q_s)^2}$$

as the series-to-parallel transformation. Again, the equivalence depends on the operating frequency.

11.11 LOCUS DIAGRAMS

Heretofore, the frequency response of a network has been exhibited by plotting separately the magnitude and the angle of a suitable network function against frequency ω. This same information can be presented in a single plot: one finds the curve (*locus diagram*) in the complex plane traced by the point representing the network function as ω varies from 0 to ∞. In this section we shall discuss locus diagrams for the input impedance or the input admittance; in some cases the variable will not be ω, but another parameter (such as resistance R).

For the series RL circuit, Fig. 11-22(a) shows the $\mathbf{Z}$-locus when ωL is fixed and R is variable; Fig. 11-22(b) shows the $\mathbf{Z}$-locus when R is fixed and L or ω is variable; and Fig. 11-22(c) shows the $\mathbf{Y}$-locus when R is fixed and L or ω is variable. This last locus is obtained from

$$\mathbf{Y} = \frac{1}{R + j\omega L} = \frac{1}{\sqrt{R^2 + (\omega L)^2}} \underline{/\tan^{-1}(-\omega L/R)}$$

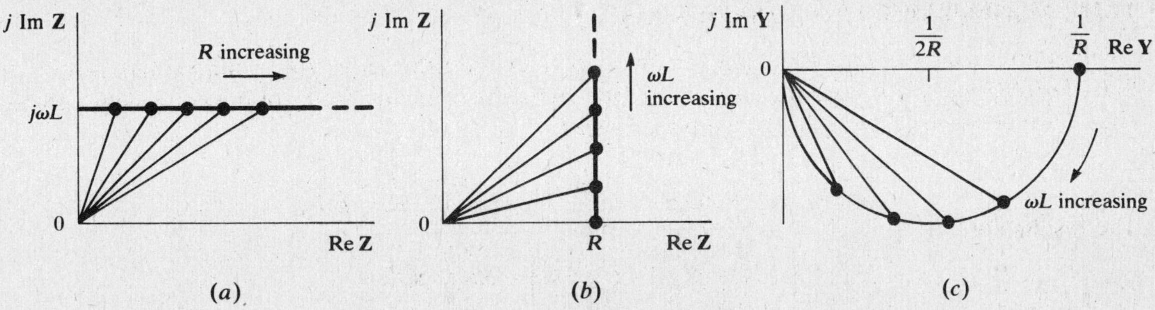

Fig. 11-22

Note that for $\omega L = 0$, $\mathbf{Y} = (1/R)\underline{/0^\circ}$; and for $\omega L \to \infty$, $\mathbf{Y} \to 0\underline{/-90^\circ}$. When $\omega L = R$,

$$\mathbf{Y} = \frac{1}{R\sqrt{2}}\underline{/-45^\circ}$$

A few other points will confirm the semicircular locus, with the center at $1/2R$ and the radius $1/2R$. Either Fig. 11-22(b) or 11-22(c) gives the frequency response of the circuit.

A parallel RC circuit has the **Y**- and **Z**-loci shown in Fig. 11-23; these are derived from

$$\mathbf{Y} = \frac{1}{R} + j\omega C \qquad \text{and} \qquad \mathbf{Z} = \frac{R}{\sqrt{1+(\omega CR)^2}}\underline{/\tan^{-1}(-\omega CR)}$$

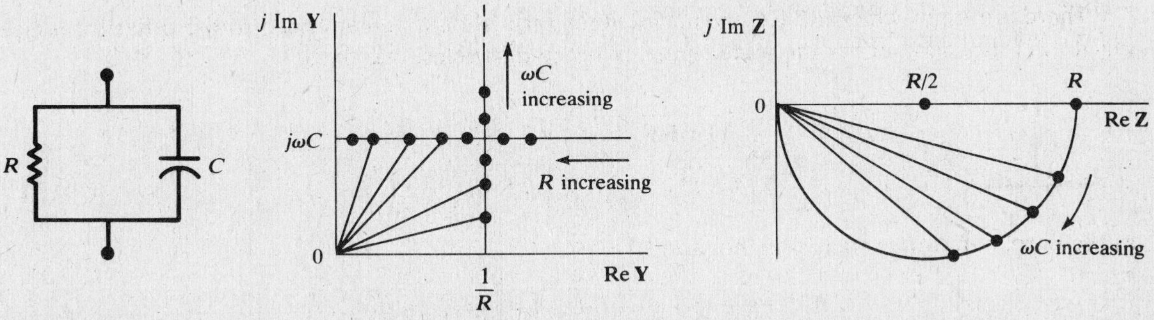

Fig. 11-23

For the RLC series circuit, the **Y**-locus, with ω as the variable, may be determined by writing

$$\mathbf{Y} = G + jB = \frac{1}{R+jX} = \frac{R-jX}{R^2+X^2}$$

whence

$$G = \frac{R}{R^2+X^2} \qquad B = -\frac{X}{R^2+X^2}$$

Both G and B depend on ω via X. Eliminating X between the two expressions yields the equation of the locus in the form

$$G^2 + B^2 = \frac{G}{R} \qquad \text{or} \qquad \left(G - \frac{1}{2R}\right)^2 + B^2 = \left(\frac{1}{2R}\right)^2$$

i.e. the circle shown in Fig. 11-24. Note the points on the locus corresponding to $\omega = \omega_l$, $\omega = \omega_0$, and $\omega = \omega_h$.

For the practical "tank" circuit examined in Section 11.9, the **Y**-locus may be constructed by combining the C-branch locus and the RL-branch locus. To illustrate the addition, the points

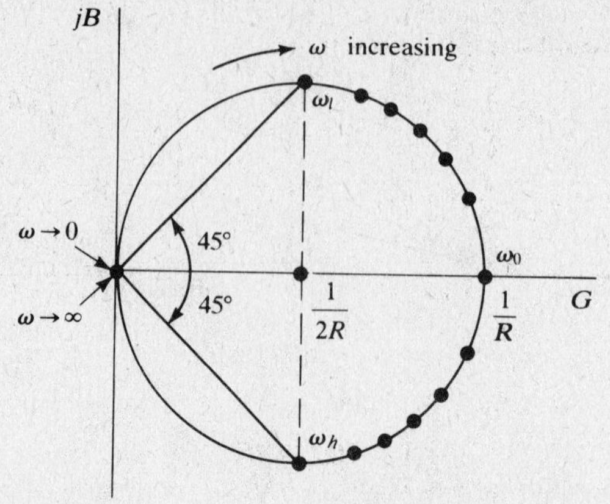

Fig. 11-24

corresponding to frequencies $\omega_1 < \omega_2 < \omega_3$ are marked on the individual loci and on the sum, shown in Fig. 11-25(*c*). It is seen that $|\mathbf{Y}|_{\min}$ occurs at a frequency greater than ω_a; that is, the resonance is high-impedance but not maximum-impedance. This comes about because G varies with ω (see Section 11.9), and varies in such a way that forcing $B = 0$ does not automatically minimize $G^2 + B^2$. The separation of the resonance and minimum-admittance frequencies is governed by the Q of the coil. Higher Q_{ind} corresponds to lower values of R. It is seen from Fig. 11-25(*b*) that low R results in a larger semicircle, which when combined with the $\mathbf{Y}_C$-locus, gives a higher ω_a and a lower minimum-admittance frequency. When $Q_{\text{ind}} \geq 10$, the two frequencies may be taken as coincident.

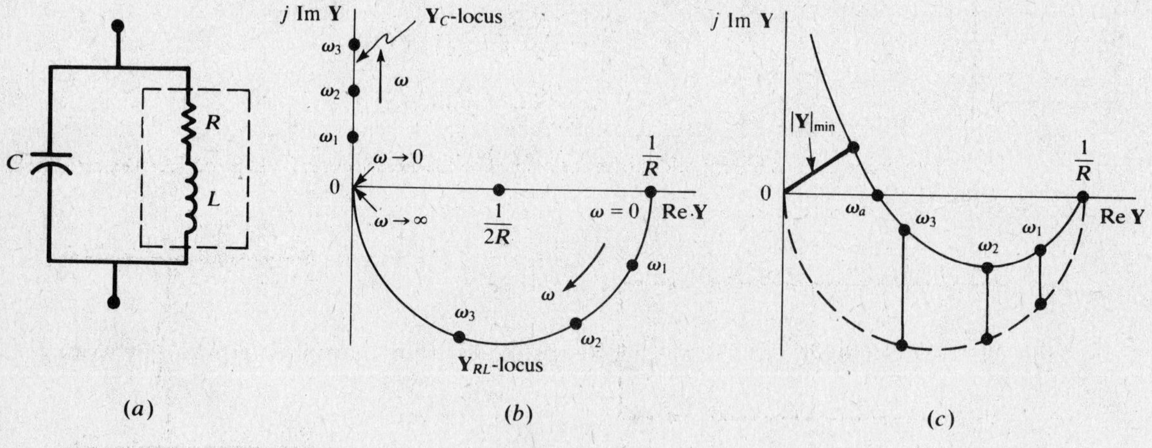

Fig. 11-25

The case of the two-branch RC and RL circuit shown in Fig. 11-26(*a*) can be examined by adding the admittance loci of the two branches. For fixed $\mathbf{V} = V\underline{/0^\circ}$, this amounts to adding the loci of the two branch currents. Consider C variable without limit, and R_1, R_2, L, and ω constant. Then current $\mathbf{I}_L$ is fixed as shown in Fig. 11-26(*b*). The semicircular locus of $\mathbf{I}_C$ is added to $\mathbf{I}_L$ to result in the locus of $\mathbf{I}_T$.

Resonance of the circuit corresponds to $\theta_T = 0$. This may occur for two values of the real, positive parameter C [the case illustrated in Fig. 11-26(*b*)], for one value, or for no value—depending on the number of real positive roots of the equation $\text{Im}\,\mathbf{Y}_T(C) = 0$.

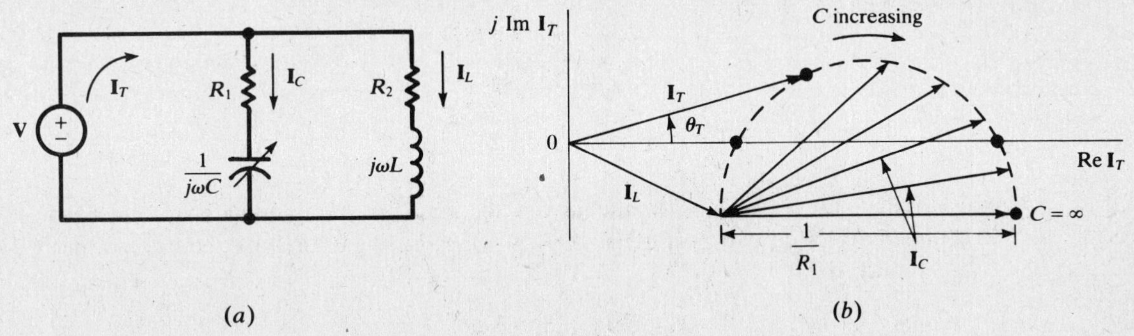

Fig. 11-26

Solved Problems

11.1 In the two-port network shown in Fig. 11-27, $R_1 = 7$ kΩ and $R_2 = 3$ kΩ. Obtain the voltage ratio $\mathbf{V}_2/\mathbf{V}_1$ (*a*) at no-load, (*b*) for $R_L = 20$ kΩ.

(*a*) At no-load, voltage division gives

$$\frac{\mathbf{V}_2}{\mathbf{V}_1} = \frac{R_2}{R_1 + R_2} = \frac{3}{7+3} = 0.30$$

(*b*) With $R_L = 20$ kΩ,

$$R_p = \frac{R_2 R_L}{R_2 + R_L} = \frac{60}{23} \quad \text{k}\Omega$$

and

$$\frac{\mathbf{V}_2}{\mathbf{V}_1} = \frac{R_p}{R_1 + R_p} = \frac{60}{221} = 0.27$$

The voltage ratio is independent of frequency. The load resistance, 20 kΩ, reduced the ratio from 0.30 to 0.27.

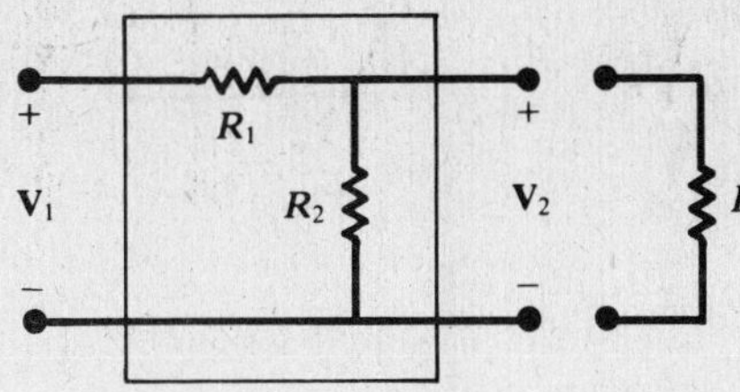

Fig. 11-27

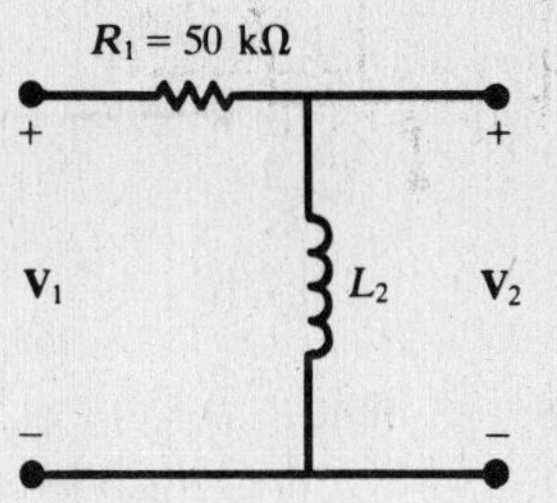

Fig. 11-28

11.2 (*a*) Find L_2 in the high-pass circuit shown in Fig. 11-28, if $|\mathbf{H}_v(\omega)| = 0.50$ at a frequency of 50 MHz. (*b*) At what frequency is $|\mathbf{H}_v| = 0.90$?

(*a*) From Section 11.3, with $\omega_x \equiv R_1/L_2$,

$$|\mathbf{H}_v(\omega)| = \frac{1}{\sqrt{1 + (\omega_x/\omega)^2}}$$

Then,

$$0.50 = \frac{1}{\sqrt{1 + (f_x/50)^2}} \quad \text{or} \quad f_x = 50\sqrt{3} \text{ MHz}$$

and $$L_2 = \frac{R_1}{2\pi f_x} = \frac{50 \times 10^3}{2\pi(50\sqrt{3} \times 10^6)} = 91.9\ \mu\text{H}$$

(*b*) $$0.90 = \frac{1}{1 + (50\sqrt{3}/f)^2} \qquad \text{or} \qquad f = 179\ \text{MHz}$$

11.3 A voltage divider, useful for high-frequency applications, can be made with two capacitors C_1 and C_2 in the generalized two-port network shown in Fig. 11-11. Under open-circuit, find C_2 if $C_1 = 0.01\ \mu\text{F}$ and $|\mathbf{H}_v| = 0.20$.

From Table 11-1,

$$\mathbf{H}_v = \frac{\mathbf{Z}_2}{\mathbf{Z}_1 + \mathbf{Z}_2} = \frac{1/j\omega C_2}{\dfrac{1}{j\omega C_1} + \dfrac{1}{j\omega C_2}} = \frac{C_1}{C_1 + C_2}$$

Hence $$0.20 = \frac{0.01}{0.01 + C_2} \qquad \text{or} \qquad C_2 = 0.04\ \mu\text{F}$$

The voltage ratio is seen to be frequency-independent under open-circuit.

11.4 Find the frequency at which $|\mathbf{H}_v| = 0.50$ for the low-pass RC network shown in Fig. 11-29.

$$\mathbf{H}_v(\omega) = \frac{1}{1 + j(\omega/\omega_x)} \qquad \text{where} \qquad \omega_x \equiv \frac{1}{R_1 C_2}$$

Then $$(0.50)^2 = \frac{1}{1 + (\omega/\omega_x)^2} \qquad \text{from which} \qquad \frac{\omega}{\omega_x} = \sqrt{3}$$

and $$\omega = \sqrt{3}\left(\frac{1}{R_1 C_2}\right) = 8660\ \text{rad/s} \qquad \text{or} \qquad f = 1378\ \text{Hz}$$

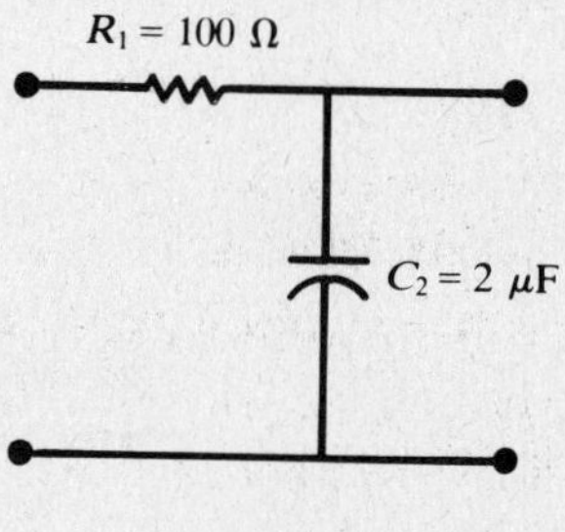

Fig. 11-29

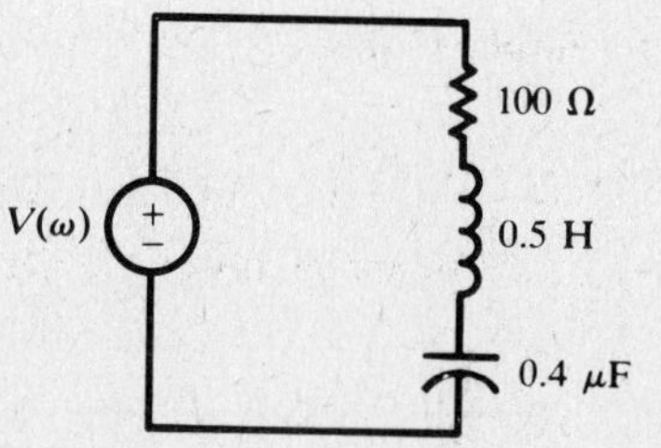

Fig. 11-30

11.5 For the series RLC circuit shown in Fig. 11-30, find the resonant frequency $\omega_0 = 2\pi f_0$. Also obtain the half-power frequencies and the bandwidth β.

$$\mathbf{Z}_{\text{in}}(\omega) = R + j\left(\omega L - \frac{1}{\omega C}\right)$$

At resonance, $\mathbf{Z}_{\text{in}}(\omega) = R$ and $\omega_0 = 1/\sqrt{LC}$.

$$\omega_0 = \frac{1}{\sqrt{0.5(0.4 \times 10^{-6})}} = 2236.1\ \text{rad/s} \qquad f_0 = \frac{\omega_0}{2\pi} = 355.9\ \text{Hz}$$

The power formula

$$P = I_{\text{eff}}^2 R = \frac{V_{\text{eff}}^2 R}{|\mathbf{Z}_{\text{in}}|^2}$$

shows that $P_{\text{max}} = V_{\text{eff}}^2/R$, achieved at $\omega = \omega_0$, and that $P = \frac{1}{2}P_{\text{max}}$ when $|\mathbf{Z}_{\text{in}}|^2 = 2R^2$; i.e., when

$$\omega L - \frac{1}{\omega C} = \pm R \qquad \text{or} \qquad \omega^2 \mp \frac{R}{L}\omega - \frac{1}{LC} = 0$$

Corresponding to the upper sign, there is a single real positive root:

$$\omega_h = \frac{R}{2L} + \sqrt{\left(\frac{R}{2L}\right)^2 + \frac{1}{LC}} = 2338.3 \text{ rad/s} \qquad \text{or} \qquad f_h = 372.1 \text{ Hz}$$

and corresponding to the lower sign, the single real positive root

$$\omega_l = -\frac{R}{2L} + \sqrt{\left(\frac{R}{2L}\right)^2 + \frac{1}{LC}} = 2138.3 \text{ rad/s} \qquad \text{or} \qquad f_l = 340.3 \text{ Hz}$$

11.6 Derive the Q of (a) the series RLC circuit, (b) the parallel RLC circuit.

(a) In the time domain, the instantaneous stored energy in the circuit is given by

$$W_s = \tfrac{1}{2}Li^2 + \frac{q^2}{2C}$$

For a maximum,

$$\frac{dW_s}{dt} = Li\frac{di}{dt} + \frac{q}{C}\frac{dq}{dt} = i\left(L\frac{di}{dt} + \frac{q}{C}\right) = i(v_L + v_C) = 0$$

Thus the maximum stored energy is W_s at $i = 0$ or W_s at $v_L + v_C = 0$, whichever is the larger. Now, the capacitor voltage, and therefore the charge, lags the current by 90°; hence $i = 0$ implies $q = \pm Q_{\text{max}}$ and

$$W_s|_{i=0} = \frac{Q_{\text{max}}^2}{2C} = \tfrac{1}{2}CV_{C\text{max}}^2 = \tfrac{1}{2}C\left(\frac{I_{\text{max}}}{\omega C}\right)^2 = \frac{I_{\text{max}}^2}{2C\omega^2}$$

On the other hand, $v_L + v_C = 0$ implies $v_L = v_C = 0$ and $i = \pm I_{\text{max}}$ (see the phasor diagram, Fig. 11-31), so that

$$W_s|_{v_L+v_C=0} = \tfrac{1}{2}LI_{\text{max}}^2$$

It follows that

$$W_{s\text{max}} = \begin{cases} I_{\text{max}}^2/2C\omega^2 & (\omega \le \omega_0) \\ LI_{\text{max}}^2/2 & (\omega \ge \omega_0) \end{cases}$$

The energy dissipated per cycle (in the resistor) is $W_d = I_{\text{max}}^2 R\pi/\omega$. Consequently,

$$Q = 2\pi\frac{W_{s\text{max}}}{W_d} = \begin{cases} 1/\omega CR & (\omega \le \omega_0) \\ \omega L/R & (\omega \ge \omega_0) \end{cases}$$

(b) For the parallel combination with applied voltage $v(t)$,

$$W_s = \tfrac{1}{2}Li_L^2 + \frac{1}{2C}q_C^2$$

and

$$\frac{dW_s}{dt} = Li_L\frac{di_L}{dt} + \frac{q_C}{C}i_C = v(i_L + i_C) = 0$$

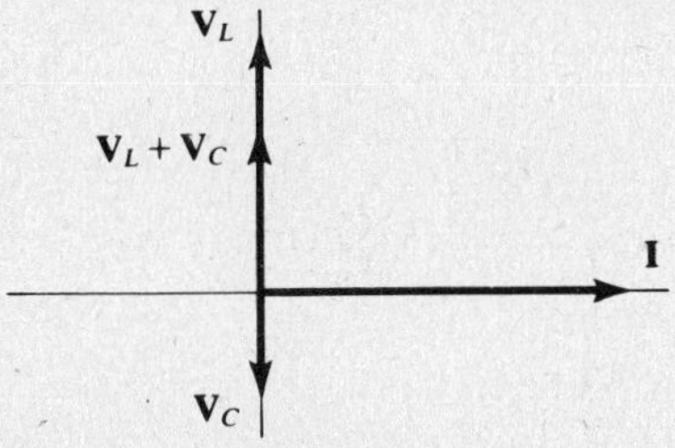

Fig. 11-31

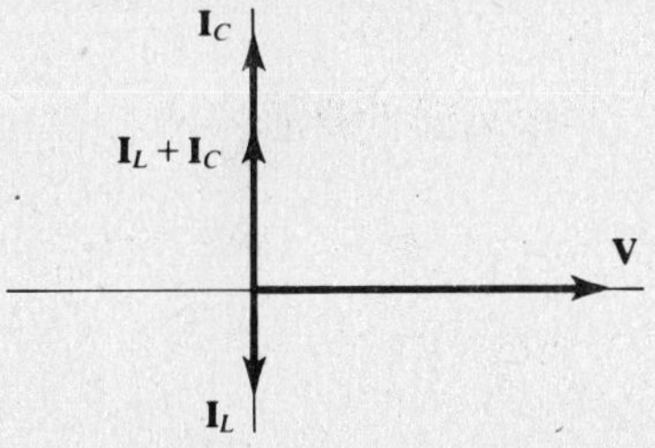

Fig. 11-32

If $v = 0$, then $q_C = 0$ and

$$i_L = \pm I_{L\max} = \pm \frac{V_{\max}}{\omega L}$$

giving
$$W_{s|v=0} = \frac{V_{\max}^2}{2L\omega^2}$$

If $i_L + i_C = 0$, then (see Fig. 11-32) $i_L = i_C = 0$ and $q_C = \pm CV_{\max}$, giving

$$W_s|_{i_L+i_C=0} = \tfrac{1}{2}CV_{\max}^2$$

Therefore,

$$W_{s\max} = \begin{cases} V_{\max}^2/2L\omega^2 & (\omega \le \omega_a) \\ CV_{\max}^2/2 & (\omega \ge \omega_a) \end{cases}$$

The energy dissipated per cycle in R is $W_d = V_{\max}^2\pi/R\omega$. Consequently,

$$Q = 2\pi \frac{W_{s\max}}{W_d} = \begin{cases} R/L\omega & (\omega \le \omega_a) \\ \omega CR & (\omega \ge \omega_a) \end{cases}$$

11.7 A three-element series circuit contains $R = 10\ \Omega$, $L = 5$ mH, and $C = 12.5\ \mu$F. Plot the magnitude and angle of $\mathbf{Z}$ as functions of ω for values of ω from $0.8\omega_0$ through $1.2\omega_0$.

$\omega_0 = 1/\sqrt{LC} = 4000$ rad/s. At ω_0,

$$X_L = (4000)(5 \times 10^{-3}) = 20\ \Omega \qquad X_C = \frac{1}{(4000)(12.5 \times 10^{-6})} = 20\ \Omega$$

$$\mathbf{Z} = 10 + j(X_L - X_C) = 10 + j0\ \ \Omega$$

The values of the reactances at other frequencies are readily obtained. A tabulation of reactances and impedances appears in Fig. 11-33(*a*), and Fig. 11-33(*b*) shows the required plots.

ω	X_L	X_C	$\mathbf{Z}$	
3200	16	25	$10 - j9$	$13.4\angle{-42^\circ}$
3600	18	22.2	$10 - j4.2$	$10.8\angle{-22.8^\circ}$
4000	20	20	10	$10\angle{0^\circ}$
4400	22	18.2	$10 + j3.8$	$10.7\angle{20.8^\circ}$
4800	24	16.7	$10 + j7.3$	$12.4\angle{36.2^\circ}$

(*a*)

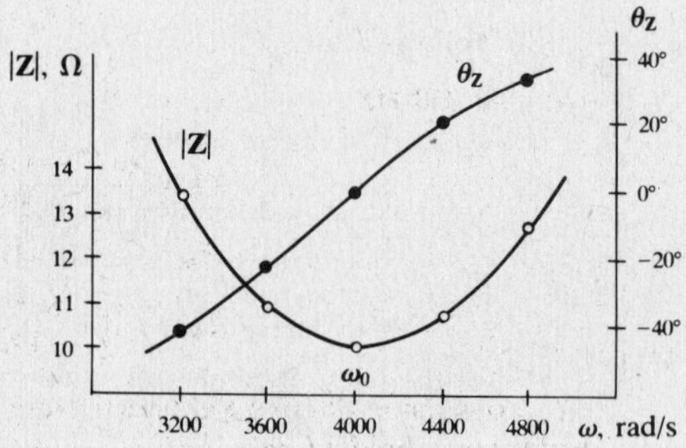

(*b*)

Fig. 11-33

11.8 Show that $\omega_0 = \sqrt{\omega_l\omega_h}$ for the series RLC circuit.

By the results of Problem 11.5,

$$\omega_l\omega_h = \left(\sqrt{\left(\frac{R}{2L}\right)^2 + \frac{1}{LC}} - \frac{R}{2L}\right)\left(\sqrt{\left(\frac{R}{2L}\right)^2 + \frac{1}{LC}} + \frac{R}{2L}\right) = \frac{1}{LC} = \omega_0^2$$

11.9 Compute the quality factor of an RLC series circuit, with $R = 20\ \Omega$, $L = 50$ mH, and $C = 1\ \mu$F, using (*a*) $Q = \omega_0 L/R$, (*b*) $Q = 1/\omega_0 CR$, (*c*) $Q = \omega_0/\beta$.

$$\omega_0 = \frac{1}{\sqrt{0.05 \times 10^{-6}}} = 4472 \text{ rad/s}$$

$$\omega_l = -\frac{R}{2L} + \sqrt{\left(\frac{R}{2L}\right)^2 + \frac{1}{LC}} = 4276.6 \text{ rad/s} \qquad \omega_h = \frac{R}{2L} + \sqrt{\left(\frac{R}{2L}\right)^2 + \frac{1}{LC}} = 4676.6 \text{ rad/s}$$

and $\beta = \omega_h - \omega_l = 400$ rad/s.

(*a*)
$$Q = \frac{\omega_0 L}{R} = \frac{4472(0.050)}{20} = 11.2$$

(*b*)
$$Q = \frac{1}{\omega_0 CR} = \frac{1}{4472(10^{-6})20} = 11.2$$

(*c*)
$$Q = \frac{\omega_0}{\beta} = \frac{4472}{400} = 11.2$$

11.10 A coil is represented by a series combination of $L = 50$ mH and $R = 15\ \Omega$. Calculate the quality factor at (*a*) 10 kHz, (*b*) 50 kHz.

(*a*)
$$Q_{\text{coil}} = \frac{\omega L}{R} = \frac{2\pi(10 \times 10^3)(50 \times 10^{-3})}{15} = 209$$

(*b*)
$$Q_{\text{coil}} = 209\left(\frac{50}{10}\right) = 1047$$

11.11 Convert the circuit constants of Problem 11.10 to the parallel form (*a*) at 10 kHz, (*b*) at 250 Hz.

(*a*)
$$R_p = R_s\left[1 + \left(\frac{\omega L_s}{R_s}\right)^2\right] = R_s[1 + Q_s^2] = 15[1 + (209)^2] = 655 \text{ k}\Omega$$

or, since $Q_s \gg 10$, $R_p \approx R_s Q_s^2 = 15(209)^2 = 655$ kΩ.

$$L_p = L_s\left(1 + \frac{1}{Q_s^2}\right) \approx L_s = 50 \text{ mH}$$

(*b*) At 250 Hz,

$$Q_s = \frac{2\pi(250)(50 \times 10^{-3})}{15} = 5.24$$

$$R_p = R_s[1 + Q_s^2] = 15[1 + (5.24)^2] = 426.9\ \Omega$$

$$L_p = L_s\left[1 + \frac{1}{Q_s^2}\right] = (50 \times 10^{-3})\left[1 + \frac{1}{(5.24)^2}\right] = 51.8 \text{ mH}$$

Conversion of circuit elements from series to parallel can be carried out at a specific frequency, the equivalence holding only at that frequency. Note that in (*b*), where $Q_s < 10$, L_p differs significantly from L_s.

11.12 For the circuit shown in Fig. 11-34, (*a*) obtain the voltage transfer function $\mathbf{H}_v(\omega)$, and (*b*) find the frequency at which the function is real.

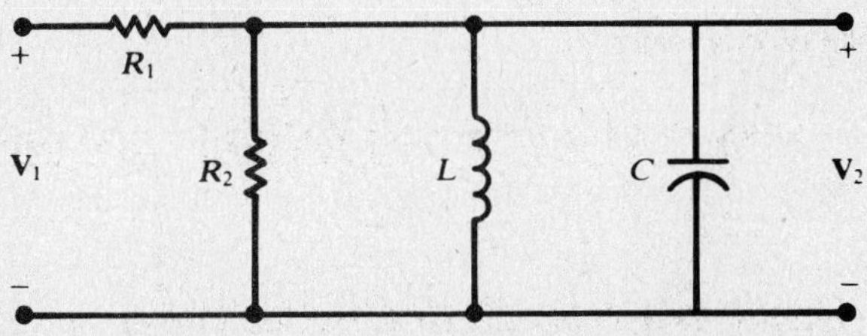

Fig. 11-34

(*a*) Let $\mathbf{Z}_2$ and $\mathbf{Y}_2$ represent the impedance and admittance of the R_2LC parallel tank.

$$\mathbf{H}_v(\omega) = \frac{\mathbf{Z}_2}{R_1 + \mathbf{Z}_2} = \frac{1}{1 + R_1\mathbf{Y}_2} = \frac{1}{1 + R_1\left(\dfrac{1}{R_2} + \dfrac{1}{j\omega L} + j\omega C\right)}$$

$$= \frac{1}{1 + \dfrac{R_1}{R_2} + jR_1\left(\omega C - \dfrac{1}{\omega L}\right)}$$

(*b*) The transfer function is real when $\mathbf{Y}_2$ is real; i.e., when

$$\omega = \omega_a \equiv \frac{1}{\sqrt{LC}}$$

At $\omega = \omega_a$, not only are $|\mathbf{Z}_2|$ and $|\mathbf{H}_v|$ maximized, but $|\mathbf{Z}_{in}| = |R_1 + \mathbf{Z}_2|$ also is maximized (because R_1 is real and positive—see the locus diagram, Fig. 11-35).

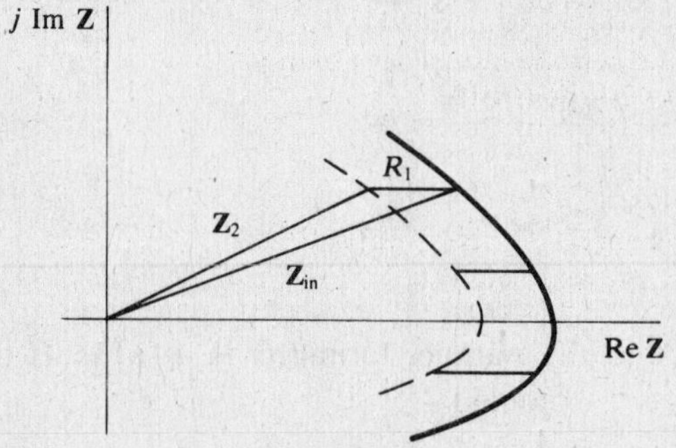

Fig. 11-35

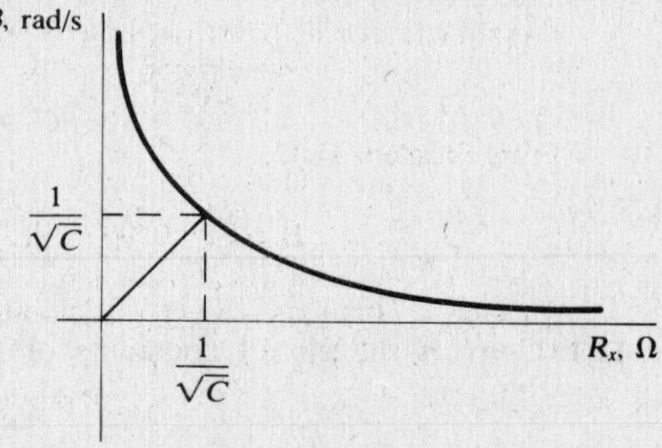

Fig. 11-36

11.13 Obtain the bandwidth β for the circuit of Fig. 11-34 and plot β against the parameter

$$R_x \equiv \frac{R_1R_2}{R_1 + R_2}$$

Here, the half-power frequencies are determined by the condition $|\mathbf{H}_v(\omega)| = 0.707\,|\mathbf{H}_v|_{max}$, or, from Problem 11.12(*a*),

$$R_1\left(\omega C - \frac{1}{\omega L}\right) = \pm\left(1 + \frac{R_1}{R_2}\right) \qquad \text{or} \qquad R_x\left(\omega C - \frac{1}{\omega L}\right) = \pm 1$$

But (see Section 11.8) this is just the equation for the half-power frequencies of an R_xLC parallel circuit. Hence,

$$\beta = \frac{\omega_a}{Q_a} = \frac{1}{CR_x}$$

The hyperbolic graph is shown in Fig. 11-36.

11.14 In the circuit of Fig. 11-34, let $R_1 = R_2 = 2\ \text{k}\Omega$, $L = 10$ mH, and $C = 40$ nF. Find the resonant frequency and bandwidth, and compare with the results for $R_1 = 0$ (i.e., a pure parallel circuit).

$$\omega_a = \frac{1}{\sqrt{(10 \times 10^{-3})(40 \times 10^{-9})}} = 5 \times 10^4\ \text{rad/s}$$

or $f_a = 7958$ Hz. With $R_x = 2^2/4 = 1\ \text{k}\Omega$, Problem 11.13 gives

$$\beta = \frac{1}{(40 \times 10^{-9})(1 \times 10^3)} = 2.5 \times 10^4\ \text{rad/s}$$

The results of Problems 11.12 and 11.13 cannot be applied as $R_1 \to 0$, for, in the limit, the voltage ratio is identically unity and so cannot provide any information about the residual R_2LC parallel circuit. (Note that $\beta \to \infty$ as $R_x \to 0$.) Instead, we must go over to the input impedance function, as in Section 11.8, whereby

$$\omega_a = \frac{1}{\sqrt{LC}} = 5 \times 10^4 \text{ rad/s}$$

as previously, and

$$\beta = \frac{1}{CR_2} = 1.25 \times 10^4 \text{ rad/s}$$

11.15 For the circuit of Fig. 11-34, $R_1 = 5\ \text{k}\Omega$ and $C = 10$ nF. If $\mathbf{V}_2/\mathbf{V}_1 = 0.8\underline{/0^\circ}$ at 15 kHz, calculate R_2, L, and the bandwidth.

An angle of zero on the voltage ratio $\mathbf{H}_v$ indicates that the circuit as a whole, and the parallel tank by itself, is at resonance (see Problem 11.14). Then,

$$\omega_a = \frac{1}{\sqrt{LC}} \qquad L = \frac{1}{\omega_a^2 C} = \frac{1}{[2\pi(15 \times 10^3)]^2(10 \times 10^{-9})} = 11.26 \text{ mH}$$

From Problem 11.12,

$$\mathbf{H}_v(\omega_a) = 0.8\underline{/0^\circ} = \frac{1}{1 + (R_1/R_2)} \qquad \text{whence} \qquad R_2 = \frac{R_1}{0.25} = 20\ \text{k}\Omega$$

Then $R_x = (5)(20)/25 = 4\ \text{k}\Omega$, and Problem 11.3 gives

$$\beta = \frac{1}{(10 \times 10^{-9})(4 \times 10^3)} = 2.5 \times 10^4 \text{ rad/s}$$

11.16 Compare the resonant frequency of the circuit shown in Fig. 11-37 for $R = 0$ to that for $R = 50\ \Omega$.

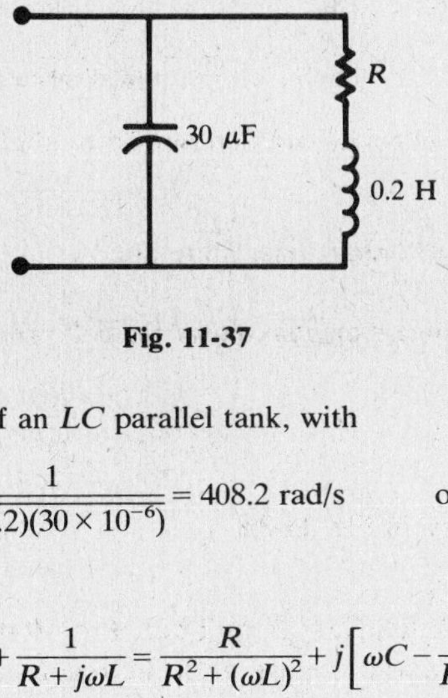

Fig. 11-37

For $R = 0$, the circuit is that of an LC parallel tank, with

$$\omega_a = \frac{1}{\sqrt{LC}} = \frac{1}{\sqrt{(0.2)(30 \times 10^{-6})}} = 408.2 \text{ rad/s} \qquad \text{or} \qquad f_a = 65 \text{ Hz}$$

For $R = 50\ \Omega$,

$$\mathbf{Y}_{\text{in}} = j\omega C + \frac{1}{R + j\omega L} = \frac{R}{R^2 + (\omega L)^2} + j\left[\omega C - \frac{\omega L}{R^2 + (\omega L)^2}\right]$$

For resonance, Im $\mathbf{Y}_{\text{in}}$ is zero, so that

$$\omega_a = \frac{1}{\sqrt{LC}}\sqrt{1 - \frac{R^2 C}{L}}$$

Clearly, as $R \to 0$, this expression reduces to that given for the pure LC tank. Substituting the numerical values produces a value of 0.791 for the radical; hence

$$\omega_a = 408.2(0.791) = 322.9 \text{ rad/s} \qquad \text{or} \qquad f_a = 51.4 \text{ Hz}$$

11.17 Measurements on a practical inductor at 10 MHz give $L = 8.0\ \mu\text{H}$ and $Q_{ind} = 40$. (*a*) Find the ideal capacitance C for parallel resonance at 10 MHz and calculate the corresponding bandwidth β. (*b*) Repeat if a practical capacitor, with a *dissipation factor* $D = Q_{cap}^{-1} = 0.005$ at 10 MHz, is used instead of an ideal capacitance.

(*a*) From Section 11.9,

$$\omega_a = \frac{1}{\sqrt{LC}} \frac{1}{\sqrt{1+Q_{ind}^{-2}}} \quad \text{or} \quad C = \frac{1}{\omega_a^2 L(1+Q_{ind}^{-2})} = \frac{1}{[2\pi(10\times 10^6)]^2(8.0\times 10^{-6})\left(1+\dfrac{1}{1600}\right)} = 31.6\ \text{pF}$$

Using Section 11.10 to convert the series RL branch of Fig. 11-19 to parallel at the resonant frequency,

$$R_p = R(1+Q_{ind}^2) = \frac{\omega_a L}{Q_{ind}}(1+Q_{ind}^2)$$

Then, from Section 11.8,

$$\beta = \frac{\omega_a}{Q_a} = \frac{\omega_a^2 L}{R_p} = \frac{\omega_a Q_{ind}}{1+Q_{ind}^2} = \frac{2\pi(10\times 10^6)(40)}{1+1600}\ \text{rad/s}$$

or 0.25 MHz.

(*b*) The circuit is shown in Fig. 11-38; part (*a*) gives the resistance of the practical inductor as

$$R = \frac{\omega_a L}{Q_{ind}} = 4\pi\ \Omega$$

Also, from the given dissipation factor, it is known that

$$\frac{1}{\omega_a C R_C} = 0.005$$

The input admittance is

$$\mathbf{Y}_{in} = \frac{1}{R_C} + j\omega C + \frac{1}{R+j\omega L} = \left[\frac{1}{R_C} + \frac{R}{R^2+(\omega L)^2}\right] + j\left[\omega C - \frac{\omega L}{R^2+(\omega L)^2}\right]$$

which differs from the input admittance for part (*a*) only in the real part. Since the imaginary part involves the same L and the same R, and must vanish at the same frequency, C must be the same as in part (*a*); namely, $C = 31.6$ pF.

For fixed C, bandwidth is inversely proportional to resistance. With the practical capacitor, the net parallel resistance is

$$R' = \frac{R_p R_C}{R_p + R_C}$$

where R_p is as calculated in part (*a*). Therefore,

$$\frac{\beta}{0.25\ \text{MHz}} = \frac{R_p}{R'} = 1 + \frac{R_p}{R_C} = 1 + \frac{(\omega_a L/Q_{ind})(1+Q_{ind}^2)}{1/\omega_a C(0.005)}$$

$$= 1 + \frac{(1+Q_{ind}^2)(0.005)}{Q_{ind}(1+Q_{ind}^{-2})}$$

$$= 1 + \frac{(1+1600)(0.005)}{40\left(1+\dfrac{1}{1600}\right)} = 1.2$$

and so $\beta = 0.30$ MHz.

A lossy capacitor has the same effect as any loading resistor placed across the tank; the Q_a is reduced and the bandwidth increased, while f_a is unchanged.

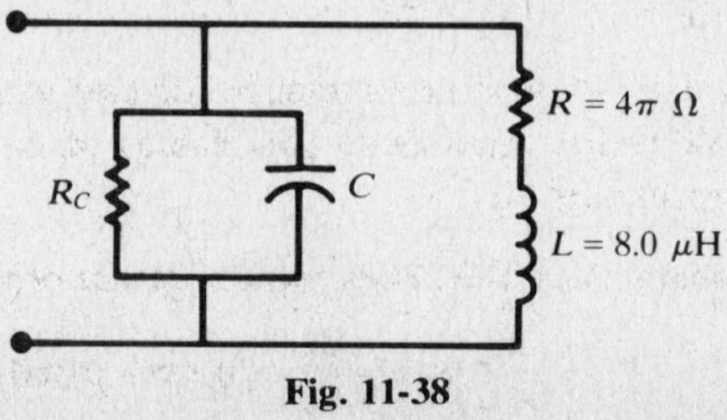

Fig. 11-38

11.18 A lossy capacitor, in the series-circuit model, consists of $R = 25\ \Omega$ and $C = 20$ pF. Obtain the equivalent parallel model at 50 kHz.

From Section 11.10, or by letting $L \to 0$ in Problem 11.6(*a*),

$$Q_s = \frac{1}{\omega C_s R_s} = \frac{1}{2\pi(50 \times 10^3)(20 \times 10^{-12})(25)} = 6370$$

For this large Q_s-value,

$$R_p \approx R_s Q_s^2 = 1010\ \text{M}\Omega \qquad C_p \approx C_s = 20\ \text{pF}$$

11.19 A variable-frequency source of $\mathbf{V} = 100\underline{/0^\circ}$ V is applied to a series *RL* circuit having $R = 20\ \Omega$ and $L = 10$ mH. Compute $\mathbf{I}$ for $\omega = 0, 500, 1000, 2000, 5000$ rad/s. Plot all currents on the same phasor diagram and note the locus of the currents.

$$\mathbf{Z} = R + jX_L = R + j\omega L$$

Table 11-2 exhibits the required computations. With the phasor voltage at the angle zero, the locus of $\mathbf{I}$ as ω varies is the semicircle shown in Fig. 11-39. Since $\mathbf{I} = \mathbf{VY}$, with constant $\mathbf{V}$, Fig. 11-39 is essentially the same as Fig. 11-22(*c*), the admittance locus diagram for the series *RL* circuit.

Table 11-2

ω, rad/s	X_L, Ω	R, Ω	$\mathbf{Z}$, Ω	$\mathbf{I}$, A
0	0	20	$20\underline{/0^\circ}$	$5\underline{/0^\circ}$
500	5	20	$20.6\underline{/14.04^\circ}$	$4.85\underline{/-14.04^\circ}$
1000	10	20	$22.4\underline{/26.57^\circ}$	$4.46\underline{/-26.57^\circ}$
2000	20	20	$28.3\underline{/45^\circ}$	$3.54\underline{/-45^\circ}$
5000	50	20	$53.9\underline{/68.20^\circ}$	$1.86\underline{/-68.20^\circ}$

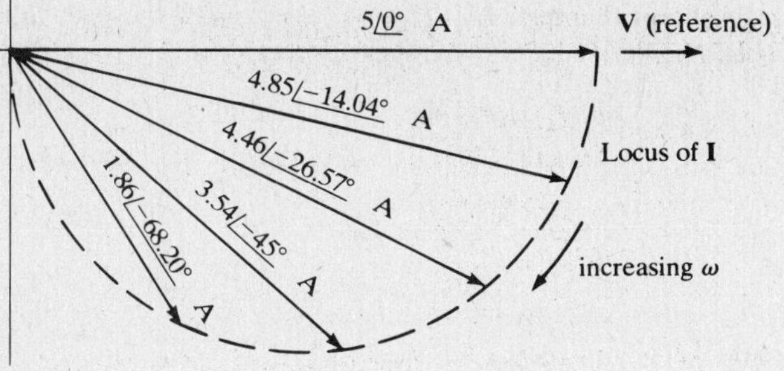

Fig. 11-39

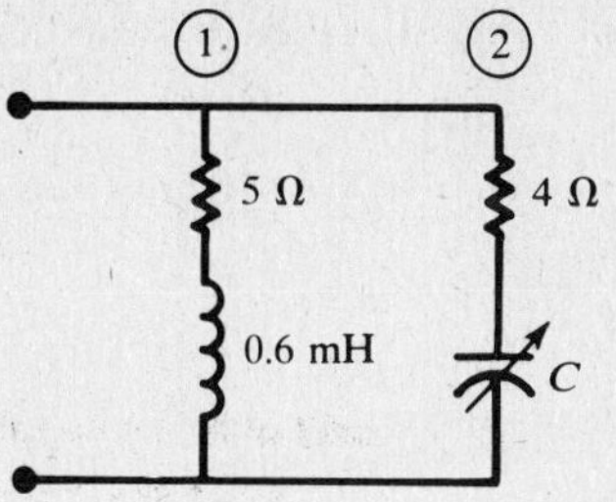

Fig. 11-40

11.20 The circuit shown in Fig. 11-40 is in resonance for two values of C when the frequency of the driving voltage is 5000 rad/s. Find these two values of C and construct the admittance locus diagram which illustrates this fact.

At the given frequency, $X_L = 3\ \Omega$. Then the admittance of this fixed branch is

$$\mathbf{Y}_1 = \frac{1}{5 + j3} = 0.147 - j0.088 \quad \text{S}$$

The semicircular admittance locus of branch 2 has the radius $r = 1/2R = 0.125$ S. The total admittance is the sum of the fixed admittance $\mathbf{Y}_1$ and the variable admittance $\mathbf{Y}_2$. In Fig. 11-41, the semicircular locus is added to the fixed complex number $\mathbf{Y}_1$. The circuit resonance occurs at points a and b, where $\mathbf{Y}_T$ is real.

$$\mathbf{Y}_T = 0.147 - j0.088 + \frac{1}{4 - jX_C}$$

which is real if

$$X_C^2 - 11.36X_C + 16 = 0$$

or $X_{C_1} = 9.71\ \Omega$, $X_{C_2} = 1.65\ \Omega$. With $\omega = 5000$ rad/s,

$$C_1 = 20.6\ \mu\text{F} \qquad C_2 = 121\ \mu\text{F}$$

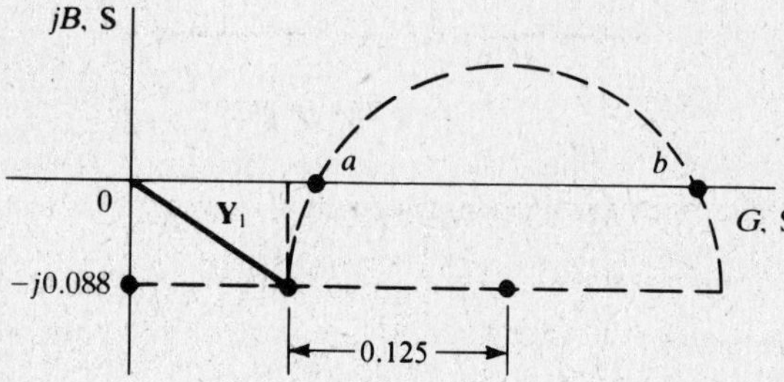

Fig. 11-41

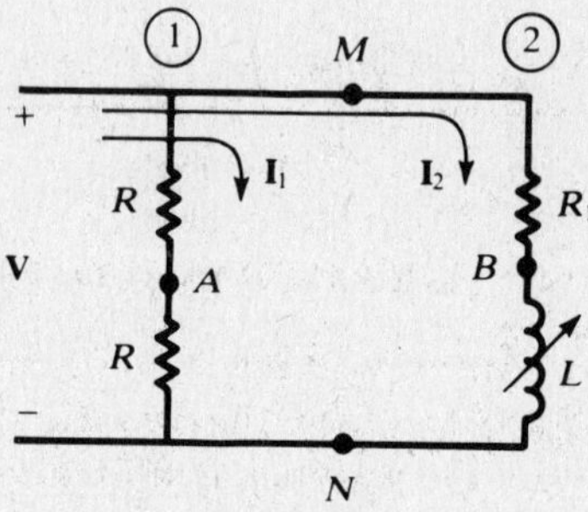

Fig. 11-42

11.21 Show by locus diagrams that the magnitude of the voltage between points A and B in Fig. 11-42 is always one-half the magnitude of the applied voltage $\mathbf{V}$ as L is varied.

Branch-1 current $\mathbf{I}_1$ passes through two equal resistors R. Thus A is the midpoint on the phasor $\mathbf{V}$, as shown in Fig. 11-43.

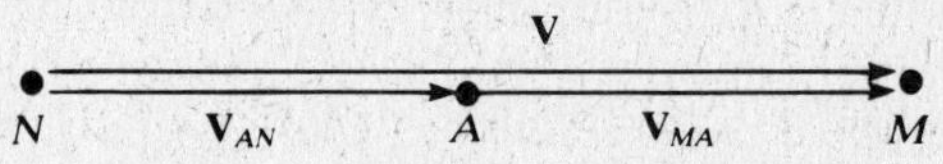

Fig. 11-43

Branch 2 has a semicircular $\mathbf{Y}$-locus [see Fig. 11-22(*c*)]. Then the current locus is also semicircular, as shown in Fig. 11-44(*a*). The voltage phasor diagram, Fig. 11-44(*b*), consists of the voltage across the inductance, $\mathbf{V}_{BN}$, and the voltage across R_1, $\mathbf{V}_{MB}$. The two voltages add vectorially,

$$\mathbf{V} = \mathbf{V}_{MN} = \mathbf{V}_{BN} + \mathbf{V}_{MB}$$

Because $\mathbf{I}_2$ lags $\mathbf{V}_{BN}$ by 90°, $\mathbf{V}_{BN}$ and $\mathbf{V}_{MB}$ are perpendicular for all values of L in Fig. 11-44(*b*). As L

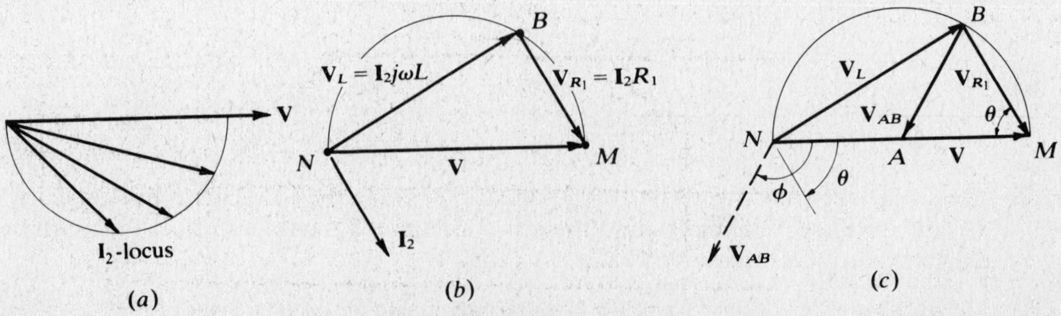

Fig. 11-44

varies from 0 to ∞, point B moves from N toward M along the semicircle. Figures 11-43 and 11-44(b) are superimposed in Fig. 11-44(c). It is clear that $\mathbf{V}_{AB}$ is a radius of the semicircle and therefore

$$|\mathbf{V}_{AB}| = \tfrac{1}{2}|\mathbf{V}|$$

Further, the angle ϕ by which $\mathbf{V}_{AB}$ lags $\mathbf{V}$ is equal to 2θ, where $\theta = \tan^{-1} \omega L/R_1$.

Supplementary Problems

11.22 A high-pass RL circuit has $R_1 = 50$ kΩ and $L_2 = 0.2$ mH. (a) Find ω if the magnitude of the voltage transfer function is $|\mathbf{H}_{v\infty}| = 0.90$. ($b$) With a load $R = 1$ MΩ across L_2, find $|\mathbf{H}_v|$ at $\omega = 7.5 \times 10^8$ rad/s.
Ans. (a) 5.16×10^8 rad/s; (b) 0.908

11.23 Obtain $\mathbf{H}_{v\infty}$ for a high-pass RL circuit at $\omega = 2.5\omega_x$, $R = 2$ kΩ, $L = 0.05$ H. *Ans.* $0.928\underline{/21.80^\circ}$

11.24 A low-pass RC circuit under no-load has $R_1 = 5$ kΩ. (a) Find C_2 if $|\mathbf{H}_v| = 0.5$ at 10 kHz. (b) Obtain $\mathbf{H}_v$ at 5 kHz. (c) What value of C_2 results in $|\mathbf{H}_v| = 0.90$ at 8 kHz? (d) With C_2 as in (a), find a new value for R_1 to result in $|\mathbf{H}_v| = 0.90$ at 8 kHz. *Ans.* (a) 5.51 μF; (b) $0.756\underline{/-40.89^\circ}$; ($c$) 1.93 μF; (d) 1749 Ω

11.25 A simple voltage divider would consist of R_1 and R_2. If stray capacitance C_s is present, then the divider would generally be frequency-dependent. Show, however, that $\mathbf{V}_2/\mathbf{V}_1$ is independent of frequency for the circuit of Fig. 11-45 if the compensating capacitance C_1 has a certain value. *Ans.* $C_1 = (R_2/R_1)C_s$

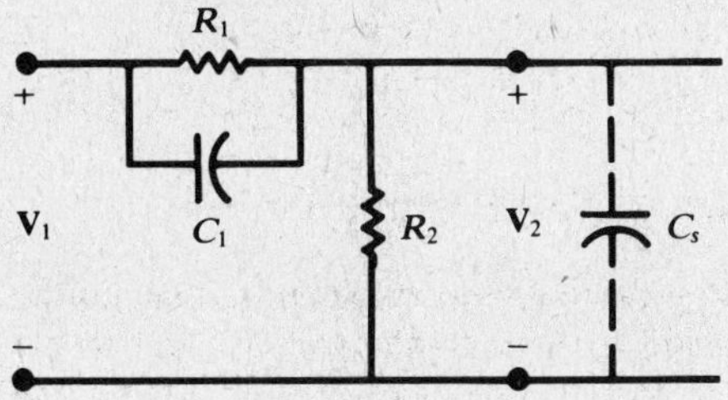

Fig. 11-45

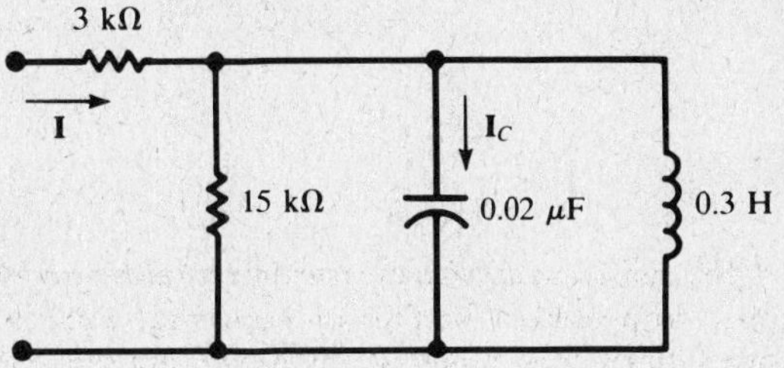

Fig. 11-46

11.26 Assume that a sinusoidal voltage source with a variable frequency and $V_{max} = 50$ V is applied to the circuit shown in Fig. 11-46. (a) At what frequency f is $|\mathbf{I}|$ a minimum? (b) Calculate this minimum current. (c) What is $|\mathbf{I}_C|$ at this frequency? *Ans.* (a) 2.05 kHz; (b) 2.78 mA; (c) 10.8 mA

11.27 A 20-μF capacitor is in parallel with a practical inductor represented by $L = 1$ mH in series with $R = 7$ Ω. Find the resonant frequency, in rad/s and in Hz, of the parallel circuit.
Ans. 1000 rad/s, 159.2 Hz

11.28 What must be the relationship between the values of R_L and R_C if the network shown in Fig. 11-47 is to be resonant at all frequencies? *Ans.* $R_L = R_C = 5$ Ω

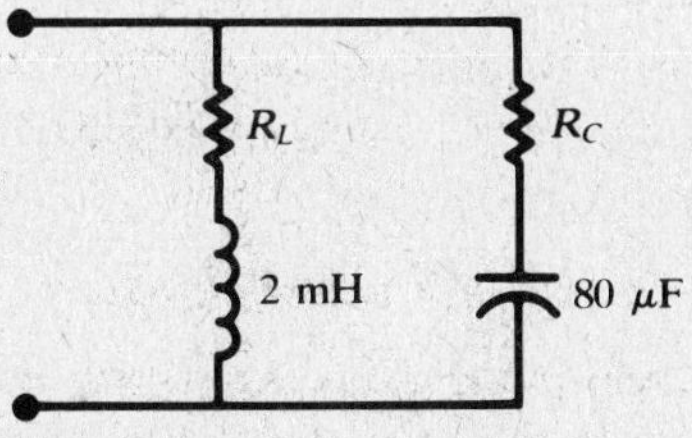

Fig. 11-47

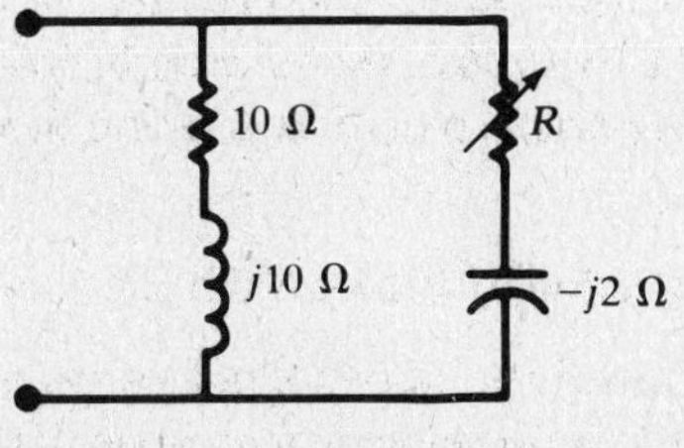

Fig. 11-48

11.29 For the parallel network shown in Fig. 11-48, (*a*) find the value of R for resonance; (*b*) convert the RC branch to a parallel equivalent. *Ans.* (*a*) 6.0 Ω; (*b*) $R_p = 6.67\ \Omega$, $X_{Cp} = 20\ \Omega$

11.30 For the network of Fig. 11-49(*a*), find R for resonance. Obtain the values of R', X_L, and X_C in the parallel equivalent of Fig. 11-49(*b*). *Ans.* $R = 12.25\ \Omega$, $R' = 7.75\ \Omega$, $X_L = 25\ \Omega$, $X_C = 25\ \Omega$

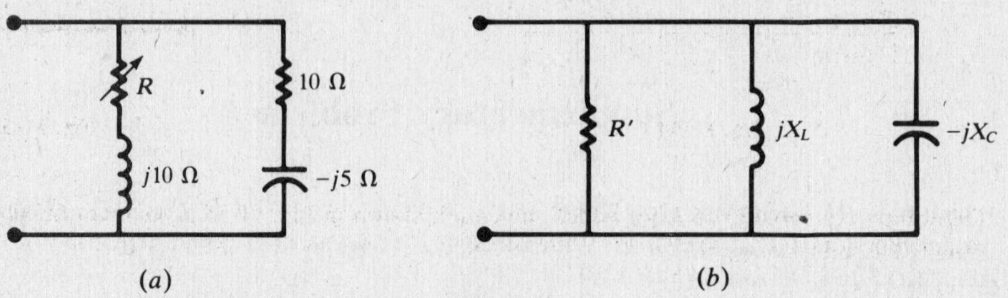

Fig. 11-49

11.31 Branch 1 of a two-branch parallel circuit has an impedance $\mathbf{Z}_1 = 8 + j6\ \ \Omega$ at $\omega = 5000$ rad/s. Branch 2 contains $R = 8.34\ \Omega$ in series with a variable capacitance C. (*a*) Find C for resonance. (*b*) Sketch the admittance locus diagram. *Ans.* (*a*) 24 μF. (*b*) See Fig. 11-50.

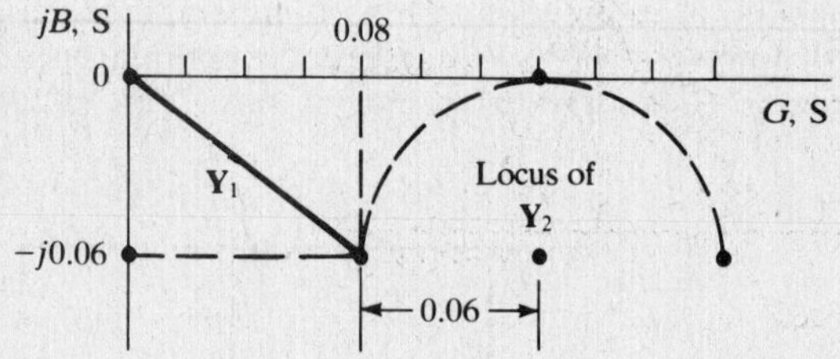

Fig. 11-50

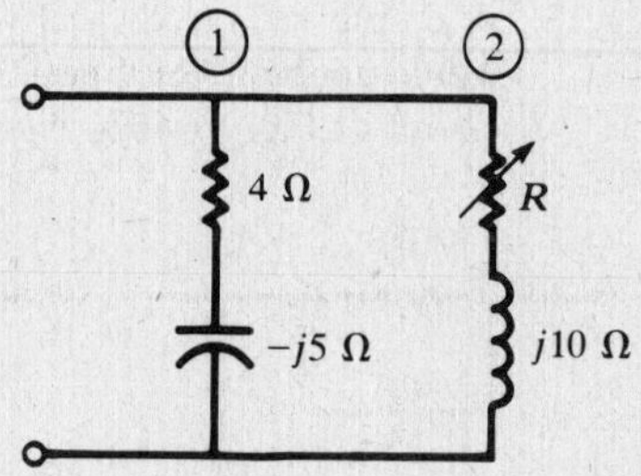

Fig. 11-51

11.32 Find R for resonance of the network shown in Fig. 11-51. Sketch the admittance locus diagram.
Ans. Resonance cannot be achieved by varying R. See Fig. 11-52.

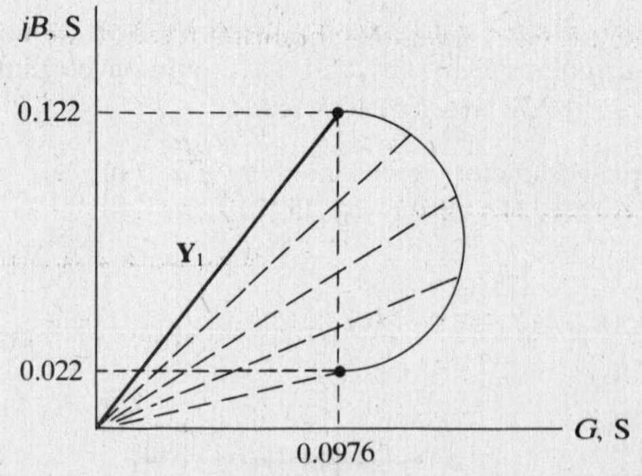

Fig. 11-52

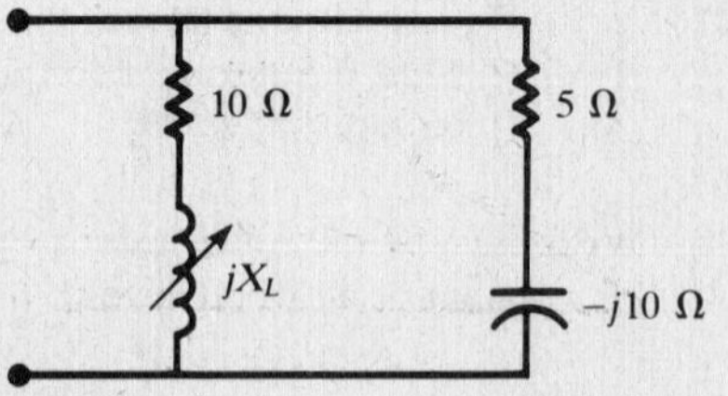

Fig. 11-53

11.33 In Problem 11.32, for what values of the inductive reactance will it be possible to obtain resonance at some value of the variable resistance R? *Ans.* $X_L \leq 8.2\ \Omega$

11.34 (*a*) Construct the admittance locus diagram for the circuit shown in Fig. 11-53. (*b*) For what value of resistance in the RL branch is resonance possible for only one value of X_L?
Ans. (*a*) See Fig. 11-54. (*b*) 6.25 Ω.

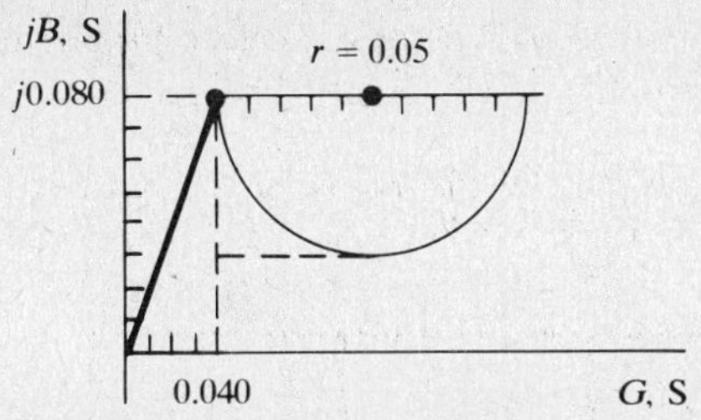

Fig. 11-54

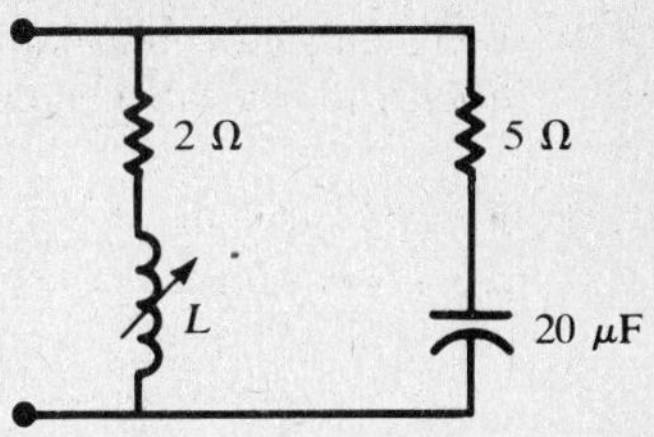

Fig. 11-55

11.35 Determine the value(s) of L for which the circuit shown in Fig. 11-55 is resonant at 5000 rad/s.
Ans. 2.43 mH, 66.0 μH

11.36 A three-branch parallel circuit has fixed elements in two branches; in the third branch, one element is variable. The voltage-current phasor diagram is shown in Fig. 11-56. Identify all the elements if $\omega = 5000$ rad/s.
Ans. Branch 1: $R = 8.05\ \Omega$, $L = 0.431$ mH
Branch 2: $R = 4.16\ \Omega$, $C = 27.7\ \mu$F
Branch 3: $L = 2.74$ mH, variable R

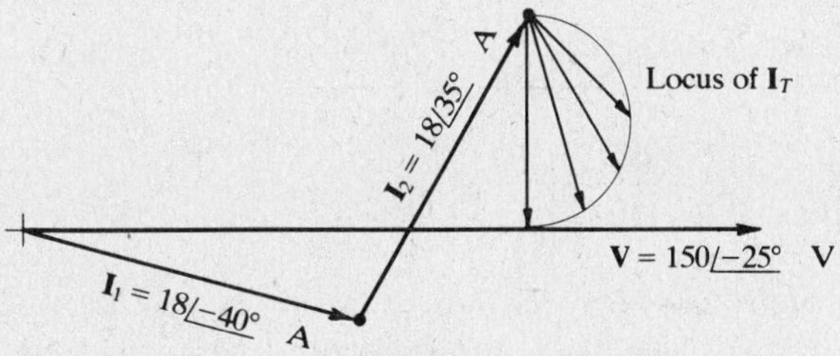

Fig. 11-56

11.37 Describe the circuit which corresponds to each locus in Fig. 11-57, if there is only one variable element in each circuit.
Ans. (*a*) A two-branch parallel circuit. Branch 1: fixed R and X_C; branch 2: fixed R and variable X_C.
(*b*) A three-branch parallel circuit. Branch 1: fixed R and X_C; branch 2: fixed X_C; branch 3: fixed R and variable X_L.
(*c*) A two-branch parallel circuit. Branch 1: fixed R and X_C; branch 2: fixed X_L and variable R.

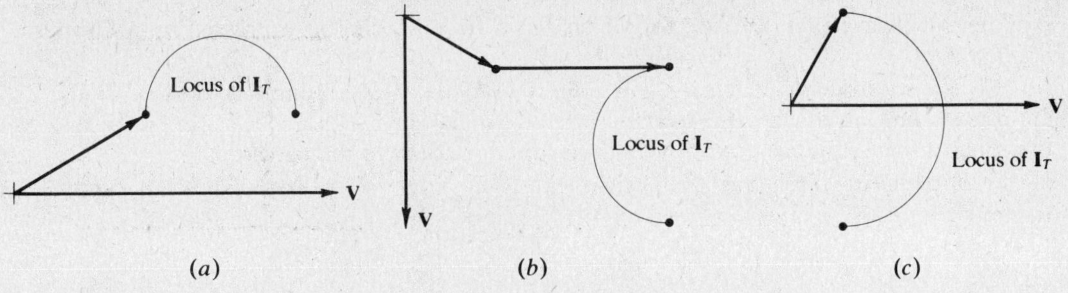

Fig. 11-57

Chapter 12

Fourier Method of Waveform Analysis

12.1 INTRODUCTION

In the circuits examined previously, the response was obtained for excitations having either constant or sinusoidal form. In such cases a single expression described the forcing function for all time; e.g., $v =$ constant or $v = V \sin \omega t$, as shown in Fig. 12-1(*a*) and (*b*).

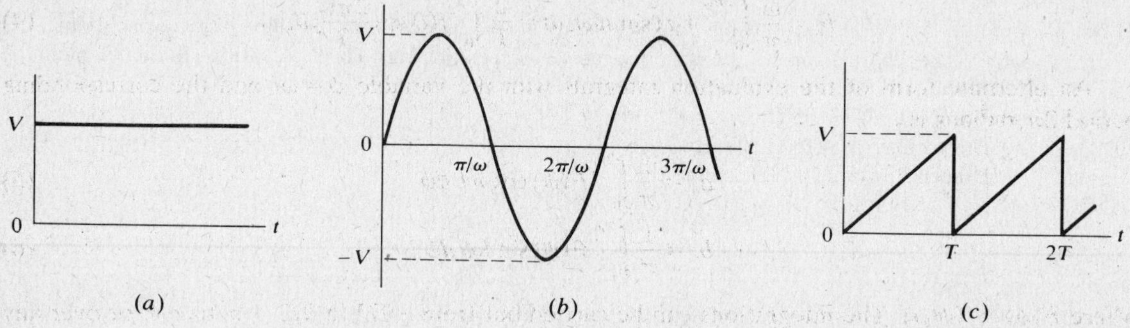

Fig. 12-1

Certain periodic waveforms, of which the sawtooth in Fig. 12-1(*c*) is an example, can be only locally defined by single functions. Thus, the sawtooth is expressed by $f(t) = (V/T)t$ in the interval $0 < t < T$ and by $f(t) = (V/T)(t - T)$ in the interval $T < t < 2T$. While such piecemeal expressions describe the waveform satisfactorily, they do not permit the determination of the circuit response. Now, if a periodic function can be expressed as the sum of a finite or infinite number of sinusoidal functions, the responses of linear networks to nonsinusoidal excitations can be determined by applying the superposition theorem. The Fourier method provides the means for solving this type of problem.

12.2 TRIGONOMETRIC FOURIER SERIES

Any periodic waveform, i.e. one for which $f(t) = f(t + T)$, can be expressed by a Fourier series provided that

(1) if it is discontinuous, there are only a finite number of discontinuities in the period T;

(2) it has a finite average value over the period T;

(3) it has a finite number of positive and negative maxima in the period T.

When these *Dirichlet conditions* are satisfied, the Fourier series exists and can be written in trigonometric form:

$$f(t) = \tfrac{1}{2}a_0 + a_1 \cos \omega t + a_2 \cos 2\omega t + a_3 \cos 3\omega t + \cdots$$
$$+ b_1 \sin \omega t + b_2 \sin 2\omega t + b_3 \sin 3\omega t + \cdots \qquad (1)$$

The Fourier coefficients, a's and b's, are determined for a given waveform by the evaluation integrals. We obtain the cosine coefficient evaluation integral by multiplying both sides of (*1*) by $\cos n\omega t$ and integrating over a full period. The period of the fundamental, $2\pi/\omega$, is the period of the series since each term in the series has a frequency which is an integral multiple of the fundamental frequency.

$$\int_0^{2\pi/\omega} f(t) \cos n\omega t\, dt = \int_0^{2\pi/\omega} \tfrac{1}{2}a_0 \cos n\omega t\, dt + \int_0^{2\pi/\omega} a_1 \cos \omega t \cos n\omega t\, dt + \cdots$$
$$+ \int_0^{2\pi/\omega} a_n \cos^2 n\omega t\, dt + \cdots + \int_0^{2\pi/\omega} b_1 \sin \omega t \cos n\omega t\, dt$$
$$+ \int_0^{2\pi/\omega} b_2 \sin 2\omega t \cos n\omega t\, dt + \cdots \tag{2}$$

The definite integrals on the right side of (*2*) are all zero except that involving $\cos^2 n\omega t$, which has the value $(\pi/\omega)a_n$. Then

$$a_n = \frac{\omega}{\pi}\int_0^{2\pi/\omega} f(t) \cos n\omega t\, dt = \frac{2}{T}\int_0^T f(t) \cos \frac{2\pi n t}{T}\, dt \tag{3}$$

Multiplying (*1*) by $\sin n\omega t$ and integrating as above results in the sine coefficient evaluation integral.

$$b_n = \frac{\omega}{\pi}\int_0^{2\pi/\omega} f(t) \sin n\omega t\, dt = \frac{2}{T}\int_0^T f(t) \sin \frac{2\pi n t}{T}\, dt \tag{4}$$

An alternate form of the evaluation integrals with the variable $\psi = \omega t$ and the corresponding period 2π radians is

$$a_n = \frac{1}{\pi}\int_0^{2\pi} F(\psi) \cos n\psi\, d\psi \tag{5}$$

$$b_n = \frac{1}{\pi}\int_0^{2\pi} F(\psi) \sin n\psi\, d\psi \tag{6}$$

where $F(\psi) = f(\psi/\omega)$. The integrations can be carried out from $-T/2$ to $T/2$, $-\pi$ to $+\pi$, or over any other full period that might simplify the calculation. The constant a_0 is obtained from (*3*) or (*5*) with $n = 0$; however, since $\frac{1}{2}a_0$ is the average value of the function, it can frequently be determined by inspection of the waveform. The series with coefficients obtained from the above evaluation integrals converges uniformly to the function at all points of continuity and converges to the mean value at points of discontinuity.

EXAMPLE 12.1 Find the Fourier series for the waveform shown in Fig. 12-2.

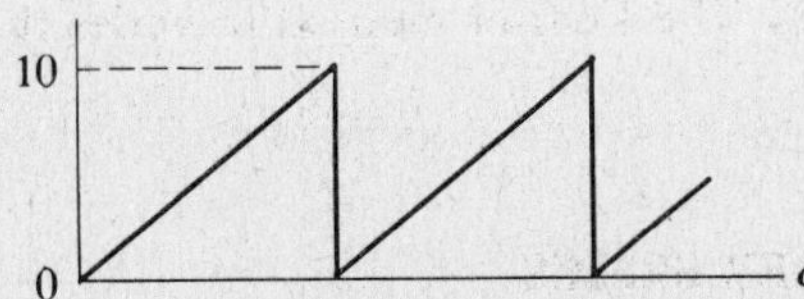

Fig. 12-2

The waveform is periodic, of period $2\pi/\omega$ in t or 2π in ωt. It is continuous for $0 < \omega t < 2\pi$ and given therein by $f(t) = (10/2\pi)\omega t$, with discontinuities at $\omega t = n2\pi$ where $n = 0, 1, 2, \ldots$. The Dirichlet conditions are satisfied. The average value of the function is 5, by inspection, and thus $\frac{1}{2}a_0 = 5$. For $n > 0$, (*5*) gives

$$a_n = \frac{1}{\pi}\int_0^{2\pi}\left(\frac{10}{2\pi}\right)\omega t \cos n\omega t\, d(\omega t) = \frac{10}{2\pi^2}\left[\frac{\omega t}{n}\sin n\omega t + \frac{1}{n^2}\cos n\omega t\right]_0^{2\pi}$$
$$= \frac{10}{2\pi^2 n^2}(\cos n2\pi - \cos 0) = 0$$

Thus the series contains no cosine terms. Using (*6*), we obtain

$$b_n = \frac{1}{\pi}\int_0^{2\pi}\left(\frac{10}{2\pi}\right)\omega t \sin n\omega t\, d(\omega t) = \frac{10}{2\pi^2}\left[-\frac{\omega t}{n}\cos n\omega t + \frac{1}{n^2}\sin n\omega t\right]_0^{2\pi} = -\frac{10}{\pi n}$$

Using these sine-term coefficients and the average term, the series is

$$f(t) = 5 - \frac{10}{\pi}\sin\omega t - \frac{10}{2\pi}\sin 2\omega t - \frac{10}{3\pi}\sin 3\omega t - \cdots = 5 - \frac{10}{\pi}\sum_{n=1}^{\infty}\frac{\sin n\omega t}{n}$$

The sine and cosine terms of like frequency can be combined as a single sine or cosine term with a phase angle. Two alternate forms of the trigonometric series result.

$$f(t) = \tfrac{1}{2}a_0 + \sum c_n \cos(n\omega t - \theta_n) \qquad (7)$$

and

$$f(t) = \tfrac{1}{2}a_0 + \sum c_n \sin(n\omega t + \phi_n) \qquad (8)$$

where $c_n = \sqrt{a_n^2 + b_n^2}$, $\theta_n = \tan^{-1}(b_n/a_n)$, and $\phi_n = \tan^{-1}(a_n/b_n)$. In (*7*) and (*8*), c_n is the harmonic amplitude, and the harmonic phase angles are θ_n or ϕ_n.

12.3 EXPONENTIAL FOURIER SERIES

If we express each of the sine and cosine terms in the trigonometric series by its exponential equivalent, the result is a series of exponential terms:

$$f(t) = \frac{a_0}{2} + a_1\left(\frac{e^{j\omega t} + e^{-j\omega t}}{2}\right) + a_2\left(\frac{e^{j2\omega t} + e^{-j2\omega t}}{2}\right) + \cdots$$
$$+ b_1\left(\frac{e^{j\omega t} - e^{-j\omega t}}{2j}\right) + b_2\left(\frac{e^{j2\omega t} - e^{-j2\omega t}}{2j}\right) + \cdots \qquad (9)$$

Rearranging,

$$f(t) = \cdots + \left(\frac{a_2}{2} - \frac{b_2}{2j}\right)e^{-j2\omega t} + \left(\frac{a_1}{2} - \frac{b_1}{2j}\right)e^{-j\omega t}$$
$$+ \frac{a_0}{2} + \left(\frac{a_1}{2} + \frac{b_1}{2j}\right)e^{j\omega t} + \left(\frac{a_2}{2} + \frac{b_2}{2j}\right)e^{j2\omega t} + \cdots \qquad (10)$$

We now define new complex constants $\mathbf{A}_n$ such that $A_0 = \frac{1}{2}a_0$ and, for $n = 1, 2, 3, \ldots$,

$$\mathbf{A}_n = \tfrac{1}{2}(a_n - jb_n) \qquad \mathbf{A}_{-n} = \tfrac{1}{2}(a_n + jb_n) \qquad (11)$$

and rewrite (*10*) as

$$f(t) = \cdots + \mathbf{A}_{-2}e^{-j2\omega t} + \mathbf{A}_{-1}e^{-j\omega t} + A_0 + \mathbf{A}_1 e^{j\omega t} + \mathbf{A}_2 e^{j2\omega t} + \cdots \qquad (12)$$

To obtain the evaluation integral for the $\mathbf{A}_n$ coefficients, we multiply (*12*) on both sides by $e^{-jn\omega t}$ and integrate over the full period:

$$\int_0^{2\pi} f(t)e^{-jn\omega t}\,d(\omega t) = \cdots + \int_0^{2\pi}\mathbf{A}_{-2}e^{-j2\omega t}e^{-jn\omega t}\,d(\omega t) + \int_0^{2\pi}\mathbf{A}_{-1}e^{-j\omega t}e^{-jn\omega t}\,d(\omega t)$$
$$+ \int_0^{2\pi} A_0 e^{-jn\omega t}\,d(\omega t) + \int_0^{2\pi}\mathbf{A}_1 e^{j\omega t}e^{-jn\omega t}\,d(\omega t) + \cdots$$
$$+ \int_0^{2\pi}\mathbf{A}_n e^{jn\omega t}\,e^{-jn\omega t}\,d(\omega t) + \cdots \qquad (13)$$

The definite integrals on the right side of (*13*) are all zero except $\int_0^{2\pi}\mathbf{A}_n\,d(\omega t)$, which has the value $2\pi\mathbf{A}_n$. Then

$$\mathbf{A}_n = \frac{1}{2\pi}\int_0^{2\pi} f(t)\,e^{-jn\omega t}\,d(\omega t) \qquad \text{or} \qquad \mathbf{A}_n = \frac{1}{T}\int_0^{T} f(t)\,e^{-j2\pi nt/T}\,dt \qquad (14)$$

Just as with the a_n and b_n evaluation integrals, the limits of integration in (*14*) may be the endpoints of any convenient full period and not necessarily 0 to 2π or 0 to T. Note that, $f(t)$ being real, $\mathbf{A}_{-n} = \mathbf{A}_n^*$, so that only positive n need be considered in (*14*). Furthermore, we have, inverse to (*11*),

$$a_n = 2\,\mathrm{Re}\,\mathbf{A}_n \qquad b_n = -2\,\mathrm{Im}\,\mathbf{A}_n \qquad (15)$$

EXAMPLE 12.2 Find the exponential Fourier series for the waveform shown in Fig. 12-2. Using the coefficients of this exponential series, obtain a_n and b_n of the trigonometric series and compare with Example 12.1.

In the interval $0 < \omega t < 2\pi$ the function is given by $f(t) = (10/2\pi)\omega t$. By inspection, the average value of the function is $A_0 = 5$. Substituting $f(t)$ in (14), we obtain the coefficients $\mathbf{A}_n$.

$$\mathbf{A}_n = \frac{1}{2\pi}\int_0^{2\pi}\left(\frac{10}{2\pi}\right)\omega t\, e^{-jn\omega t}\, d(\omega t) = \frac{10}{(2\pi)^2}\left[\frac{e^{-jn\omega t}}{(-jn)^2}(-jn\omega t - 1)\right]_0^{2\pi} = j\frac{10}{2\pi n}$$

Inserting the coefficients $\mathbf{A}_n$ in (12), the exponential form of the Fourier series for the given waveform is

$$f(t) = \cdots - j\frac{10}{4\pi}e^{-j2\omega t} - j\frac{10}{2\pi}e^{-j\omega t} + 5 + j\frac{10}{2\pi}e^{j\omega t} + j\frac{10}{4\pi}e^{j2\omega t} + \cdots \tag{16}$$

The trigonometric series coefficients are, by (15),

$$a_n = 0 \qquad b_n = -\frac{10}{\pi n}$$

and so

$$f(t) = 5 - \frac{10}{\pi}\sin\omega t - \frac{10}{2\pi}\sin 2\omega t - \frac{10}{3\pi}\sin 3\omega t - \cdots$$

which is the same as in Example 12.1.

12.4 WAVEFORM SYMMETRY

The series obtained in Example 12.1 contained only sine terms in addition to a constant term. Other waveforms will have only cosine terms; and sometimes only odd harmonics are present in the series, whether the series contains sine, cosine or both types of terms. This is the result of certain types of symmetry exhibited by the waveform. Knowledge of such symmetry results in reduced calculations in determining the Fourier series. For this reason the following definitions are important.

1. A function $f(x)$ is said to be *even* if $f(x) = f(-x)$.

The function $f(x) = 2 + x^2 + x^4$ is an example of even functions since the functional values for x and $-x$ are equal. The cosine is an even function, since it can be expressed as the power series

$$\cos x = 1 - \frac{x^2}{2!} + \frac{x^4}{4!} - \frac{x^6}{6!} + \frac{x^8}{8!} - \cdots$$

The sum or product of two or more even functions is an even function, and with the addition of a constant the even nature of the function is still preserved.

In Fig. 12-3, the waveforms shown represent even functions of x. They are symmetrical with respect to the vertical axis, as indicated by the construction in Fig. 12-3(a).

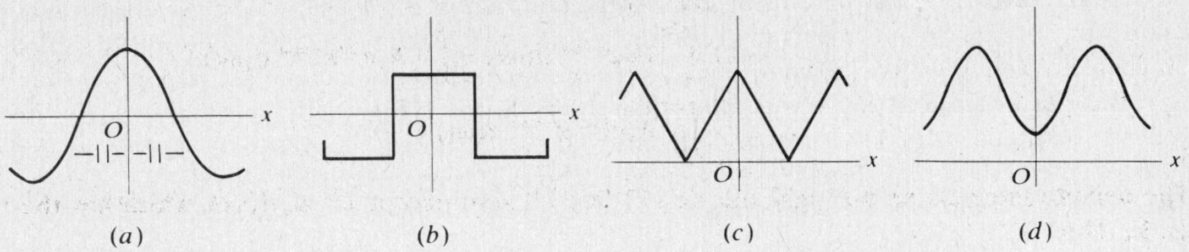

Fig. 12-3

2. A function $f(x)$ is said to be *odd* if $f(x) = -f(-x)$.

The function $f(x) = x + x^3 + x^5$ is an example of odd functions since the values of the function for x and $-x$ are of opposite sign. The sine is an odd function, since it can be expressed as the power series

$$\sin x = x - \frac{x^3}{3!} + \frac{x^5}{5!} - \frac{x^7}{7!} + \frac{x^9}{9!} - \cdots$$

The sum of two or more odd functions is an odd function, but the addition of a constant removes the odd nature of the function. The product of two odd functions is an even function.

The waveforms shown in Fig. 12-4 represent odd functions of x. They are symmetrical with respect to the origin, as indicated by the construction in Fig. 12-4(a).

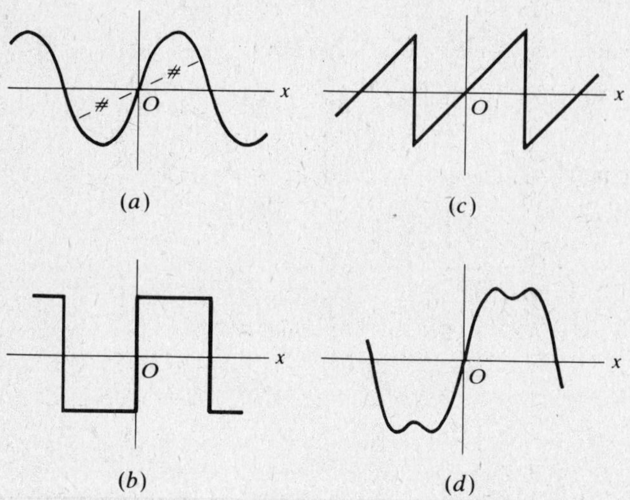

Fig. 12-4

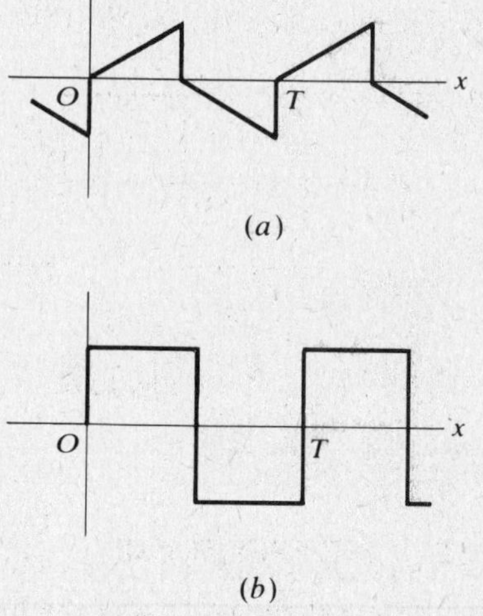

Fig. 12-5

3. A periodic function $f(x)$ is said to have *half-wave symmetry* if $f(x) = -f(x + T/2)$ where T is the period. Two waveforms with half-wave symmetry are shown in Fig. 12-5.

When the type of symmetry of a waveform is established, the following conclusions are reached. If the waveform is even, all terms of its Fourier series are cosine terms, including a constant if the waveform has a nonzero average value. Hence there is no need of evaluating the integral for the coefficients b_n, since no sine terms can be present. If the waveform is odd, the series contains only sine terms. The wave may be odd only after its average value is subtracted, in which case its Fourier representation will simply contain that constant and a series of sine terms. If the waveform has half-wave symmetry, only odd harmonics are present in the series. This series will contain both sine and cosine terms unless the function is also odd or even. In any case, a_n and b_n are equal to zero for $n = 2, 4, 6, \ldots$ for any waveform with half-wave symmetry. Half-wave symmetry, too, may be present only after subtraction of the average value.

Certain waveforms can be odd or even, depending upon the location of the vertical axis. The square wave of Fig. 12-6(a) meets the condition of an even function, i.e. $f(x) = f(-x)$. A shift of the vertical axis to the position shown in Fig. 12-6(b) produces an odd function, i.e. $f(x) = -f(-x)$. With the vertical axis placed at any points other than those shown in Fig. 12-6, the square wave is neither even nor odd, and its series contains both sine and cosine terms. Thus, in the analysis of periodic functions, the vertical axis should be conveniently chosen to result in either an even or odd function, if the type of waveform makes this possible.

The shifting of the horizontal axis may simplify the series representation of the function. As an example, the waveform of Fig. 12-7(a) does not meet the requirements of an odd function until the average value is removed as shown in Fig. 12-7(b). Thus its series will contain a constant term and only sine terms.

The above symmetry considerations can be used to check the coefficients of the exponential Fourier series. An even waveform contains only cosine terms in its trigonometric series and therefore, by (*11*), the exponential Fourier coefficients must be pure real numbers. Similarly, an odd function whose trigonometric series consists of sine terms has pure imaginary coefficients in its exponential series.

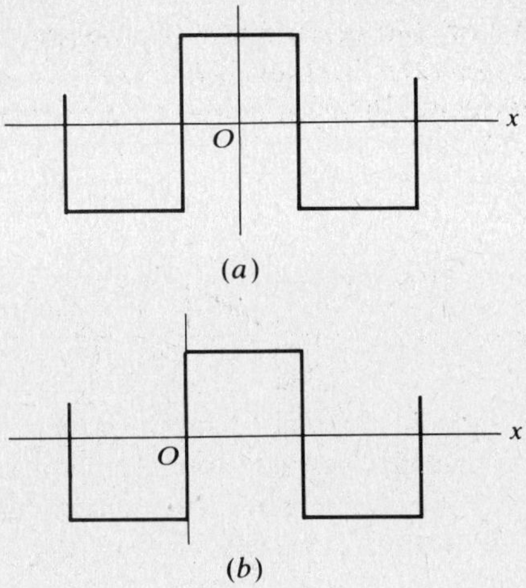

Fig. 12-6

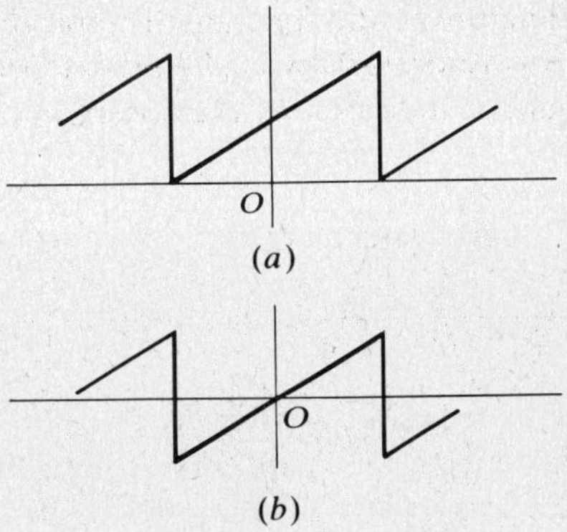

Fig. 12-7

12.5 LINE SPECTRUM

A plot showing each of the harmonic amplitudes in the wave is called the *line spectrum*. The lines decrease rapidly for waves with rapidly convergent series. Waves with discontinuities, such as the sawtooth and square wave, have spectra with slowly decreasing amplitudes since their series have strong high harmonics. Their 10th harmonics will often have amplitudes of significant value as compared to the fundamental. In contrast, the series for waveforms without discontinuities and with a generally smooth appearance will converge rapidly, and only a few terms are required to generate the wave. Such rapid convergence will be evident from the line spectrum where the harmonic amplitudes decrease rapidly, so that any above the 5th or 6th are insignificant.

The harmonic content and the line spectrum of a wave are part of the very nature of that wave and never change, regardless of the method of analysis. Shifting the origin gives the trigonometric series a completely different appearance, and the exponential series coefficients also change greatly. However, the same harmonics always appear in the series, and their amplitudes,

$$c_0 = |\tfrac{1}{2}a_0| \qquad \text{and} \qquad c_n = \sqrt{a_n^2 + b_n^2} \quad (n \geq 1) \tag{17}$$

or

$$c_0 = |A_0| \qquad \text{and} \qquad c_n = |\mathbf{A}_n| + |\mathbf{A}_{-n}| = 2|\mathbf{A}_n| \quad (n \geq 1) \tag{18}$$

remain the same. Note that when the exponential form is used, the amplitude of the nth harmonic combines the contributions of frequencies $+n\omega$ and $-n\omega$.

EXAMPLE 12.3 In Fig. 12-8, the sawtooth wave of Example 12.1 and its line spectrum are shown. Since there were only sine terms in the trigonometric series, the harmonic amplitudes are given directly by $\frac{1}{2}a_0$ and $|b_n|$. The same line spectrum is obtained from the exponential Fourier series, (*16*).

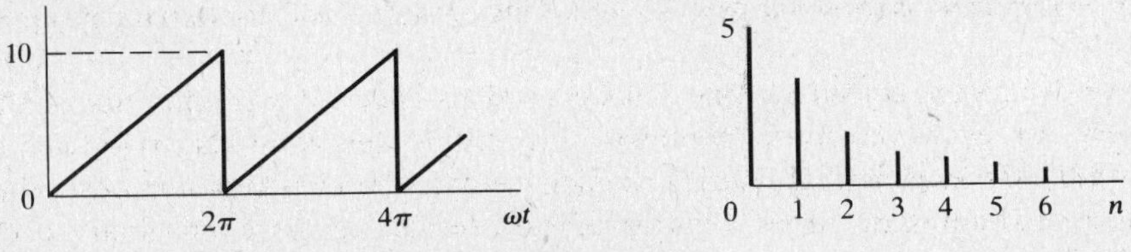

Fig. 12-8

12.6 WAVEFORM SYNTHESIS

Synthesis is a combination of parts so as to form a whole. Fourier synthesis is the recombination of the terms of the trigonometric series, usually the first four or five, to produce the original wave. Often it is only after synthesizing a wave that the student is convinced that the Fourier series does in fact represent the periodic wave for which it was obtained.

The trigonometric series for the sawtooth wave of Fig. 12-8 is

$$f(t) = 5 - \frac{10}{\pi}\sin\omega t - \frac{10}{2\pi}\sin 2\omega t - \frac{10}{3\pi}\sin 3\omega t - \cdots$$

These four terms are plotted and added in Fig. 12-9. Although the result is not a perfect sawtooth wave, it appears that with more terms included the sketch will more nearly resemble a sawtooth. Since this wave has discontinuities, its series is not rapidly convergent and, consequently, the synthesis using only four terms does not produce a very good result. The next term, at the frequency 4ω, has amplitude $10/4\pi$, which is certainly significant compared to the fundamental amplitude, $10/\pi$. As each term is added in the synthesis, the irregularities of the resultant are reduced and the approximation to the original wave is improved. This is what was meant when we said earlier that *the series converges to the function at all points of continuity and to the mean value at points of discontinuity.* In Fig. 12-9, at 0 and 2π it is clear that a value of 5 will remain, since all sine terms are zero at these points. These are the points of discontinuity; and the value of the function when they are approached from the left is 10, and from the right 0, with the mean value 5.

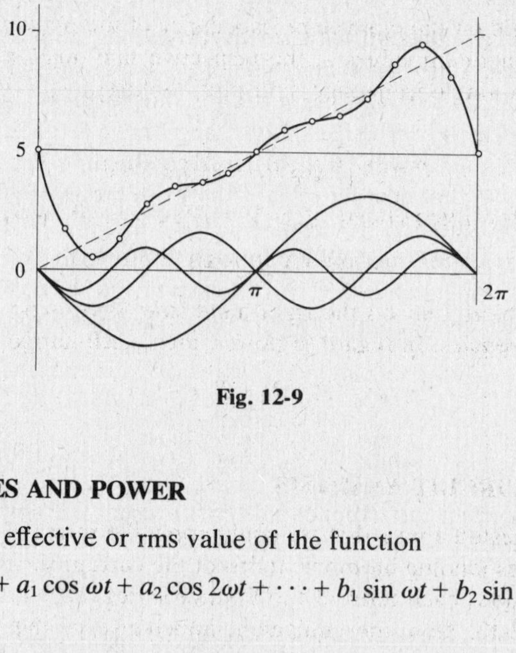

Fig. 12-9

12.7 EFFECTIVE VALUES AND POWER

From Appendix A, the effective or rms value of the function

$$f(t) = \tfrac{1}{2}a_0 + a_1\cos\omega t + a_2\cos 2\omega t + \cdots + b_1\sin\omega t + b_2\sin 2\omega t + \cdots$$

is

$$F_{\text{rms}} = \sqrt{(\tfrac{1}{2}a_0)^2 + \tfrac{1}{2}a_1^2 + \tfrac{1}{2}a_2^2 + \cdots + \tfrac{1}{2}b_1^2 + \tfrac{1}{2}b_2^2 + \cdots} = \sqrt{c_0^2 + \tfrac{1}{2}c_1^2 + \tfrac{1}{2}c_2^2 + \tfrac{1}{2}c_3^2 + \cdots} \tag{19}$$

where (*17*) has been used.

Considering a linear network with an applied voltage which is periodic, we would expect that the resulting current would contain the same harmonic terms as the voltage but with harmonic amplitudes of different relative magnitude, since the impedance varies with $n\omega$. It is possible that some harmonics would not appear in the current; for example, in a pure LC parallel circuit, one of the harmonic frequencies might coincide with the resonant frequency, making the impedance at that frequency infinite. In general we may write

$$v = V_0 + \sum V_n \sin(n\omega t + \phi_n) \qquad \text{and} \qquad i = I_0 + \sum I_n \sin(n\omega t + \psi_n) \tag{20}$$

with corresponding effective values of

$$V_{\text{rms}} = \sqrt{V_0^2 + \tfrac{1}{2}V_1^2 + \tfrac{1}{2}V_2^2 + \cdots} \qquad \text{and} \qquad I_{\text{rms}} = \sqrt{I_0^2 + \tfrac{1}{2}I_1^2 + \tfrac{1}{2}I_2^2 + \cdots} \tag{21}$$

The average power P follows from integration of the instantaneous power, which is given by the product of v and i:

$$p = vi = \left[V_0 + \sum V_n \sin(n\omega t + \phi_n)\right]\left[I_0 + \sum I_n \sin(n\omega t + \psi_n)\right] \tag{22}$$

Since v and i both have period T, their product must have an integral number of its periods in T. (Recall that for a single sine wave of applied voltage, the product vi has a period half that of the voltage wave.) The average may therefore be calculated over one period of the voltage wave:

$$P = \frac{1}{T}\int_0^T \left[V_0 + \sum V_n \sin(n\omega t + \phi_n)\right]\left[I_0 + \sum I_n \sin(n\omega t + \psi_n)\right] dt \tag{23}$$

Examination of the possible terms in the product of the two infinite series shows them to be of the following types: the product of two constants, the product of a constant and a sine function, the product of two sine functions of different frequencies, and sine functions squared. After integration, the product of the two constants is still V_0I_0 and the sine functions squared with the limits applied appear as $(V_nI_n/2)\cos(\phi_n - \psi_n)$; all other products upon integration over the period T are zero. Then the average power is

$$P = V_0I_0 + \tfrac{1}{2}V_1I_1\cos\theta_1 + \tfrac{1}{2}V_2I_2\cos\theta_2 + \tfrac{1}{2}V_3I_3\cos\theta_3 + \cdots \tag{24}$$

where $\theta_n = \phi_n - \psi_n$ is the angle on the equivalent impedance of the network at the angular frequency $n\omega$, and V_n and I_n are the maximum values of the respective sine functions.

In the special case of a single-frequency sinusoidal voltage, $V_0 = V_2 = V_3 = \cdots = 0$, and (*24*) reduces to the familiar

$$P = \tfrac{1}{2}V_1I_1\cos\theta_1 = V_{\text{eff}}I_{\text{eff}}\cos\theta$$

Compare Section 9.2. Also, for a dc voltage, $V_1 = V_2 = V_3 = \cdots = 0$, and (*24*) becomes

$$P = V_0I_0 = VI$$

Thus, (*24*) is quite general. Note that on the right-hand side there is no term that involves voltage and current of different frequencies. In regard to power, then, each harmonic acts independently, and

$$P = P_0 + P_1 + P_2 + \cdots$$

12.8 APPLICATIONS IN CIRCUIT ANALYSIS

It has already been suggested above that we could apply the terms of a voltage series to a linear network and obtain the corresponding harmonic terms of the current series. This result is obtained by superposition. Thus we consider each term of the Fourier series representing the voltage as a single source, as shown in Fig. 12-10. Now the equivalent impedance of the network at each harmonic frequency $n\omega$ is used to compute the current at that harmonic. The sum of these individual responses is the total response i, in series form, to the applied voltage.

EXAMPLE 12.4 A series RL circuit in which $R = 5\ \Omega$ and $L = 20$ mH (Fig. 12-11) has an applied voltage $v = 100 + 50\sin\omega t + 25\sin 3\omega t$ (V), with $\omega = 500$ rad/s. Find the current and the average power.

Compute the equivalent impedance of the circuit at each frequency found in the voltage function. Then obtain the respective currents.

At $\omega = 0$, $Z_0 = R = 5\ \Omega$ and

$$I_0 = \frac{V_0}{R} = \frac{100}{5} = 20\ \text{A}$$

At $\omega = 500$ rad/s, $\mathbf{Z}_1 = 5 + j(500)(20 \times 10^{-3}) = 5 + j10 = 11.15\underline{/63.4^\circ}\ \Omega$ and

$$i_1 = \frac{V_{1\max}}{Z_1}\sin(\omega t - \theta_1) = \frac{50}{11.15}\sin(\omega t - 63.4^\circ) = 4.48\sin(\omega t - 63.4^\circ) \quad \text{(A)}$$

At $3\omega = 1500$ rad/s, $\mathbf{Z}_3 = 5 + j30 = 30.4\angle 80.54^\circ \;\; \Omega$ and

$$i_3 = \frac{V_{3\max}}{Z_3}\sin(3\omega t - \theta_3) = \frac{25}{30.4}\sin(3\omega t - 80.54^\circ) = 0.823\sin(3\omega t - 80.54^\circ) \quad \text{(A)}$$

The sum of the harmonic currents is the required total response; it is a Fourier series of the type (*8*).

$$i = 20 + 4.48\sin(\omega t - 63.4^\circ) + 0.823\sin(3\omega t - 80.54^\circ) \quad \text{(A)}$$

This current has the effective value

$$I_{\text{eff}} = \sqrt{20^2 + (4.48^2/2) + (0.823^2/2)} = \sqrt{410.6} = 20.25 \text{ A}$$

which results in a power in the 5-Ω resistor of

$$P = I_{\text{eff}}^2 R = (410.6)5 = 2053 \text{ W}$$

As a check, we compute the total average power by calculating first the power contributed by each harmonic and then adding the results.

At $\omega = 0$: $\quad P_0 = V_0 I_0 = 100(20) = 2000$ W

At $\omega = 500$ rad/s: $\quad P_1 = \frac{1}{2}V_1 I_1 \cos\theta_1 = \frac{1}{2}(50)(4.48)\cos 63.4^\circ = 50.1$ W

At $3\omega = 1500$ rad/s: $\quad P_3 = \frac{1}{2}V_3 I_3 \cos\theta_3 = \frac{1}{2}(25)(0.823)\cos 80.54^\circ = 1.69$ W

Then $\quad P = 2000 + 50.1 + 1.69 = 2052$ W

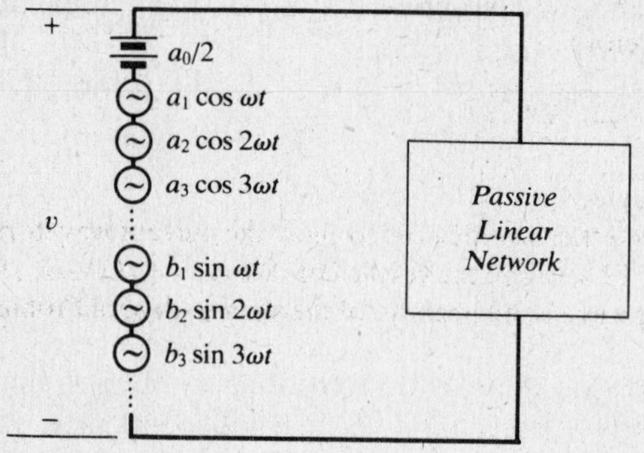

Fig. 12-10

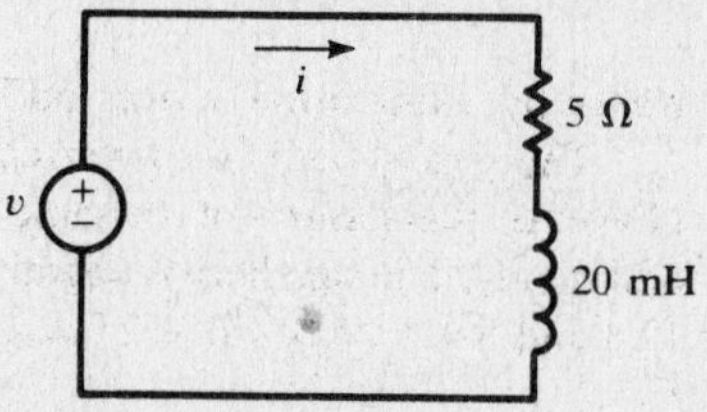

Fig. 12-11

Another Method

The Fourier series expression for the voltage across the resistor is

$$v_R = Ri = 100 + 22.4\sin(\omega t - 63.4^\circ) + 4.11\sin(3\omega t - 80.54^\circ) \quad \text{(V)}$$

and
$$V_{R\text{eff}} = \sqrt{100^2 + \tfrac{1}{2}(22.4)^2 + \tfrac{1}{2}(4.11)^2} = \sqrt{10\,259} = 101.3 \text{ V}$$

Then the power delivered by the source is $P = V_{R\text{eff}}^2/R = (10\,259)/5 = 2052$ W.

In Example 12.4 the driving voltage was given as a trigonometric Fourier series in t, and the computations were in the time domain. (The complex impedance was used only as a shortcut; Z_n and θ_n could have been obtained directly from R, L, and $n\omega$.) If, instead, the voltage is represented by an exponential Fourier series,

$$v(t) = \sum_{-\infty}^{+\infty} \mathbf{V}_n e^{jn\omega t}$$

then we have to do with a superposition of *phasors* $\mathbf{V}_n$ (rotating counterclockwise if $n > 0$, clockwise if $n < 0$), and so frequency-domain methods are called for. This is illustrated in Example 12.5.

EXAMPLE 12.5 A voltage represented by the triangular wave shown in Fig. 12-12 is applied to a pure capacitor C. Determine the resulting current.

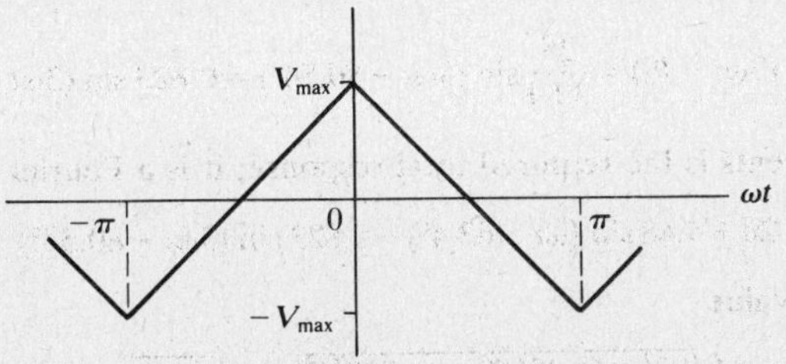

Fig. 12-12

In the interval $-\pi < \omega t < 0$ the voltage function is $v = V_{max} + (2V_{max}/\pi)\omega t$; and for $0 < \omega t < \pi$, $v = V_{max} - (2V_{max}/\pi)\omega t$. Then the coefficients of the exponential series are determined by the evaluation integral

$$\mathbf{V}_n = \frac{1}{2\pi}\int_{-\pi}^{0} [V_{max} + (2V_{max}/\pi)\omega t]e^{-jn\omega t}\, d(\omega t) + \frac{1}{2\pi}\int_{0}^{\pi} [V_{max} - (2V_{max}/\pi)\omega t]e^{-jn\omega t}\, d(\omega t)$$

from which $\mathbf{V}_n = 4V_{max}/\pi^2 n^2$ for odd n, and $\mathbf{V}_n = 0$ for even n.

The phasor current produced by $\mathbf{V}_n$ (n odd) is

$$\mathbf{I}_n = \frac{\mathbf{V}_n}{\mathbf{Z}_n} = \frac{4V_{max}/\pi^2 n^2}{1/jn\omega C} = j\frac{4V_{max}\omega C}{\pi^2 n}$$

with an implicit time factor $e^{jn\omega t}$. The resultant current is therefore

$$i(t) = \sum_{-\infty}^{+\infty} \mathbf{I}_n e^{jn\omega t} = j\frac{4V_{max}\omega C}{\pi^2}\sum_{-\infty}^{+\infty}\frac{e^{jn\omega t}}{n}$$

where the summation is over odd n only.

The series could be converted to the trigonometric form and then synthesized to show the current waveform. However, this series is of the same form as the result in Problem 12.8, where the coefficients are $\mathbf{A}_n = -j(2V/n\pi)$ for odd n only. The sign here is negative, indicating that our current wave is the negative of the square wave of Problem 12.8 and has a peak value $2V_{max}\omega C/\pi$.

Solved Problems

12.1 Find the trigonometric Fourier series for the square wave shown in Fig. 12-13 and plot the line spectrum.

In the interval $0 < \omega t < \pi$, $f(t) = V$; and for $\pi < \omega t < 2\pi$, $f(t) = -V$. The average value of the wave is zero; hence $a_0/2 = 0$. The cosine coefficients are obtained by writing the evaluation integral with the functions inserted as follows.

$$a_n = \frac{1}{\pi}\left\{\int_0^{\pi} V\cos n\omega t\, d(\omega t) + \int_{\pi}^{2\pi}(-V)\cos n\omega t\, d(\omega t)\right\} = \frac{V}{\pi}\left\{\left[\frac{1}{n}\sin n\omega t\right]_0^{\pi} - \left[\frac{1}{n}\sin n\omega t\right]_{\pi}^{2\pi}\right\}$$

$$= 0 \qquad \text{for all } n$$

Thus the series contains no cosine terms. Proceeding with the evaluation integral for the sine terms,

$$b_n = \frac{1}{\pi}\left\{\int_0^{\pi} V\sin n\omega t\, d(\omega t) + \int_{\pi}^{2\pi}(-V)\sin n\omega t\, d(\omega t)\right\}$$

$$= \frac{V}{\pi}\left\{\left[-\frac{1}{n}\cos n\omega t\right]_0^{\pi} + \left[\frac{1}{n}\cos n\omega t\right]_{\pi}^{2\pi}\right\}$$

$$= \frac{V}{\pi n}(-\cos n\pi + \cos 0 + \cos n2\pi - \cos n\pi) = \frac{2V}{\pi n}(1 - \cos n\pi)$$

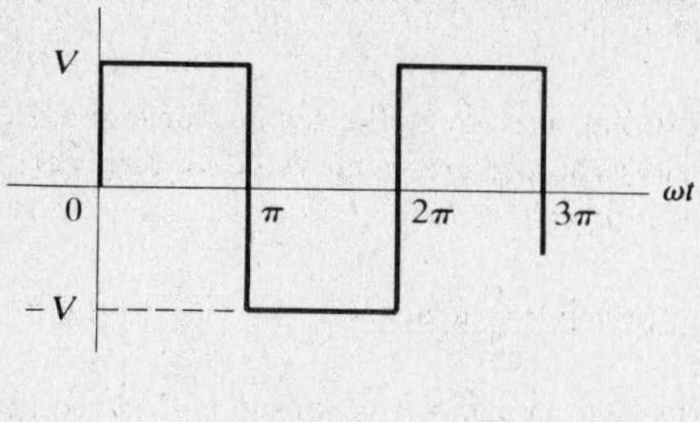

Fig. 12-13

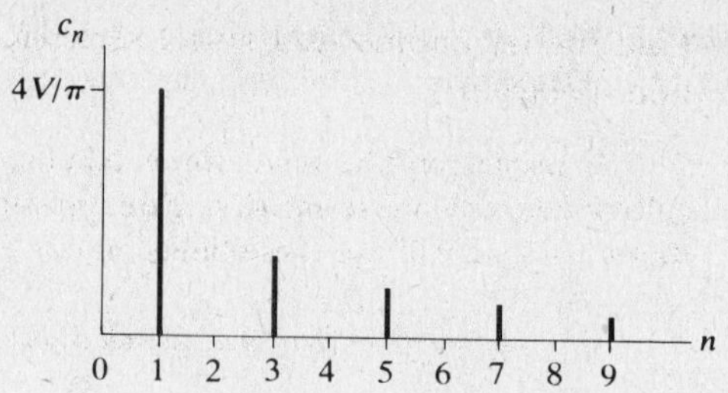

Fig. 12-14

Then $b_n = 4V/\pi n$ for $n = 1, 3, 5, \ldots$, and $b_n = 0$ for $n = 2, 4, 6, \ldots$. The series for the square wave is

$$f(t) = \frac{4V}{\pi} \sin \omega t + \frac{4V}{3\pi} \sin 3\omega t + \frac{4V}{5\pi} \sin 5\omega t + \cdots$$

The line spectrum for this series is shown in Fig. 12-14. The series contains only odd-harmonic sine terms, as could have been anticipated by examination of the waveform for symmetry. Since the wave in Fig. 12-13 is odd, its series contains only sine terms; and since it also has half-wave symmetry, only odd harmonics are present.

12.2 Find the trigonometric Fourier series for the triangular wave shown in Fig. 12-15 and plot the line spectrum.

The wave is an even function since $f(t) = f(-t)$, and if its average value, $V/2$, is subtracted, it also has half-wave symmetry, i.e. $f(t) = -f(t + \pi)$. For $-\pi < \omega t < 0$, $f(t) = V + (V/\pi)\omega t$; and for $0 < \omega t < \pi$, $f(t) = V - (V/\pi)\omega t$. Since even waveforms have only cosine terms, all $b_n = 0$. For $n \geq 1$,

$$\begin{aligned} a_n &= \frac{1}{\pi}\int_{-\pi}^{0} [V + (V/\pi)\omega t] \cos n\omega t \, d(\omega t) + \frac{1}{\pi}\int_{0}^{\pi} [V - (V/\pi)\omega t] \cos n\omega t \, d(\omega t) \\ &= \frac{V}{\pi}\left\{\int_{-\pi}^{\pi} \cos n\omega t \, d(\omega t) + \int_{-\pi}^{0} \frac{\omega t}{\pi} \cos n\omega t \, d(\omega t) - \int_{0}^{\pi} \frac{\omega t}{\pi} \cos n\omega t \, d(\omega t)\right\} \\ &= \frac{V}{\pi^2}\left\{\left[\frac{1}{n^2}\cos n\omega t + \frac{\omega t}{\pi}\sin n\omega t\right]_{-\pi}^{0} - \left[\frac{1}{n^2}\cos n\omega t + \frac{\omega t}{n}\sin n\omega t\right]_{0}^{\pi}\right\} \\ &= \frac{V}{\pi^2 n^2}\{\cos 0 - \cos(-n\pi) - \cos n\pi + \cos 0\} = \frac{2V}{\pi^2 n^2}(1 - \cos n\pi) \end{aligned}$$

As predicted from half-wave symmetry, the series contains only odd terms, since $a_n = 0$ for $n = 2, 4, 6, \ldots$. For $n = 1, 3, 5, \ldots$, $a_n = 4V/\pi^2 n^2$. Then the required Fourier series is

$$f(t) = \frac{V}{2} + \frac{4V}{\pi^2} \cos \omega t + \frac{4V}{(3\pi)^2} \cos 3\omega t + \frac{4V}{(5\pi)^2} \cos 5\omega t + \cdots$$

The coefficients decrease as $1/n^2$, and thus the series converges more rapidly than that of Problem 12.1. This fact is evident from the line spectrum shown in Fig. 12-16.

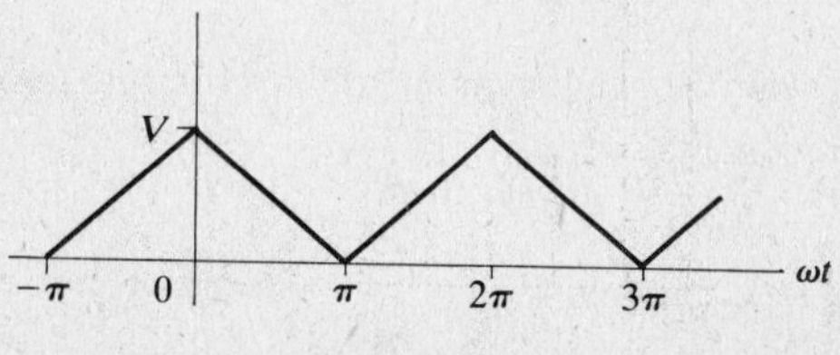

Fig. 12-15

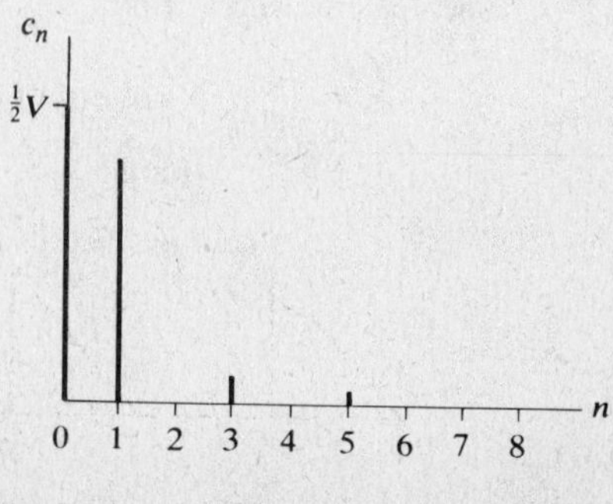

Fig. 12-16

12.3 Find the trigonometric Fourier series for the sawtooth wave shown in Fig. 12-17 and plot the line spectrum.

By inspection, the waveform is odd (and therefore has average value zero). Consequently the series will contain only sine terms. A single expression, $f(t) = (V/\pi)\omega t$, describes the wave over the period from $-\pi$ to $+\pi$, and we will use these limits on our evaluation integral for b_n.

$$b_n = \frac{1}{\pi}\int_{-\pi}^{\pi} (V/\pi)\omega t \sin n\omega t \, d(\omega t) = \frac{V}{\pi^2}\left[\frac{1}{n^2}\sin n\omega t - \frac{\omega t}{n}\cos n\omega t\right]_{-\pi}^{\pi} = -\frac{2V}{n\pi}(\cos n\pi)$$

As $\cos n\pi$ is $+1$ for even n and -1 for odd n, the signs of the coefficients alternate. The required series is

$$f(t) = \frac{2V}{\pi}\{\sin \omega t - \tfrac{1}{2}\sin 2\omega t + \tfrac{1}{3}\sin 3\omega t - \tfrac{1}{4}\sin 4\omega t + \cdots\}$$

The coefficients decrease as $1/n$, and thus the series converges slowly, as shown by the spectrum in Fig. 12-18. Except for the shift in the origin and the average term, this waveform is the same as in Fig. 12-8; compare the two spectra.

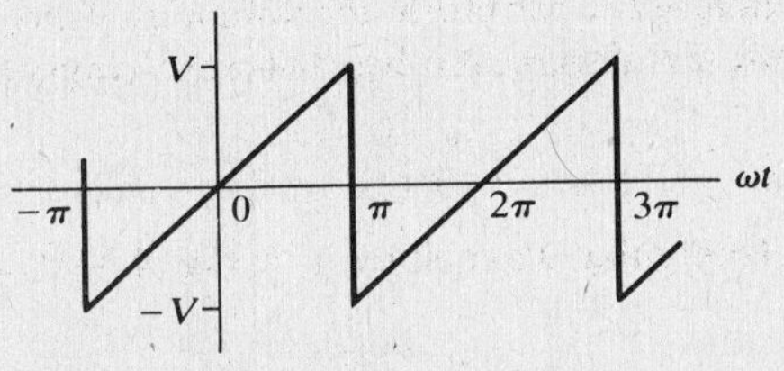

Fig. 12-17

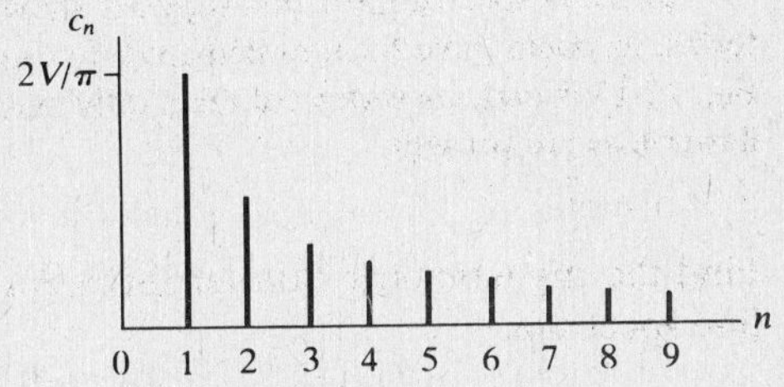

Fig. 12-18

12.4 Find the trigonometric Fourier series for the waveform shown in Fig. 12-19 and sketch the line spectrum.

In the interval $0 < \omega t < \pi$, $f(t) = (V/\pi)\omega t$; and for $\pi < \omega t < 2\pi$, $f(t) = 0$. By inspection, the average value of the wave is $V/4$. Since the wave is neither even nor odd, the series will contain both sine and cosine terms. For $n > 0$, we have

$$a_n = \frac{1}{\pi}\int_{0}^{\pi} (V/\pi)\omega t \cos n\omega t \, d(\omega t) = \frac{V}{\pi^2}\left[\frac{1}{n^2}\cos n\omega t + \frac{\omega t}{n}\sin n\omega t\right]_{0}^{\pi} = \frac{V}{\pi^2 n^2}(\cos n\pi - 1)$$

When n is even, $\cos n\pi - 1 = 0$ and $a_n = 0$. When n is odd, $a_n = -2V/(\pi^2 n^2)$. The b_n coefficients are

$$b_n = \frac{1}{\pi}\int_{0}^{\pi} (V/\pi)\omega t \sin n\omega t \, d(\omega t) = \frac{V}{\pi^2}\left[\frac{1}{n^2}\sin n\omega t - \frac{\omega t}{n}\cos n\omega t\right]_{0}^{\pi} = -\frac{V}{\pi n}(\cos n\pi) = (-1)^{n+1}\frac{V}{\pi n}$$

Then the required Fourier series is

$$f(t) = \frac{V}{4} - \frac{2V}{\pi^2}\cos \omega t - \frac{2V}{(3\pi)^2}\cos 3\omega t - \frac{2V}{(5\pi)^2}\cos 5\omega t - \cdots$$
$$+ \frac{V}{\pi}\sin \omega t - \frac{V}{2\pi}\sin 2\omega t + \frac{V}{3\pi}\sin 3\omega t - \cdots$$

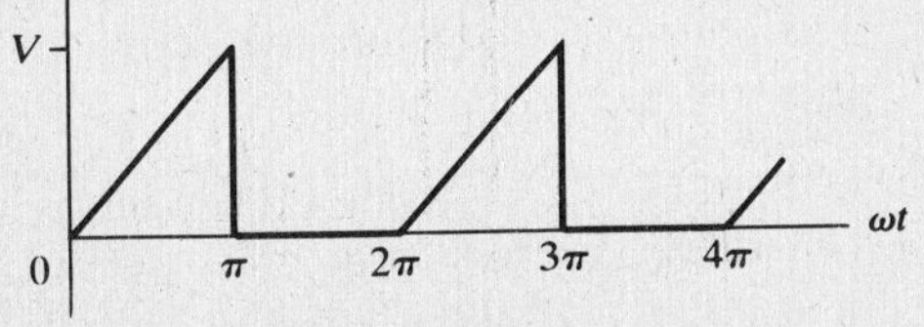

Fig. 12-19

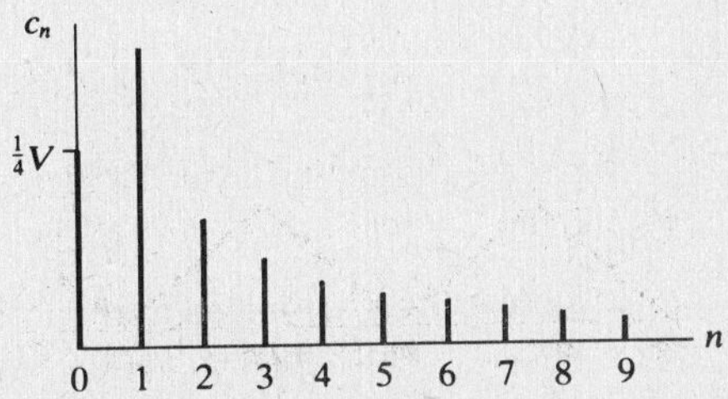

Fig. 12-20

The even-harmonic amplitudes are given directly by $|b_n|$ since there are no even-harmonic cosine terms. However, the odd-harmonic amplitudes must be computed using $c_n = \sqrt{a_n^2 + b_n^2}$. Thus

$$c_1 = \sqrt{(2V/\pi^2)^2 + (V/\pi)^2} = V(0.377) \qquad c_3 = V(0.109) \qquad c_5 = V(0.064)$$

The line spectrum is shown in Fig. 12-20.

12.5 Find the trigonometric Fourier series for the half-wave-rectified sine wave shown in Fig. 12-21 and sketch the line spectrum.

The wave shows no symmetry and we therefore expect the series to contain both sine and cosine terms. Since the average value is not obtainable by inspection, we evaluate a_0 for use in the term $a_0/2$.

$$a_0 = \frac{1}{\pi}\int_0^{\pi} V \sin \omega t \, d(\omega t) = \frac{V}{\pi}[-\cos \omega t]_0^{\pi} = \frac{2V}{\pi}$$

Next we determine a_n:

$$a_n = \frac{1}{\pi}\int_0^{\pi} V \sin \omega t \cos n\omega t \, d(\omega t)$$

$$= \frac{V}{\pi}\left[\frac{-n \sin \omega t \sin n\omega t - \cos n\omega t \cos \omega t}{-n^2+1}\right]_0^{\pi} = \frac{V}{\pi(1-n^2)}(\cos n\pi + 1)$$

With n even, $a_n = 2V/\pi(1-n^2)$; and with n odd, $a_n = 0$. However, this expression is indeterminate for $n = 1$ and therefore we must integrate separately for a_1.

$$a_1 = \frac{1}{\pi}\int_0^{\pi} V \sin \omega t \cos \omega t \, d(\omega t) = \frac{V}{\pi}\int_0^{\pi} \tfrac{1}{2} \sin 2\omega t \, d(\omega t) = 0$$

Now we evaluate b_n:

$$b_n = \frac{1}{\pi}\int_0^{\pi} V \sin \omega t \sin n\omega t \, d(\omega t) = \frac{V}{\pi}\left[\frac{n \sin \omega t \cos n\omega t - \sin n\omega t \cos \omega t}{-n^2+1}\right]_0^{\pi} = 0$$

Here again the expression is indeterminate for $n = 1$, and b_1 is evaluated separately.

$$b_1 = \frac{1}{\pi}\int_0^{\pi} V \sin^2 \omega t \, d(\omega t) = \frac{V}{\pi}\left[\frac{\omega t}{2} - \frac{\sin 2\omega t}{4}\right]_0^{\pi} = \frac{V}{2}$$

Then the required Fourier series is

$$f(t) = \frac{V}{\pi}\left\{1 + \frac{\pi}{2}\sin \omega t - \frac{2}{3}\cos 2\omega t - \frac{2}{15}\cos 4\omega t - \frac{2}{35}\cos 6\omega t - \cdots\right\}$$

The spectrum, Fig. 12-22, shows the strong fundamental term in the series and the rapidly decreasing amplitudes of the higher harmonics.

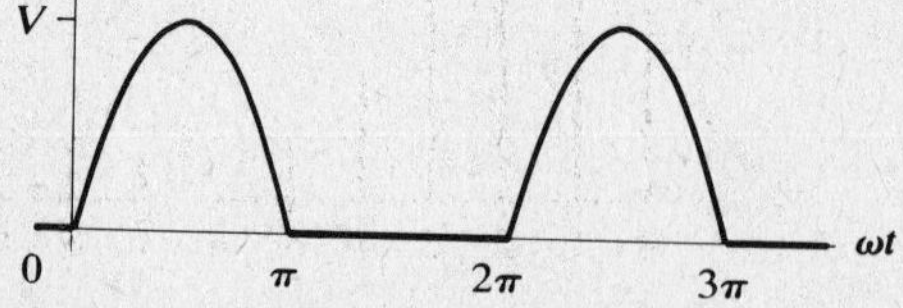

Fig. 12-21

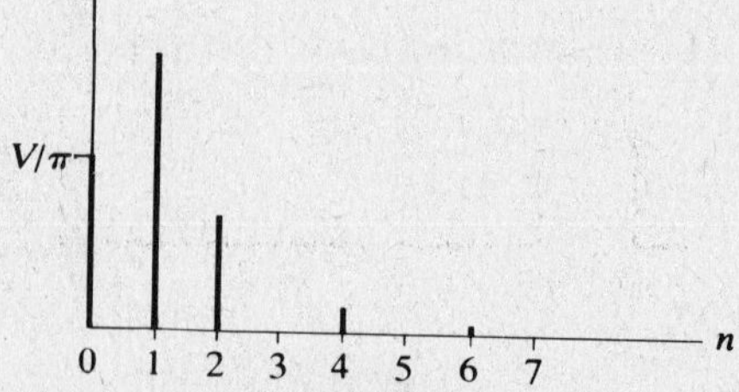

Fig. 12-22

12.6 Find the trigonometric Fourier series for the half-wave-rectified sine wave shown in Fig. 12-23, where the vertical axis is shifted from its position in Fig. 12-21.

The function is described in the interval $-\pi < \omega t < 0$ by $f(t) = -V \sin \omega t$. The average value is the same as that in Problem 12.5, i.e. $\frac{1}{2}a_0 = V/\pi$. For the coefficients a_n, we have

$$a_n = \frac{1}{\pi}\int_{-\pi}^{0} (-V \sin \omega t) \cos n\omega t \, d(\omega t) = \frac{V}{\pi(1-n^2)}(1 + \cos n\pi)$$

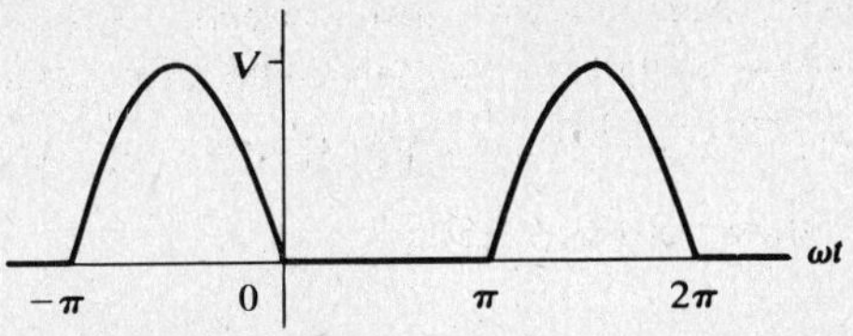

Fig. 12-23

For n even, $a_n = 2V/\pi(1-n^2)$; and for n odd, $a_n = 0$, except that $n = 1$ must be examined separately.

$$a_1 = \frac{1}{\pi}\int_{-\pi}^{0} (-V \sin \omega t) \cos \omega t \, d(\omega t) = 0$$

For the coefficients b_n, we obtain

$$b_n = \frac{1}{\pi}\int_{-\pi}^{0} (-V \sin \omega t) \sin n\omega t \, d(\omega t) = 0$$

except for $n = 1$.

$$b_1 = \frac{1}{\pi}\int_{-\pi}^{0} (-V) \sin^2 \omega t \, d(\omega t) = -\frac{V}{2}$$

Thus the series is

$$f(t) = \frac{V}{\pi}\left\{1 - \frac{\pi}{2}\sin \omega t - \frac{2}{3}\cos 2\omega t - \frac{2}{15}\cos 4\omega t - \frac{2}{35}\cos 6\omega t - \cdots\right\}$$

This series is identical to that of Problem 12.5, except for the fundamental term, which has a negative coefficient in this series. The spectrum would obviously be identical to that of Fig. 12-22.

Another Method

When the sine wave $V \sin \omega t$ is subtracted from the graph of Fig. 12-21, the graph of Fig. 12-23 results.

12.7 Obtain the trigonometric Fourier series for the repeating rectangular pulse shown in Fig. 12-24 and plot the line spectrum.

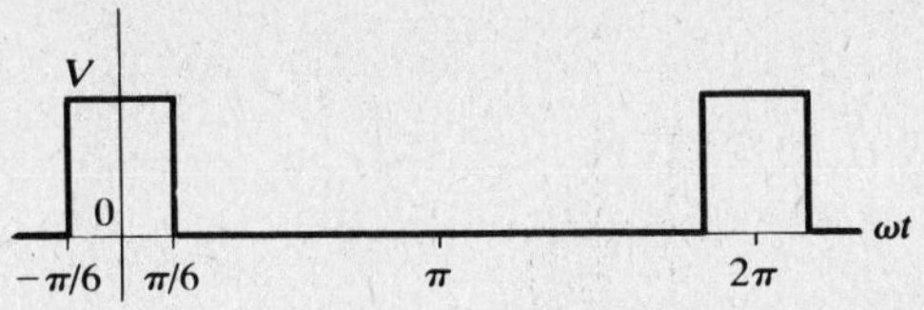

Fig. 12-24

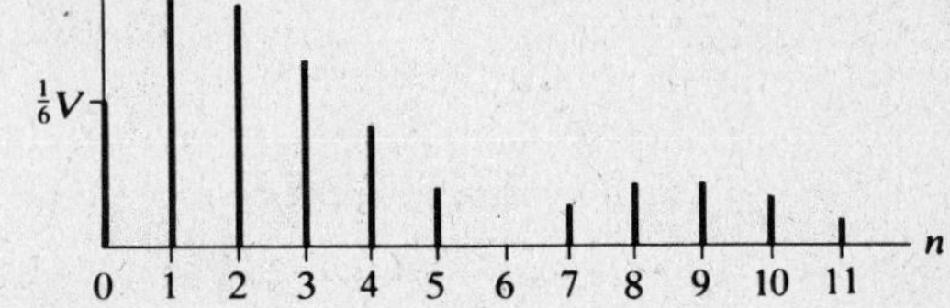

Fig. 12-25

With the vertical axis positioned as shown, the wave is even and the series will contain only cosine terms and a constant term. In the period from $-\pi$ to $+\pi$ used for the evaluation integrals, the function is zero except from $-\pi/6$ to $+\pi/6$.

$$a_0 = \frac{1}{\pi}\int_{-\pi/6}^{\pi/6} V \, d(\omega t) = \frac{V}{3} \qquad a_n = \frac{1}{\pi}\int_{-\pi/6}^{\pi/6} V \cos n\omega t \, d(\omega t) = \frac{2V}{n\pi}\sin\frac{n\pi}{6}$$

Since $\sin n\pi/6 = 1/2, \sqrt{3}/2, 1, \sqrt{3}/2, 1/2, 0, -1/2, \ldots$ for $n = 1, 2, 3, 4, 5, 6, 7, \ldots$, respectively the series is

$$f(t)=\frac{V}{6}+\frac{2V}{\pi}\left\{\frac{1}{2}\cos\omega t+\frac{\sqrt{3}}{2}\left(\frac{1}{2}\right)\cos 2\omega t+1\left(\frac{1}{3}\right)\cos 3\omega t+\frac{\sqrt{3}}{2}\left(\frac{1}{4}\right)\cos 4\omega t\right.$$
$$\left.+\frac{1}{2}\left(\frac{1}{5}\right)\cos 5\omega t-\frac{1}{2}\left(\frac{1}{7}\right)\cos 7\omega t-\cdots\right\}$$

or

$$f(t)=\frac{V}{6}+\frac{2V}{\pi}\sum_{n=1}^{\infty}\frac{1}{n}\sin(n\pi/6)\cos n\omega t$$

The line spectrum, shown in Fig. 12-25, decreases very slowly for this wave, since the series converges very slowly to the function. Of particular interest is the fact that the 8th, 9th and 10th harmonic amplitudes exceed the 7th. With the simple waves considered previously, the higher-harmonic amplitudes were progressively lower.

12.8 Find the exponential Fourier series for the square wave shown in Figs. 12-13 and 12-26, and sketch the line spectrum. Obtain the trigonometric series coefficients from those of the exponential series and compare with Problem 12.1.

In the interval $-\pi<\omega t<0$, $f(t)=-V$; and for $0<\omega t<\pi$, $f(t)=V$. The wave is odd; therefore, $A_0=0$ and the $\mathbf{A}_n$ will be pure imaginaries.

$$\mathbf{A}_n=\frac{1}{2\pi}\left\{\int_{-\pi}^{0}(-V)e^{-jn\omega t}\,d(\omega t)+\int_{0}^{\pi}Ve^{-jn\omega t}\,d(\omega t)\right\}$$
$$=\frac{V}{2\pi}\left\{-\left[\frac{1}{(-jn)}e^{-jn\omega t}\right]_{-\pi}^{0}+\left[\frac{1}{(-jn)}e^{-jn\omega t}\right]_{0}^{\pi}\right\}$$
$$=\frac{V}{(-j2\pi n)}(-e^{0}+e^{jn\pi}+e^{-jn\pi}-e^{0})=j\frac{V}{n\pi}(e^{jn\pi}-1)$$

For n even, $e^{jn\pi}=+1$ and $\mathbf{A}_n=0$; for n odd, $e^{jn\pi}=-1$ and $\mathbf{A}_n=-j(2V/n\pi)$ (half-wave symmetry). The required Fourier series is

$$f(t)=\cdots+j\frac{2V}{3\pi}e^{-j3\omega t}+j\frac{2V}{\pi}e^{-j\omega t}-j\frac{2V}{\pi}e^{j\omega t}-j\frac{2V}{3\pi}e^{j3\omega t}-\cdots$$

The graph in Fig. 12-27 shows amplitudes for both positive and negative frequencies. Combining the values at $+n$ and $-n$ yields the same line spectrum as plotted in Fig. 12-14.

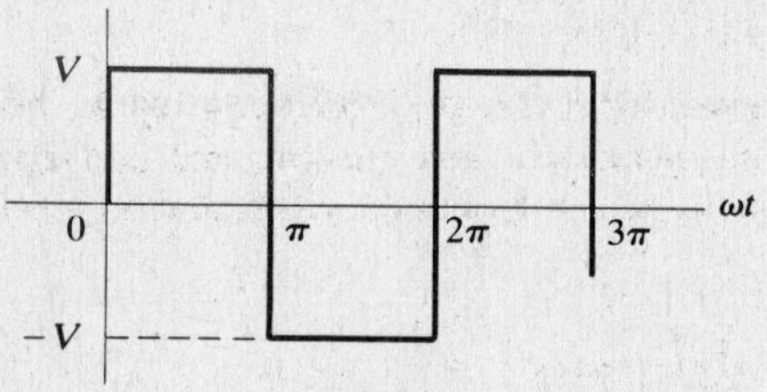

Fig. 12-26

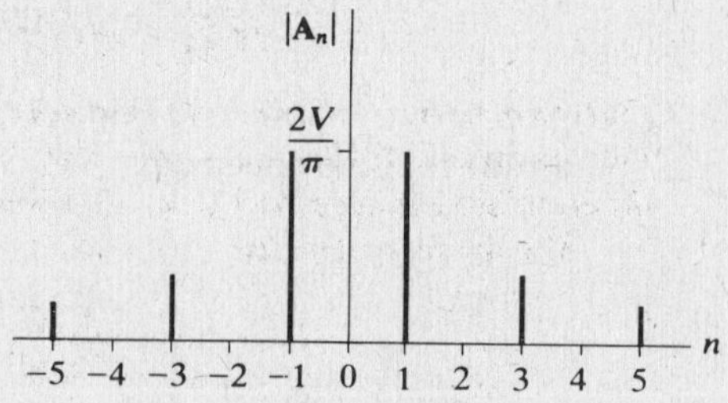

Fig. 12-27

The trigonometric-series cosine coefficients are

$$a_n=2\,\mathrm{Re}\,\mathbf{A}_n=0$$

and

$$b_n=-2\,\mathrm{Im}\,\mathbf{A}_n=\frac{4V}{n\pi}\qquad\text{for odd } n \text{ only}$$

These agree with the coefficients obtained in Problem 12.1.

12.9 Find the exponential Fourier series for the triangular wave shown in Figs. 12-15 and 12-28, and sketch the line spectrum.

In the interval $-\pi<\omega t<0$, $f(t)=V+(V/\pi)\omega t$; and for $0<\omega t<\pi$, $f(t)=V-(V/\pi)\omega t$. The wave is even and therefore the $\mathbf{A}_n$ coefficients will be pure real. By inspection the average value is $V/2$.

$$\mathbf{A}_n = \frac{1}{2\pi}\left\{\int_{-\pi}^{0}[V+(V/\pi)\omega t]e^{-jn\omega t}\,d(\omega t) + \int_0^{\pi}[V-(V/\pi)\omega t]e^{-jn\omega t}\,d(\omega t)\right\}$$
$$= \frac{V}{2\pi^2}\left\{\int_{-\pi}^{0}\omega t\,e^{-jn\omega t}\,d(\omega t) + \int_0^{\pi}(-\omega t)e^{-jn\omega t}\,d(\omega t) + \int_{-\pi}^{\pi}\pi\,e^{-jn\omega t}\,d(\omega t)\right\}$$
$$= \frac{V}{2\pi^2}\left\{\left[\frac{e^{-jn\omega t}}{(-jn)^2}(-jn\omega t - 1)\right]_{-\pi}^{0} - \left[\frac{e^{-jn\omega t}}{(-jn)^2}(-jn\omega t - 1)\right]_{0}^{\pi}\right\} = \frac{V}{\pi^2 n^2}[1 - e^{jn\pi}]$$

For even n, $e^{jn\pi} = +1$ and $\mathbf{A}_n = 0$; for odd n, $\mathbf{A}_n = 2V/\pi^2 n^2$. Thus the series is

$$f(t) = \cdots + \frac{2V}{(-3\pi)^2}e^{-j3\omega t} + \frac{2V}{(-\pi)^2}e^{-j\omega t} + \frac{V}{2} + \frac{2V}{(\pi)^2}e^{j\omega t} + \frac{2V}{(3\pi)^2}e^{j3\omega t} + \cdots$$

The harmonic amplitudes

$$c_0 = \frac{V}{2} \qquad c_n = 2|\mathbf{A}_n| = \begin{cases} 0 & (n = 2, 4, 6, \ldots) \\ 4V/\pi^2 n^2 & (n = 1, 3, 5, \ldots) \end{cases}$$

are exactly as plotted in Fig. 12-16.

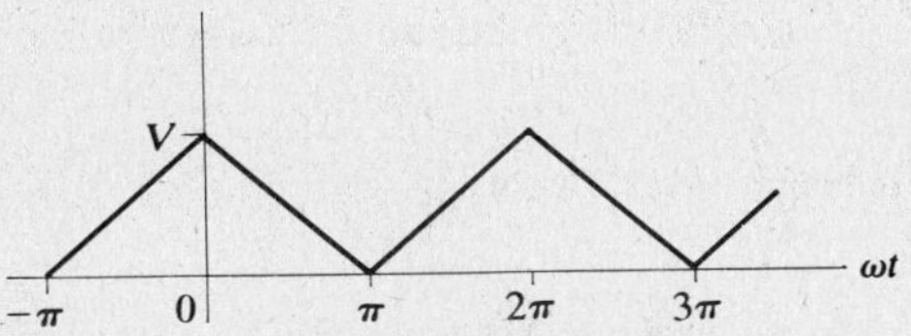

Fig. 12-28

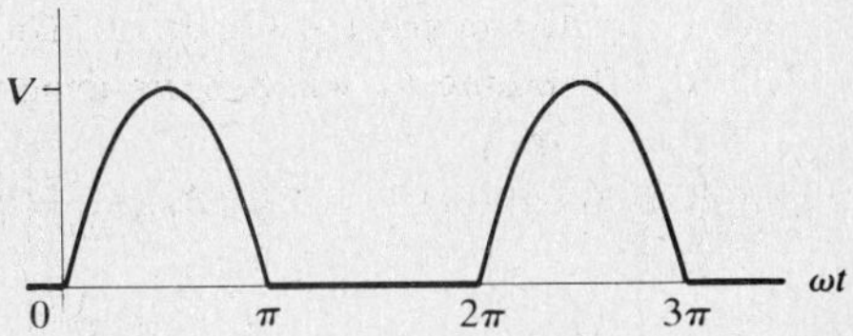

Fig. 12-29

12.10 Find the exponential Fourier series for the half-wave-rectified sine wave shown in Figs. 12-21 and 12-29, and sketch the line spectrum.

In the interval $0 < \omega t < \pi$, $f(t) = V \sin \omega t$; and from π to 2π, $f(t) = 0$. Then

$$\mathbf{A}_n = \frac{1}{2\pi}\int_0^{\pi} V \sin \omega t\, e^{-jn\omega t}\,d(\omega t)$$
$$= \frac{V}{2\pi}\left[\frac{e^{-jn\omega t}}{(1-n^2)}(-jn \sin \omega t - \cos \omega t)\right]_0^{\pi} = \frac{V(e^{-jn\pi}+1)}{2\pi(1-n^2)}$$

For even n, $\mathbf{A}_n = V/\pi(1-n^2)$; for odd n, $\mathbf{A}_n = 0$. However, for $n = 1$, the expression for $\mathbf{A}_n$ becomes indeterminate. L'Hospital's rule may be applied; i.e. the numerator and denominator are separately differentiated with respect to n, after which n is allowed to approach 1, with the result that $\mathbf{A}_1 = -j(V/4)$.

The average value is

$$A_0 = \frac{1}{2\pi}\int_0^{\pi} V \sin \omega t\, d(\omega t) = \frac{V}{2\pi}\left[-\cos \omega t\right]_0^{\pi} = \frac{V}{\pi}$$

Then the exponential Fourier series is

$$f(t) = \cdots - \frac{V}{15\pi}e^{-j4\omega t} - \frac{V}{3\pi}e^{-j2\omega t} + j\frac{V}{4}e^{-j\omega t} + \frac{V}{\pi} - j\frac{V}{4}e^{j\omega t} - \frac{V}{3\pi}e^{j2\omega t} - \frac{V}{15\pi}e^{j4\omega t} - \cdots$$

The harmonic amplitudes,

$$c_0 = A_0 = \frac{V}{\pi} \qquad c_n = 2|\mathbf{A}_n| = \begin{cases} 2V/\pi(n^2-1) & (n = 2, 4, 6, \ldots) \\ V/2 & (n = 1) \\ 0 & (n = 3, 5, 7, \ldots) \end{cases}$$

are exactly as plotted in Fig. 12-22.

12.11 Find the average power in a resistance $R = 10\ \Omega$, if the current in Fourier series form is $i = 10 \sin \omega t + 5 \sin 3\omega t + 2 \sin 5\omega t$ (A).

The current has an effective value $I_{\text{eff}} = \sqrt{\frac{1}{2}(10)^2 + \frac{1}{2}(5)^2 + \frac{1}{2}(2)^2} = \sqrt{64.5} = 8.03$ A. Then the average power is $P = I_{\text{eff}}^2 R = (64.5)10 = 645$ W.

Another Method

The total power is the sum of the harmonic powers, which are given by $\frac{1}{2}V_{\text{max}}I_{\text{max}} \cos \theta$. But the voltage across the resistor and the current are in phase for all harmonics, and $\theta_n = 0$. Then

$$v_R = Ri = 100 \sin \omega t + 50 \sin 3\omega t + 20 \sin 5\omega t$$

and $P = \frac{1}{2}(100)(10) + \frac{1}{2}(50)(5) + \frac{1}{2}(20)(2) = 645$ W.

12.12 Find the average power supplied to a network if the applied voltage and resulting current are

$$v = 50 + 50 \sin 5 \times 10^3 t + 30 \sin 10^4 t + 20 \sin 2 \times 10^4 t \quad \text{(V)}$$
$$i = 11.2 \sin (5 \times 10^3 t + 63.4°) + 10.6 \sin (10^4 t + 45°) + 8.97 \sin (2 \times 10^4 t + 26.6°) \quad \text{(A)}$$

The total average power is the sum of the harmonic powers:

$$P = (50)(0) + \tfrac{1}{2}(50)(11.2) \cos 63.4° + \tfrac{1}{2}(30)(10.6) \cos 45° + \tfrac{1}{2}(20)(8.97) \cos 26.6° = 317.7 \text{ W}$$

12.13 Obtain the constants of the two-element series circuit with the applied voltage and resultant current given in Problem 12.12.

The voltage series contains a constant term 50 but there is no corresponding term in the current series, thus indicating that one of the elements is a capacitor. Since power is delivered to the circuit, the other element must be a resistor.

$$I_{\text{eff}} = \sqrt{\tfrac{1}{2}(11.2)^2 + \tfrac{1}{2}(10.6)^2 + \tfrac{1}{2}(8.97)^2} = 12.6 \text{ A}$$

The average power is $P = I_{\text{eff}}^2 R$, from which $R = P/I_{\text{eff}}^2 = 317.7/159.2 = 2\ \Omega$.
At $\omega = 10^4$ rad/s, the current leads the voltage by 45°. Hence,

$$1 = \tan 45° = \frac{1}{\omega CR} \qquad \text{or} \qquad C = \frac{1}{(10^4)(2)} = 50\ \mu\text{F}$$

Therefore the two-element series circuit consists of a resistor of 2 Ω and a capacitor of 50 μF.

12.14 The voltage wave shown in Fig. 12-30 is applied to a series circuit of $R = 2$ kΩ and $L = 10$ H. Use the trigonometric Fourier series to obtain the voltage across the resistor. Plot the line spectra of the applied voltage and v_R to show the effect of the inductance on the harmonics. $\omega = 377$ rad/s.

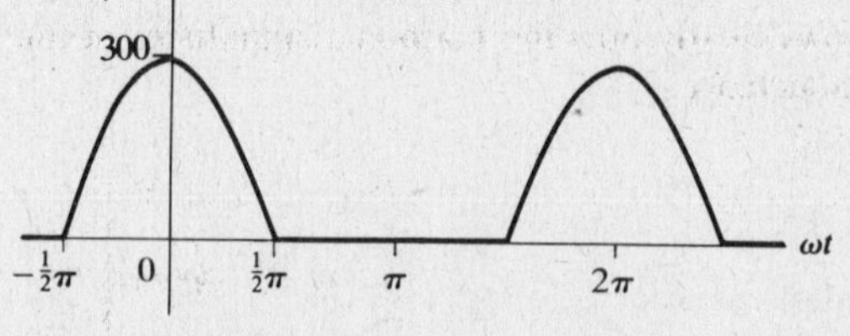

Fig. 12-30

The applied voltage has average value V_{max}/π, as in Problem 12.5. The wave function is even and hence the series contains only cosine terms, with coefficients obtained by the following evaluation integral:

$$a_n = \frac{1}{\pi}\int_{-\pi/2}^{\pi/2} 300 \cos \omega t \cos n\omega t \, d(\omega t) = \frac{600}{\pi(1-n^2)} \cos n\pi/2 \quad \text{V}$$

Here, $\cos n\pi/2$ has the value −1 for $n = 2, 6, 10, \ldots$, and +1 for $n = 4, 8, 12, \ldots$. For n odd, $\cos n\pi/2 = 0$. However, for $n = 1$, the expression is indeterminate and must be evaluated separately.

Table 12-1

n	$n\omega$, rad/s	R, kΩ	$n\omega L$, kΩ	Z_n, kΩ	θ_n
0	0	2	0	2	0°
1	377	2	3.77	4.26	62°
2	754	2	7.54	7.78	75.1°
4	1508	2	15.08	15.2	82.45°
6	2262	2	22.62	22.6	84.92°

$$a_1 = \frac{1}{\pi}\int_{-\pi/2}^{\pi/2} 300\cos^2 \omega t\, d(\omega t) = \frac{300}{\pi}\left[\frac{\omega t}{2} + \frac{\sin 2\omega t}{4}\right]_{-\pi/2}^{\pi/2} = \frac{300}{2}\ \text{V}$$

Thus,

$$v = \frac{300}{\pi}\left\{1 + \frac{\pi}{2}\cos \omega t + \frac{2}{3}\cos 2\omega t - \frac{2}{15}\cos 4\omega t + \frac{2}{35}\cos 6\omega t - \cdots\right\} \quad \text{(V)}$$

In Table 12-1, the total impedance of the series circuit is computed for each harmonic in the voltage expression. The Fourier coefficients of the current series are the voltage series coefficients divided by the Z_n; the current terms lag the voltage terms by the phase angles θ_n.

$$I_0 = \frac{300/\pi}{2}\ \text{mA}$$

$$i_1 = \frac{300/2}{4.26}\cos(\omega t - 62°) \quad \text{(mA)}$$

$$i_2 = \frac{600/3\pi}{7.78}\cos(2\omega t - 75.1°) \quad \text{(mA)}$$

. . . .

Then the current series is

$$i = \frac{300}{2\pi} + \frac{300}{(2)(4.26)}\cos(\omega t - 62°) + \frac{600}{3\pi(7.78)}\cos(2\omega t - 75.1°)$$
$$- \frac{600}{15\pi(15.2)}\cos(4\omega t - 82.45°) + \frac{600}{35\pi(22.6)}\cos(6\omega t - 84.92°) - \cdots \quad \text{(mA)}$$

and the voltage across the resistor is

$$v_R = Ri = 95.5 + 70.4\cos(\omega t - 62°) + 16.4\cos(2\omega t - 75.1°)$$
$$- 1.67\cos(4\omega t - 82.45°) + 0.483\cos(6\omega t - 84.92°) - \cdots \quad \text{(V)}$$

Figure 12-31 shows clearly how the harmonic amplitudes of the applied voltage have been reduced by the 10-H series inductance.

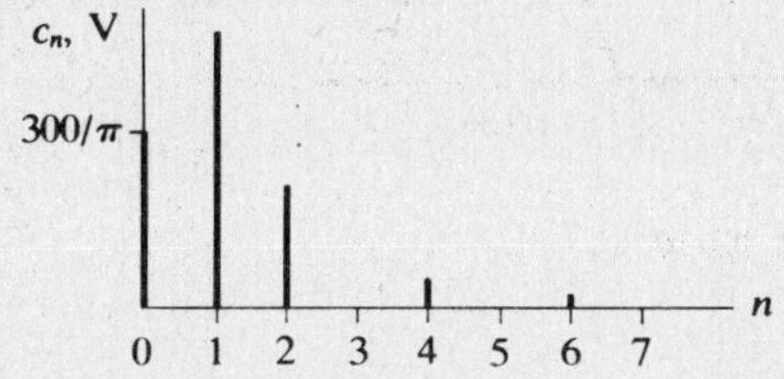

(*a*) Spectrum of v

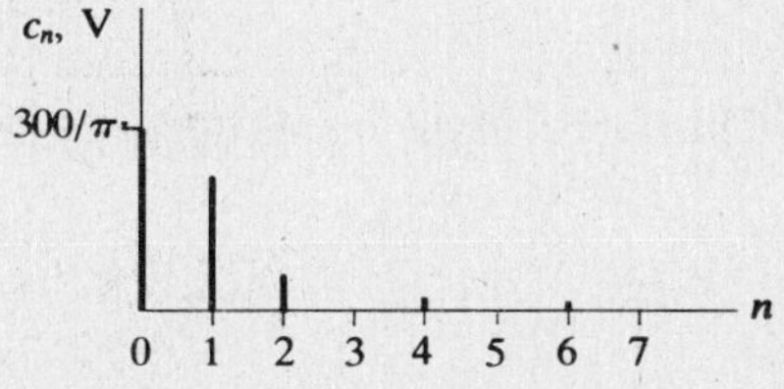

(*b*) Spectrum of v_R

Fig. 12-31

12.15 The current in a 10-mH inductance has the waveform shown in Fig. 12-32. Obtain the trigonometric series for the voltage across the inductance, given that $\omega = 500$ rad/s.

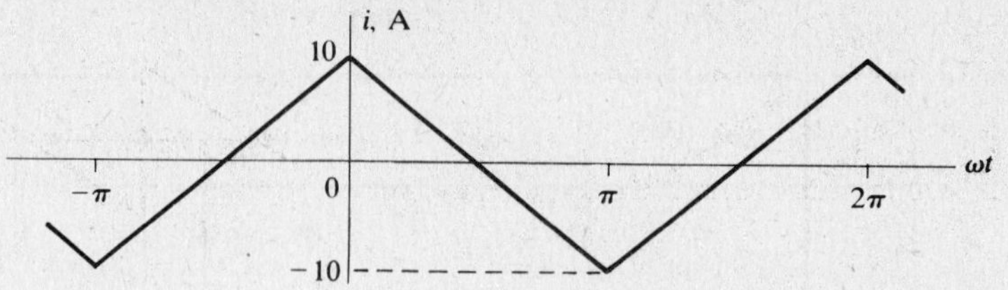

Fig. 12-32

The derivative of the waveform of Fig. 12-32 is graphed in Fig. 12-33. This is just Fig. 12-13 with $V = -20/\pi$. Hence, from Problem 12.1,

$$\frac{di}{d(\omega t)} = -\frac{80}{\pi^2}[\sin \omega t + \tfrac{1}{3}\sin 3\omega t + \tfrac{1}{5}\sin 5\omega t + \cdots] \quad \text{(A)}$$

and so

$$v_L = L\omega\frac{di}{d(\omega t)} = -\frac{400}{\pi^2}[\sin \omega t + \tfrac{1}{3}\sin 3\omega t + \tfrac{1}{5}\sin 5\omega t + \cdots] \quad \text{(V)}$$

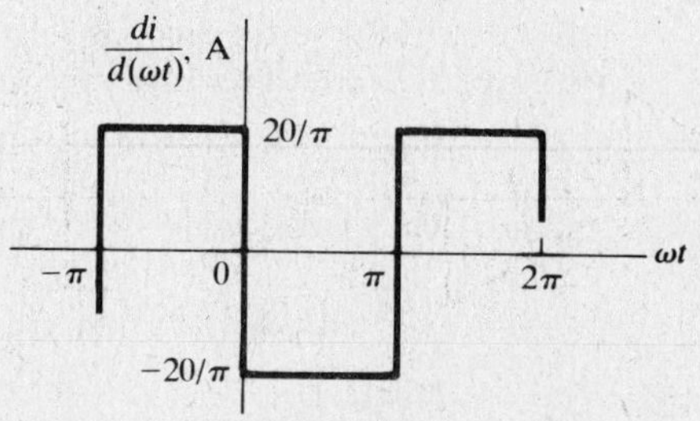

Fig. 12-33

Supplementary Problems

12.16 Synthesize the waveform for which the trigonometric Fourier series is

$$f(t) = \frac{8V}{\pi^2}\{\sin \omega t - \tfrac{1}{9}\sin 3\omega t + \tfrac{1}{25}\sin 5\omega t - \tfrac{1}{49}\sin 7\omega t + \cdots\}$$

12.17 Synthesize the waveform if its Fourier series is

$$f(t) = 5 - \frac{40}{\pi^2}(\cos \omega t + \tfrac{1}{9}\cos 3\omega t + \tfrac{1}{25}\cos 5\omega t + \cdots) + \frac{20}{\pi}(\sin \omega t - \tfrac{1}{2}\sin 2\omega t + \tfrac{1}{3}\sin 3\omega t - \tfrac{1}{4}\sin 4\omega t + \cdots)$$

12.18 Synthesize the waveform for the given Fourier series.

$$f(t) = V\left\{\frac{1}{2\pi} - \frac{1}{\pi}\cos \omega t - \frac{1}{3\pi}\cos 2\omega t + \frac{1}{2\pi}\cos 3\omega t - \frac{1}{15\pi}\cos 4\omega t - \frac{1}{6\pi}\cos 6\omega t + \cdots \right.$$
$$\left. + \frac{1}{4}\sin \omega t - \frac{2}{3\pi}\sin 2\omega t + \frac{4}{15\pi}\sin 4\omega t - \cdots \right\}$$

12.19 Find the trigonometric Fourier series for the sawtooth wave shown in Fig. 12-34 and plot the line spectrum. Compare with Example 12.1.

Ans. $f(t) = \dfrac{V}{2} + \dfrac{V}{\pi}\{\sin \omega t + \tfrac{1}{2}\sin 2\omega t + \tfrac{1}{3}\sin 3\omega t + \cdots\}$

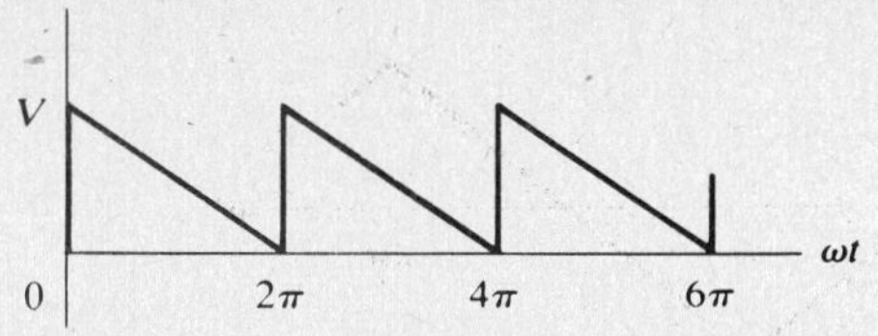

Fig. 12-34

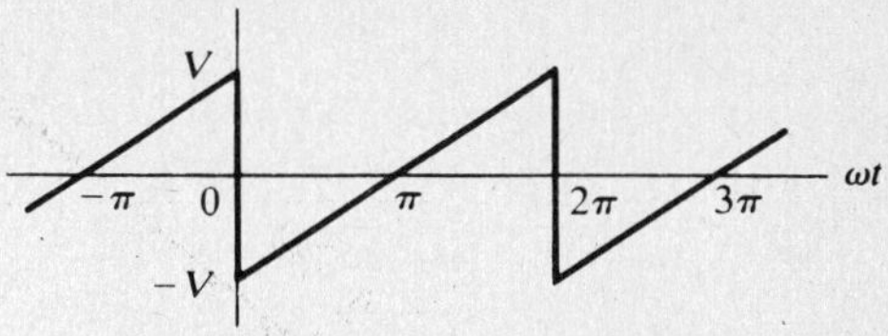

Fig. 12-35

12.20 Find the trigonometric Fourier series for the sawtooth wave shown in Fig. 12-35 and plot the spectrum. Compare with the result of Problem 12.3.

Ans. $f(t) = \dfrac{-2V}{\pi}\{\sin \omega t + \frac{1}{2}\sin 2\omega t + \frac{1}{3}\sin 3\omega t + \frac{1}{4}\sin 4\omega t + \cdots\}$

12.21 Find the trigonometric Fourier series for the waveform shown in Fig. 12-36 and plot the line spectrum.

Ans. $f(t) = \dfrac{4V}{\pi^2}\{\cos \omega t + \frac{1}{9}\cos 3\omega t + \frac{1}{25}\cos 5\omega t + \cdots\} - \dfrac{2V}{\pi}\{\sin \omega t + \frac{1}{3}\sin 3\omega t + \frac{1}{5}\sin 5\omega t + \cdots\}$

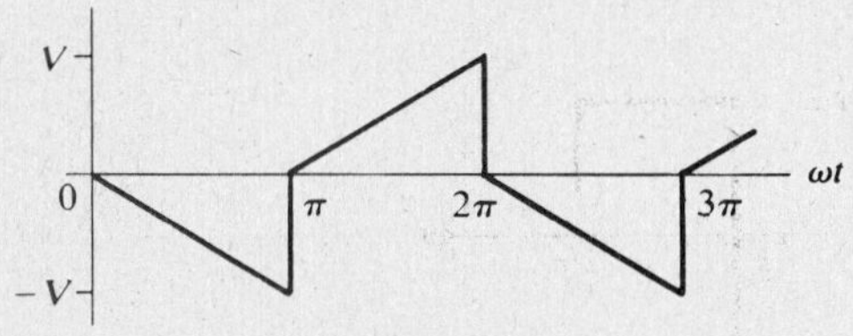

Fig. 12-36

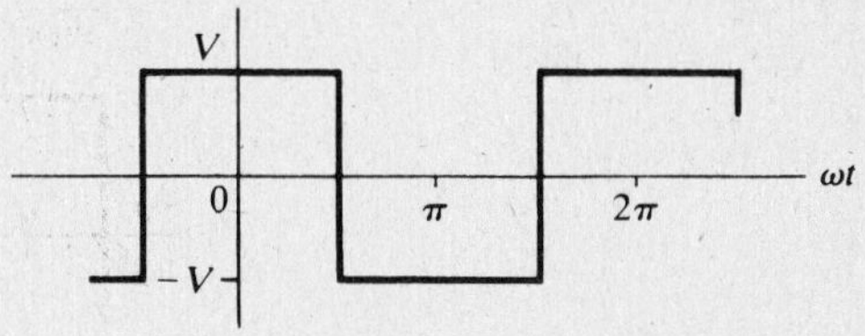

Fig. 12-37

12.22 Find the trigonometric Fourier series of the square wave shown in Fig. 12-37 and plot the line spectrum. Compare with the result of Problem 12.1.

Ans. $f(t) = \dfrac{4V}{\pi}\{\cos \omega t - \frac{1}{3}\cos 3\omega t + \frac{1}{5}\cos 5\omega t - \frac{1}{7}\cos 7\omega t + \cdots\}$

12.23 Find the trigonometric Fourier series for the waveforms shown in Fig. 12-38. Plot the line spectrum of each and compare.

Ans. (a) $f(t) = \dfrac{5}{12} + \sum_{n=1}^{\infty}\left\{\dfrac{10}{n\pi}\left(\sin\dfrac{n\pi}{12}\right)\cos n\omega t + \dfrac{10}{n\pi}\left(1 - \cos\dfrac{n\pi}{12}\right)\sin n\omega t\right\}$

(b) $f(t) = \dfrac{50}{6} + \sum_{n=1}^{\infty}\left\{\dfrac{10}{n\pi}\left(\sin\dfrac{n5\pi}{3}\right)\cos n\omega t + \dfrac{10}{n\pi}\left(1 - \cos\dfrac{n5\pi}{3}\right)\sin n\omega t\right\}$

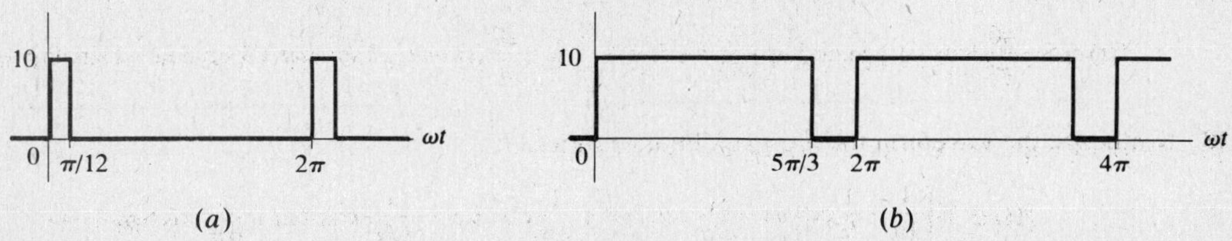

Fig. 12-38

12.24 Find the trigonometric Fourier series for the half-wave-rectified sine wave shown in Fig. 12-39 and plot the line spectrum. Compare the answer with the results of Problems 12.5 and 12.6.

Ans. $f(t) = \dfrac{V}{\pi}\left\{1 + \dfrac{\pi}{2}\cos \omega t + \dfrac{2}{3}\cos 2\omega t - \dfrac{2}{15}\cos 4\omega t + \dfrac{2}{35}\cos 6\omega t - \cdots\right\}$

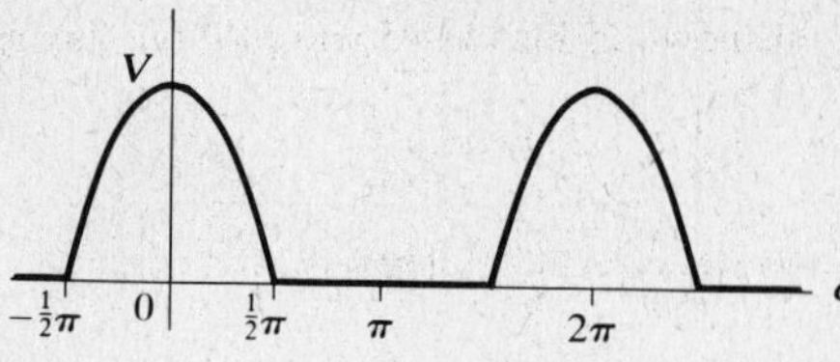

Fig. 12-39

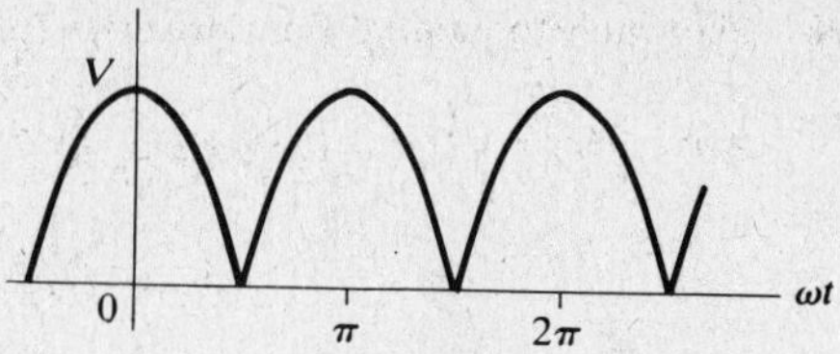

Fig. 12-40

12.25 Find the trigonometric Fourier series for the full-wave-rectified sine wave shown in Fig. 12-40 and plot the spectrum.

Ans. $f(t) = \dfrac{2V}{\pi}\{1 + \frac{2}{3}\cos 2\omega t - \frac{2}{15}\cos 4\omega t + \frac{2}{35}\cos 6\omega t - \cdots\}$

12.26 The waveform in Fig. 12-41 is that of Fig. 12-40 with the origin shifted. Find the Fourier series and show that the two spectra are identical.

Ans. $f(t) = \dfrac{2V}{\pi}\{1 - \frac{2}{3}\cos 2\omega t - \frac{2}{15}\cos 4\omega t - \frac{2}{35}\cos 6\omega t - \cdots\}$

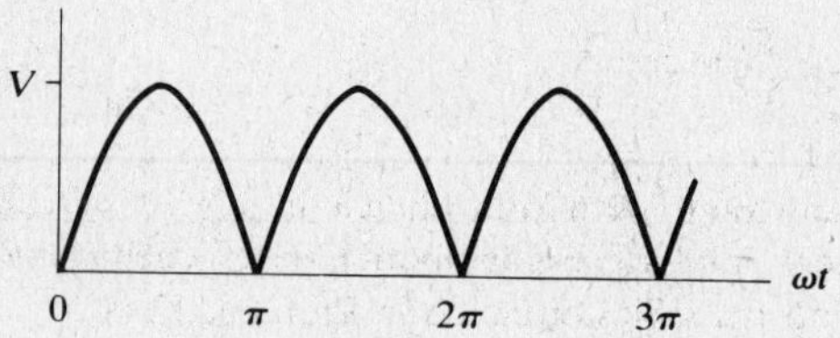

Fig. 12-41

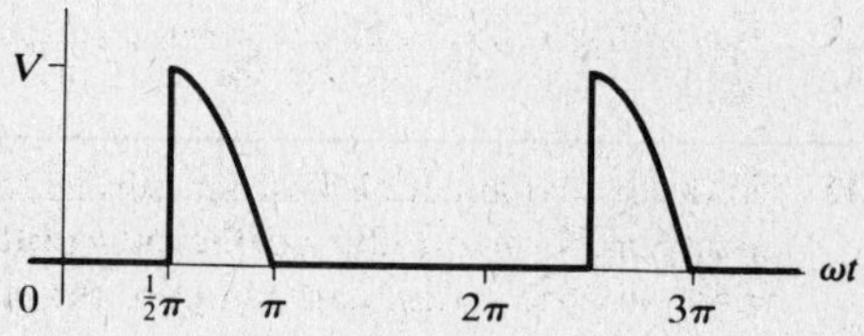

Fig. 12-42

12.27 Find the trigonometric Fourier series for the waveform shown in Fig. 12-42.

Ans.
$$f(t) = \frac{V}{2\pi} - \frac{V}{2\pi}\cos\omega t + \sum_{n=2}^{\infty} \frac{V}{\pi(1-n^2)}(\cos n\pi + n\sin n\pi/2)\cos n\omega t + \frac{V}{4}\sin\omega t + \sum_{n=2}^{\infty}\left[\frac{-nV\cos n\pi/2}{\pi(1-n^2)}\right]\sin n\omega t$$

12.28 Find the trigonometric Fourier series for the waveform shown in Fig. 12-43. Add this series termwise to that of Problem 12.27 and compare the sum with the series obtained in Problem 12.5.

Ans.
$$f(t) = \frac{V}{2\pi} + \frac{V}{2\pi}\cos\omega t + \sum_{n=2}^{\infty}\frac{V[n\sin n\pi/2 - 1]}{\pi(n^2-1)}\cos n\omega t + \frac{V}{4}\sin\omega t + \sum_{n=2}^{\infty}\frac{nV\cos n\pi/2}{\pi(1-n^2)}\sin n\omega t$$

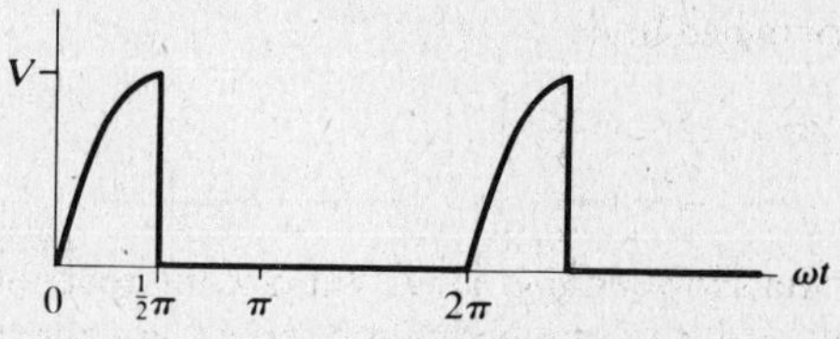

Fig. 12-43

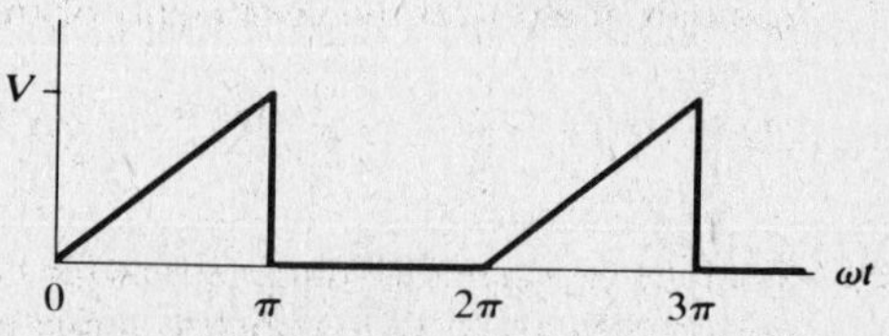

Fig. 12-44

12.29 Find the exponential Fourier series for the waveform shown in Fig. 12-44 and plot the line spectrum. Convert the coefficients obtained here into the trigonometric series coefficients, write the trigonometric series, and compare it with the result of Problem 12.4.

Ans.
$$f(t) = V\Big\{\cdots - \left(\frac{1}{9\pi^2} - j\frac{1}{6\pi}\right)e^{-j3\omega t} - j\frac{1}{4\pi}e^{-j2\omega t} - \left(\frac{1}{\pi^2} - j\frac{1}{2\pi}\right)e^{-j\omega t} + \frac{1}{4} - \left(\frac{1}{\pi^2} + j\frac{1}{2\pi}\right)e^{j\omega t} + j\frac{1}{4\pi}e^{j2\omega t} - \left(\frac{1}{9\pi^2} + j\frac{1}{6\pi}\right)e^{j3\omega t} - \cdots\Big\}$$

12.30 Find the exponential Fourier series for the waveform shown in Fig. 12-45 and plot the line spectrum.

Ans. $f(t) = V\left\{\cdots + \left(\frac{1}{9\pi^2} + j\frac{1}{6\pi}\right)e^{-j3\omega t} + j\frac{1}{4\pi}e^{-j2\omega t} + \left(\frac{1}{\pi^2} + j\frac{1}{2\pi}\right)e^{-j\omega t} + \frac{1}{4} + \left(\frac{1}{\pi^2} - j\frac{1}{2\pi}\right)e^{j\omega t} - j\frac{1}{4\pi}e^{j2\omega t} + \left(\frac{1}{9\pi^2} - j\frac{1}{6\pi}\right)e^{j3\omega t} + \cdots\right\}$

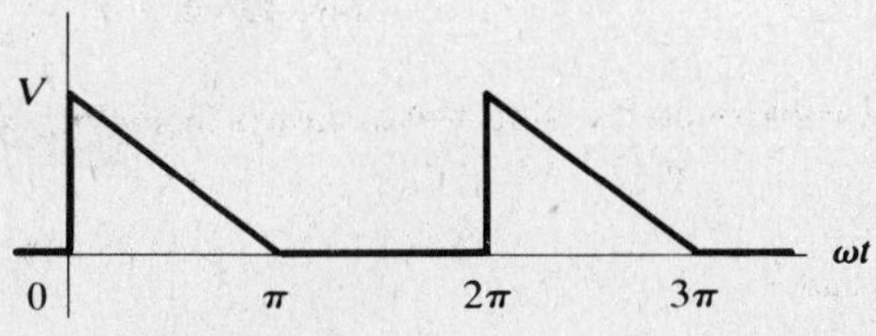

Fig. 12-45

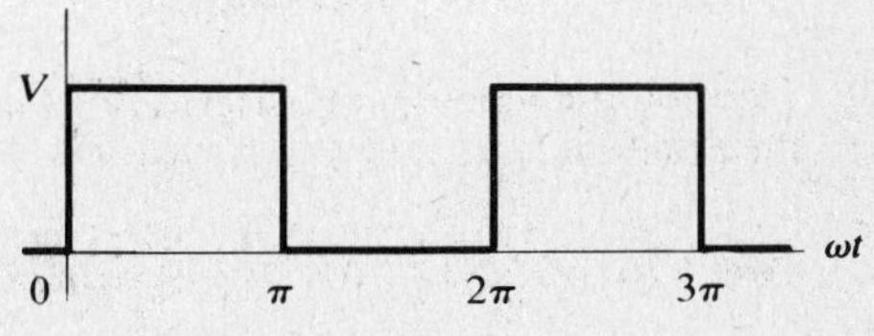

Fig. 12-46

12.31 Find the exponential Fourier series for the square wave shown in Fig. 12-46 and plot the line spectrum. Add the exponential series of Problems 12.29 and 12.30 and compare the sum to the series obtained here.

Ans. $f(t) = V\left\{\cdots + j\frac{1}{3\pi}e^{-j3\omega t} + j\frac{1}{\pi}e^{-j\omega t} + \frac{1}{2} - j\frac{1}{\pi}e^{j\omega t} - j\frac{1}{3\pi}e^{j3\omega t} - \cdots\right\}$

12.32 Find the exponential Fourier series for the sawtooth waveform shown in Fig. 12-47 and plot the spectrum. Convert the coefficients obtained here into the trigonometric series coefficients, write the trigonometric series, and compare the results with the series obtained in Problem 12.19.

Ans. $f(t) = V\left\{\cdots + j\frac{1}{4\pi}e^{-j2\omega t} + j\frac{1}{2\pi}e^{-j\omega t} + \frac{1}{2} - j\frac{1}{2\pi}e^{j\omega t} - j\frac{1}{4\pi}e^{j2\omega t} - \cdots\right\}$

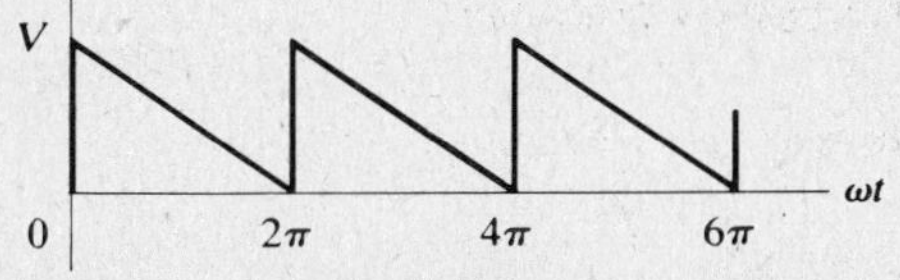

Fig. 12-47

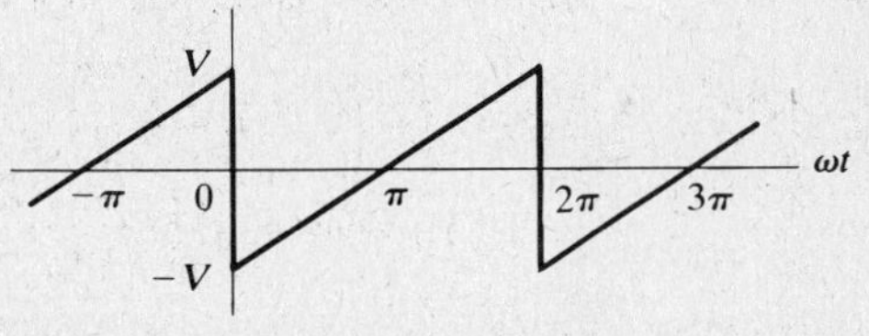

Fig. 12-48

12.33 Find the exponential Fourier series for the waveform shown in Fig. 12-48 and plot the spectrum. Convert the trigonometric series coefficients found in Problem 12.20 into exponential series coefficients and compare them with the coefficients of the series obtained here.

Ans. $f(t) = V\left\{\cdots - j\frac{1}{2\pi}e^{-j2\omega t} - j\frac{1}{\pi}e^{-j\omega t} + j\frac{1}{\pi}e^{j\omega t} + j\frac{1}{2\pi}e^{j2\omega t} + \cdots\right\}$

12.34 Find the exponential Fourier series for the waveform shown in Fig. 12-49 and plot the spectrum. Convert the coefficients to trigonometric series coefficients, write the trigonometric series, and compare it with that obtained in Problem 12.21.

Ans. $f(t) = V\left\{\cdots + \left(\frac{2}{9\pi^2} - j\frac{1}{3\pi}\right)e^{-j3\omega t} + \left(\frac{2}{\pi^2} - j\frac{1}{\pi}\right)e^{-j\omega t} + \left(\frac{2}{\pi^2} + j\frac{1}{\pi}\right)e^{j\omega t} + \left(\frac{2}{9\pi^2} + j\frac{1}{3\pi}\right)e^{j3\omega t} + \cdots\right\}$

12.35 Find the exponential Fourier series for the square wave shown in Fig. 12-50 and plot the line spectrum. Convert the trigonometric series coefficients of Problem 12.22 into exponential series coefficients and compare with the coefficients in the result obtained here.

Ans. $f(t) = \frac{2V}{\pi}\{\cdots + \tfrac{1}{5}e^{-j5\omega t} - \tfrac{1}{3}e^{-j3\omega t} + e^{-j\omega t} + e^{j\omega t} - \tfrac{1}{3}e^{j3\omega t} + \tfrac{1}{5}e^{j5\omega t} - \cdots\}$

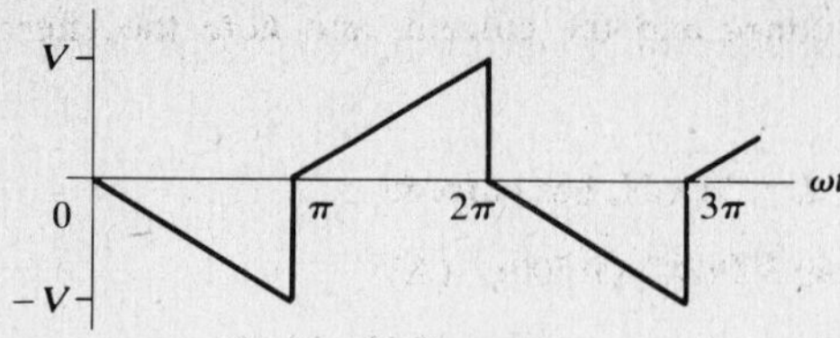

Fig. 12-49

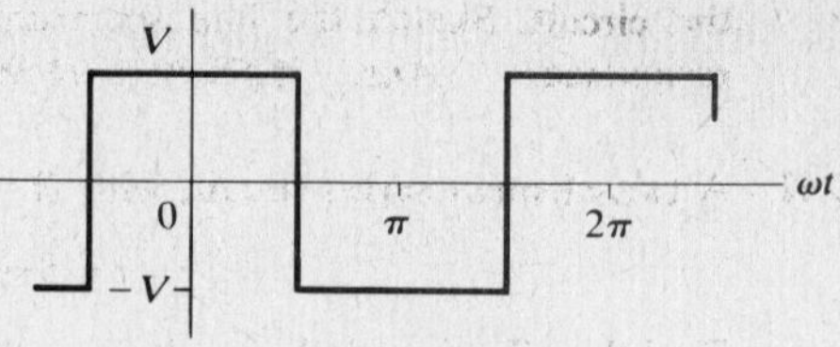

Fig. 12-50

12.36 Find the exponential Fourier series for the waveform shown in Fig. 12-51 and plot the line spectrum.

Ans. $f(t) = \cdots + \dfrac{V}{2\pi}\sin\left(\dfrac{2\pi}{6}\right)e^{-j2\omega t} + \dfrac{V}{\pi}\sin\left(\dfrac{\pi}{6}\right)e^{-j\omega t} + \dfrac{V}{6} + \dfrac{V}{\pi}\sin\left(\dfrac{\pi}{6}\right)e^{j\omega t} + \dfrac{V}{2\pi}\sin\left(\dfrac{2\pi}{6}\right)e^{j2\omega t} + \cdots$

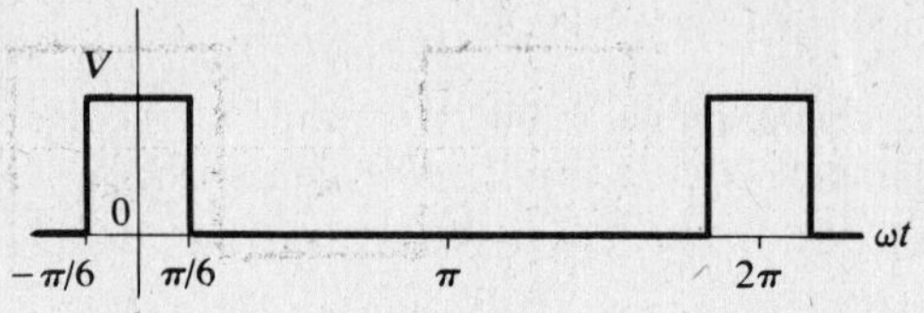

Fig. 12-51

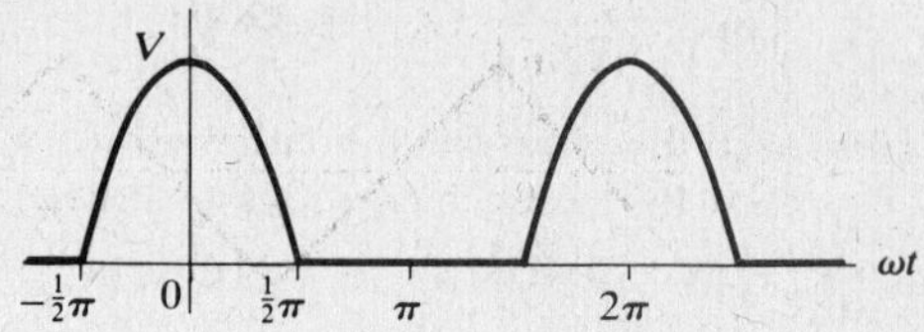

Fig. 12-52

12.37 Find the exponential Fourier series for the half-wave-rectified sine wave shown in Fig. 12-52. Convert these coefficients into the trigonometric series coefficients, write the trigonometric series, and compare it with the result of Problem 12.24.

Ans. $f(t) = \cdots - \dfrac{V}{15\pi}e^{-j4\omega t} + \dfrac{V}{3\pi}e^{-j2\omega t} + \dfrac{V}{4}e^{-j\omega t} + \dfrac{V}{\pi} + \dfrac{V}{4}e^{j\omega t} + \dfrac{V}{3\pi}e^{j2\omega t} - \dfrac{V}{15\pi}e^{j4\omega t} + \cdots$

12.38 Find the exponential Fourier series for the full-wave-rectified sine wave shown in Fig. 12-53 and plot the line spectrum.

Ans. $f(t) = \cdots - \dfrac{2V}{15\pi}e^{-j4\omega t} + \dfrac{2V}{3\pi}e^{-j2\omega t} + \dfrac{2V}{\pi} + \dfrac{2V}{3\pi}e^{j2\omega t} - \dfrac{2V}{15\pi}e^{j4\omega t} + \cdots$

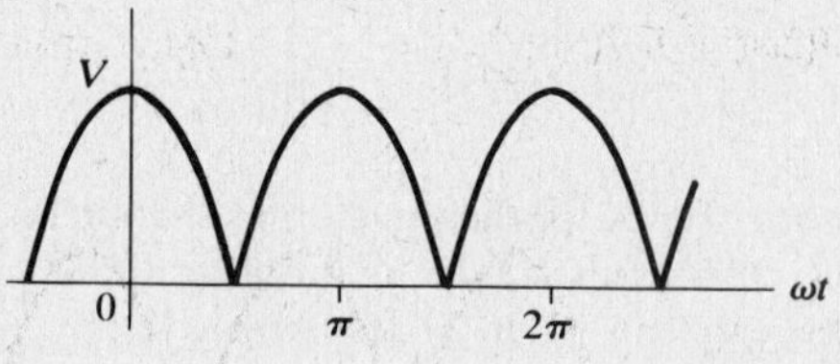

Fig. 12-53

12.39 Find the effective voltage, effective current, and average power supplied to a passive network if the applied voltage is $v = 200 + 100\cos(500t + 30°) + 75\cos(1500t + 60°)$ (V) and the resulting current is $i = 3.53\cos(500t + 75°) + 3.55\cos(1500t + 78.45°)$ (A). *Ans.* 218.5 V, 3.54 A, 250.8 W

12.40 A voltage $v = 50 + 25\sin 500t + 10\sin 1500t + 5\sin 2500t$ (V) is applied to the terminals of a passive network and the resulting current is

$$i = 5 + 2.23\sin(500t - 26.6°) + 0.556\sin(1500t - 56.3°) + 0.186\sin(2500t - 68.2°) \quad \text{(A)}$$

Find the effective voltage, effective current, and the average power. *Ans.* 53.6 V, 5.25 A, 276.5 W

12.41 A three-element series circuit, with $R = 5\ \Omega$, $L = 5$ mH, and $C = 50\ \mu$F, has an applied voltage $v = 150\sin 1000t + 100\sin 2000t + 75\sin 3000t$ (V). Find the effective current and the average power for the circuit. Sketch the line spectrum of the voltage and the current, and note the effect of series resonance. *Ans.* 16.58 A, 1374 W

12.42 A two-element series circuit, with $R = 10\ \Omega$ and $L = 20$ mH, has current

$$i = 5 \sin 100t + 3 \sin 300t + 2 \sin 500t \quad \text{(A)}$$

Find the effective applied voltage and the average power. *Ans.* 48 V, 190 W

12.43 A pure inductance, $L = 10$ mH, has the triangular current wave shown in Fig. 12-54, where $\omega = 500$ rad/s. Obtain the exponential Fourier series for the voltage across the inductance. Compare the answer with the result of Problem 12.8.

Ans. $v_L = \dfrac{200}{\pi^2}\{\cdots - j\tfrac{1}{3}e^{-j3\omega t} - je^{-j\omega t} + je^{j\omega t} + j\tfrac{1}{3}e^{j3\omega t} + \cdots\}$ (V)

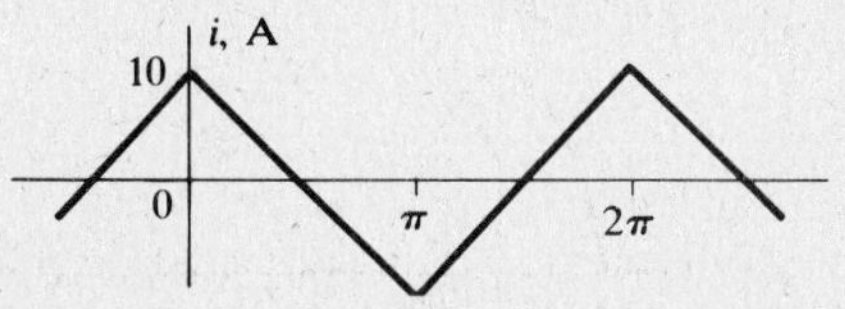

Fig. 12-54

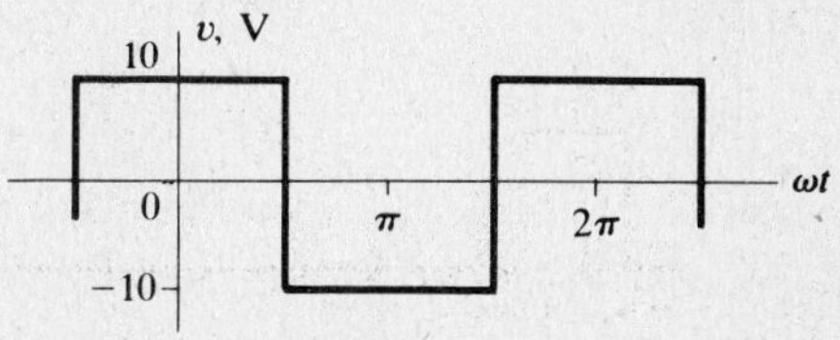

Fig. 12-55

12.44 A pure inductance, $L = 10$ mH, has an applied voltage with the waveform shown in Fig. 12-55, where $\omega = 200$ rad/s. Obtain the current series in trigonometric form and identify the current waveform.

Ans. $i = \dfrac{20}{\pi}\{\sin \omega t - \tfrac{1}{9}\sin 3\omega t + \tfrac{1}{25}\sin 5\omega t - \tfrac{1}{49}\sin 7\omega t + \cdots\}$ (A); triangular

12.45 Figure 12-56 shows a full-wave-rectified sine wave representing the voltage applied to the terminals of an LC series circuit. Use the trigonometric Fourier series to find the voltages across the inductor and the capacitor.

Ans. $$v_L = \frac{4V_m}{\pi}\left\{\frac{2\omega L}{3\left(2\omega L - \dfrac{1}{2\omega C}\right)}\cos 2\omega t - \frac{4\omega L}{15\left(4\omega L - \dfrac{1}{4\omega C}\right)}\cos 4\omega t + \cdots\right\}$$

$$v_C = \frac{4V_m}{\pi}\left\{\frac{1}{2} - \frac{1}{3(2\omega C)\left(2\omega L - \dfrac{1}{2\omega C}\right)}\cos 2\omega t + \frac{1}{15(4\omega C)\left(4\omega L - \dfrac{1}{4\omega C}\right)}\cos 4\omega t - \cdots\right\}$$

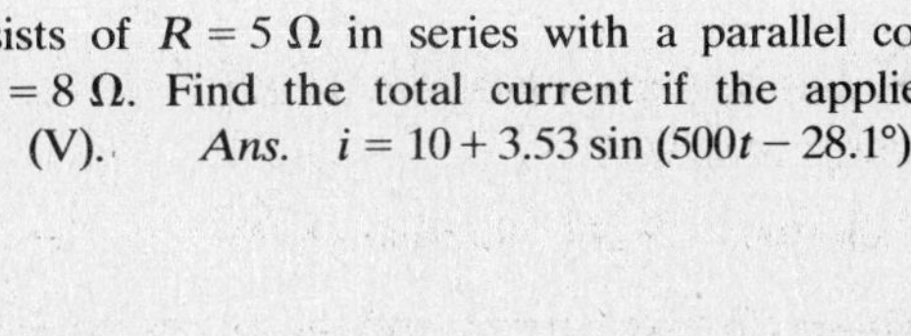

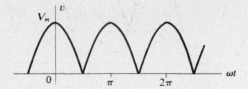

Fig. 12-56

12.46 A three-element circuit consists of $R = 5\ \Omega$ in series with a parallel combination of L and C. At $\omega = 500$ rad/s, $X_L = 2\ \Omega$, $X_C = 8\ \Omega$. Find the total current if the applied voltage is given by $v = 50 + 20 \sin 500t + 10 \sin 1000t$ (V). *Ans.* $i = 10 + 3.53 \sin (500t - 28.1°)$ (A)

Chapter 13

Complex Frequency

13.1 INTRODUCTION

Earlier chapters have dealt with the forced responses of passive circuits to excitations that were constant (dc conditions) or sinusoidal functions of time (ac conditions), the latter case including, via Fourier's theorem, general periodic functions of time. Let us now consider a driving voltage of the form

$$v = V_m e^{\sigma t} \cos(\omega t + \phi) = \text{Re}\,[V_m e^{j\phi} e^{(\sigma + j\omega)t}] \qquad (1)$$

where V_m is real and positive. This kind of function has already been encountered, though as a response, in Example 5.6. In practical circuits σ will be negative. This imposes the familiar exponential damping on the cosine function, as shown in Fig. 13-1. The coefficient of t in the exponential in (*1*), $\sigma + j\omega$, is called the *complex frequency* and given the symbol **s**. Because this chapter will utilize the cosine rather than the sine function, the "real part of" the symbol will be dropped from (*1*), and we shall write simply $v = V_m e^{j\phi} e^{st}$, the real part being understood.

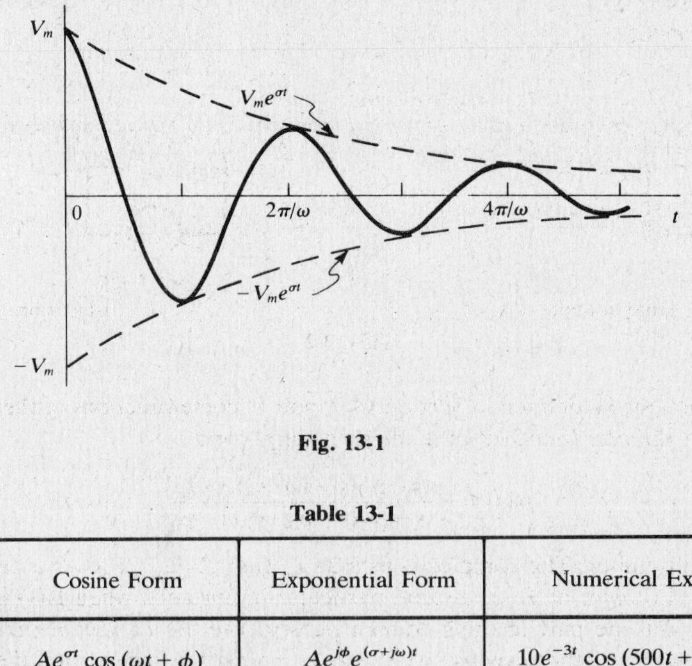

Fig. 13-1

Table 13-1

s	Cosine Form	Exponential Form	Numerical Example
$\sigma + j\omega$	$Ae^{\sigma t}\cos(\omega t + \phi)$	$Ae^{j\phi}e^{(\sigma + j\omega)t}$	$10e^{-3t}\cos(500t + 30°)$ (V)
$0 + j\omega$	$A\cos(\omega t + \phi)$	$Ae^{j\phi}e^{j\omega t}$	$10\cos(500t + 30°)$ (V)
$\sigma + j0$	$Ae^{\sigma t}\cos\phi$	$Ae^{j\phi}e^{\sigma t}$	$8.66e^{-3t}$ (V)
$0 + j0$	$A\cos\phi$	$Ae^{j\phi}$	8.66 V

The damped cosine function (*1*) includes as special cases the familiar ac sinusoid and the constant or dc case, as shown in rows 2 and 4 of Table 13-1.

Frequency has now taken on a broader meaning. The real and imaginary parts of s both have the dimensions s^{-1}. Just as ω is measured in rad/s, where the radian is dimensionless, σ is given the units Np/s, where the *neper* (after John Napier or Neper, the inventor of natural logarithms) is another dimensionless unit.

13.2 NETWORK ANALYSIS IN THE s-DOMAIN

A driving voltage of the form $v = V_m e^{st}$ applied to a passive network will result in currents and voltages throughout the network, each having the time dependence e^{st}; e.g. $I_m e^{j\psi} e^{st}$. Therefore, only the magnitude I_m and the phase angle ψ need be determined, just as was the case in the sinusoidal steady state ($\sigma = 0$). We are thus led to consider phasors $I_m\angle\psi$ in an *s-domain*. Figure 13-2 indicates the correspondence between the time domain and the s-domain; it should be compared with Fig. 7-1, which pictures the analogous relation between the time and frequency domains. In general, all frequency-domain methods, wherein $j\omega$ is replaced by s, remain valid in the s-domain. Thus, the impedances of an inductor and a capacitor are sL and 1/sC, respectively; impedances in series, or admittances in parallel, add; the usual formulas for voltage and current division apply; etc.

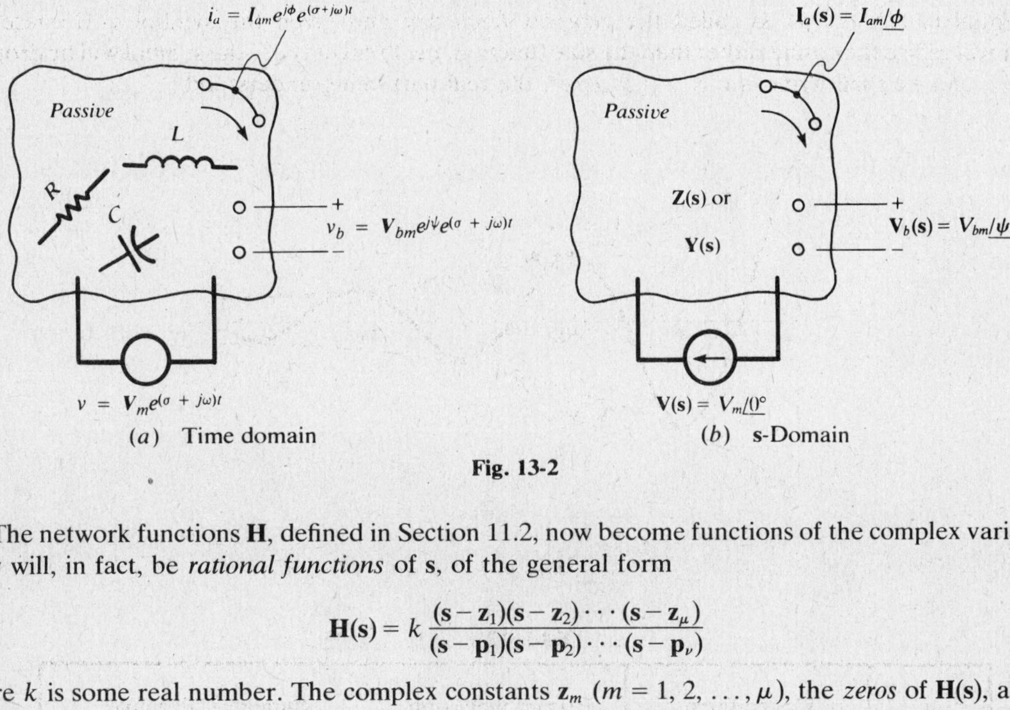

(*a*) Time domain (*b*) s-Domain

Fig. 13-2

The network functions **H**, defined in Section 11.2, now become functions of the complex variable **s**. They will, in fact, be *rational functions* of s, of the general form

$$\mathbf{H}(\mathbf{s}) = k\,\frac{(\mathbf{s}-\mathbf{z}_1)(\mathbf{s}-\mathbf{z}_2)\cdots(\mathbf{s}-\mathbf{z}_\mu)}{(\mathbf{s}-\mathbf{p}_1)(\mathbf{s}-\mathbf{p}_2)\cdots(\mathbf{s}-\mathbf{p}_\nu)} \tag{2}$$

where k is some real number. The complex constants $\mathbf{z}_m$ ($m = 1, 2, \ldots, \mu$), the *zeros* of **H**(s), and the $\mathbf{p}_n$ ($n = 1, 2, \ldots, \nu$), the *poles* of **H**(s), assume particular importance when **H**(s) is interpreted as the ratio of the *response* (in one part of the s-domain network) to the *excitation* (in another part of the network). Thus, when $\mathbf{s} = \mathbf{z}_m$, the response will be zero, no matter how great the excitation; whereas, when $\mathbf{s} = \mathbf{p}_n$, the response will be infinite, no matter how small the excitation.

EXAMPLE 13.1 Figure 13-3 shows a one-port passive network in the s-domain. The excitation is the applied voltage **V**(s); the response is current **I**(s). Their ratio is the input admittance, $\mathbf{Y}_{in}(\mathbf{s})$. By the rules for combining admittances,

$$\mathbf{Y}_{in}(\mathbf{s}) = \frac{\left(\frac{1}{2.5}\right)\left(\frac{3}{5\mathbf{s}} + \frac{\mathbf{s}}{20}\right)}{\left(\frac{1}{2.5}\right) + \left(\frac{3}{5\mathbf{s}} + \frac{\mathbf{s}}{20}\right)} = (0.4)\frac{\mathbf{s}^2 + 12}{\mathbf{s}^2 + 8\mathbf{s} + 12}$$

Restricting attention to the zeros of $\mathbf{Y}_{in}$ (the roots of $\mathbf{s}^2 + 12 = 0$), these are seen to be $\mathbf{z}_1 = j\sqrt{12}$ rad/s, $\mathbf{z}_2 = -j\sqrt{12}$ rad/s. Being pure imaginaries, these zeros correspond to sinusoidal excitation at $\omega = \pm\sqrt{12}$ rad/s; a practical source might be $v = V_m \cos \sqrt{12}\,t$ (V). At this frequency, the LC tank is at resonance, has an infinite impedance, and will limit the circuit current to zero.

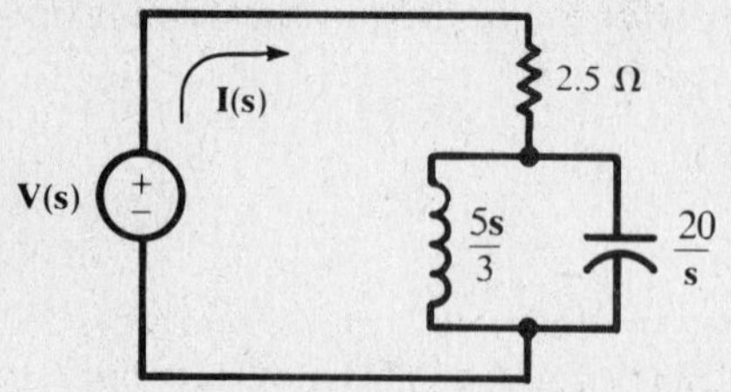

Fig. 13-3

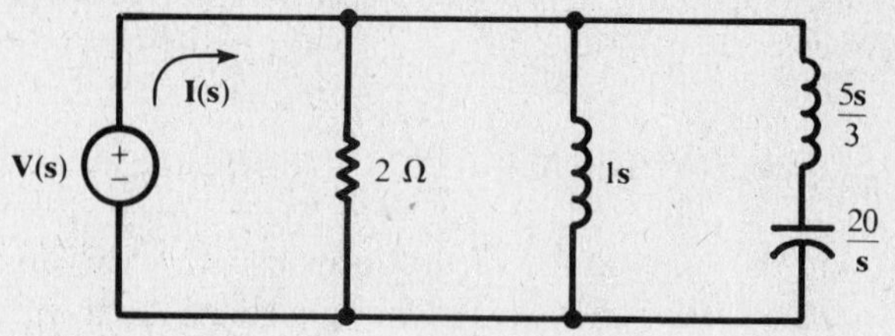

Fig. 13-4

EXAMPLE 13.2 For the network of Fig. 13-4,

$$\mathbf{Y}_{in}(\mathbf{s}) = \frac{1}{2} + \frac{1}{\mathbf{s}} + \frac{1}{\dfrac{5\mathbf{s}}{3} + \dfrac{20}{\mathbf{s}}} = (0.1)\frac{5\mathbf{s}^3 + 16\mathbf{s}^2 + 60\mathbf{s} + 120}{\mathbf{s}^3 + 12\mathbf{s}} \quad \text{(S)}$$

There are three poles (roots of $\mathbf{s}^3 + 12\mathbf{s} = 0$): $\mathbf{p}_1 = 0$, $\mathbf{p}_2 = j\sqrt{12}$ rad/s, $\mathbf{p}_3 = -j\sqrt{12}$ rad/s. A constant-voltage source applied to the network would result in an infinite current due to the inductive branch, which, to dc, appears as a short circuit. This is the first pole, $\mathbf{p}_1 = 0$. Poles $\mathbf{p}_2$ and $\mathbf{p}_3$ show that a sinusoidal source at $\sqrt{12}$ rad/s would also produce infinite current. This is due to the series-resonant LC branch, which, with zero impedance, has the same effect as a short circuit.

The zeros and poles of $\mathbf{H}$ can be plotted in the complex s-plane. Figure 13-5 shows these points for the network function of Example 13.1,

$$\mathbf{H}(\mathbf{s}) = (0.4)\frac{(\mathbf{s} - j\sqrt{12})(\mathbf{s} + j\sqrt{12})}{(\mathbf{s}+2)(\mathbf{s}+6)}$$

The zeros (marked ⊙) are $\pm j\sqrt{12}$, and the poles (marked ×) are -2, -6. The network function itself is shown boxed on the pole-zero plot.

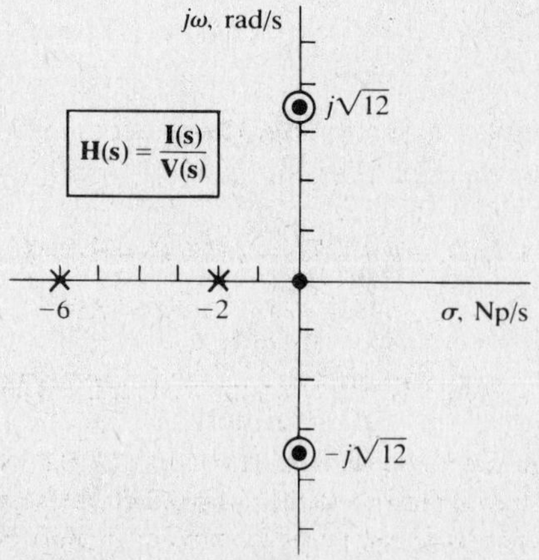

Fig. 13-5

Observe that in Fig. 13-5 both the poles and the zeros are situated symmetrically with respect to the σ-axis. This holds true in general: the two polynomials composing the numerator and denominator in (2) will have all coefficients real, and therefore any nonreal zeros of these polynomials must occur in complex conjugate pairs.

13.3 NETWORK RESPONSE FROM THE s-PLANE

Relative to a given excitation, the response of the network at complex frequency $\mathbf{s}$ is measured by $\mathbf{H}(\mathbf{s})$. Going over to polar form in (2) by setting

$$\mathbf{s}-\mathbf{z}_m \equiv N_m\underline{/\alpha_m} \quad (m = 1, 2, \ldots, \mu)$$
$$\mathbf{s}-\mathbf{p}_n = D_n\underline{/\beta_n} \quad (n = 1, 2, \ldots, \nu)$$

we have:

$$\begin{aligned}\mathbf{H}(\mathbf{s}) &= k\frac{(N_1\underline{/\alpha_1})(N_2\underline{/\alpha_2})\cdots(N_\mu\underline{/\alpha_\mu})}{(D_1\underline{/\beta_1})(D_2\underline{/\beta_2})\cdots(D_\nu\underline{/\beta_\nu})}\\ &= k\frac{N_1N_2\cdots N_\mu}{D_1D_2\cdots D_\nu}\underline{/(\alpha_1+\cdots+\alpha_\mu)-(\beta_1+\cdots+\beta_\nu)}\end{aligned} \tag{3}$$

It follows from (3) that the response of the network at an arbitrary *test point* $\mathbf{s}$ in the $(\sigma + j\omega)$-plane can be determined by measuring the lengths of the vectors from the zeros and poles to $\mathbf{s}$, as well as the angles these vectors make with the positive σ-axis.

EXAMPLE 13.3 Test the response of the network of Example 13.1 to exponential voltage excitation at 1 Np/s.

On the pole-zero plot, Fig. 13-5, locate the test point, $1 + j0$, and draw the vectors from the poles and zeros to the test point, giving Fig. 13-6.

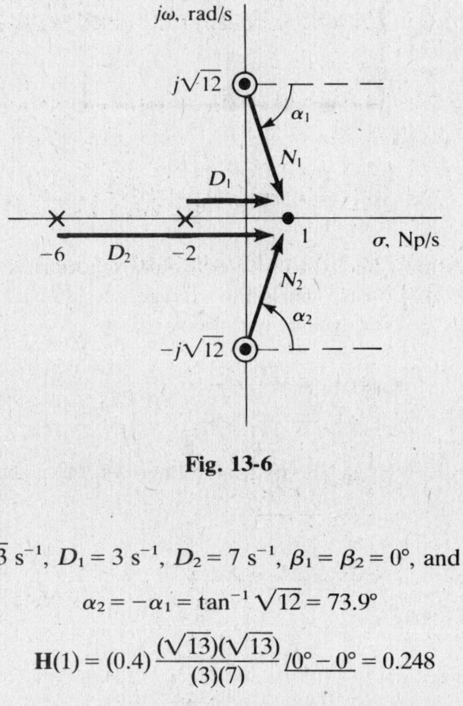

Fig. 13-6

It is seen that $N_1 = N_2 = \sqrt{13}\ \text{s}^{-1}$, $D_1 = 3\ \text{s}^{-1}$, $D_2 = 7\ \text{s}^{-1}$, $\beta_1 = \beta_2 = 0°$, and

$$\alpha_2 = -\alpha_1 = \tan^{-1}\sqrt{12} = 73.9°$$

wherefore
$$\mathbf{H}(1) = (0.4)\frac{(\sqrt{13})(\sqrt{13})}{(3)(7)}\underline{/0° - 0°} = 0.248$$

(As a check, the analytic expression for $\mathbf{Y}_{in}(\mathbf{s})$ gives $\mathbf{Y}_{in}(1) = (0.4)(13/21)$ S.)

The above result implies that, in the time domain, $i(t) = 0.248\,v(t)$, so that both voltage and current become infinite as e^{1t}. As was previously remarked, any realistic value of σ must be negative or zero.

The above geometrical method does not seem to require knowledge of the analytic expression for $\mathbf{H}$ as a rational function of $\mathbf{s}$. It is clear, however, that that expression could be constructed, to within the constant factor k, from the assumedly known poles and zeros of $\mathbf{H}$. See Problem 13.14.

13.4 THE NATURAL RESPONSE

The focus of this chapter is on the forced or steady-state response, and it is in obtaining this response that the complex-frequency methods are most helpful. However, the natural frequencies, which characterize the transient response, are easily obtained. They are, in fact, the poles of the network function.

EXAMPLE 13.4 If we excite the network of Fig. 13-7 with $\mathbf{V}(\mathbf{s})$ inserted at xx', then the input admittance function is the same as for Fig. 13-3:

$$\mathbf{Y}_{\text{in}}(\mathbf{s}) = \frac{\mathbf{I}(\mathbf{s})}{\mathbf{V}(\mathbf{s})} = (0.4)\frac{\mathbf{s}^2 + 12}{(\mathbf{s}+2)(\mathbf{s}+6)}$$

The natural frequencies are seen to be -2 Np/s and -6 Np/s. Hence, in the time domain, the natural current is of the form

$$i_n = A_1 e^{-2t} + A_2 e^{-6t}$$

where the constants A_1 and A_2 are determined by applying the initial conditions to the complete response, $i = i_n + i_f$.

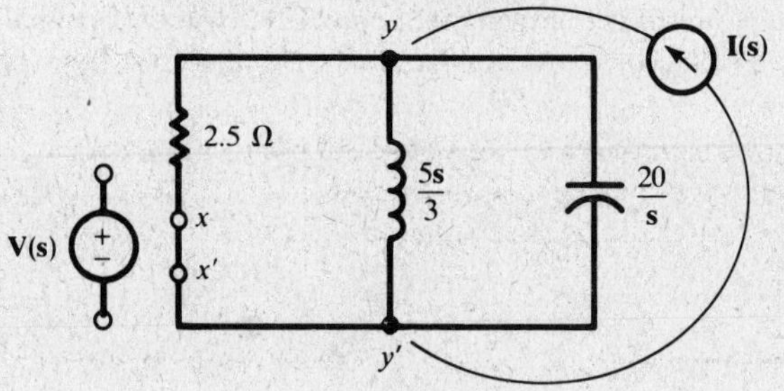

Fig. 13-7

The same natural frequencies must be found whatever the excitation of the network. Suppose that $\mathbf{I}(\mathbf{s})$ is applied to points yy' in Fig. 13-7 and that $\mathbf{V}_1(\mathbf{s})$ is the voltage across these same points. Now the pertinent network function is $\mathbf{V}_1(\mathbf{s})/\mathbf{I}(\mathbf{s})$, the input impedance seen between yy'.

$$\mathbf{Z}_{\text{in}}(\mathbf{s}) = \frac{1}{\dfrac{1}{2.5} + \dfrac{3}{5\mathbf{s}} + \dfrac{\mathbf{s}}{20}} = \frac{20\mathbf{s}}{(\mathbf{s}+2)(\mathbf{s}+6)}$$

Again the poles are at -2 Np/s and -6 Np/s. The natural voltage will be of the same form as the natural current above.

13.5 SCALING IN THE s-DOMAIN

Magnitude Scaling

Let a network have input impedance function $\mathbf{Z}_{\text{in}}(\mathbf{s})$, and let K_m be a positive real number. Then, if each resistance R in the network is replaced by $K_m R$, each inductance L by $K_m L$, and each capacitance C by C/K_m, the new input impedance function will be $K_m \mathbf{Z}_{\text{in}}(\mathbf{s})$. We say that the network has been *magnitude-scaled by a factor* K_m.

Frequency Scaling

If, instead of the above changes, we preserve each resistance R, replace each inductance L by L/K_f $(K_f > 0)$, and replace each capacitance C by C/K_f, then the new input impedance function will be $\mathbf{Z}_{\text{in}}(\mathbf{s}/K_f)$. That is, the new network has the same impedance at complex frequency $K_f\mathbf{s}$ as the old had at $\mathbf{s}$. We say that the network has been *frequency-scaled by a factor* K_f.

In practical situations, one usually restricts attention to pure imaginary values of **s**. Frequency scaling then produces the same impedance at radian frequency $K_f\omega$ as originally obtained at ω.

Solved Problems

13.1 For the time functions listed in the first column of Table 13-2, write the corresponding phasor (cosine-based) and the complex frequency **s**.

See columns 2 and 3 of the table.

Table 13-2

Time Function	Phasor	s
$i(t) = 86.6$ A	$86.6\angle 0°$ A	0
$i(t) = 15.0e^{-2\times 10^3 t}$ (A)	$15.0\angle 0°$ A	-2×10^3 Np/s
$v(t) = 25.0\cos(250t - 45°)$ (V)	$25.0\angle -45°$ V	$\pm j250$ rad/s
$v(t) = 0.50\sin(250t + 30°)$ (V)	$0.50\angle -60°$ V	$\pm j250$ rad/s
$i(t) = 5.0e^{-100t}\sin(50t + 90°)$ (A)	$5.0\angle 0°$ A	$-100 \pm j50$ s^{-1}
$i(t) = 3\cos 50t + 4\sin 50t$ (A)	$5\angle -53.13°$ A	$\pm j50$ rad/s

13.2 For each phasor and complex frequency in Table 13-3, write the corresponding time function.

See column 3 of the table.

Table 13-3

Phasor	s	Time Function
$10\angle 0°$	$+j120\pi$	$10\cos 120\pi t$
$2\angle 45°$	$-j120\pi$	$2\cos(120\pi t + 45°)$
$5\angle -90°$	$-2 \pm j50$	$5e^{-2t}\cos(50t - 90°)$
$15\angle 0°$	$-5000 \pm j1000$	$15e^{-5000t}\cos 1000t$
$100\angle 30°$	0	86.6

13.3 A phasor voltage, $10\sqrt{2}\angle 45°$ V, has an associated complex frequency $\mathbf{s} = -50 + j100$ s^{-1}. Find the voltage at $t = 10$ ms.

$$v(t) = 10\sqrt{2}e^{-50t}\cos(100t + 45°) \quad (V)$$

At $t = 10^{-2}$ s, $100t = 1$ rad $= 57.3°$, and so

$$v = 10\sqrt{2}e^{-0.5}\cos 102.3° = -1.83 \text{ V}$$

13.4 A passive network contains resistors, a 70-mH inductor, and a 25-μF capacitor. Obtain the respective **s**-domain impedances for a driving voltage (*a*) $v = 100\sin(300t + 45°)$ (V), (*b*) $v = 100e^{-100t}\cos 300t$ (V).

(*a*) Resistance is independent of frequency. At $\mathbf{s} = j300$ rad/s, the impedance of the inductor is

$$\mathbf{s}L = (j300)(70 \times 10^{-3}) = j21$$

and that of the capacitor is

$$\frac{1}{\mathbf{s}C} = -j133.3$$

(*b*) At $\mathbf{s} = -100 + j300 \quad \mathrm{s}^{-1}$,

$$\mathbf{s}L = (-100 + j300)(70 \times 10^{-3}) = -7 + j21$$

$$\frac{1}{\mathbf{s}C} = \frac{1}{(-100 + j300)(25 \times 10^{-6})} = -40 - j120$$

13.5 The three currents at the junction shown in Fig. 13-8 have positive directions as indicated. Obtain i_3, given that

$$i_1 = 6.4e^{-0.2t}\cos(100t + 12°) \quad \text{(A)} \qquad i_2 = 3.7e^{-0.2t}\sin(100t - 27°) \quad \text{(A)}$$

The two currents have the same complex frequency, $\mathbf{s} = -0.2 + j100 \quad \mathrm{s}^{-1}$, allowing them to be added or subtracted as phasors.

$$\mathbf{I}_2 = 3.7\underline{/-117°} \text{ A} \qquad \mathbf{I}_1 = 6.4\underline{/12°} \text{ A} \qquad \mathbf{I}_3 = \mathbf{I}_1 - \mathbf{I}_2 = 9.19\underline{/30.25°} \text{ A}$$

and $i_3 = 9.19e^{-0.2t}\cos(100t + 30.25°)$ (A).

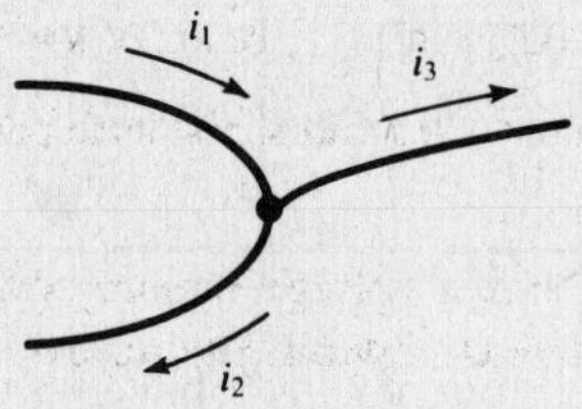

Fig. 13-8

13.6 For the circuit shown in Fig. 13-9, obtain v at $t = 0.1$ s for source current (*a*) $i = 10\cos 2t$ (A), (*b*) $i = 10e^{-t}\cos 2t$ (A).

$$\mathbf{Z}_{in}(\mathbf{s}) = 2 + \frac{2(\mathbf{s}+2)}{\mathbf{s}+4} = 4\frac{\mathbf{s}+3}{\mathbf{s}+4}$$

(*a*) At $\mathbf{s} = j2$ rad/s, $\mathbf{Z}_{in}(j2) = 3.22\underline{/7.13°}$ Ω. Then,

$$\mathbf{V} = \mathbf{I}\mathbf{Z}_{in} = (10\underline{/0°})(3.22\underline{/7.13°}) = 32.2\underline{/7.13°} \text{ V} \qquad \text{or} \qquad v = 32.2\cos(2t + 7.13°) \quad \text{(V)}$$

and $v(0.1) = 32.2\cos(18.59°) = 30.5$ V.

(*b*) At $\mathbf{s} = -1 + j2 \quad \mathrm{s}^{-1}$, $\mathbf{Z}_{in}(-1 + j2) = 3.14\underline{/11.31°}$ Ω. Then,

$$\mathbf{V} = \mathbf{I}\mathbf{Z}_{in} = 31.4\underline{/11.31°} \text{ V} \qquad \text{or} \qquad v = 31.4e^{-t}\cos(2t + 11.31°) \quad \text{(V)}$$

and $v(0.1) = 31.4e^{-0.1}\cos 22.77° = 26.2$ V.

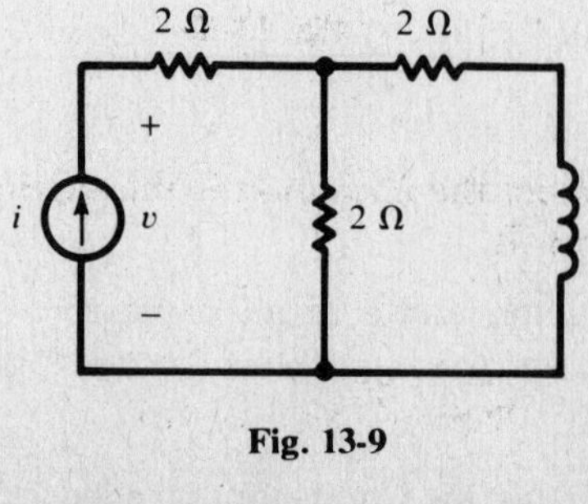

Fig. 13-9

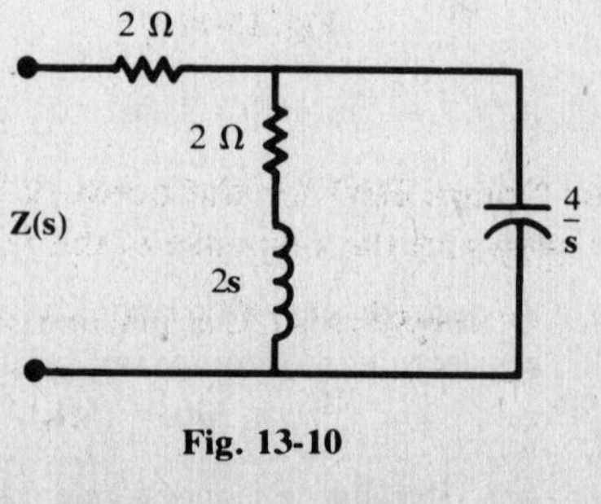

Fig. 13-10

13.7 Obtain the impedance $\mathbf{Z}_{in}(\mathbf{s})$ for the circuit shown in Fig. 13-10 at (*a*) $\mathbf{s} = 0$, (*b*) $\mathbf{s} = j4$ rad/s, (*c*) $|\mathbf{s}| = \infty$.

$$\mathbf{Z}_{in}(\mathbf{s}) = 2 + \frac{2(\mathbf{s}+1)\left(\frac{4}{\mathbf{s}}\right)}{2(\mathbf{s}+1)+\frac{4}{\mathbf{s}}} = 2\,\frac{\mathbf{s}^2+3\mathbf{s}+4}{\mathbf{s}^2+\mathbf{s}+2}$$

(*a*) $\mathbf{Z}_{in}(0) = 4\ \Omega$, the impedance offered to a constant (dc) source.

(*b*)
$$\mathbf{Z}_{in}(j4) = 2\,\frac{(j4)^2+3(j4)+4}{(j4)^2+j4+2} = 2.33\underline{/-29.05^\circ}\ \Omega$$

This is the impedance offered to a source $\sin 4t$ or $\cos 4t$.

(*c*) $\mathbf{Z}_{in}(\infty) = 2\ \Omega$. At very high frequencies the capacitance acts like a short circuit across the *RL* branch.

13.8 Express the impedance $\mathbf{Z}(\mathbf{s})$ of the parallel combination of $L = 4$ H and $C = 1$ F. At what frequencies $\mathbf{s}$ is this impedance zero or infinite?

$$\mathbf{Z}(\mathbf{s}) = \frac{(4\mathbf{s})(1/\mathbf{s})}{4\mathbf{s}+(1/\mathbf{s})} = \frac{\mathbf{s}}{\mathbf{s}^2+0.25}$$

By inspection, $\mathbf{Z}(0) = 0$ and $\mathbf{Z}(\infty) = 0$, which agrees with our earlier understanding of parallel *LC* circuits at frequencies of zero (dc) and infinity. For $|\mathbf{Z}(\mathbf{s})| = \infty$,

$$\mathbf{s}^2 + 0.25 = 0 \qquad \text{or} \qquad \mathbf{s} = \pm j0.5 \text{ rad/s}$$

A sinusoidal driving source, of frequency 0.5 rad/s, results in parallel resonance and an infinite impedance.

13.9 The circuit shown in Fig. 13-11 has a voltage source connected at terminals *ab*. The response to the excitation is the input current. Obtain the appropriate network function $\mathbf{H}(\mathbf{s})$.

$$\mathbf{H}(\mathbf{s}) = \frac{\text{response}}{\text{excitation}} = \frac{\mathbf{I}(\mathbf{s})}{\mathbf{V}(\mathbf{s})} \equiv \mathbf{Y}_{in}(\mathbf{s})$$

By the rules for combining admittances,

$$\mathbf{Y}_{in}(\mathbf{s}) = \frac{\left(\frac{1}{2}\right)\left[\frac{1}{2+(1/\mathbf{s})}+\frac{1}{1}\right]}{\frac{1}{2}+\frac{1}{2+(1/\mathbf{s})}+\frac{1}{1}} = \frac{3\mathbf{s}+1}{8\mathbf{s}+3}$$

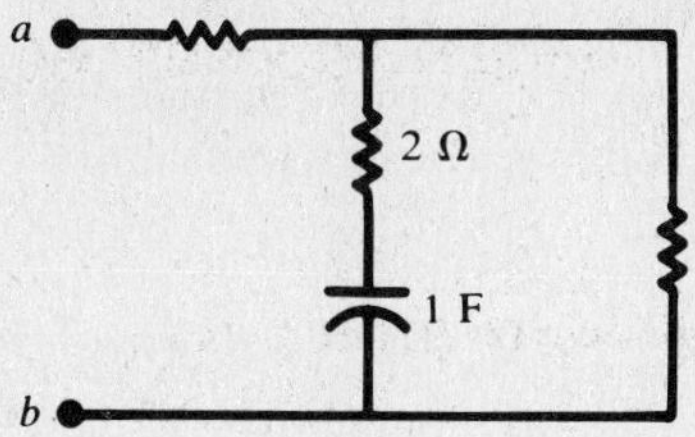

Fig. 13-11

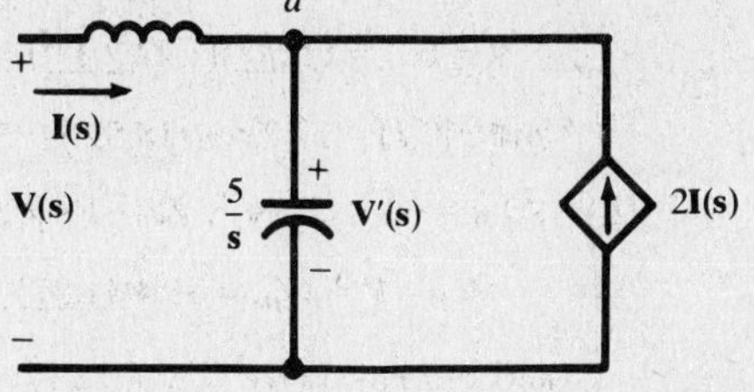

Fig. 13-12

13.10 Obtain $\mathbf{H}(\mathbf{s})$ for the network shown in Fig. 13-12, where the excitation is the driving current $\mathbf{I}(\mathbf{s})$ and the response is the voltage at the input terminals.

Applying KCL at junction *a*,

$$\mathbf{I}(\mathbf{s}) + 2\mathbf{I}(\mathbf{s}) = \frac{\mathbf{s}}{5}\mathbf{V}'(\mathbf{s}) \qquad \text{or} \qquad \mathbf{V}'(\mathbf{s}) = \frac{15}{\mathbf{s}}\mathbf{I}(\mathbf{s})$$

At the input terminals, KVL gives

$$\mathbf{V}(\mathbf{s}) = 2s\mathbf{I}(\mathbf{s}) + \mathbf{V}'(\mathbf{s}) = \left(2\mathbf{s} + \frac{15}{\mathbf{s}}\right)\mathbf{I}(\mathbf{s})$$

Then
$$\mathbf{H}(\mathbf{s}) = \frac{\mathbf{V}(\mathbf{s})}{\mathbf{I}(\mathbf{s})} = \frac{2\mathbf{s}^2 + 15}{\mathbf{s}}$$

13.11 For the two-port network shown in Fig. 13-13 find the values of R_1, R_2, and C, given that the voltage transfer function is

$$\mathbf{H}_v(\mathbf{s}) \equiv \frac{\mathbf{V}_o(\mathbf{s})}{\mathbf{V}_i(\mathbf{s})} = \frac{0.2}{\mathbf{s}^2 + 3\mathbf{s} + 2}$$

The impedance looking into xx' is

$$\mathbf{Z}' = \frac{(1/\mathbf{s}C)(R_1 + R_2)}{(1/\mathbf{s}C) + R_1 + R_2} = \frac{R_1 + R_2}{1 + (R_1 + R_2)C\mathbf{s}}$$

Then, by repeated voltage division,

$$\frac{\mathbf{V}_o}{\mathbf{V}_i} = \left(\frac{\mathbf{V}_o}{\mathbf{V}_{xx'}}\right)\left(\frac{\mathbf{V}_{xx'}}{\mathbf{V}_i}\right) = \left(\frac{R_2}{R_1 + R_2}\right)\left(\frac{\mathbf{Z}'}{\mathbf{Z}' + \mathbf{s}1}\right) = \frac{R_2/(R_1 + R_2)C}{\mathbf{s}^2 + \dfrac{1}{(R_1 + R_2)C}\mathbf{s} + \dfrac{1}{C}}$$

Equating the coefficients in this expression to those in the given expression for $\mathbf{H}_v(\mathbf{s})$, we find:

$$C = \frac{1}{2}\text{ F} \qquad R_1 = \frac{3}{5}\ \Omega \qquad R_2 = \frac{1}{15}\ \Omega$$

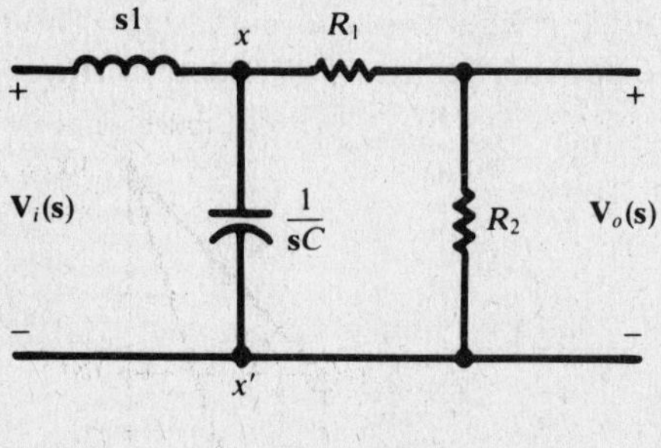

Fig. 13-13

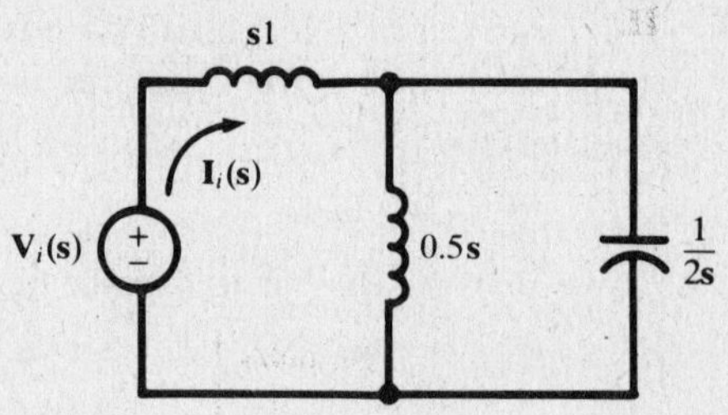

Fig. 13-14

13.12 Obtain $\mathbf{H}(\mathbf{s}) = \mathbf{V}_i(\mathbf{s})/\mathbf{I}_i(\mathbf{s})$ for the circuit shown in Fig. 13-14 and construct the pole-zero plot.

$$\mathbf{H}(\mathbf{s}) = \mathbf{Z}(\mathbf{s}) = \mathbf{s}1 + \frac{(0.5\mathbf{s})(1/2\mathbf{s})}{0.5\mathbf{s} + (1/2\mathbf{s})} = \frac{\mathbf{s}(\mathbf{s}^2 + 1.5)}{\mathbf{s}^2 + 1}$$

Poles occur at $\mathbf{s} = \pm j1$ rad/s, and zeros at $\mathbf{s} = 0$ and $\mathbf{s} = \pm j\sqrt{1.5} = \pm j1.22$ rad/s. The pole-zero plot is shown in Fig. 13-15.

13.13 Construct the pole-zero plot for the transfer admittance function

$$\mathbf{H}(\mathbf{s}) = \frac{\mathbf{I}_o(\mathbf{s})}{\mathbf{V}_i(\mathbf{s})} = \frac{\mathbf{s}^2 + 2\mathbf{s} + 17}{\mathbf{s}^2 + 3\mathbf{s} + 2}$$

In factored form,

$$\mathbf{H}(\mathbf{s}) = \frac{(\mathbf{s} + 1 + j4)(\mathbf{s} + 1 - j4)}{(\mathbf{s} + 1)(\mathbf{s} + 2)}$$

Poles exist at -1 and -2; zeros at $-1 \pm j4$. See Fig. 13-16.

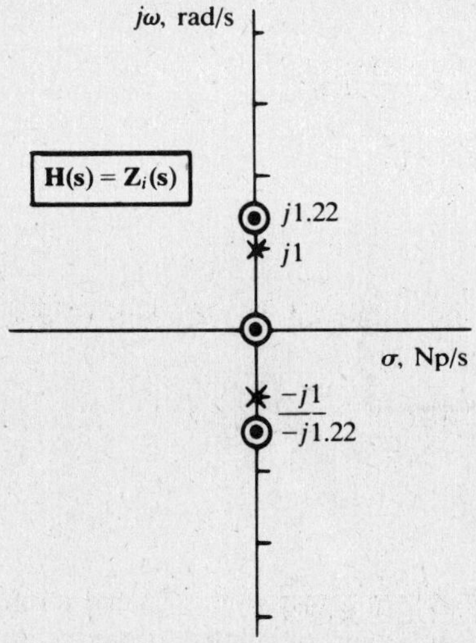

Fig. 13-15

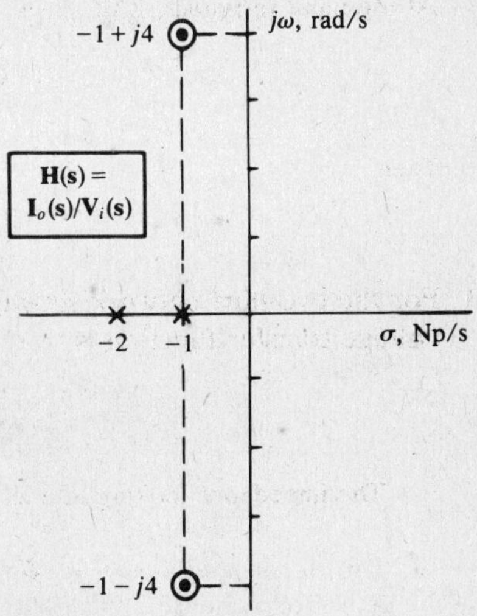

Fig. 13-16

13.14 Write the transfer function **H**(s) whose pole-zero plot is given in Fig. 13-17.

$$\mathbf{H}(\mathbf{s}) = k\frac{(\mathbf{s}+10)(\mathbf{s}+40)}{(\mathbf{s}+20-j40)(\mathbf{s}+20+j40)} = k\frac{\mathbf{s}^2+50\mathbf{s}+400}{\mathbf{s}^2+40\mathbf{s}+2000}$$

Additional information would be needed to evaluate the constant k and to ascertain the units carried by **H**.

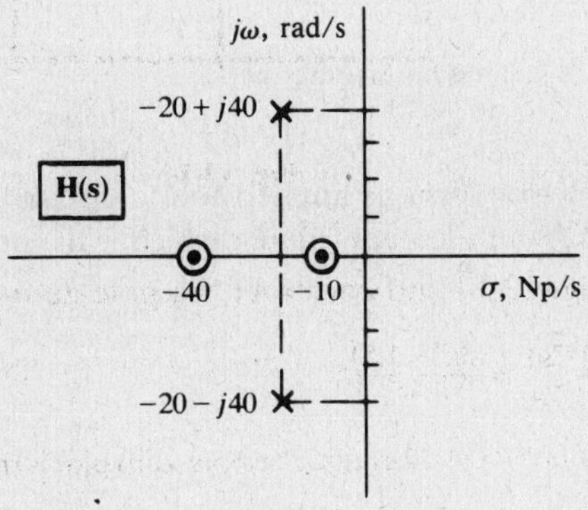

Fig. 13-17

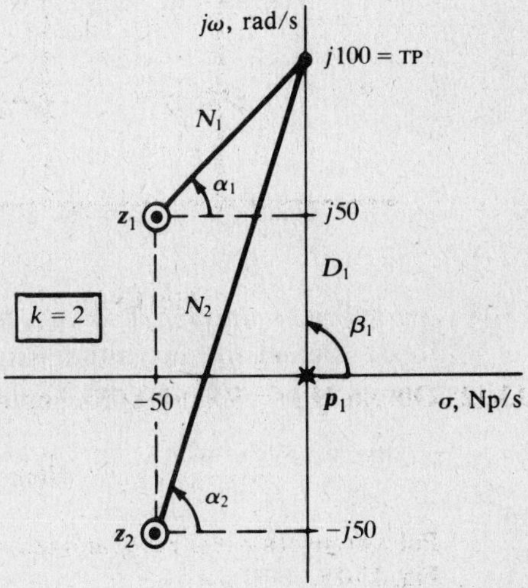

Fig. 13-18

13.15 The pole-zero plot in Fig. 13-18 shows a pole at $\mathbf{s} = 0$ and zeros at $\mathbf{s} = -50 \pm j50$. Use the geometrical method to evaluate the transfer function at the test point $j100$.

This test point corresponds to sinusoidal excitation at a frequency of 100 rad/s. We have $D_1 = 100$, $\beta_1 = 90°$; and, from the two right triangles, $N_1 = 50\sqrt{2}$ and $\alpha_1 = 45°$, $N_2 = 158.1$ and $\alpha_2 = 71.57°$. Then,

$$\mathbf{H}(j100) = (2)\frac{(50\sqrt{2})(158.1)\underline{/45° + 71.57°}}{100\underline{/90°}} = 223.6\underline{/26.57°}$$

13.16 Obtain the natural frequencies of the network shown in Fig. 13-19 by driving it with a conveniently located current source.

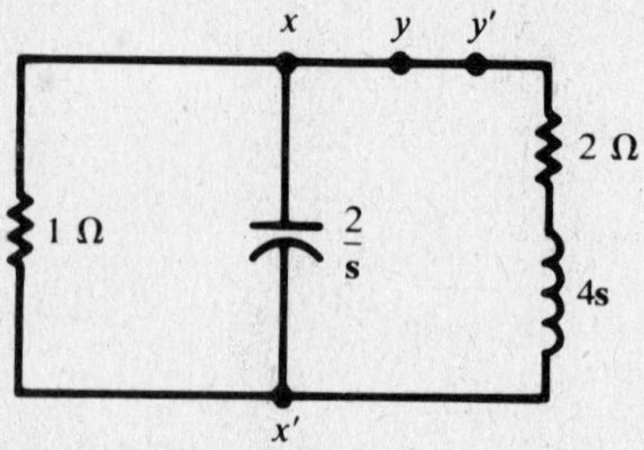

Fig. 13-19

The response to a current source connected at xx' is a voltage across these same terminals; hence the network function $\mathbf{H(s)} = \mathbf{V(s)}/\mathbf{I(s)} = \mathbf{Z(s)}$. This impedance is easiest obtained as the reciprocal of

$$\mathbf{Y(s)} = \frac{1}{1} + \frac{1}{2/\mathbf{s}} + \frac{1}{2+4\mathbf{s}} = \left(\frac{1}{2}\right)\frac{\mathbf{s}^2 + 2.5\mathbf{s} + 1.5}{\mathbf{s} + 0.5}$$

Thus
$$\mathbf{Z(s)} = (2)\frac{\mathbf{s}+0.5}{\mathbf{s}^2 + 2.5\mathbf{s} + 1.5} = (2)\frac{\mathbf{s}+0.5}{(\mathbf{s}+1)(\mathbf{s}+1.5)}$$

The natural frequencies are the poles of the network function, $\mathbf{s} = -1.0$ Np/s and $\mathbf{s} = -1.5$ Np/s.

13.17 Repeat Problem 13.16, now driving the network with a conveniently located voltage source.

The conductor at yy' in Fig. 13-19 can be opened and a voltage source inserted. Then, $\mathbf{H(s)} = \mathbf{I(s)}/\mathbf{V(s)} = \mathbf{Y(s)}$. But the admittance seen by the voltage source is

$$\mathbf{Y(s)} = \frac{\left(\frac{1}{2+4\mathbf{s}}\right)\left(\frac{1}{1} + \frac{\mathbf{s}}{2}\right)}{\frac{1}{2+4\mathbf{s}} + \frac{1}{1} + \frac{\mathbf{s}}{2}} = \left(\frac{1}{4}\right)\frac{\mathbf{s}+2}{\mathbf{s}^2 + 2.5\mathbf{s} + 1.5}$$

The denominator is the same as in Problem 13.6, giving the same natural frequencies.

13.18 A 5000-rad/s sinusoidal source, $\mathbf{V} = 100\underline{/0^\circ}$ V in phasor form, is applied to the circuit of Fig. 13-20. Obtain the magnitude-scaling factor K_m and the element values which will limit the current to 89 mA (maximum value).

At $\omega = 5000$ rad/s,

$$\mathbf{Z}_{\text{in}} = j\omega L_1 + \frac{(j\omega L_2)\left(R + \frac{1}{j\omega C}\right)}{j\omega L_2 + R + \frac{1}{j\omega C}}$$

$$= j0.250 + \frac{(j0.500)(0.40 - j0.80)}{0.40 - j0.30} = 1.124\underline{/69.15^\circ}\ \Omega$$

For $|\mathbf{V}| = 100$ V, $|\mathbf{I}| = 100/1.124 = 89.0$ A. Thus, to limit the current to 89×10^{-3} A, the impedance must be increased by the factor $K_m = 10^3$.

The scaled element values are as follows: $R = 10^3(0.4\ \Omega) = 400\ \Omega$, $L_1 = 10^3(50\ \mu\text{H}) = 50$ mH, $L_2 = 10^3(100\ \mu\text{H}) = 100$ mH, and $C = (250\ \mu\text{F})/10^3 = 0.250\ \mu$F.

13.19 Refer to Fig. 13-21. Obtain $\mathbf{H(s)} = \mathbf{V}_o/\mathbf{V}_i$ for $\mathbf{s} = j4 \times 10^6$ rad/s. Scale the network with $K_m = 10^{-3}$ and compare $\mathbf{H(s)}$ for the two networks.

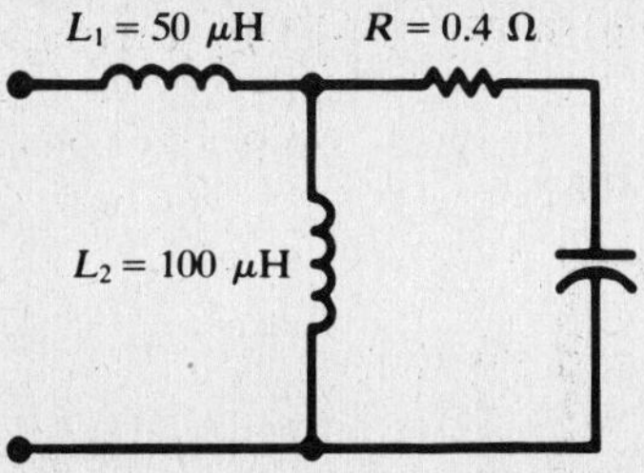

Fig. 13-20

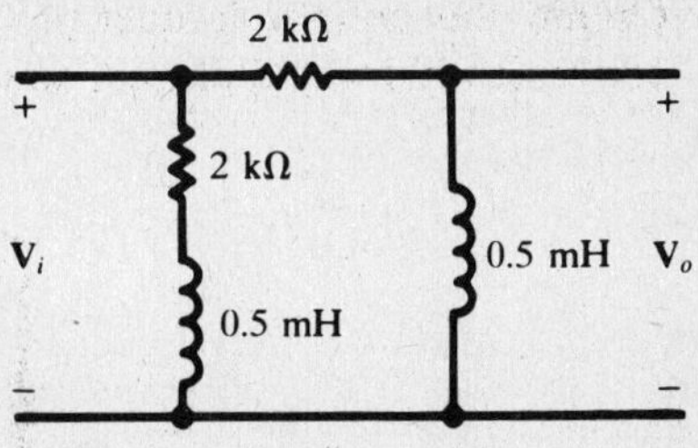

Fig. 13-21

At $\omega = 4 \times 10^6$ rad/s, $X_L = (4 \times 10^6)(0.5 \times 10^{-3}) = 2000\ \Omega$. Then,

$$\mathbf{H(s)} = \frac{\mathbf{V}_o}{\mathbf{V}_i} = \frac{j2000}{2000 + j2000} = \frac{1}{\sqrt{2}}\underline{/45^\circ}$$

After magnitude scaling, the inductive reactance is $10^{-3}(2000\ \Omega) = 2\ \Omega$ and the resistance is $10^{-3}(2\ \text{k}\Omega) = 2\ \Omega$. Thus

$$\mathbf{H(s)} = \frac{j2}{2 + j2} = \frac{1}{\sqrt{2}}\underline{/45^\circ}$$

The voltage transfer function remains unchanged by magnitude scaling. In general, any dimensionless transfer function is unaffected by magnitude scaling; a transfer function having units Ω is multiplied by K_m; and a function having units S is multiplied by $1/K_m$.

13.20 A three-element series circuit contains $R = 5\ \Omega$, $L = 4$ H, and $C = 3.91$ mF. Obtain the series resonant frequency, in rad/s, and then frequency-scale the circuit with $K_f = 1000$. Plot $|\mathbf{Z}(\omega)|$ for both circuits.

Before scaling,

$$\omega_0 = \frac{1}{\sqrt{LC}} = 8 \text{ rad/s} \qquad \text{and} \qquad \mathbf{Z}(\omega_0) = R = 5\ \Omega$$

After scaling,

$$R = 5\ \Omega \qquad L = \frac{4\ \text{H}}{1000} = 4\ \text{mH} \qquad C = \frac{3.91\ \text{mF}}{1000} = 3.91\ \mu\text{F}$$

$$\omega_0 = 1000(8\ \text{rad/s}) = 8000\ \text{rad/s} \qquad \mathbf{Z}(\omega_0) = R = 5\ \Omega$$

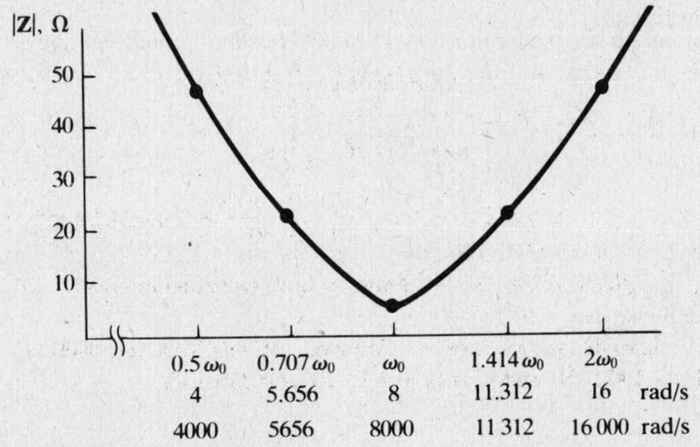

Fig. 13-22

Thus, frequency scaling by a factor 1000 results in the 5 Ω impedance value being attained at 8000 rad/s instead of 8 rad/s. Any other value of the impedance is likewise attained, after scaling, at a frequency 1000 times that at which it was attained before scaling. Consequently, the two graphs of $|\mathbf{Z}(\omega)|$ differ only in the horizontal scale—see Fig. 13-22. (The same would be true of the two graphs of $\theta_{\mathbf{Z}(\omega)}$.)

Supplementary Problems

13.21 A voltage has the s-domain representation $100\underline{/30°}$ V. Express the time function for (*a*) $\mathbf{s} = -2$ Np/s, (*b*) $\mathbf{s} = -1 + j5\ \ \mathrm{s}^{-1}$. *Ans.* (*a*) $86.6e^{-2t}$ (V); (*b*) $100e^{-t}\cos(5t + 30°)$ (V)

13.22 Give the complex frequencies associated with the current $i(t) = 5.0 + 10e^{-3t}\cos(50t + 90°)$ (A). *Ans.* $0, -3 \pm j50\ \ \mathrm{s}^{-1}$

13.23 A phasor current $25\underline{/40°}$ A has complex frequency $\mathbf{s} = -2 + j3\ \ \mathrm{s}^{-1}$. What is the magnitude of $i(t)$ at $t = 0.2$ s? *Ans.* 4.51 A

13.24 Calculate the impedance $\mathbf{Z}(\mathbf{s})$ for the circuit shown in Fig. 13-23, at (*a*) $\mathbf{s} = 0$, (*b*) $\mathbf{s} = j1$ rad/s, (*c*) $\mathbf{s} = j2$ rad/s, (*d*) $|\mathbf{s}| = \infty$. *Ans.* (*a*) 1 Ω; (*b*) $1.58\underline{/18.43°}$ Ω; (*c*) $1.84\underline{/12.53°}$ Ω; (*d*) 2 Ω

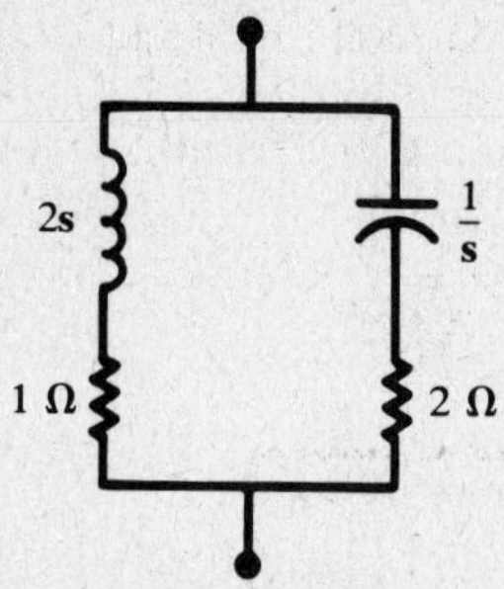

Fig. 13-23

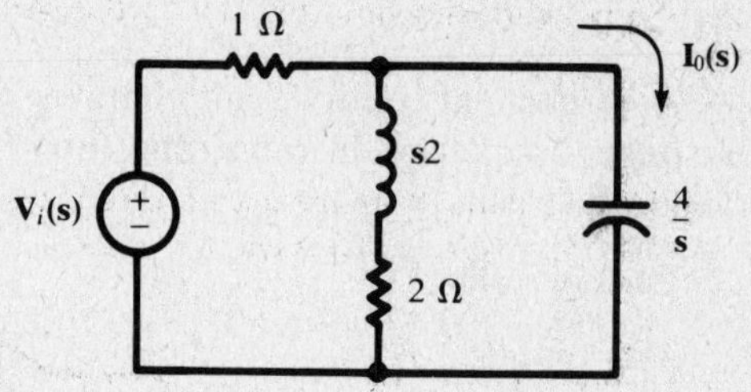

Fig. 13-24

13.25 The voltage source in the s-domain circuit shown in Fig. 13-24 has the time-domain expression

$$v_i(t) = 10e^{-t}\cos 2t \quad \text{(V)}$$

Obtain $i_o(t)$. *Ans.* $7.07e^{-t}\cos(2t + 98.13°)$ (A)

13.26 In the time domain, a series circuit of R, L, and C has an applied voltage v_i and element voltages v_R, v_L, and v_C. Obtain the voltage transfer functions (*a*) $\mathbf{V}_R(\mathbf{s})/\mathbf{V}_i(\mathbf{s})$, (*b*) $\mathbf{V}_C(\mathbf{s})/\mathbf{V}_i(\mathbf{s})$.

Ans. (*a*) $\dfrac{R\mathbf{s}/L}{\mathbf{s}^2 + \dfrac{R}{L}\mathbf{s} + \dfrac{1}{LC}}$; (*b*) $\dfrac{1/LC}{\mathbf{s}^2 + \dfrac{R}{L}\mathbf{s} + \dfrac{1}{LC}}$

13.27 Obtain the network function $\mathbf{H}(\mathbf{s})$ for the circuit shown in Fig. 13-25. The response is the voltage $\mathbf{V}_i(\mathbf{s})$.

Ans. $\dfrac{(\mathbf{s} + 7 - j2.65)(\mathbf{s} + 7 + j2.65)}{(\mathbf{s} + 2)(\mathbf{s} + 4)}$

13.28 Construct the s-plane plot for the transfer function of Problem 13.27. Evaluate $\mathbf{H}(j3)$ from the plot. *Ans.* See Fig. 13-26.

$$\frac{(7.02)(9.0)\underline{/2.86° + 38.91°}}{(3.61)(5.0)\underline{/56.31° + 36.87°}} = 3.50\underline{/-51.41°}\ \ \Omega$$

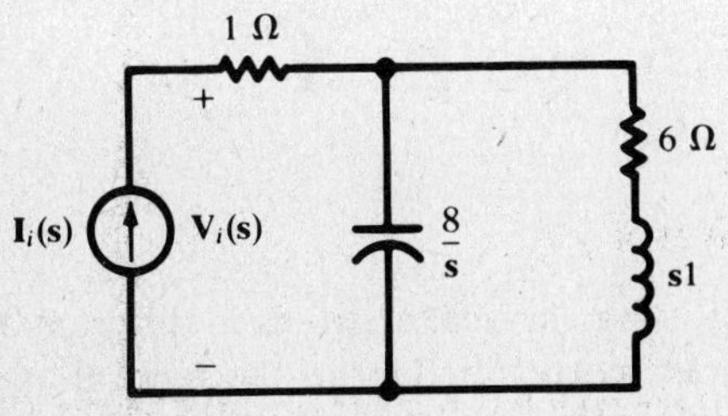

Fig. 13-25

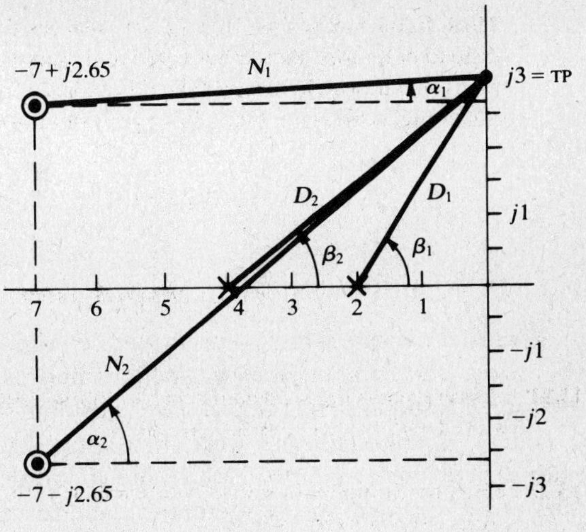

Fig. 13-26

13.29 A two-branch parallel circuit has a resistance of 20 Ω in one branch and the series combination of $R = 10\ \Omega$ and $L = 0.1$ H in the other. First, apply an excitation, $\mathbf{I}_i(\mathbf{s})$, and obtain the natural frequency from the denominator of the network function. Try different locations for applying the current source. Second, insert a voltage source, $\mathbf{V}_i(\mathbf{s})$, and obtain the natural frequency. *Ans.* −300 Np/s in all cases

13.30 In the network shown in Fig. 13-27, the switch is closed at $t = 0$. At $t = 0^+$, $i = 0$ and

$$\frac{di}{dt} = 25 \text{ A/s}$$

Obtain the natural frequencies and the complete current, $i = i_n + i_f$.
Ans. −8.5 Np/s, −23.5 Np/s; $i = -2.25e^{-8.5t} - 0.25e^{-23.5t} + 2.5$ (A)

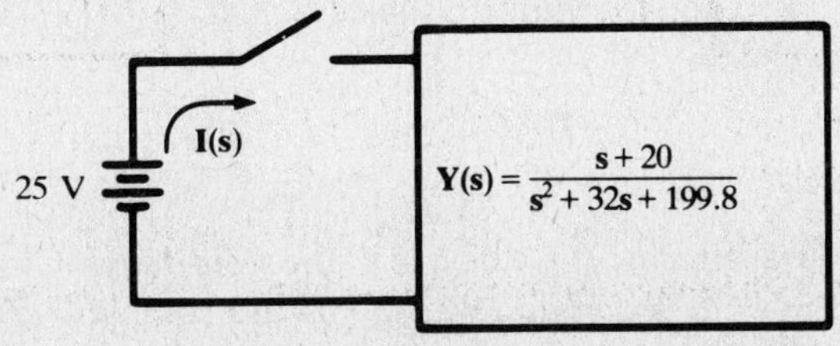

Fig. 13-27

13.31 A series *RLC* circuit contains $R = 1\ \Omega$, $L = 2$ H, and $C = 0.25$ F. Simultaneously apply magnitude and frequency scaling, with $K_m = 2000$ and $K_f = 10^4$. What are the scaled element values?
Ans. 2000 Ω, 0.4 H, 12.5 μF

13.32 At a certain frequency ω_1, a voltage $\mathbf{V}_1 = 25\underline{/0^\circ}$ V applied to a passive network results in a current $\mathbf{I}_1 = 3.85\underline{/-30^\circ}$ A. The network elements are magnitude-scaled with $K_m = 10$. Obtain the current which results from a second voltage source, $\mathbf{V}_2 = 10\underline{/45^\circ}$ V, replacing the first, if the second source frequency is $\omega_2 = 10^3\omega_1$. *Ans.* $0.154\underline{/15^\circ}$ A

Chapter 14

The Laplace Transform Method

14.1 INTRODUCTION

Transient voltages and currents were obtained in Chapter 5 using classical methods in differential equations. The complementary functions and particular solutions, corresponding to the natural and the forced responses, respectively, were combined into single expressions which involved one or more constants. Initial conditions were then applied to establish the values of the constants. The present chapter provides a method in which the conditions are applied at the beginning of the problem rather than at the end, and where the natural and forced responses are obtained at the same time.

As was seen in Chapter 12, particularly in Example 12.5, the concept of *frequency domain* could be broadened to include not just one-frequency phasors, representing time functions of the form $A\cos(\omega t + \phi)$, but superpositions of infinitely many such phasors having different frequencies. Such superpositions in the frequency domain could represent arbitrary periodic voltages or currents in the time domain. Quite similarly, the Laplace transform method may be viewed as generalizing the **s**-*domain* of Chapter 13, wherein the phasors represented exponentially damped sinusoids, $Ae^{j\phi}e^{\mathbf{s}t}$. In fact, formula (*2*) of Section 14.2 will show that, by means of an infinite superposition (integral) over **s** of "phasors" $\mathbf{F}(\mathbf{s})\,d\mathbf{s}/2\pi j$, each with its implied time factor $e^{\mathbf{s}t}$, almost any function of time can be represented for $t > 0$.

14.2 THE LAPLACE TRANSFORM

Let $f(t)$ be a time function which is zero for $t \leq 0$ and which is (subject to some mild conditions) arbitrarily defined for $t > 0$. Then the *direct Laplace transform* of $f(t)$, denoted $\mathscr{L}[f(t)]$, is defined by

$$\mathscr{L}[f(t)] = \mathbf{F}(\mathbf{s}) = \int_{0^+}^{\infty} f(t)e^{-\mathbf{s}t}\,dt \qquad (1)$$

Thus, the operation $\mathscr{L}[\ ]$ transforms $f(t)$, which is in the *time domain*, into $\mathbf{F}(\mathbf{s})$, which is in the *complex frequency domain*, or simply the **s**-*domain*, where **s** is the complex variable $\sigma + j\omega$. While it appears that the integration could prove difficult, it will soon be apparent that application of the Laplace transform method utilizes tables which cover all functions likely to be encountered in elementary circuit theory.

There is a uniqueness in the transform pairs; that is, if $f_1(t)$ and $f_2(t)$ have the same **s**-domain image $\mathbf{F}(\mathbf{s})$, then $f_1(t) = f_2(t)$. This permits going back in the other direction, from the **s**-domain to the time domain, a process called the *inverse Laplace transform*, $\mathscr{L}^{-1}[\mathbf{F}(\mathbf{s})] = f(t)$. The inverse Laplace transform can also be expressed as an integral, the *complex inversion integral*:

$$\mathscr{L}^{-1}[\mathbf{F}(\mathbf{s})] = f(t) = \frac{1}{2\pi j}\int_{\sigma_0 - j\infty}^{\sigma_0 + j\infty} \mathbf{F}(\mathbf{s})e^{\mathbf{s}t}\,d\mathbf{s} \qquad (2)$$

In (*2*) the path of integration is a straight line parallel to the $j\omega$-axis, such that all the poles of $\mathbf{F}(\mathbf{s})$ lie to the left of the line. Here again, the integration need not actually be performed, unless it is a question of adding to existing tables of transform pairs.

It should be remarked that taking the direct Laplace transform of a physical quantity introduces an extra time unit in the result. For instance, if $i(t)$ is a current in A, then $\mathbf{I}(\mathbf{s})$ has the units A · s (or C). Because the extra unit s will be removed in taking the inverse Laplace transform, we shall generally omit to cite units in the s-domain, shall still call $\mathbf{I(s)}$ a "current," indicate it by an arrow, etc.

14.3 SELECTED LAPLACE TRANSFORMS

The Laplace transform of the unit step function is easily obtained:

$$\mathscr{L}[u(t)] = \int_0^\infty (1)e^{-\mathbf{s}t}\,dt = -\frac{1}{\mathbf{s}}[e^{-\mathbf{s}t}]_0^\infty = \frac{1}{\mathbf{s}}$$

From the linearity of the Laplace transform, it follows that $v(t) = Vu(t)$ in the time domain has the s-domain image $\mathbf{V(s)} = V/\mathbf{s}$.

The exponential decay function, which appeared often in the transients of Chapter 5, is another time function which is readily transformed.

$$\mathscr{L}[Ae^{-at}] = \int_0^\infty Ae^{-at}e^{-\mathbf{s}t}\,dt = \frac{-A}{a+\mathbf{s}}[e^{-(a+\mathbf{s})t}]_0^\infty = \frac{A}{\mathbf{s}+a}$$

or, inversely,

$$\mathscr{L}^{-1}\left[\frac{A}{\mathbf{s}+a}\right] = Ae^{-at}$$

The transform of a sine function is also easily obtained.

$$\mathscr{L}[\sin \omega t] = \int_0^\infty (\sin \omega t)e^{-\mathbf{s}t}\,dt = \left[\frac{-\mathbf{s}(\sin \omega t)e^{-\mathbf{s}t} - e^{-\mathbf{s}t}\,\omega \cos \omega t}{\mathbf{s}^2+\omega^2}\right]_0^\infty = \frac{\omega}{\mathbf{s}^2+\omega^2}$$

It will be useful now to obtain the transform of a derivative, $df(t)/dt$.

$$\mathscr{L}\left[\frac{df(t)}{dt}\right] = \int_0^\infty \frac{df(t)}{dt}e^{-\mathbf{s}t}\,dt$$

Integrating by parts,

$$\mathscr{L}\left[\frac{df(t)}{dt}\right] = [e^{-\mathbf{s}t}f(t)]_{0^+}^\infty - \int_0^\infty f(t)(-\mathbf{s}e^{-\mathbf{s}t})\,dt = -f(0^+) + \mathbf{s}\int_0^\infty f(t)e^{-\mathbf{s}t}\,dt = -f(0^+) + \mathbf{sF(s)}$$

A small collection of transform pairs, including those obtained above, is given in Table 14-1. The last five lines of the table present some general properties of the Laplace transform.

EXAMPLE 14.1 Consider a series RL circuit, with $R = 5\ \Omega$ and $L = 2.5$ mH. At $t = 0$, when the current in the circuit is 2 A, a source of 50 V is applied. The time-domain circuit is shown in Fig. 14-1.

Time Domain ***s-Domain***

(i) $Ri + L\dfrac{di}{dt} = v$ ⟶ (ii) $R\mathbf{I(s)} + L[-i(0^+) + \mathbf{sI(s)}] = \mathbf{V(s)}$

(iii) $5\mathbf{I(s)} + (2.5\times 10^{-3})[-2 + \mathbf{sI(s)}] = \dfrac{50}{\mathbf{s}}$

↓ (*classical methods*)

(iv) $\mathbf{I(s)} = \dfrac{10}{\mathbf{s}} + \dfrac{(-8)}{\mathbf{s}+2000}$

(v) $10\mathscr{L}^{-1}\left[\dfrac{1}{\mathbf{s}}\right] = 10$

(vi) $(-8)\mathscr{L}^{-1}\left[\dfrac{1}{\mathbf{s}+2000}\right] = (-8)e^{-2000t}$

(vii) $i(t) = 10 - 8e^{-2000t}$ (A) ⟵

Table 14-1. Laplace Transform Pairs

	$f(t)$	$\mathbf{F(s)}$
1.	1	$\frac{1}{s}$
2.	t	$\frac{1}{s^2}$
3.	e^{-at}	$\frac{1}{s+a}$
4.	te^{-at}	$\frac{1}{(s+a)^2}$
5.	$\sin \omega t$	$\frac{\omega}{s^2+\omega^2}$
6.	$\cos \omega t$	$\frac{s}{s^2+\omega^2}$
7.	$\sin (\omega t + \theta)$	$\frac{s \sin \theta + \omega \cos \theta}{s^2+\omega^2}$
8.	$\cos (\omega t + \theta)$	$\frac{s \cos \theta - \omega \sin \theta}{s^2+\omega^2}$
9.	$e^{-at} \sin \omega t$	$\frac{\omega}{(s+a)^2+\omega^2}$
10.	$e^{-at} \cos \omega t$	$\frac{s+a}{(s+a)^2+\omega^2}$
11.	$\sinh \omega t$	$\frac{\omega}{s^2-\omega^2}$
12.	$\cosh \omega t$	$\frac{s}{s^2-\omega^2}$
13.	$\frac{df}{dt}$	$\mathbf{sF(s)} - f(0^+)$
14.	$\int_0^t f(\tau)\, d\tau$	$\frac{\mathbf{F(s)}}{s}$
15.	$f(t-t_1)$	$e^{-t_1 s}\mathbf{F(s)}$
16.	$c_1 f_1(t) + c_2 f_2(t)$	$c_1\mathbf{F}_1(s) + c_2\mathbf{F}_2(s)$
17.	$\int_0^t f_1(\tau) f_2(t-\tau)\, d\tau$	$\mathbf{F}_1(s)\,\mathbf{F}_2(s)$

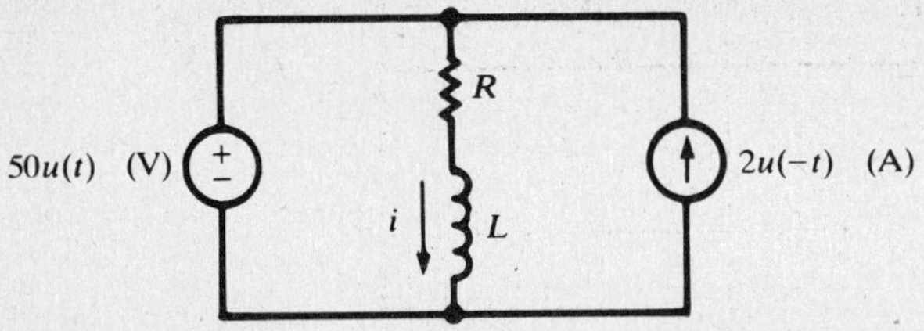

Fig. 14-1

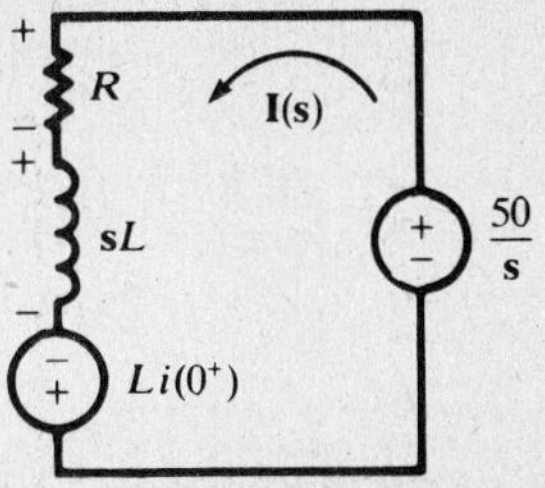

Fig. 14-2

Kirchhoff's voltage law, applied to the circuit for $t>0$, yields the familiar differential equation (i). This equation is transformed, term by term, into the s-domain equation (ii). The unknown current $i(t)$ becomes $\mathbf{I}(\mathbf{s})$, while the known voltage $v = 50u(t)$ is transformed to $50/\mathbf{s}$. Also, di/dt is transformed into $-i(0^+) + \mathbf{sI}(\mathbf{s})$, in which $i(0^+)$ is 2 A. Equation (iii) is solved for $\mathbf{I}(\mathbf{s})$, and the solution is put in the form (iv) by the techniques of Section 14.5. Then lines 1, 3, and 16 of Table 14-1 are applied to obtain the inverse Laplace transform of $\mathbf{I}(\mathbf{s})$, which is $i(t)$.

A circuit can be drawn in the s-domain, as shown in Fig. 14-2. The initial current appears in the circuit as a voltage source, $Li(0^+)$. The s-domain current establishes the voltage terms $R\mathbf{I}(\mathbf{s})$ and $\mathbf{s}L\mathbf{I}(\mathbf{s})$ in (ii) just as a phasor current $\mathbf{I}$ and an impedance $\mathbf{Z}$ create a phasor voltage $\mathbf{IZ}$.

14.4 INITIAL-VALUE AND FINAL-VALUE THEOREMS

Taking the limit as $\mathbf{s} \to \infty$ (through real values) of the direct Laplace transform of the derivative, $df(t)/dt$,

$$\lim_{s\to\infty} \mathscr{L}\left[\frac{df(t)}{dt}\right] = \lim_{s\to\infty} \int_0^\infty \frac{df(t)}{dt} e^{-st}\, dt = \lim_{s\to\infty} \{s\mathbf{F}(s) - f(0^+)\}$$

But e^{-st} in the integrand approaches zero as $s \to \infty$. Thus,

$$\lim_{s\to\infty} \{s\mathbf{F}(s) - f(0^+)\} = 0$$

Since $f(0^+)$ is a constant, we may write

$$f(0^+) = \lim_{s\to\infty} \{s\mathbf{F}(s)\}$$

which is the statement of the initial-value theorem.

EXAMPLE 14.2 In Example 14.1,

$$\lim_{s\to\infty} \{s\mathbf{I}(s)\} = \lim_{s\to\infty} \left\{10 - \frac{8s}{s+2000}\right\} = 10 - 8 = 2$$

which is indeed the initial current, $i(0^+) = 2$ A.

The final-value theorem is also developed from the direct Laplace transform of the derivative, but now the limit is taken as $\mathbf{s} \to 0$ (through real values).

$$\lim_{s\to 0} \mathscr{L}\left[\frac{df(t)}{dt}\right] = \lim_{s\to 0} \int_0^\infty \frac{df(t)}{dt} e^{-st}\, dt = \lim_{s\to 0} \{s\mathbf{F}(s) - f(0^+)\}$$

But
$$\lim_{s\to 0} \int_0^\infty \frac{df(t)}{dt} e^{-st}\, dt = \int_0^\infty df(t) = f(\infty) - f(0^+)$$

and $f(0^+)$ is a constant. Therefore,

$$f(\infty) - f(0^+) = -f(0^+) + \lim_{s\to 0} \{s\mathbf{F}(s)\}$$

or

$$f(\infty) = \lim_{s \to 0} \{s\mathbf{F}(s)\}$$

This is the statement of the final-value theorem. The theorem may be applied only when all poles of $\mathbf{sF(s)}$ have negative real parts. This excludes the transforms of such functions as e^t and cos t, which become infinite or indeterminate as $t \to \infty$.

14.5 PARTIAL-FRACTIONS EXPANSIONS

The unknown quantity in a problem in circuit analysis can be either a current $i(t)$ or a voltage $v(t)$. In the s-domain, it is $\mathbf{I(s)}$ or $\mathbf{V(s)}$; for the circuits considered in this book, this will be a rational function of the form

$$\mathbf{R(s)} = \frac{\mathbf{P(s)}}{\mathbf{Q(s)}}$$

where the polynomial $\mathbf{Q(s)}$ is of higher degree than $\mathbf{P(s)}$. Furthermore, $\mathbf{R(s)}$ is real for real values of $\mathbf{s}$, so that any nonreal poles of $\mathbf{R(s)}$, i.e. nonreal roots of $\mathbf{Q(s)} = 0$, must occur in complex conjugate pairs.

In a *partial-fractions expansion*, the function $\mathbf{R(s)}$ is broken down into a sum of simpler rational functions, its so-called *principal parts*, with each pole of $\mathbf{R(s)}$ contributing a principal part.

Case 1*: $\mathbf{s = a}$ *is a simple pole. When $\mathbf{s = a}$ is a nonrepeated root of $\mathbf{Q(s)} = 0$, the corresponding principal part of $\mathbf{R(s)}$ is

$$\frac{\mathbf{A}}{\mathbf{s - a}} \qquad \text{where} \qquad \mathbf{A} = \lim_{\mathbf{s \to a}} \{(\mathbf{s - a})\mathbf{R(s)}\}$$

If $\mathbf{a}$ is real, so will be $\mathbf{A}$; if $\mathbf{a}$ is complex, then $\mathbf{a}^*$ is also a simple pole and the numerator of its principal part is $\mathbf{A}^*$. Notice that if $\mathbf{a} = 0$, $\mathbf{A}$ is the final value of $r(t)$.

Case 2*: $\mathbf{s = b}$ *is a double pole. When $\mathbf{s = b}$ is a double root of $\mathbf{Q(s)} = 0$, the corresponding principal part of $\mathbf{R(s)}$ is

$$\frac{\mathbf{B_1}}{\mathbf{s - b}} + \frac{\mathbf{B_2}}{(\mathbf{s - b})^2}$$

where the constants $\mathbf{B_2}$ and $\mathbf{B_1}$ may be found as

$$\mathbf{B_2} = \lim_{\mathbf{s \to b}} \{(\mathbf{s - b})^2\mathbf{R(s)}\} \qquad \text{and} \qquad \mathbf{B_1} = \lim_{\mathbf{s \to b}} \left\{(\mathbf{s - b})\left[\mathbf{R(s)} - \frac{\mathbf{B_2}}{(\mathbf{s - b})^2}\right]\right\}$$

$\mathbf{B_1}$ may be zero. Similar to Case 1, $\mathbf{B_1}$ and $\mathbf{B_2}$ are real if $\mathbf{b}$ is real, and these constants for the double pole $\mathbf{b}^*$ are the conjugates of those for $\mathbf{b}$.

The principal part at a higher-order pole can be obtained by analogy to Case 2; we shall assume, however, that $\mathbf{R(s)}$ has no such poles. Once the partial-functions expansion of $\mathbf{R(s)}$ is known, Table 14-1 can be used to invert each term and thus to obtain the time function $r(t)$.

EXAMPLE 14.3 Find the time-domain current $i(t)$ if its Laplace transform is

$$\mathbf{I(s)} = \frac{\mathbf{s} - 10}{\mathbf{s}^4 + \mathbf{s}^2}$$

Factoring the denominator,

$$\mathbf{I(s)} = \frac{\mathbf{s} - 10}{\mathbf{s}^2(\mathbf{s} - j)(\mathbf{s} + j)}$$

we see that the poles of $\mathbf{I(s)}$ are $\mathbf{s} = 0$ (double pole) and $\mathbf{s} = \pm j$ (simple poles).

The principal part at $\mathbf{s} = 0$ is

$$\frac{B_1}{\mathbf{s}} + \frac{B_2}{\mathbf{s}^2} = \frac{1}{\mathbf{s}} - \frac{10}{\mathbf{s}^2}$$

since

$$B_2 = \lim_{s\to 0}\left\{\frac{s-10}{(s-j)(s+j)}\right\} = -10$$

$$B_1 = \lim_{s\to 0}\left\{s\left[\frac{s-10}{s^2(s^2+1)}+\frac{10}{s^2}\right]\right\} = \lim_{s\to 0}\left\{\frac{10s+1}{s^2+1}\right\} = 1$$

The principal part at $\mathbf{s} = +j$ is

$$\frac{\mathbf{A}}{\mathbf{s}-j} = -\frac{0.5+j5}{\mathbf{s}-j}$$

since

$$\mathbf{A} = \lim_{s\to j}\left\{\frac{s-10}{s^2(s+j)}\right\} = -(0.5+j5)$$

It follows at once that the principal part at $\mathbf{s} = -j$ is

$$-\frac{0.5-j5}{\mathbf{s}+j}$$

The partial-fractions expansion of $\mathbf{I}(\mathbf{s})$ is therefore

$$\mathbf{I}(\mathbf{s}) = \frac{1}{\mathbf{s}} - 10\frac{1}{\mathbf{s}^2} - (0.5+j5)\frac{1}{\mathbf{s}-j} - (0.5-j5)\frac{1}{\mathbf{s}+j}$$

and term-by-term inversion using Table 14-1 gives

$$\begin{aligned} i(t) &= 1 - 10t - (0.5+j5)e^{jt} - (0.5-j5)e^{-jt} \\ &= 1 - 10t - (\cos t - 10\sin t) \end{aligned}$$

Heaviside Expansion Formula

If all poles of $\mathbf{R}(\mathbf{s})$ *are simple*, the partial-fractions expansion and termwise inversion can be accomplished in a single step:

$$\mathcal{L}^{-1}\left[\frac{\mathbf{P}(\mathbf{s})}{\mathbf{Q}(\mathbf{s})}\right] = \sum_{k=1}^{n}\frac{\mathbf{P}(\mathbf{a}_k)}{\mathbf{Q}'(\mathbf{a}_k)}e^{\mathbf{a}_k t} \qquad (3)$$

where $\mathbf{a}_1, \mathbf{a}_2, \ldots, \mathbf{a}_n$ are the poles and $\mathbf{Q}'(\mathbf{a}_k)$ is $d\mathbf{Q}(\mathbf{s})/ds$ evaluated at $\mathbf{s} = \mathbf{a}_k$.

14.6 s-DOMAIN CIRCUITS

Table 14-2 exhibits the elements needed to construct the s-domain image of a given time-domain circuit. The first three lines of the table were in effect developed in Example 14.1. As for the capacitor, we have, for $t > 0$,

$$v_C(t) = V_0 + \frac{1}{C}\int_0^t i(\tau)\,d\tau$$

so that, from Table 14-1,

$$\mathbf{V}_C(\mathbf{s}) = \frac{V_0}{\mathbf{s}} + \frac{\mathbf{I}(\mathbf{s})}{C\mathbf{s}}$$

EXAMPLE 14.4 In the circuit shown in Fig. 14-3(*a*), an initial current i_1 is established while the switch is in position *1*. At $t = 0$ it is moved to position *2*, introducing both a capacitor with initial charge Q_0 and a constant-voltage source V_2.

The s-domain circuit is shown in Fig. 14-3(*b*). The s-domain equation is

$$R\mathbf{I}(\mathbf{s}) + \mathbf{s}L\mathbf{I}(\mathbf{s}) - Li(0^+) + \frac{\mathbf{I}(\mathbf{s})}{\mathbf{s}C} + \frac{V_0}{\mathbf{s}C} = \frac{V_2}{\mathbf{s}}$$

in which $V_0 = Q_0/C$ and $i(0^+) = i_1 = V_1/R$.

Table 14-2

Time Domain	s-Domain	s-Domain Voltage Term
$i \rightarrow$ R	$\mathbf{I(s)} \rightarrow$ R	$R\,\mathbf{I(s)}$
$i \rightarrow$ L $\rightarrow i(0^+)$	$\mathbf{I(s)} \rightarrow$ $\mathbf{s}L$ $-+$ $L\,i(0^+)$	$\mathbf{s}L\,\mathbf{I(s)} - L\,i(0^+)$
$i \rightarrow$ L $\leftarrow i(0^+)$	$\mathbf{I(s)} \rightarrow$ $\mathbf{s}L$ $+-$ $L\,i(0^+)$	$\mathbf{s}L\,\mathbf{I(s)} + L\,i(0^+)$
$i \rightarrow$ C $+V_0^-$	$\mathbf{I(s)} \rightarrow$ $\frac{1}{\mathbf{s}C}$ $+-$ $\frac{V_0}{\mathbf{s}}$	$\frac{\mathbf{I(s)}}{\mathbf{s}C} + \frac{V_0}{\mathbf{s}}$
$i \rightarrow$ C $-V_0^+$	$\mathbf{I(s)} \rightarrow$ $\frac{1}{\mathbf{s}C}$ $-+$ $\frac{V_0}{\mathbf{s}}$	$\frac{\mathbf{I(s)}}{\mathbf{s}C} - \frac{V_0}{\mathbf{s}}$

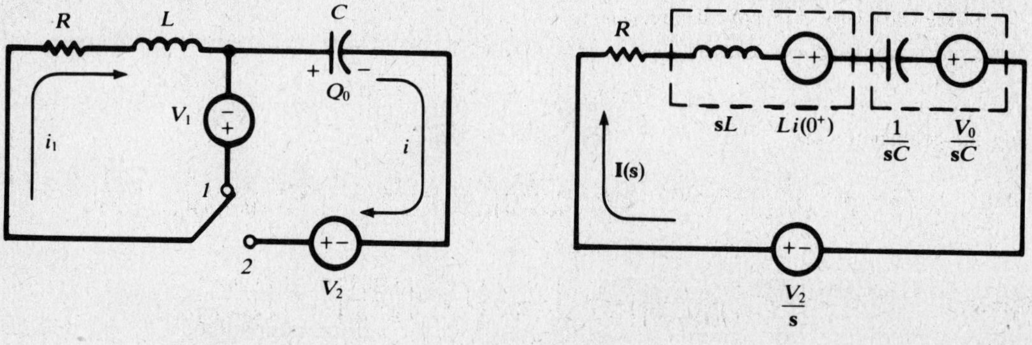

Fig. 14-3

Solved Problems

14.1 Find the Laplace transform of $e^{-at} \cos \omega t$, where a is a constant.

Applying the defining equation $\mathscr{L}[f(t)] = \int_0^\infty f(t)e^{-\mathbf{s}t}\,dt$ to the given function, we obtain

$$\mathscr{L}[e^{-at} \cos \omega t] = \int_0^\infty \cos \omega t\, e^{-(\mathbf{s}+a)t}\,dt$$

$$= \left[\frac{-(\mathbf{s}+a)\cos \omega t\, e^{-(\mathbf{s}+a)t} + e^{-(\mathbf{s}+a)t}\,\omega \sin \omega t}{(\mathbf{s}+a)^2 + \omega^2}\right]_0^\infty$$

$$= \frac{\mathbf{s}+a}{(\mathbf{s}+a)^2 + \omega^2}$$

14.2 If $\mathscr{L}[f(t)] = \mathbf{F}(\mathbf{s})$, show that $\mathscr{L}[e^{-at}f(t)] = \mathbf{F}(\mathbf{s}+a)$. Apply this result to Problem 14.1.

By definition, $\mathscr{L}[f(t)] = \int_0^\infty f(t)e^{-\mathbf{s}t}\,dt = \mathbf{F}(\mathbf{s})$. Then

$$\mathscr{L}[e^{-at}f(t)] = \int_0^\infty [e^{-at}f(t)]e^{-\mathbf{s}t}\,dt = \int_0^\infty f(t)e^{-(\mathbf{s}+a)t}\,dt = \mathbf{F}(\mathbf{s}+a) \qquad (1)$$

Applying (1) to line 6 of Table 14-1 gives

$$\mathscr{L}[e^{-at}\cos\omega t] = \frac{\mathbf{s}+a}{(\mathbf{s}+a)^2+\omega^2}$$

as determined in Problem 14.1.

14.3 Find the Laplace transform of $f(t) = 1 - e^{-at}$, where a is a constant.

$$\mathscr{L}[1-e^{-at}] = \int_0^\infty (1-e^{-at})e^{-\mathbf{s}t}\,dt = \int_0^\infty e^{-\mathbf{s}t}\,dt - \int_0^\infty e^{-(\mathbf{s}+a)t}\,dt$$

$$= \left[-\frac{1}{\mathbf{s}}e^{-\mathbf{s}t} + \frac{1}{\mathbf{s}+a}e^{-(\mathbf{s}+a)t}\right]_0^\infty = \frac{1}{\mathbf{s}} - \frac{1}{\mathbf{s}+a} = \frac{a}{\mathbf{s}(\mathbf{s}+a)}$$

Another Method

$$\mathscr{L}\left[a\int_0^t e^{-a\tau}\,d\tau\right] = a\,\frac{1/(\mathbf{s}+a)}{\mathbf{s}} = \frac{a}{\mathbf{s}(\mathbf{s}+a)}$$

14.4 Find

$$\mathscr{L}^{-1}\left[\frac{1}{\mathbf{s}(\mathbf{s}^2-a^2)}\right]$$

Using the method of partial fractions,

$$\frac{1}{\mathbf{s}(\mathbf{s}^2-a^2)} = \frac{A}{\mathbf{s}} + \frac{B}{\mathbf{s}+a} + \frac{C}{\mathbf{s}-a}$$

and the coefficients are

$$A = \left.\frac{1}{\mathbf{s}^2-a^2}\right|_{\mathbf{s}=0} = -\frac{1}{a^2} \qquad B = \left.\frac{1}{\mathbf{s}(\mathbf{s}-a)}\right|_{\mathbf{s}=-a} = \frac{1}{2a^2} \qquad C = \left.\frac{1}{\mathbf{s}(\mathbf{s}+a)}\right|_{\mathbf{s}=a} = \frac{1}{2a^2}$$

Hence

$$\mathscr{L}^{-1}\left[\frac{1}{\mathbf{s}(\mathbf{s}^2-a^2)}\right] = \mathscr{L}^{-1}\left[\frac{-1/a^2}{\mathbf{s}}\right] + \mathscr{L}^{-1}\left[\frac{1/2a^2}{\mathbf{s}+a}\right] + \mathscr{L}^{-1}\left[\frac{1/2a^2}{\mathbf{s}-a}\right]$$

The corresponding time functions are found in Table 14-1:

$$\mathscr{L}^{-1}\left[\frac{1}{\mathbf{s}(\mathbf{s}^2-a^2)}\right] = -\frac{1}{a^2} + \frac{1}{2a^2}e^{-at} + \frac{1}{2a^2}e^{at}$$

$$= -\frac{1}{a^2} + \frac{1}{a^2}\left(\frac{e^{at}+e^{-at}}{2}\right) = \frac{1}{a^2}(\cosh at - 1)$$

Another Method

By lines 11 and 14 of Table 14-1,

$$\mathscr{L}^{-1}\left[\frac{1/(\mathbf{s}^2-a^2)}{\mathbf{s}}\right] = \int_0^t \frac{\sinh a\tau}{a}\,d\tau = \left[\frac{\cosh a\tau}{a^2}\right]_0^t = \frac{1}{a^2}(\cosh at - 1)$$

14.5 Find

$$\mathscr{L}^{-1}\left[\frac{\mathbf{s}+1}{\mathbf{s}(\mathbf{s}^2+4\mathbf{s}+4)}\right]$$

Using the method of partial fractions, we have

$$\frac{s+1}{s(s+2)^2} = \frac{A}{s} + \frac{B_1}{s+2} + \frac{B_2}{(s+2)^2}$$

Then
$$A = \frac{s+1}{(s+2)^2}\bigg|_{s=0} = \frac{1}{4} \qquad B_2 = \frac{s+1}{s}\bigg|_{s=-2} = \frac{1}{2}$$

and
$$B_1 = (s+2)\frac{s+2}{2s(s+2)^2}\bigg|_{s=-2} = -\frac{1}{4}$$

Hence
$$\mathscr{L}^{-1}\left[\frac{s+1}{s(s^2+4s+4)}\right] = \mathscr{L}^{-1}\left[\frac{\frac{1}{4}}{s}\right] + \mathscr{L}^{-1}\left[\frac{-\frac{1}{4}}{s+2}\right] + \mathscr{L}^{-1}\left[\frac{\frac{1}{2}}{(s+2)^2}\right]$$

The corresponding time functions are found in Table 14-1:

$$\mathscr{L}^{-1}\left[\frac{s+1}{s(s^2+4s+4)}\right] = \frac{1}{4} - \frac{1}{4}e^{-2t} + \frac{1}{2}te^{-2t}$$

14.6 In the series RC circuit of Fig. 14-4 the capacitor has initial charge 2.5 mC. At $t=0$, the switch is closed and a constant-voltage source $V = 100$ V is applied. Use the Laplace transform method to find the current.

The time-domain equation for the given circuit after the switch is closed is

$$Ri(t) + \frac{1}{C}\left[Q_0 + \int_0^t i(\tau)\,d\tau\right] = V$$

or
$$10i(t) + \frac{1}{50\times 10^{-6}}\left[(-2.5\times 10^{-3}) + \int_0^t i(\tau)\,d\tau\right] = V \qquad (1)$$

Q_0 is opposite in polarity to the charge which the source will deposit on the capacitor. Taking the Laplace transform of the terms in (1), we obtain the s-domain equation

$$10\mathbf{I}(s) - \frac{2.5\times 10^{-3}}{50\times 10^{-6}s} + \frac{\mathbf{I}(s)}{50\times 10^{-6}s} = \frac{100}{s} \qquad (2)$$

or
$$\mathbf{I}(s) = \frac{15}{s+(2\times 10^3)} \qquad (3)$$

The time function is now obtained by taking the inverse Laplace transform of (3):

$$i(t) = \mathscr{L}^{-1}\left[\frac{15}{s+(2\times 10^3)}\right] = 15e^{-2\times 10^3 t} \quad \text{(A)} \qquad (4)$$

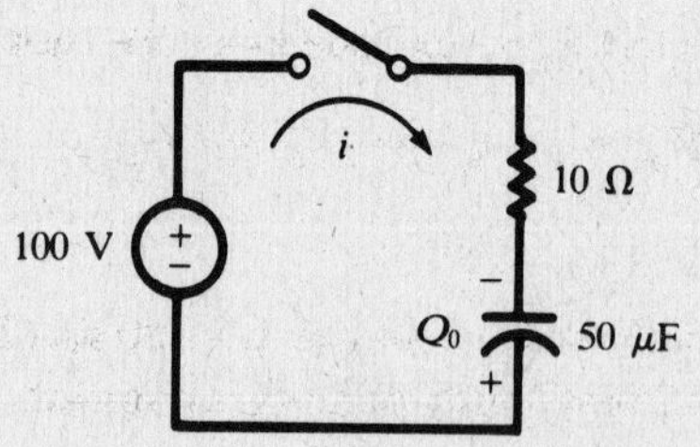

Fig. 14-4

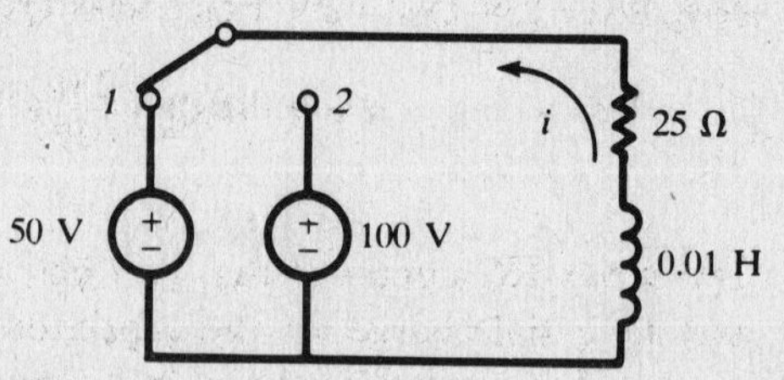

Fig. 14-5

14.7 In the RL circuit shown in Fig. 14-5, the switch is in position 1 long enough to establish steady-state conditions, and at $t=0$ it is switched to position 2. Find the resulting current.

Assume the direction of the current as shown in the diagram. The initial current is then $i_0 = -50/25 = -2$ A.

The time-domain equation is

$$25i + 0.01\frac{di}{dt} = 100 \qquad (1)$$

Taking the Laplace transform of (*1*),

$$25\mathbf{I(s)} + 0.01\,\mathbf{sI(s)} - 0.01\,i(0^+) = 100/\mathbf{s} \tag{2}$$

Substituting for $i(0^+)$,

$$25\mathbf{I(s)} + 0.01\,\mathbf{sI(s)} + 0.01(2) = 100/\mathbf{s} \tag{3}$$

and

$$\mathbf{I(s)} = \frac{100}{\mathbf{s}(0.01\mathbf{s}+25)} - \frac{0.02}{0.01\mathbf{s}+25} = \frac{10^4}{\mathbf{s}(\mathbf{s}+2500)} - \frac{2}{\mathbf{s}+2500} \tag{4}$$

Applying the method of partial fractions,

$$\frac{10^4}{\mathbf{s}(\mathbf{s}+2500)} = \frac{A}{\mathbf{s}} + \frac{B}{\mathbf{s}+2500} \tag{5}$$

with

$$A = \left.\frac{10^4}{\mathbf{s}+2500}\right|_{\mathbf{s}=0} = 4 \qquad \text{and} \qquad B = \left.\frac{10^4}{\mathbf{s}}\right|_{\mathbf{s}=-2500} = -4$$

Then

$$\mathbf{I(s)} = \frac{4}{\mathbf{s}} - \frac{4}{\mathbf{s}+2500} - \frac{2}{\mathbf{s}+2500} = \frac{4}{\mathbf{s}} - \frac{6}{\mathbf{s}+2500} \tag{6}$$

Taking the inverse Laplace transform of (*6*), we obtain $i = 4 - 6e^{-2500t}$ (A).

14.8 In the series *RL* circuit of Fig. 14-6 an exponential voltage $v = 50e^{-100t}$ (V) is applied by closing the switch at $t = 0$. Find the resulting current.

The time-domain equation for the given circuit is

$$Ri + L\frac{di}{dt} = v \tag{1}$$

In the s-domain, (*1*) has the form

$$R\mathbf{I(s)} + \mathbf{s}L\,\mathbf{I(s)} - Li(0^+) = \mathbf{V(s)} \tag{2}$$

Substituting the circuit constants and the transform of the source, $\mathbf{V(s)} = 50/(\mathbf{s}+100)$, in (*2*),

$$10\mathbf{I(s)} + \mathbf{s}(0.2)\mathbf{I(s)} = \frac{50}{\mathbf{s}+100} \qquad \text{or} \qquad \mathbf{I(s)} = \frac{250}{(\mathbf{s}+100)(\mathbf{s}+50)} \tag{3}$$

By the Heaviside expansion formula,

$$\mathscr{L}^{-1}[\mathbf{I(s)}] = \mathscr{L}^{-1}\left[\frac{\mathbf{P(s)}}{\mathbf{Q(s)}}\right] = \sum_{n=1,2} \frac{\mathbf{P}(\mathbf{a}_n)}{\mathbf{Q}'(\mathbf{a}_n)}\, e^{\mathbf{a}_n t}$$

Here, $\mathbf{P(s)} = 250$, $\mathbf{Q(s)} = \mathbf{s}^2 + 150\mathbf{s} + 5000$, $\mathbf{Q}'(\mathbf{s}) = 2\mathbf{s} + 150$, $\mathbf{a}_1 = -100$, and $\mathbf{a}_2 = -50$. Then

$$i = \mathscr{L}^{-1}[\mathbf{I(s)}] = \frac{250}{-50}e^{-100t} + \frac{250}{50}e^{-50t} = -5e^{-100t} + 5e^{-50t} \quad \text{(A)}$$

14.9 The series *RC* circuit of Fig. 14-7 has a sinusoidal voltage source $v = 180 \sin(2000t + \phi)$ (V) and an initial charge on the capacitor $Q_0 = 1.25$ mC with polarity as shown. Determine the current if the switch is closed at a time corresponding to $\phi = 90°$.

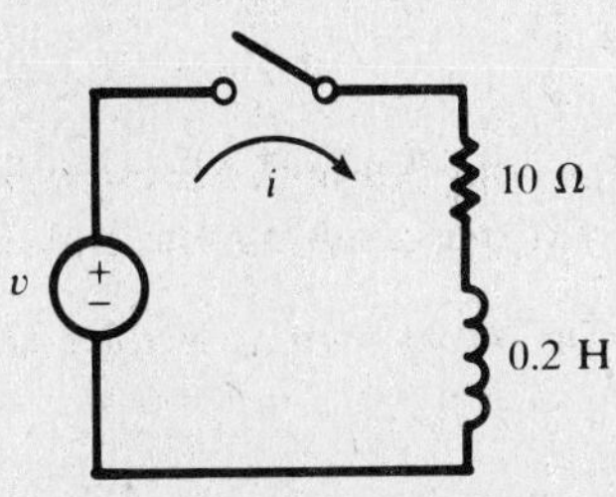

Fig. 14-6

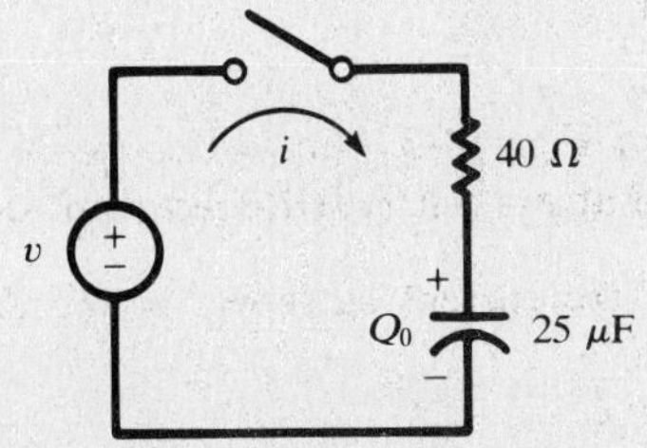

Fig. 14-7

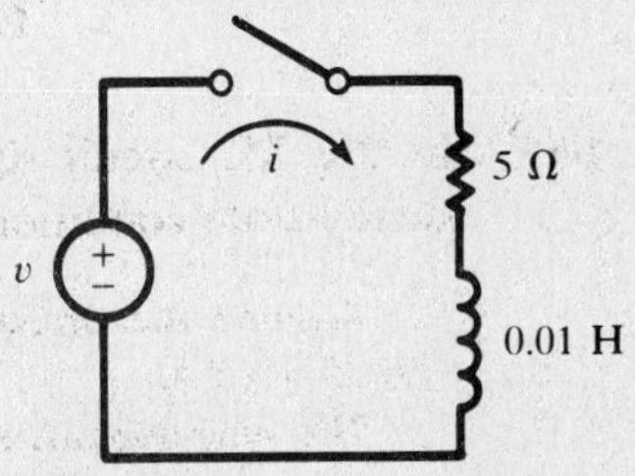

Fig. 14-8

The time-domain equation of the circuit is

$$40i(t) + \frac{1}{25\times 10^{-6}}\left(1.25\times 10^{-3} + \int_0^t i(\tau)\,d\tau\right) = 180\cos 2000t \tag{1}$$

The Laplace transform of (1) gives the s-domain equation

$$40\mathbf{I}(\mathbf{s}) + \frac{1.25\times 10^{-3}}{25\times 10^{-6}\mathbf{s}} + \frac{4\times 10^4}{\mathbf{s}}\mathbf{I}(\mathbf{s}) = \frac{180\mathbf{s}}{\mathbf{s}^2 + 4\times 10^6} \tag{2}$$

or

$$\mathbf{I}(\mathbf{s}) = \frac{4.5\mathbf{s}^2}{(\mathbf{s}^2 + 4\times 10^6)(\mathbf{s}+10^3)} - \frac{1.25}{\mathbf{s}+10^3} \tag{3}$$

Applying the Heaviside expansion formula to the first term on the right in (3), we have $\mathbf{P}(\mathbf{s}) = 4.5\mathbf{s}^2$, $\mathbf{Q}(\mathbf{s}) = \mathbf{s}^3 + 10^3\mathbf{s}^2 + 4\times 10^6\mathbf{s} + 4\times 10^9$, $\mathbf{Q}'(\mathbf{s}) = 3\mathbf{s}^2 + 2\times 10^3\mathbf{s} + 4\times 10^6$, $\mathbf{a}_1 = -j2\times 10^3$, $\mathbf{a}_2 = j2\times 10^3$, and $\mathbf{a}_3 = -10^3$. Then

$$\begin{aligned} i &= \frac{\mathbf{P}(-j2\times 10^3)}{\mathbf{Q}'(-j2\times 10^3)}e^{-j2\times 10^3 t} + \frac{\mathbf{P}(j2\times 10^3)}{\mathbf{Q}'(j2\times 10^3)}e^{j2\times 10^3 t} + \frac{\mathbf{P}(-10^3)}{\mathbf{Q}'(-10^3)}e^{-10^3 t} - 1.25e^{-10^3 t} \\ &= (1.8 - j0.9)e^{-j2\times 10^3 t} + (1.8 + j0.9)e^{j2\times 10^3 t} - 0.35e^{-10^3 t} \\ &= -1.8\sin 2000t + 3.6\cos 2000t - 0.35e^{-10^3 t} \\ &= 4.02\sin(2000t + 116.6^\circ) - 0.35e^{-10^3 t} \quad \text{(A)} \end{aligned} \tag{4}$$

At $t = 0$ the current is given by the instantaneous voltage, consisting of the source voltage and the charged capacitor voltage, divided by the resistance. Thus

$$i_0 = \left(180\sin 90^\circ - \frac{1.25\times 10^{-3}}{25\times 10^{-6}}\right)\bigg/ 40 = 3.25\ \text{A}$$

The same result is obtained if we set $t = 0$ in (4).

14.10 In the series RL circuit of Fig. 14-8 the source is $v = 100\sin(500t + \phi)$ (V). Determine the resulting current if the switch is closed at a time corresponding to $\phi = 0$.

The s-domain equation of a series RL circuit is

$$R\mathbf{I}(\mathbf{s}) + sL\mathbf{I}(\mathbf{s}) - Li(0^+) = \mathbf{V}(\mathbf{s}) \tag{1}$$

The transform of the source with $\phi = 0$ is

$$\mathbf{V}(\mathbf{s}) = \frac{(100)(500)}{\mathbf{s}^2 + (500)^2}$$

Since there is no initial current in the inductance, $Li(0^+) = 0$. Substituting the circuit constants into (1),

$$5\mathbf{I}(\mathbf{s}) + 0.01\mathbf{s}\mathbf{I}(\mathbf{s}) = \frac{5\times 10^4}{\mathbf{s}^2 + 25\times 10^4} \quad \text{or} \quad \mathbf{I}(\mathbf{s}) = \frac{5\times 10^6}{(\mathbf{s}^2 + 25\times 10^4)(\mathbf{s}+500)} \tag{2}$$

Expanding (2) by partial fractions,

$$\mathbf{I}(\mathbf{s}) = 5\left(\frac{-1+j}{\mathbf{s}+j500}\right) + 5\left(\frac{-1-j}{\mathbf{s}-j500}\right) + \frac{10}{\mathbf{s}+500} \tag{3}$$

The inverse Laplace transform of (3) is

$$i = 10\sin 500t - 10\cos 500t + 10e^{-500t} = 10e^{-500t} + 14.14\sin(500t - 45^\circ) \quad \text{(A)}$$

14.11 Rework Problem 14.10 by writing the voltage function as

$$v = 100e^{j500t} \quad \text{(V)} \tag{1}$$

Now $\mathbf{V}(\mathbf{s}) = 100/(\mathbf{s} - j500)$, and the s-domain equation is

$$5\mathbf{I}(\mathbf{s}) + 0.01\mathbf{s}\mathbf{I}(\mathbf{s}) = \frac{100}{\mathbf{s} - j500} \quad \text{or} \quad \mathbf{I}(\mathbf{s}) = \frac{10^4}{(\mathbf{s} - j500)(\mathbf{s}+500)}$$

Using partial fractions,

$$\mathbf{I}(\mathbf{s}) = \frac{10 - j10}{\mathbf{s} - j500} + \frac{-10 + j10}{\mathbf{s} + 500}$$

and inverting,

$$\begin{aligned} i &= (10 - j10)e^{j500t} + (-10 + j10)e^{-500t} \\ &= 14.14\,e^{j(500t - \pi/4)} + (-10 + j10)e^{-500t} \quad \text{(A)} \end{aligned} \qquad (2)$$

The actual voltage is the imaginary part of (*1*); hence the actual current is the imaginary part of (*2*).

$$i = 14.14 \sin (500t - \pi/4) + 10e^{-500t} \quad \text{(A)}$$

14.12 In the series *RLC* circuit shown in Fig. 14-9, there is no initial charge on the capacitor. If the switch is closed at $t = 0$, determine the resulting current.

The time-domain equation of the given circuit is

$$Ri + L\frac{di}{dt} + \frac{1}{C}\int_0^t i(\tau)\, d\tau = V \qquad (1)$$

Because $i(0^+) = 0$, the Laplace transform of (*1*) is

$$R\mathbf{I}(\mathbf{s}) + \mathbf{s}L\mathbf{I}(\mathbf{s}) + \frac{1}{\mathbf{s}C}\mathbf{I}(\mathbf{s}) = \frac{V}{\mathbf{s}} \qquad (2)$$

or

$$2\mathbf{I}(\mathbf{s}) + 1\mathbf{s}\mathbf{I}(\mathbf{s}) + \frac{1}{0.5\mathbf{s}}\mathbf{I}(\mathbf{s}) = \frac{50}{\mathbf{s}} \qquad (3)$$

Hence

$$\mathbf{I}(\mathbf{s}) = \frac{50}{\mathbf{s}^2 + 2\mathbf{s} + 2} = \frac{50}{(\mathbf{s} + 1 + j)(\mathbf{s} + 1 - j)} \qquad (4)$$

Expanding (*4*) by partial fractions,

$$\mathbf{I}(\mathbf{s}) = \frac{j25}{(\mathbf{s} + 1 + j)} - \frac{j25}{(\mathbf{s} + 1 - j)} \qquad (5)$$

and the inverse Laplace transform of (*5*) gives

$$i = j25\{e^{(-1-j)t} - e^{(-1+j)t}\} = 50e^{-t} \sin t \quad \text{(A)}$$

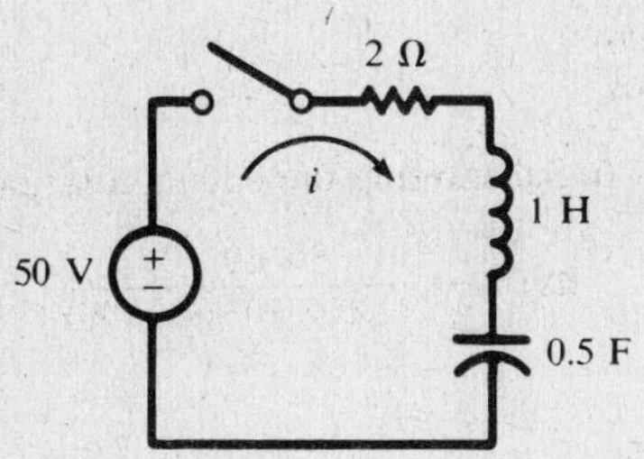

Fig. 14-9

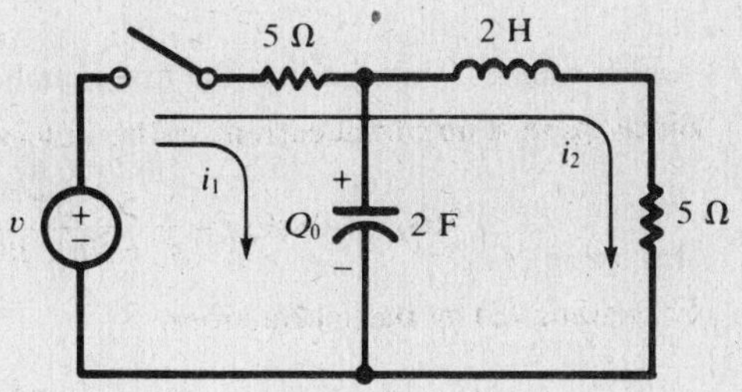

Fig. 14-10

14.13 In the two-mesh network of Fig. 14-10, the two loop currents are selected as shown. Write the s-domain equations in matrix form and construct the corresponding circuit.

Writing the set of equations in the time domain,

$$5i_1 + \frac{1}{2}\left[Q_0 + \int_0^t i_1(\tau)\, d\tau\right] + 5i_2 = v \quad \text{and} \quad 10i_2 + 2\frac{di_2}{dt} + 5i_1 = v \qquad (1)$$

Taking the Laplace transform of (*1*) to obtain the corresponding s-domain equations,

$$5\mathbf{I}_1(\mathbf{s}) + \frac{Q_0}{2\mathbf{s}} + \frac{1}{2\mathbf{s}}\mathbf{I}_1(\mathbf{s}) + 5\mathbf{I}_2(\mathbf{s}) = \mathbf{V}(\mathbf{s}) \qquad 10\mathbf{I}_2(\mathbf{s}) + 2\mathbf{s}\mathbf{I}_2(\mathbf{s}) - 2i_2(0^+) + 5\mathbf{I}_1(\mathbf{s}) = \mathbf{V}(\mathbf{s}) \qquad (2)$$

When this set of s-domain equations is written in matrix form,

$$\begin{bmatrix} 5+(1/2\mathbf{s}) & 5 \\ 5 & 10+2\mathbf{s} \end{bmatrix} \begin{bmatrix} \mathbf{I}_1(\mathbf{s}) \\ \mathbf{I}_2(\mathbf{s}) \end{bmatrix} = \begin{bmatrix} \mathbf{V}(\mathbf{s}) - (Q_0/2\mathbf{s}) \\ \mathbf{V}(\mathbf{s}) + 2i_2(0^+) \end{bmatrix}$$

the required s-domain circuit can be determined by examination of the $\mathbf{Z}(\mathbf{s})$, $\mathbf{I}(\mathbf{s})$, and $\mathbf{V}(\mathbf{s})$ matrices (see Fig. 14-11).

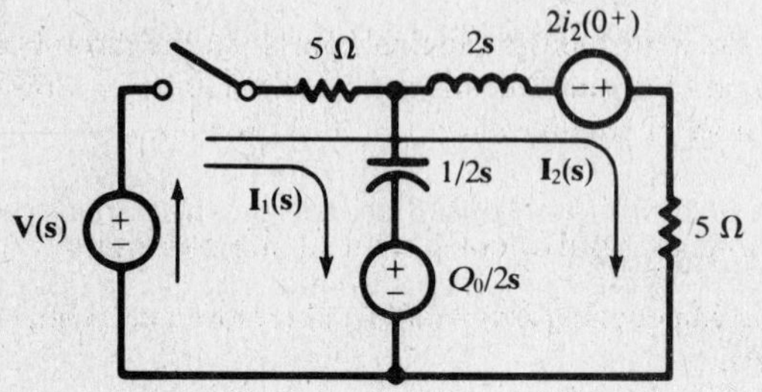

Fig. 14-11

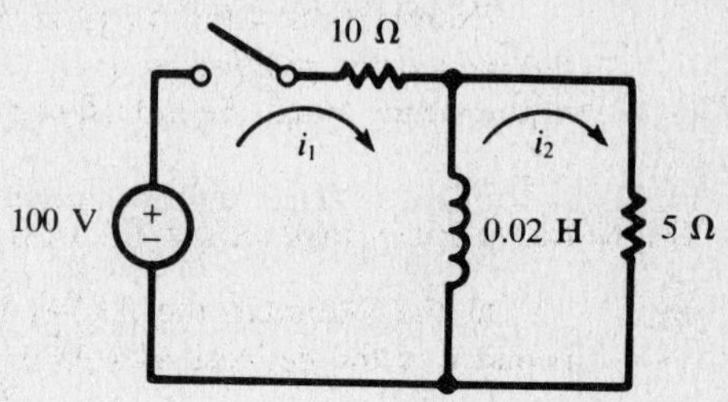

Fig. 14-12

14.14 In the two-mesh network of Fig. 14-12, find the currents which result when the switch is closed.

The time-domain equations for the network are

$$10i_1 + 0.02\frac{di_1}{dt} - 0.02\frac{di_2}{dt} = 100$$

$$0.02\frac{di_2}{dt} + 5i_2 - 0.02\frac{di_1}{dt} = 0 \qquad (1)$$

Taking the Laplace transform of set (*1*),

$$(10+0.02\mathbf{s})\mathbf{I}_1(\mathbf{s}) - 0.02\mathbf{s}\mathbf{I}_2(\mathbf{s}) = 100/\mathbf{s} \qquad (5+0.02\mathbf{s})\mathbf{I}_2(\mathbf{s}) - 0.02\mathbf{s}\mathbf{I}_1(\mathbf{s}) = 0 \qquad (2)$$

From the second equation in set (*2*) we find

$$\mathbf{I}_2(\mathbf{s}) = \mathbf{I}_1(\mathbf{s})\left(\frac{\mathbf{s}}{\mathbf{s}+250}\right) \qquad (3)$$

which when substituted into the first equation gives

$$\mathbf{I}_1(\mathbf{s}) = 6.67\left\{\frac{\mathbf{s}+250}{\mathbf{s}(\mathbf{s}+166.7)}\right\} = \frac{10}{\mathbf{s}} - \frac{3.33}{\mathbf{s}+166.7} \qquad (4)$$

Inverting (*4*),

$$i_1 = 10 - 3.33e^{-166.7t} \quad \text{(A)}$$

Finally, substitute (*4*) into (*3*) and obtain

$$\mathbf{I}_2(\mathbf{s}) = 6.67\left(\frac{1}{\mathbf{s}+166.7}\right) \qquad \text{whence} \qquad i_2 = 6.67e^{-166.7t} \quad \text{(A)}$$

14.15 Apply the initial- and final-value theorems in Problem 14.14.

The initial value of i_1 is given by

$$i_1(0^+) = \lim_{s\to\infty}[s\mathbf{I}_1(s)] = \lim_{s\to\infty}\left[6.67\left(\frac{s+250}{s+166.7}\right)\right] = 6.67 \text{ A}$$

and the final value is

$$i_1(\infty) = \lim_{s\to 0}[s\mathbf{I}_1(s)] = \lim_{s\to 0}\left[6.67\left(\frac{s+250}{s+166.7}\right)\right] = 10 \text{ A}$$

The initial value of i_2 is given by

$$i_2(0^+) = \lim_{s\to\infty} [s\mathbf{I}_2(s)] = \lim_{s\to\infty} \left[6.67\left(\frac{s}{s+166.7}\right)\right] = 6.67 \text{ A}$$

and the final value is

$$i_2(\infty) = \lim_{s\to 0} [s\mathbf{I}_2(s)] = \lim_{s\to 0} \left[6.67\left(\frac{s}{s+166.7}\right)\right] = 0$$

Examination of Fig. 14-12 verifies each of the above initial and final values. At the instant of closing, the inductance presents an infinite impedance and the currents are $i_1 = i_2 = 100/(10+5) = 6.67$ A. Then, in the steady state, the inductance appears as a short circuit; hence $i_1 = 10$ A, $i_2 = 0$.

14.16 Solve for i_1 in Problem 14.14 by determining an equivalent circuit in the s-domain.

In the s-domain the 0.02-H inductor has impedance $\mathbf{Z(s)} = 0.02\mathbf{s}$. Therefore, the equivalent impedance of the network as seen from the source is

$$\mathbf{Z(s)} = 10 + \frac{(0.02\mathbf{s})(5)}{0.02\mathbf{s}+5} = 15\left(\frac{\mathbf{s}+166.7}{\mathbf{s}+250}\right)$$

and the s-domain equivalent circuit is as shown in Fig. 14-13. The current is then

$$\mathbf{I}_1(\mathbf{s}) = \frac{\mathbf{V(s)}}{\mathbf{Z(s)}} = \frac{100}{\mathbf{s}}\left\{\frac{\mathbf{s}+250}{15(\mathbf{s}+166.7)}\right\} = 6.67\left\{\frac{\mathbf{s}+250}{\mathbf{s}(\mathbf{s}+166.7)}\right\}$$

This expression is identical with (4) of Problem 14.14, and so the same time function i_1 is obtained.

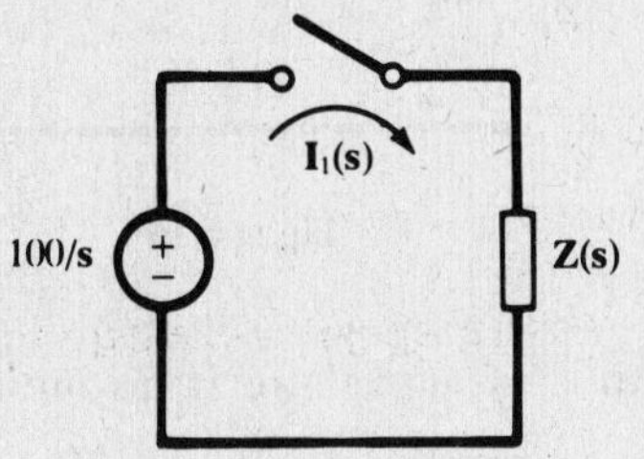

Fig. 14-13

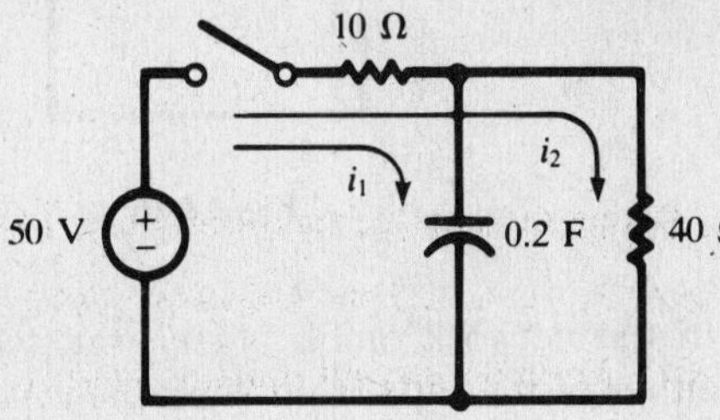

Fig. 14-14

14.17 In the two-mesh network shown in Fig. 14-14 there is no initial charge on the capacitor. Find the loop currents i_1 and i_2 which result when the switch is closed at $t = 0$.

The time-domain equations for the circuit are

$$10i_1 + \frac{1}{0.2}\int_0^t i_1\, d\tau + 10i_2 = 50 \qquad 50i_2 + 10i_1 = 50$$

The corresponding s-domain equations are

$$10\mathbf{I}_1(\mathbf{s}) + \frac{1}{0.2\mathbf{s}}\mathbf{I}_1(\mathbf{s}) + 10\mathbf{I}_2(\mathbf{s}) = \frac{50}{\mathbf{s}} \qquad 50\mathbf{I}_2(\mathbf{s}) + 10\mathbf{I}_1(\mathbf{s}) = \frac{50}{\mathbf{s}}$$

Solving,

$$\mathbf{I}_1(\mathbf{s}) = \frac{5}{\mathbf{s}+0.625} \qquad \mathbf{I}_2(\mathbf{s}) = \frac{1}{\mathbf{s}} - \frac{1}{\mathbf{s}+0.625}$$

which invert to

$$i_1 = 5e^{-0.625t} \text{ (A)} \qquad i_2 = 1 - e^{-0.625t} \text{ (A)}$$

14.18 Referring to Problem 14.17, obtain the equivalent impedance of the s-domain network and determine the total current and the branch currents using the current-division rule.

The s-domain impedance as seen by the voltage source is

$$\mathbf{Z(s)} = 10 + \frac{40(1/0.2\mathbf{s})}{40 + 1/0.2\mathbf{s}} = \frac{80\mathbf{s} + 50}{8\mathbf{s} + 1} = 10\left(\frac{\mathbf{s} + 5/8}{\mathbf{s} + 1/8}\right) \tag{1}$$

The equivalent circuit is shown in Fig. 14-15; the resulting current is

$$\mathbf{I(s)} = \frac{\mathbf{V(s)}}{\mathbf{Z(s)}} = 5\frac{\mathbf{s} + 1/8}{\mathbf{s}(\mathbf{s} + 5/8)} \tag{2}$$

Expanding $\mathbf{I(s)}$ in partial fractions,

$$\mathbf{I(s)} = \frac{1}{\mathbf{s}} + \frac{4}{\mathbf{s} + 5/8} \qquad \text{from which} \qquad i = 1 + 4e^{-5t/8} \quad \text{(A)}$$

Now the branch currents $\mathbf{I}_1(\mathbf{s})$ and $\mathbf{I}_2(\mathbf{s})$ can be obtained by the current-division rule. Referring to Fig. 14-16, we have

$$\mathbf{I}_1(\mathbf{s}) = \mathbf{I(s)}\left(\frac{40}{40 + 1/0.2\mathbf{s}}\right) = \frac{5}{\mathbf{s} + 5/8} \qquad \text{and} \qquad i_1 = 5e^{-0.625t} \quad \text{(A)}$$

$$\mathbf{I}_2(\mathbf{s}) = \mathbf{I(s)}\left(\frac{1/0.2\mathbf{s}}{40 + 1/0.2\mathbf{s}}\right) = \frac{1}{\mathbf{s}} - \frac{1}{\mathbf{s} + 5/8} \qquad \text{and} \qquad i_2 = 1 - e^{-0.625t} \quad \text{(A)}$$

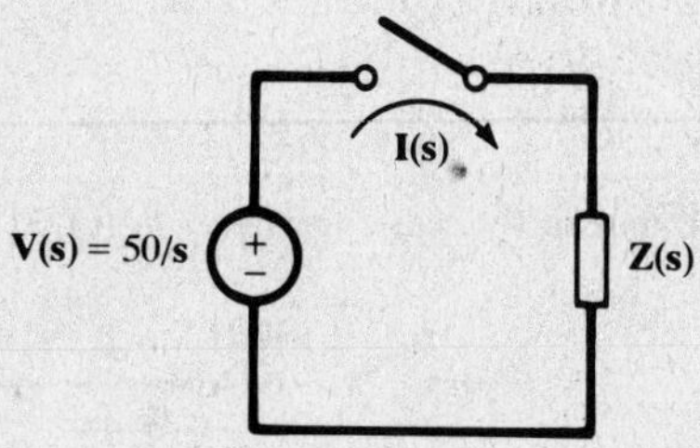

Fig. 14-15

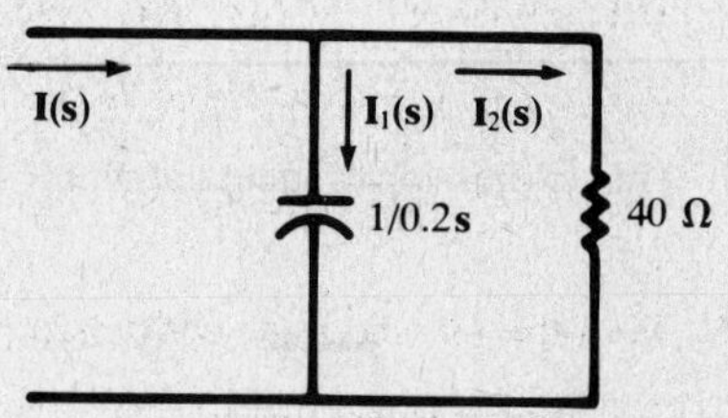

Fig. 14-16

14.19 In the network of Fig. 14-17, the switch is closed at $t = 0$ and there is no initial charge on either of the capacitors. Find the resulting current i.

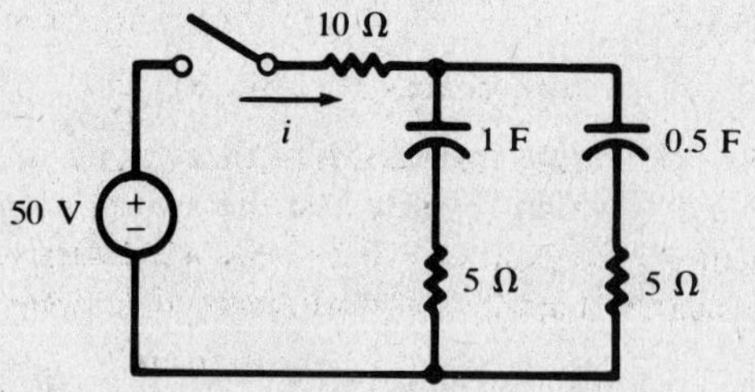

Fig. 14-17

The network has an equivalent impedance in the s-domain

$$\mathbf{Z(s)} = 10 + \frac{(5 + 1/\mathbf{s})(5 + 1/0.5\mathbf{s})}{10 + 1/\mathbf{s} + 1/0.5\mathbf{s}} = \frac{125\mathbf{s}^2 + 45\mathbf{s} + 2}{\mathbf{s}(10\mathbf{s} + 3)}$$

Hence the current is

$$\mathbf{I(s)} = \frac{\mathbf{V(s)}}{\mathbf{Z(s)}} = \frac{50}{\mathbf{s}}\frac{\mathbf{s}(10\mathbf{s} + 3)}{(125\mathbf{s}^2 + 45\mathbf{s} + 2)} = \frac{4(\mathbf{s} + 0.3)}{(\mathbf{s} + 0.308)(\mathbf{s} + 0.052)}$$

Expanding $\mathbf{I(s)}$ in partial fractions,

$$\mathbf{I(s)} = \frac{1/8}{\mathbf{s} + 0.308} + \frac{31/8}{\mathbf{s} + 0.052} \qquad \text{and} \qquad i = \frac{1}{8}e^{-0.308t} + \frac{31}{8}e^{-0.052t} \quad \text{(A)}$$

14.20 Apply the initial- and final-value theorems to the s-domain current of Problem 14.19.

$$i(0^+) = \lim_{s\to\infty} [s\mathbf{I}(s)] = \lim_{s\to\infty} \left[\frac{1}{8}\left(\frac{s}{s+0.308}\right) + \frac{31}{8}\left(\frac{s}{s+0.052}\right)\right] = 4 \text{ A}$$

$$i(\infty) = \lim_{s\to 0} [s\mathbf{I}(s)] = \lim_{s\to 0} \left[\frac{1}{8}\left(\frac{s}{s+0.308}\right) + \frac{31}{8}\left(\frac{s}{s+0.052}\right)\right] = 0$$

Examination of Fig. 14-17 shows that initially the total circuit resistance is $R = 10 + 5(5)/10 = 12.5\ \Omega$ and thus $i(0^+) = 50/12.5 = 4$ A. Then, in the steady state, both capacitors are charged to 50 V and the current is zero.

Supplementary Problems

14.21 Find the Laplace transform of each function.

(a) $f(t) = At$ (c) $f(t) = e^{-at}\sin\omega t$ (e) $f(t) = \cosh\omega t$
(b) $f(t) = te^{-at}$ (d) $f(t) = \sinh\omega t$ (f) $f(t) = e^{-at}\sinh\omega t$

Ans. (a)–(e) See Table 14-1.

(f) $\dfrac{\omega}{(\mathbf{s}+a)^2 - \omega^2}$

14.22 Find the inverse Laplace transform of each function.

(a) $\mathbf{F(s)} = \dfrac{\mathbf{s}}{(\mathbf{s}+2)(\mathbf{s}+1)}$ (d) $\mathbf{F(s)} = \dfrac{3}{\mathbf{s}(\mathbf{s}^2+6\mathbf{s}+9)}$ (g) $\mathbf{F(s)} = \dfrac{2\mathbf{s}}{(\mathbf{s}^2+4)(\mathbf{s}+5)}$

(b) $\mathbf{F(s)} = \dfrac{1}{\mathbf{s}^2+7\mathbf{s}+12}$ (e) $\mathbf{F(s)} = \dfrac{\mathbf{s}+5}{\mathbf{s}^2+2\mathbf{s}+5}$

(c) $\mathbf{F(s)} = \dfrac{5\mathbf{s}}{\mathbf{s}^2+3\mathbf{s}+2}$ (f) $\mathbf{F(s)} = \dfrac{2\mathbf{s}+4}{\mathbf{s}^2+4\mathbf{s}+13}$

Ans. (a) $2e^{-2t} - e^{-t}$ (d) $\frac{1}{3} - \frac{1}{3}e^{-3t} - te^{-3t}$ (g) $\frac{10}{29}\cos 2t + \frac{4}{29}\sin 2t - \frac{10}{29}e^{-5t}$

(b) $e^{-3t} - e^{-4t}$ (e) $e^{-t}(\cos 2t + 2\sin 2t)$

(c) $10e^{-2t} - 5e^{-t}$ (f) $2e^{-2t}\cos 3t$

14.23 A series *RL* circuit, with $R = 10\ \Omega$ and $L = 0.2$ H, has a constant voltage $V = 50$ V applied at $t = 0$. Find the resulting current using the Laplace transform method. *Ans.* $i = 5 - 5e^{-50t}$ (A)

14.24 In the series *RL* circuit of Fig. 14-18 the switch is in position 1 long enough to establish the steady state and is switched to position 2 at $t = 0$. Find the current. *Ans.* $i = 5e^{-50t}$ (A)

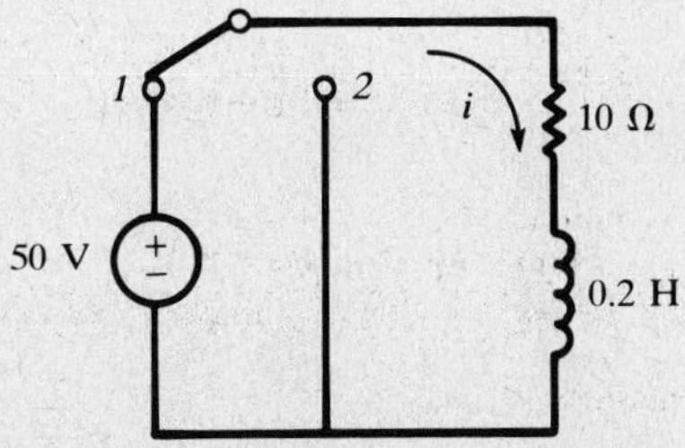

Fig. 14-18

14.25 In the circuit shown in Fig. 14-19, switch *1* is closed at $t = 0$ and then, at $t = t' = 4$ ms, switch *2* is opened. Find the current in the intervals $0 < t < t'$ and $t > t'$.
Ans. $i = 2(1 - e^{-500t})$ (A), $i = 1.06e^{-1500(t-t')} + 0.667$ (A)

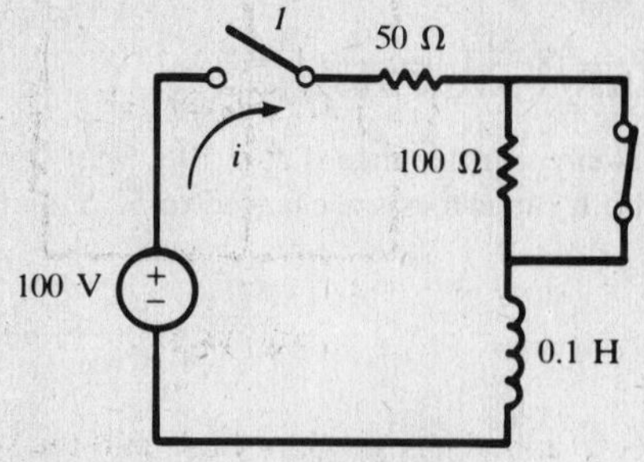

Fig. 14-19

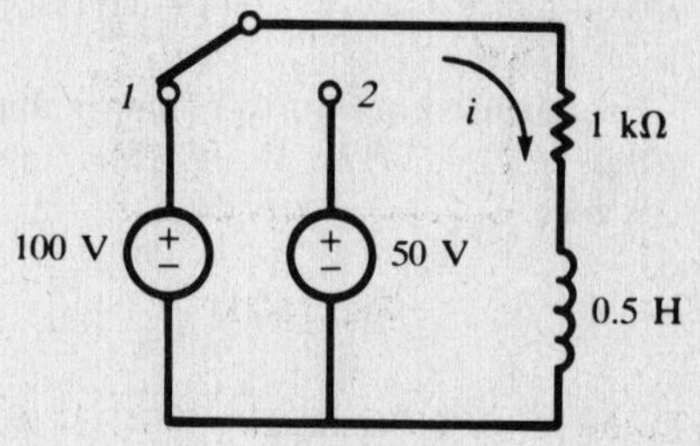

Fig. 14-20

14.26 In the series *RL* circuit shown in Fig. 14-20, the switch is closed on position *1* at $t = 0$ and then, at $t = t' = 50\ \mu$s, it is moved to position *2*. Find the current in the intervals $0 < t < t'$ and $t > t'$.
Ans. $i = 0.1(1 - e^{-2000t})$ (A), $i = 0.06e^{-2000(t-t')} - 0.05$ (A)

14.27 A series *RC* circuit, with $R = 10\ \Omega$ and $C = 4\ \mu$F, has an initial charge $Q_0 = 800\ \mu$C on the capacitor at the time the switch is closed, applying a constant-voltage source $V = 100$ V. Find the resulting current transient if the charge is (*a*) of the same polarity as that deposited by the source, and (*b*) of the opposite polarity. *Ans.* (*a*) $i = -10e^{-25\times10^3 t}$ (A); (*b*) $i = 30e^{-25\times10^3 t}$ (A)

14.28 A series *RC* circuit, with $R = 1$ kΩ and $C = 20\ \mu$F, has an initial charge Q_0 on the capacitor at the time the switch is closed, applying a constant-voltage source $V = 50$ V. If the resulting current is $i = 0.075e^{-50t}$ (A), find the charge Q_0 and its polarity.
Ans. 500 μC, opposite polarity to that deposited by source

14.29 In the *RC* circuit shown in Fig. 14-21, the switch is closed on position *1* at $t = 0$ and then, at $t = t' = \tau$ (the time constant, Section 5.4) is moved to position *2*. Find the transient current in the intervals $0 < t < t'$ and $t > t'$. *Ans.* $i = 0.5e^{-200t}$ (A), $i = -0.516e^{-200(t-t')}$ (A)

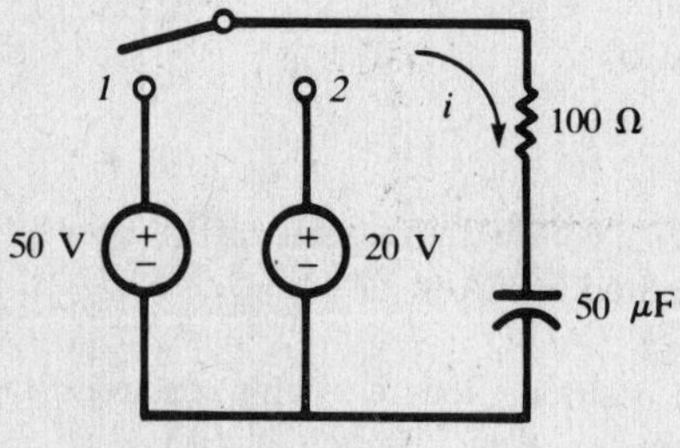

Fig. 14-21

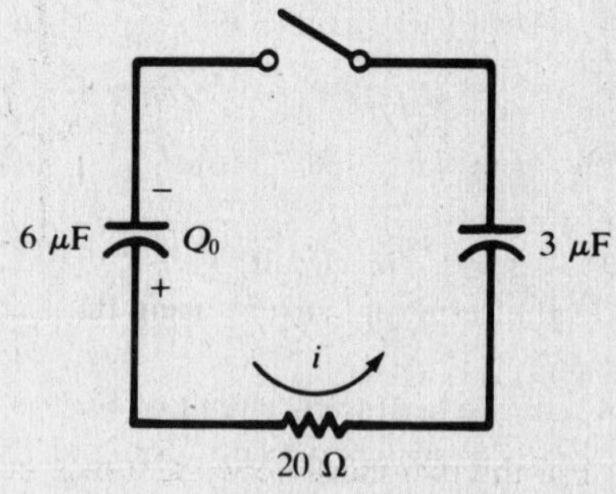

Fig. 14-22

14.30 In the circuit of Fig. 14-22, $Q_0 = 300\ \mu$C at the time the switch is closed. Find the resulting current transient. *Ans.* $i = 2.5e^{-2.5\times10^4 t}$ (A)

14.31 In the circuit shown in Fig. 14-23, the capacitor has an initial charge $Q_0 = 25\ \mu$C and the sinusoidal voltage source is $v = 100 \sin(1000t + \phi)$ (V). Find the resulting current if the switch is closed at a time corresponding to $\phi = 30°$. *Ans.* $i = 0.1535e^{-4000t} + 0.0484 \sin(1000t + 106°)$ (A)

14.32 A series *RLC* circuit, with $R = 5\ \Omega$, $L = 0.1$ H, and $C = 500\ \mu$F, has a constant voltage $V = 10$ V applied at $t = 0$. Find the resulting current. *Ans.* $i = 0.72e^{-25t} \sin 139t$ (A)

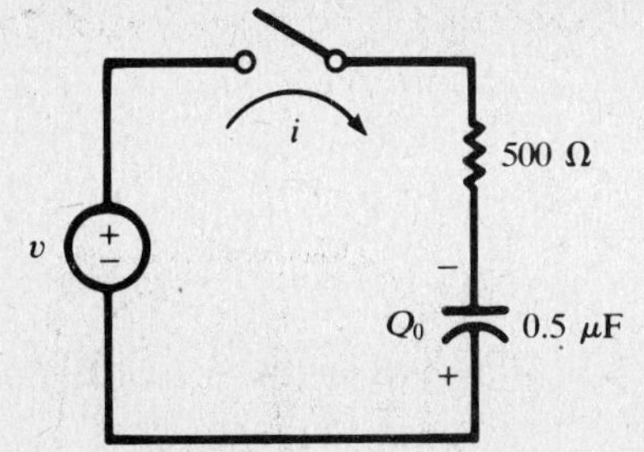

Fig. 14-23

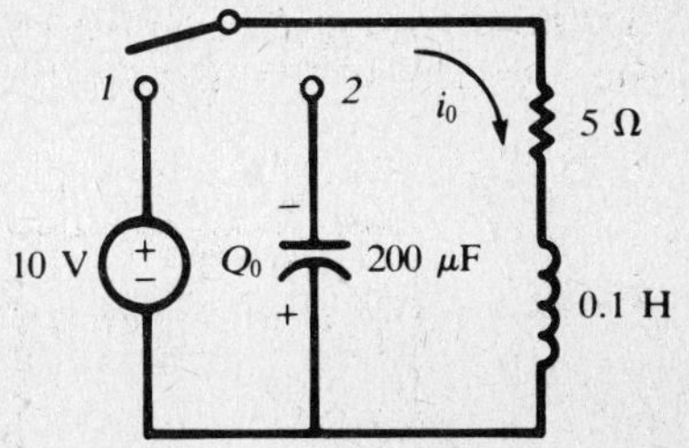

Fig. 14-24

14.33 In the series RLC circuit of Fig. 14-24, the capacitor has an initial charge $Q_0 = 1$ mC and the switch is in position *1* long enough to establish the steady state. Find the transient current which results when the switch is moved from position *1* to *2* at $t = 0$. *Ans.* $i = e^{-25t}(2\cos 222t - 0.45\sin 222t)$ (A)

14.34 A series RLC circuit, with $R = 5\ \Omega$, $L = 0.2$ H, and $C = 1$ F, has a voltage source $v = 10e^{-100t}$ (V) applied at $t = 0$. Find the resulting current. *Ans.* $i = -0.666\,e^{-100t} + 0.670\,e^{-24.8t} - 0.004\,e^{-0.2t}$ (A)

14.35 A series RLC circuit, with $R = 200\ \Omega$, $L = 0.5$ H, and $C = 100\ \mu$F, has a sinusoidal voltage source $v = 300\sin(500t + \phi)$ (V). Find the resulting current if the switch is closed at a time corresponding to $\phi = 30°$. *Ans.* $i = 0.517\,e^{-341.4t} - 0.197e^{-58.6t} + 0.983\sin(500t - 19°)$ (A)

14.36 A series RLC circuit, with $R = 5\ \Omega$, $L = 0.1$ H, and $C = 500\ \mu$F, has a sinusoidal voltage source $v = 100\sin 250t$ (V). Find the resulting current if the switch is closed at $t = 0$.
Ans. $i = e^{-25t}(5.42\cos 139t + 1.89\sin 139t) + 5.65\sin(250t - 73.6°)$ (A)

14.37 In the two-mesh network of Fig. 14-25, the currents are selected as shown in the diagram. Write the time-domain equations, transform them into the corresponding s-domain equations, and obtain the currents i_1 and i_2. *Ans.* $i_1 = 2.5(1 + e^{-10^5 t})$ (A), $i_2 = 5e^{-10^5 t}$ (A)

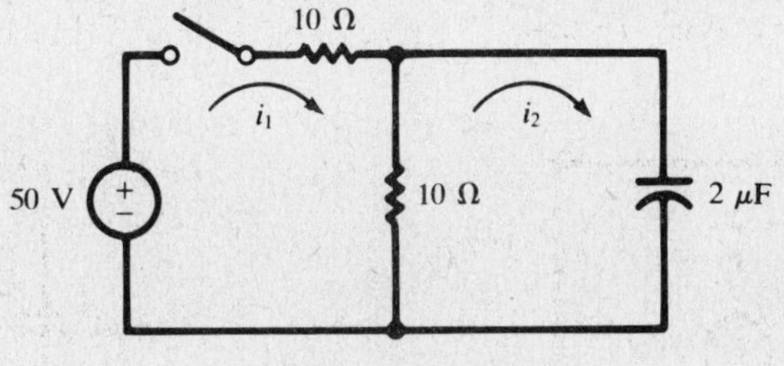

Fig. 14-25

14.38 For the two-mesh network shown in Fig. 14-26, find the currents i_1 and i_2 which result when the switch is closed at $t = 0$. *Ans.* $i_1 = 0.101\,e^{-100t} + 9.899\,e^{-9950t}$ (A), $i_2 = -5.05\,e^{-100t} + 5 + 0.05\,e^{-9950t}$ (A)

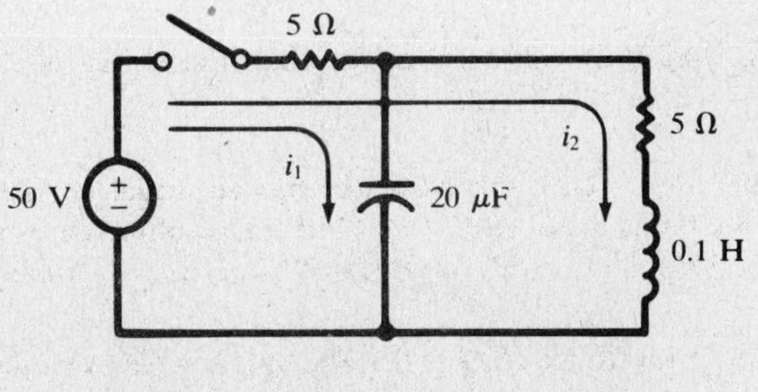

Fig. 14-26

14.39 In the network shown in Fig. 14-27, the 100-V source passes a continuous current in the first loop while the switch is open. Find the currents after the switch is closed at $t = 0$.
Ans. $i_1 = 1.67e^{-6.67t} + 5$ (A), $i_2 = -0.555e^{-6.67t} + 5$ (A)

14.40 The two-mesh network shown in Fig. 14-28 contains a sinusoidal voltage source $v = 100\sin(200t + \phi)$ (V). The switch is closed at an instant when the voltage is increasing at its maximum rate. Find the resulting mesh currents, with directions as shown in the diagram.
Ans. $i_1 = 3.01e^{-100t} + 8.96\sin(200t - 63.4°)$ (A), $i_2 = 1.505e^{-100t} + 4.48\sin(200t - 63.4°)$ (A)

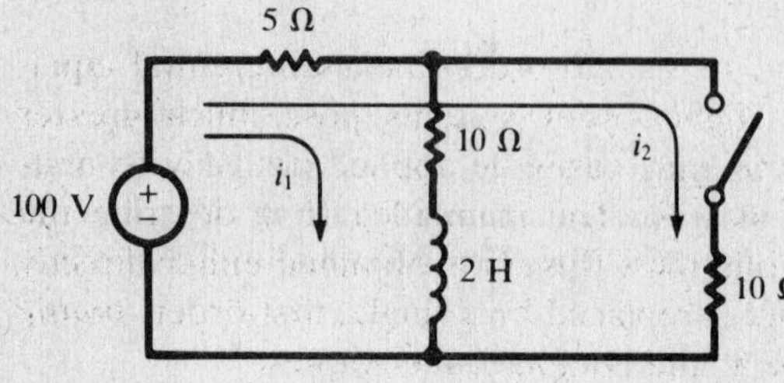

Fig. 14-27

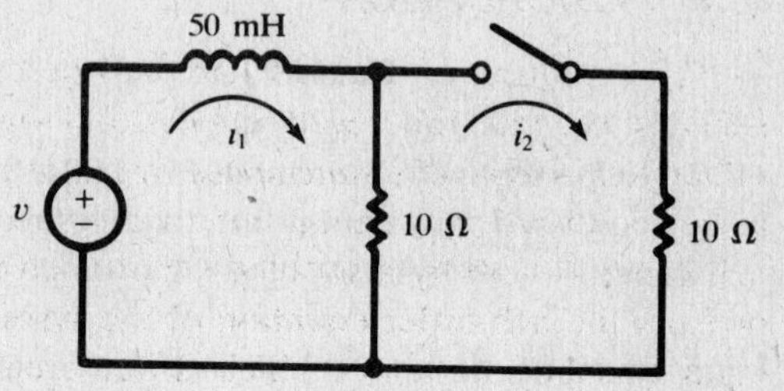

Fig. 14-28

Chapter 15

State Variable Analysis

15.1 INTRODUCTION

The solution of systems (circuits) describable by first- or second-order linear differential equations is by this time well understood by the reader. Higher-order systems pose much greater difficulties. However, *state space*, or *state variable*, *analysis* may easily be applied not only to first- and second-order systems, but to higher-order systems as well. The main idea is to describe the nth-order system by a set of n simultaneous first-order equations; this set is obtained either directly or from the nth-order equation of the system. Then the set is replaced by a single, first-order, *matrix* equation, the solution of which is well adapted to numerical and computer techniques.

15.2 NETWORK TOPOLOGY

While state equations for many simple networks may be written by inspection, their formulation for more complicated networks requires some concepts in network topology, or *planar graph theory*. Only connected graphs will be considered here.

We are already familiar with the terms *node*, *branch*, *path* (a sequence of connected branches in which no node occurs more than once), *loop* (a closed path), and *mesh* (a loop without interior loops; a "window" of the network). The following definitions will also be required.

***Tree*:** A set of branches that connects all nodes of the network but contains no loop. (More precisely, this structure is called a *spanning subtree*.) The number of branches in a tree is one less than the number of nodes in the network. Any tree provides a unique path between any two nodes. A given network may have a number of distinct trees.

Cotree*:** The complement of a tree; that is, the set of all branches of the network that do not belong to the given tree. A branch of the cotree is called a ***link (of the given tree).

If a tree is augmented by any single link, the resulting structure has exactly one loop. From this fact, it can be proved that the set of loops generated by bringing the links into the tree one at a time constitutes a *basic set* of loops. The number of loops in a basic set equals the number of meshes in the network, and a basic set shares with the set of meshes the property that *any loop in the network can be constructed from loops belonging to the set.* This being the case, a judiciously chosen basic set might well provide network equations in simpler form than would be obtained directly from the meshes—a possibility already mentioned in the discussion of the mesh current method in Chapters 4 and 8. The subject of Section 15.3 is, in effect, the selection of such a basic set.

EXAMPLE 15.1 Figure 15-1(a) depicts a simple network; its graph is given in Fig. 15-1(b). The graph has 16 trees, 12 of the type shown in Fig. 15-2(a) and 4 of the type in Fig. 15-2(b). Beside either tree in Fig. 15-2 is indicated the basic set of loops associated with it. Note that the basic set in Fig. 15-2(b) coincides with the set of meshes. This is exceptional; normally, the set of meshes cannot be obtained from any tree.

15.3 STATE EQUATIONS IN NORMAL FORM

Following is a sequence of rules whereby the first-order state equations for a given network may be obtained. There do exist networks, of complicated topology, for which these rules fail and another approach (e.g., mesh currents) must be employed.

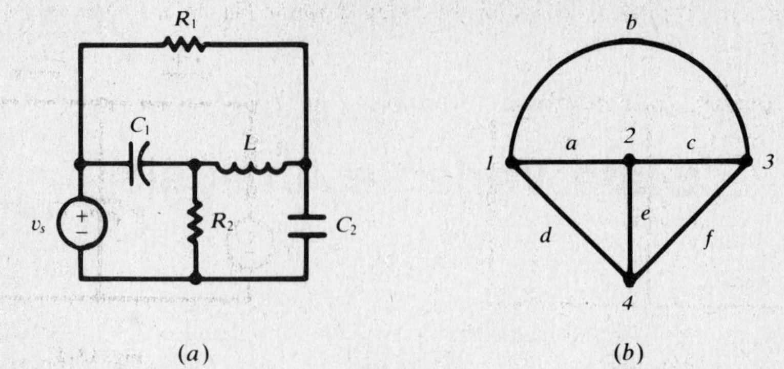

(a) (b)

Fig. 15-1

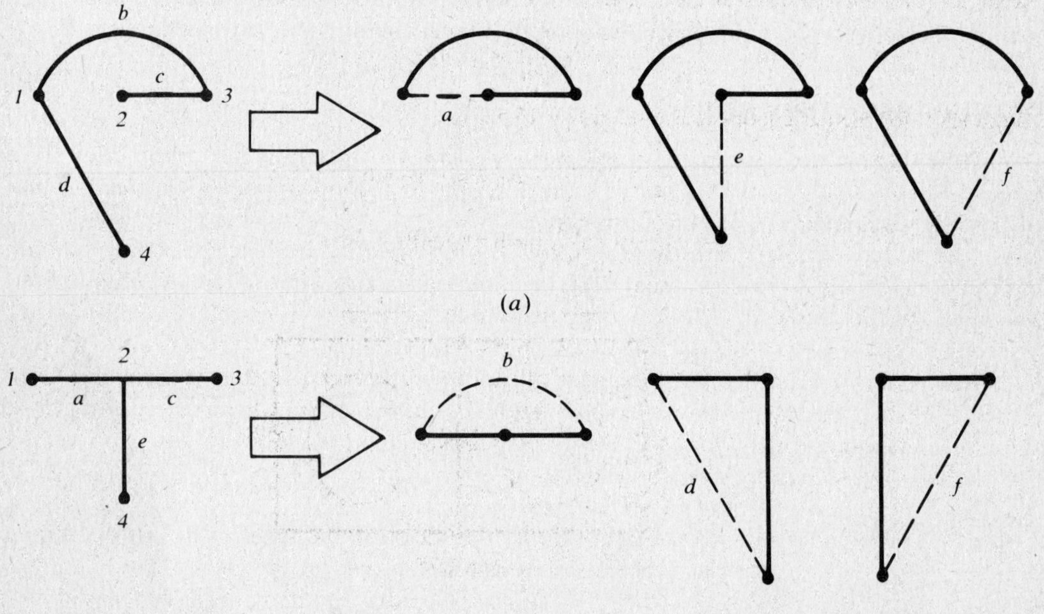

(a)

(b)

Fig. 15-2

1. In the graph corresponding to the circuit, choose a *normal tree*, one that contains all voltage sources and the maximum possible number of capacitors. All current sources and inductors are left for the cotree. Resistors may be in either the tree or the cotree. As far as possible, control voltages should also be in the tree, control currents in the cotree.

2. To each capacitor assign a voltage and mark its polarity; to each inductor assign a current and specify its direction. These capacitor voltages and inductor currents are the state variables.

3. Using KCL, write a node equation at each capacitor. Using KVL, write a loop equation for each basic loop composed of an inductor link and a path in the normal tree.

4. If resistor voltages occur in the KVL equations of Rule 3, use KCL to equate V_R/R to a sum of link currents. Similarly, if resistor currents occur in the KCL equations, use KVL to set $i_R R$ equal to a sum of basic loop voltages.

5. Substitute the expressions of Rule 4 into the equations of Rule 3, thereby obtaining the set of normal equations.

EXAMPLE 15.2 Write the normal equations for the circuit shown in Fig. 15-3.

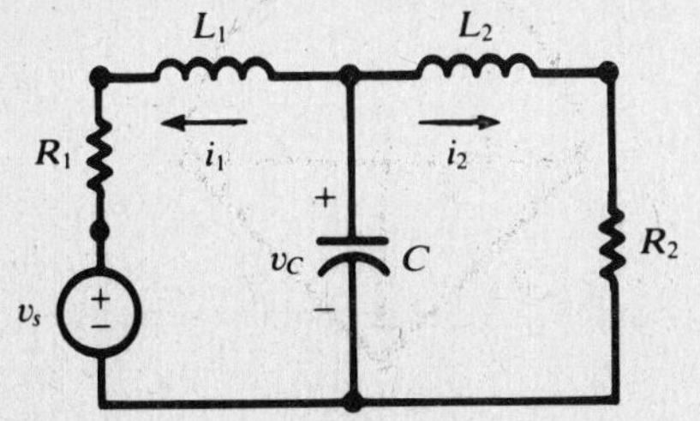

Fig. 15-3

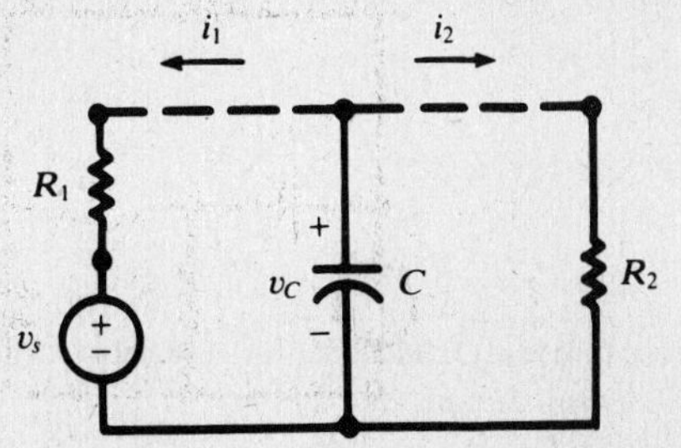

Fig. 15-4

1. Select a tree with all the capacitors and none of the inductors (Fig. 15-4).
2. Use the tree-branch capacitor voltage v_C and the link inductor currents i_1 and i_2 as state variables. The reference direction for each state variable is shown in Fig. 15-4.
3. Write KCL for the capacitor at the + end:

$$C\frac{dv_C}{dt} + i_1 + i_2 = 0 \tag{1}$$

Write KVL for the basic loops indicated in Fig. 15-5:

$$L_1\frac{di_1}{dt} + v_{R_1} + v_s - v_C = 0 \tag{2}$$

$$L_2\frac{di_2}{dt} + v_{R_2} - v_C = 0 \tag{3}$$

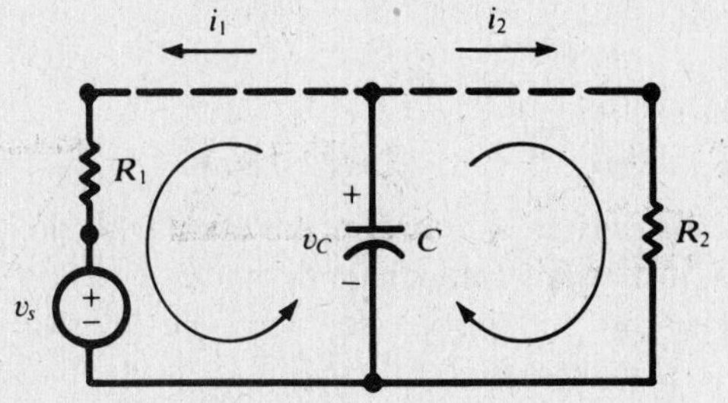

Fig. 15-5

4. In (*1*), only state variables occur, and no further simplification is needed. In (*2*) and (*3*), express v_{R_1} and v_{R_2} in terms of the state variables:

$$v_{R_1} = R_1 i_1 \qquad v_{R_2} = R_2 i_2$$

5. Now (*1*), (*2*), and (*3*) become

$$\begin{aligned} \frac{dv_C}{dt} &= \quad 0 - \frac{1}{C}i_1 - \frac{1}{C}i_2 \\ \frac{di_1}{dt} &= \frac{1}{L_1}v_C - \frac{R_1}{L_1}i_1 + 0 - \frac{v_s}{L_1} \\ \frac{di_2}{dt} &= \frac{1}{L_2}v_C + 0 - \frac{R_2}{L_2}i_2 \end{aligned} \tag{4}$$

or, in matrix form,

$$\frac{d}{dt}\begin{bmatrix} v_C \\ i_1 \\ i_2 \end{bmatrix} = \begin{bmatrix} 0 & -\dfrac{1}{C} & -\dfrac{1}{C} \\ \dfrac{1}{L_1} & -\dfrac{R_1}{L_1} & 0 \\ \dfrac{1}{L_2} & 0 & -\dfrac{R_2}{L_2} \end{bmatrix}\begin{bmatrix} v_C \\ i_1 \\ i_2 \end{bmatrix} + \begin{bmatrix} 0 \\ -\dfrac{v_s}{L_1} \\ 0 \end{bmatrix}$$

The normal tree in this case yields the meshes as the basic set of loops. It is instructive to compare the normal equations (4) with the equations given by the mesh current method, which are

$$L_1 \frac{di_1}{dt} + R_1 i_1 + v_s + \frac{1}{C}\int (i_1 + i_2)\, dt = 0$$

$$L_2 \frac{di_2}{dt} + R_2 i_2 + \frac{1}{C}\int (i_1 + i_2)\, dt = 0$$

These two integro-differential equations are equivalent to, but certainly more complex than, the three first-order differential equations (4).

15.4 STATE MATRIX DIFFERENTIAL EQUATION

The behavior or response of most continuous linear systems, such as passive electrical networks, can be described by classical means in the form of a differential equation in the time domain:

$$\frac{d^n y}{dt^n} + b_{n-1}\frac{d^{n-1}y}{dt^{n-1}} + \cdots + b_2 \frac{d^2 y}{dt^2} + b_1 \frac{dy}{dt} + b_0 y = f(t) \tag{5}$$

Here, $f(t)$ represents the forcing function. By setting

$$y \equiv x_1 \qquad \frac{dy}{dt} \equiv x_2 \qquad \frac{d^2y}{dt^2} \equiv x_3 \qquad \cdots \qquad \frac{d^{n-1}y}{dt^{n-1}} \equiv x_n$$

we may replace (5) by the first-order system (going over to the dot notation for the time derivative)

$$\begin{aligned}
\dot{x}_1 &= 0x_1 + 1x_2 + 0x_3 + 0x_4 + \cdots + 0x_n \\
\dot{x}_2 &= 0x_1 + 0x_2 + 1x_3 + 0x_4 + \cdots + 0x_n \\
&\cdots\cdots\cdots\cdots\cdots\cdots\cdots\cdots \\
\dot{x}_{n-1} &= 0x_1 + 0x_2 + 0x_3 + 0x_4 + \cdots + 1x_n \\
\dot{x}_n &= -b_0 x_1 - b_1 x_2 - b_2 x_3 - b_3 x_4 - \cdots - b_{n-1}x_n + f(t)
\end{aligned} \tag{6}$$

The method of Section 15.3 is designed to produce a system of the form (6) directly.

Denoting by $\mathbf{A}$ the $n \times n$ matrix of coefficients of the x's, and by $\mathbf{f}(t)$ the $n \times 1$ matrix (column vector) having $f(t)$ in the last row and zeros elsewhere, we rewrite (6) as a single, first-order, matrix differential equation for the *state vector* $\mathbf{x}(t)$:

$$\dot{\mathbf{x}} = \mathbf{A}\mathbf{x} + \mathbf{f}(t) \tag{7}$$

In (7), the matrix $\mathbf{A}$ is called the *characteristic* or *system matrix* and $\mathbf{f}(t)$ is the *forcing vector.* In what follows we shall allow $\mathbf{A}$ to be an arbitrary constant matrix and $\mathbf{f}(t)$ to have n distinct time functions as its components. The state matrix differential equation (7) can then represent linear systems having more than one forcing function.

15.5 SOLUTION OF THE MATRIX EQUATION

The solution of (7) proceeds by analogy with that of an ordinary linear differential equation of first order.

EXAMPLE 15.3 Solve

$$\frac{dx}{dt} - ax = f(t) \tag{8}$$

where a is a constant and where $x(0^+)$ is prescribed.

Either classical or Laplace transtorm techniques are available. Choosing the latter,

$$\mathbf{s}\mathbf{X}(\mathbf{s}) - x(0^+) - a\mathbf{X}(s) = \mathbf{F}(\mathbf{s})$$

or

$$\mathbf{X}(\mathbf{s}) = \frac{1}{\mathbf{s} - a}x(0^+) + \frac{1}{\mathbf{s} - a}\mathbf{F}(\mathbf{s}) \tag{9}$$

Then, since

$$e^{at} = \mathscr{L}^{-1}\left[\frac{1}{s-a}\right] \tag{10}$$

we obtain, using line 17 of Table 14-1,

$$x(t) = e^{at}x(0^+) + \int_0^t e^{a(t-\tau)}f(\tau)\,d\tau \tag{11}$$

The Laplace transform of (7) is

$$s\mathbf{X}(s) - \mathbf{x}(0^+) = \mathbf{A}\mathbf{X}(s) + \mathbf{F}(s) \qquad \text{or} \qquad (s\mathbf{I} - \mathbf{A})\mathbf{X}(s) = \mathbf{x}(0^+) + \mathbf{F}(s)$$

where the vectors $\mathbf{X}(s)$ and $\mathbf{F}(s)$ have as their kth components ($k = 1, 2, \ldots, n$) the ordinary Laplace transforms of $x_k(t)$ and $f_k(t)$, respectively, and where $s\mathbf{I} - \mathbf{A}$ is an $n \times n$ matrix function of s. Using the inverse matrix, $(s\mathbf{I} - \mathbf{A})^{-1}$, we solve for $\mathbf{X}(s)$:

$$\mathbf{X}(s) = (s\mathbf{I} - \mathbf{A})^{-1}\mathbf{x}(0^+) + (s\mathbf{I} - \mathbf{A})^{-1}\mathbf{F}(s) \tag{9a}$$

Then, defining the *exponential function of the matrix* $\mathbf{A}t$ by

$$e^{\mathbf{A}t} \equiv \mathscr{L}^{-1}[(s\mathbf{I} - \mathbf{A})^{-1}] = \mathbf{I} + \mathbf{A}t + \mathbf{A}^2\frac{t^2}{2!} + \mathbf{A}^3\frac{t^3}{3!} + \cdots \tag{10a}$$

we can invert (9a), obtaining

$$\mathbf{x}(t) = e^{\mathbf{A}t}\mathbf{x}(0^+) + \int_0^t e^{\mathbf{A}(t-\tau)}\mathbf{f}(\tau)\,d\tau \tag{11a}$$

The infinite series in (10a) converges for every $\mathbf{A}$ and all t, defining $e^{\mathbf{A}t}$ as an $n \times n$ matrix, the *state transition matrix* for the system. To emphasize its time dependence, we write

$$\boldsymbol{\phi}(t) \equiv e^{\mathbf{A}t}$$

for the state transition matrix, and note the property $\boldsymbol{\phi}(t+u) = \boldsymbol{\phi}(t)\boldsymbol{\phi}(u)$. As we know from Chapter 14, the time behavior of $\boldsymbol{\phi}(t)$, and hence that of the solution (11a), will be determined by the poles of its Laplace transform, $(s\mathbf{I} - \mathbf{A})^{-1}$, in the s-domain. Now, from Appendix C,

$$(s\mathbf{I} - \mathbf{A})^{-1} = \frac{\operatorname{adj}(s\mathbf{I} - \mathbf{A})}{\det(s\mathbf{I} - \mathbf{A})}$$

so that these poles are the roots of the equation

$$\det(s\mathbf{I} - \mathbf{A}) = 0 \tag{12}$$

which is a polynomial equation of degree n in s. We call (12) the *characteristic equation* of $\mathbf{A}$, and the n roots (counted according to multiplicity) are called the *eigenvalues* of $\mathbf{A}$. See Appendix C.

Natural and Forced Responses

The first term on the right-hand side of (11a) represents the zero-input (natural) response due to the initial conditions; if the system is stable (i.e., if all the eigenvalues of $\mathbf{A}$ have negative real parts), this solution is a pure transient. The second term represents the zero-state response due to the forcing functions.

EXAMPLE 15.4 Consider the system described by $\dot{x} + 3x = f(t)$, $x(0^+) = 4$.

(a) Let $f(t) = e^{-2t}$ (steady-state value is 0). Then,

$$\begin{aligned} x(t) &= e^{-3t}(4) + \int_0^t e^{-3(t-\tau)}e^{-2\tau}\,d\tau \\ &= \underbrace{4e^{-3t}}_{\substack{\textit{zero-input}\\ \textit{response}}} + \underbrace{(-e^{-3t} + e^{-2t})}_{\substack{\textit{zero-state}\\ \textit{response}}} \end{aligned}$$

(*b*) Let $f(t) = u(t)$ (unit step input). Then,

$$x(t) = 4e^{-3t} + \int_0^t e^{-3(t-\tau)}\,d\tau$$
$$= 4e^{-3t} + \left(-\frac{1}{3}e^{-3t} + \frac{1}{3}\right)$$

EXAMPLE 15.5 Replace Example 15.4(*b*) by $\dot{x} - 3x = u(t)$, $x(0^+) = 4$. Then,

$$x(t) = 4e^{3t} + \int_0^t e^{+3(t-\tau)}u(\tau)\,d\tau$$
$$= \underbrace{4e^{3t}}_{\substack{\text{zero-input}\\ \text{response}}} + \underbrace{\left(\frac{1}{3}e^{3t} - \frac{1}{3}\right)}_{\substack{\text{zero-state}\\ \text{response}}}$$

Here, the system is unstable, because the sole eigenvalue, $\mathbf{s} = 3$, has positive real part. The solution increases without limit as $t \to \infty$; so there is no transient, in the sense of a function that decays to zero, and no steady state.

Solved Problems

15.1 The series *RLC* circuit shown in Fig. 15-6(*a*) has zero input. Obtain the normal equations.

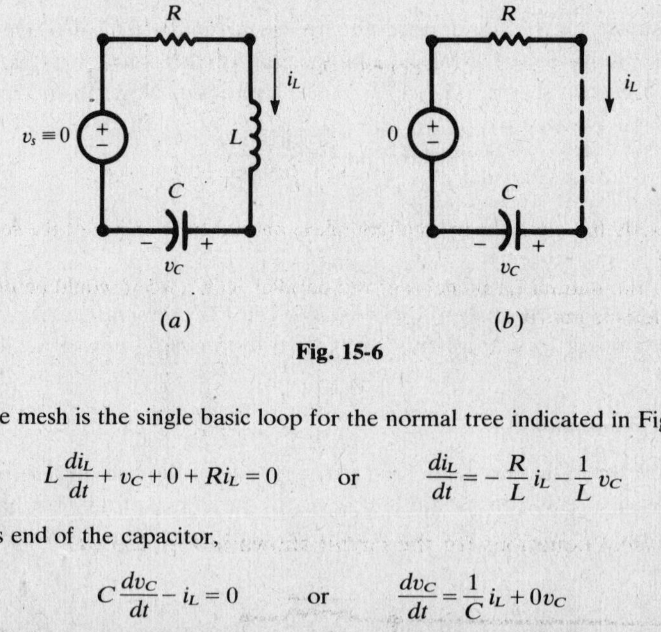

Fig. 15-6

Here, the single mesh is the single basic loop for the normal tree indicated in Fig. 15-6(*b*). By KVL,

$$L\frac{di_L}{dt} + v_C + 0 + Ri_L = 0 \qquad \text{or} \qquad \frac{di_L}{dt} = -\frac{R}{L}i_L - \frac{1}{L}v_C \qquad (1)$$

By KCL at the plus end of the capacitor,

$$C\frac{dv_C}{dt} - i_L = 0 \qquad \text{or} \qquad \frac{dv_C}{dt} = \frac{1}{C}i_L + 0v_C \qquad (2)$$

Equations (*1*) and (*2*) are the normal equations in the state variables i_L and v_C.

15.2 The parallel *RLC* circuit shown in Fig. 15-7(*a*) has zero input. Obtain the normal equations in i_L and v_C.

A normal tree is shown in Fig. 15-7(*b*); the zero-voltage link has infinite resistance. By KVL,

$$L\frac{di_L}{dt} - v_C = 0 \qquad \text{or} \qquad \frac{di_L}{dt} = 0i_L + \frac{1}{L}v_C \qquad (1)$$

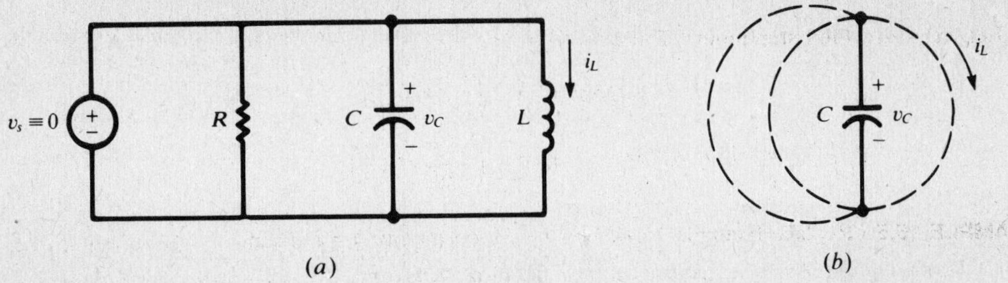

Fig. 15-7

By KCL at the plus end of the capacitor,

$$C\frac{dv_C}{dt} + i_L + \frac{v_C}{R} + 0 = 0 \qquad \text{or} \qquad \frac{dv_C}{dt} = -\frac{1}{C}i_L - \frac{1}{CR}v_C \tag{2}$$

Equations (*1*) and (*2*) are the desired normal equations.

15.3 Find the eigenvalues for the system of Problem 15.1, and verify that they are the natural frequencies of the circuit.

As found in (*1*) and (*2*) of Problem 15.1, the characteristic matrix of the system is

$$\mathbf{A} = \begin{bmatrix} -\frac{R}{L} & -\frac{1}{L} \\ \frac{1}{C} & 0 \end{bmatrix}$$

The eigenvalues are the roots of the characteristic equation,

$$\det(\mathbf{sI} - \mathbf{A}) = \begin{vmatrix} \mathbf{s} + \frac{R}{L} & \frac{1}{L} \\ -\frac{1}{C} & \mathbf{s} \end{vmatrix} = 0$$

or

$$\mathbf{s}^2 + \frac{R}{L}\mathbf{s} + \frac{1}{LC} = 0$$

But this is exactly the equation for the (complex) natural frequencies of the series *RLC* circuit, as found in Section 5.7.

Similarly, the natural frequencies of the parallel *RLC* circuit would be obtained as the eigenvalues of the characteristic matrix

$$\mathbf{A} = \begin{bmatrix} 0 & \frac{1}{L} \\ -\frac{1}{C} & -\frac{1}{CR} \end{bmatrix}$$

of Problem 15.2.

15.4 Write the normal equations for the circuit shown in Fig. 15-8(*a*).

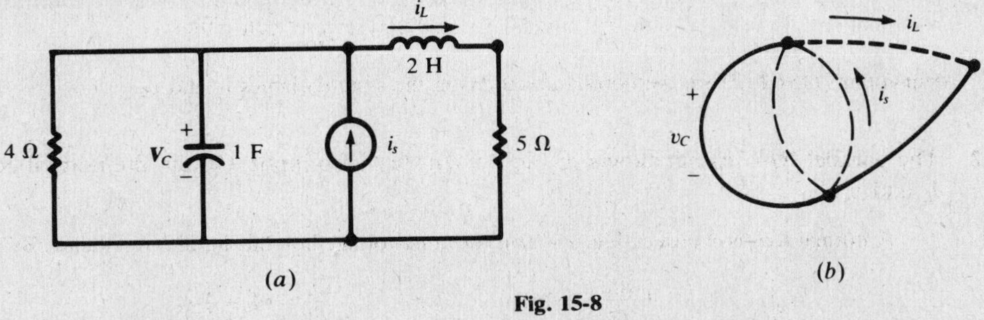

Fig. 15-8

The normal tree is shown in Fig. 15-8(*b*), with the capacitor in the tree and the inductor in the cotree. By KVL,

$$2\frac{di_L}{dt}+5i_L-v_C=0 \qquad \text{or} \qquad \frac{di_L}{dt}=-2.5\,i_L+0.5v_C \tag{1}$$

By KCL at the plus end of the capacitor,

$$1\frac{dv_C}{dt}+i_L-i_s+\frac{v_C}{4}=0 \qquad \text{or} \qquad \frac{dv_C}{dt}=-\,i_L-0.25v_C+i_s \tag{2}$$

Rewriting the normal equations (*1*) and (*2*) in matrix form,

$$\frac{d}{dt}\begin{bmatrix} i_L \\ v_C \end{bmatrix}=\begin{bmatrix} -2.5 & 0.5 \\ -1 & -0.25 \end{bmatrix}\begin{bmatrix} i_L \\ v_C \end{bmatrix}+\begin{bmatrix} 0 \\ i_s \end{bmatrix}$$

15.5 Write the normal equations for the circuit in Fig. 15-9(*a*).

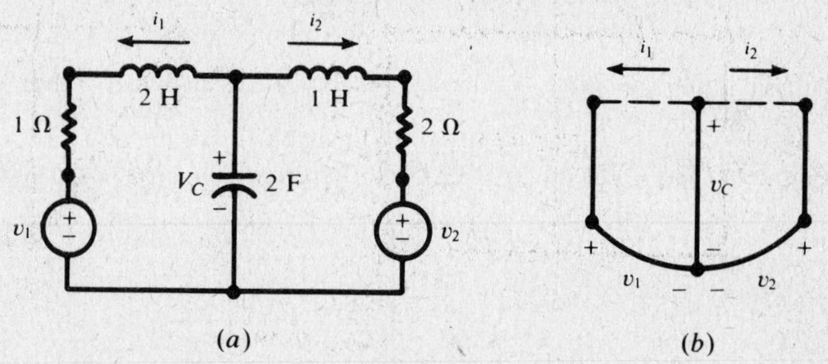

Fig. 15-9

The normal tree is shown in Fig. 15-9(*b*). By KVL in loop 1,

$$2\frac{di_1}{dt}+1i_1+v_1-v_C=0 \qquad \text{or} \qquad \frac{di_1}{dt}=-0.5\,i_1+0i_2+0.5\,v_C-0.5\,v_1 \tag{1}$$

and in loop 2,

$$1\frac{di_2}{dt}+2i_2+v_2-v_C=0 \qquad \text{or} \qquad \frac{di_2}{dt}=0i_1-2i_2+v_C-v_2 \tag{2}$$

By KCL at the plus end of the capacitor,

$$2\frac{dv_C}{dt}+i_1+i_2=0 \qquad \text{or} \qquad \frac{dv_C}{dt}=-0.5i_1-0.5i_2+0v_C \tag{3}$$

In matrix form, the normal equations are

$$\frac{d}{dt}\begin{bmatrix} i_1 \\ i_2 \\ v_C \end{bmatrix}=\begin{bmatrix} -0.5 & 0 & 0.5 \\ 0 & -2 & 1 \\ -0.5 & -0.5 & 0 \end{bmatrix}\begin{bmatrix} i_1 \\ i_2 \\ v_C \end{bmatrix}+\begin{bmatrix} -0.5v_1 \\ -v_2 \\ 0 \end{bmatrix}$$

Observe that the components of the forcing vector are not the forcing functions v_1 and v_2, but linear combinations of these functions. Such will generally be the case.

15.6 The circuit shown in Fig. 15-10(*a*) has input $v_s(t)=5e^{-2t}$ (V). Obtain the normal equations.

It is clear that the two capacitors and the voltage source cannot all be placed in a (normal) tree. Thus, in Fig. 15-10(*b*), the 0.2-F capacitor is assigned to the cotree. By KVL,

$$0.2\frac{di}{dt}-v+2i=0 \qquad \text{or} \qquad \frac{di}{dt}=-10i+5v \tag{1}$$

By KCL at the plus end of the 0.3-F capacitor,

$$0.3\frac{dv}{dt} + i - 0.2\frac{d(v_s - v)}{dt} - \frac{v_s - v}{3} = 0$$

or, since $\dot{v}_s = -2v_s$,

$$\frac{dv}{dt} = -2i - \frac{2}{3}v - \frac{2}{15}v_s \tag{2}$$

In matrix form, the normal equations are

$$\frac{d}{dt}\begin{bmatrix} i \\ v \end{bmatrix} = \begin{bmatrix} -10 & 5 \\ -2 & -2/3 \end{bmatrix}\begin{bmatrix} i \\ v \end{bmatrix} + \begin{bmatrix} 0 \\ -2v_s/15 \end{bmatrix}$$

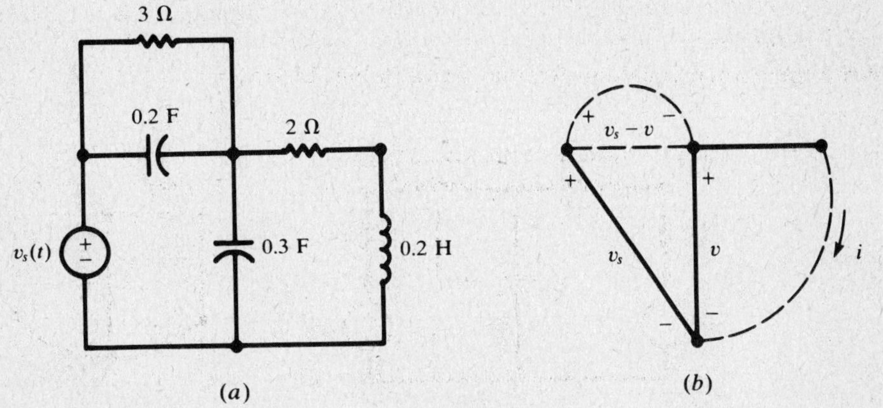

Fig. 15-10

15.7 A linear system is described by the matrix

$$\mathbf{A} = \begin{bmatrix} 1 & 2 \\ -3 & 4 \end{bmatrix}$$

Use the power-series expansion to calculate the state transition matrix at $t = 0.1$ s.

By (10*a*) of Section 15.5,

$$\boldsymbol{\phi}(t) = e^{\mathbf{A}t} = \mathbf{I} + \mathbf{A}t + \mathbf{A}^2\frac{t^2}{2!} + \mathbf{A}^3\frac{t^3}{3!} + \cdots$$

We will use only the first three terms to approximate $\boldsymbol{\phi}(0.1)$, because, as will become evident, higher-order terms become insignificant for the given values.

$$\mathbf{I} = \begin{bmatrix} 1 & 0 \\ 0 & 1 \end{bmatrix} \qquad \mathbf{A}t = \begin{bmatrix} 1 & 2 \\ -3 & 4 \end{bmatrix}(0.1) = \begin{bmatrix} 0.1 & 0.2 \\ -0.3 & 0.4 \end{bmatrix}$$

$$\mathbf{A}^2 = \mathbf{AA} = \begin{bmatrix} 1 & 2 \\ -3 & 4 \end{bmatrix}\begin{bmatrix} 1 & 2 \\ -3 & 4 \end{bmatrix} = \begin{bmatrix} (1-6) & (2+8) \\ (-3-12) & (-6+16) \end{bmatrix} = \begin{bmatrix} -5 & 10 \\ -15 & 10 \end{bmatrix}$$

$$\mathbf{A}^2\frac{t^2}{2!} = \begin{bmatrix} -5 & 10 \\ -15 & 10 \end{bmatrix}\left(\frac{0.01}{2}\right) = \begin{bmatrix} -0.025 & 0.05 \\ -0.075 & 0.05 \end{bmatrix}$$

Thus

$$\boldsymbol{\phi}(0.1) \approx \begin{bmatrix} 1 & 0 \\ 0 & 1 \end{bmatrix} + \begin{bmatrix} 0.1 & 0.2 \\ -0.3 & 0.4 \end{bmatrix} + \begin{bmatrix} -0.025 & 0.05 \\ -0.075 & 0.05 \end{bmatrix} = \begin{bmatrix} 1.075 & 0.25 \\ -0.375 & 1.45 \end{bmatrix}$$

15.8 Refer to Problem 15.7. By means of the Laplace transform, find an exact expression for $\boldsymbol{\phi}(t)$, and use it to check the approximate calculation of $\boldsymbol{\phi}(0.1)$.

We have

$$\boldsymbol{\phi}(t) = \mathscr{L}^{-1}[(s\mathbf{I} - \mathbf{A})^{-1}]$$

For the given matrix **A**,

$$\mathbf{sI} - \mathbf{A} = \begin{bmatrix} \mathbf{s}-1 & -2 \\ 3 & \mathbf{s}-4 \end{bmatrix}$$

and so

$$(\mathbf{sI}-\mathbf{A})^{-1} = \frac{\text{adj}\,(\mathbf{sI}-\mathbf{A})}{\det\,(\mathbf{sI}-\mathbf{A})} = \frac{\begin{bmatrix} \mathbf{s}-4 & 2 \\ -3 & \mathbf{s}-1 \end{bmatrix}}{\mathbf{s}^2-5\mathbf{s}+10}$$

$$= \begin{bmatrix} \dfrac{\mathbf{s}-4}{\mathbf{s}^2-5\mathbf{s}+10} & \dfrac{2}{\mathbf{s}^2-5\mathbf{s}+10} \\ \dfrac{-3}{\mathbf{s}^2-5\mathbf{s}+10} & \dfrac{\mathbf{s}-1}{\mathbf{s}^2-5\mathbf{s}+10} \end{bmatrix} \qquad (1)$$

The inverse Laplace transform of the matrix (*1*) is obtained by taking the inverse Laplace transform of each element. All elements have the same poles (the eigenvalues of **A**),

$$\mathbf{s}_1 = \frac{5}{2} + j\frac{\sqrt{15}}{2} \qquad \mathbf{s}_2 = \frac{5}{2} - j\frac{\sqrt{15}}{2} = \mathbf{s}_1^*$$

and we may use the Heaviside formula, (*3*) of Chapter 14, to find

$$\boldsymbol{\phi}(t) = \begin{bmatrix} \phi_{11} & \phi_{12} \\ \phi_{21} & \phi_{22} \end{bmatrix}$$

where

$$\phi_{11} = \mathcal{L}^{-1}\left[\frac{\mathbf{s}-4}{\mathbf{s}^2-5\mathbf{s}+10}\right] = \left(\frac{\mathbf{s}_1-4}{2\mathbf{s}_1-5}e^{\mathbf{s}_1 t}\right) + \text{(complex conjugate)}$$

$$= \frac{e^{5t/2}}{\sqrt{15}}\left(\sqrt{15}\cos\frac{\sqrt{15}}{2}t - 3\sin\frac{\sqrt{15}}{2}t\right)$$

$$\phi_{12} = \mathcal{L}^{-1}\left[\frac{2}{\mathbf{s}^2-5\mathbf{s}+10}\right] = \frac{e^{5t/2}}{\sqrt{15}}\left(4\sin\frac{\sqrt{15}}{2}t\right)$$

$$\phi_{21} = \mathcal{L}^{-1}\left[\frac{-3}{\mathbf{s}^2-5\mathbf{s}+10}\right] = \frac{e^{5t/2}}{\sqrt{15}}\left(-6\sin\frac{\sqrt{15}}{2}t\right)$$

$$\phi_{22} = \mathcal{L}^{-1}\left[\frac{\mathbf{s}-1}{\mathbf{s}^2-5\mathbf{s}+10}\right] = \frac{e^{5t/2}}{\sqrt{15}}\left(\sqrt{15}\cos\frac{\sqrt{15}}{2}t + 3\sin\frac{\sqrt{15}}{2}t\right)$$

Substitution of $t = 0.1$ gives

$$\boldsymbol{\phi}(0.1) = \begin{bmatrix} 1.068 & 0.256 \\ -0.384 & 1.452 \end{bmatrix}$$

which shows that the approximation in Problem 15.7 was quite good.

15.9 The differential equation of a certain electromechanical system is

$$\ddot{y} + 3\dot{y} + 2y = f(t)$$

Use the state-variable method to obtain the complete response of the system to the driving force $f(t) = u(t)$ (unit step function), given the initial conditions

$$y(0^+) = 0 \qquad \dot{y}(0^+) = 1$$

Choosing the state variables $x_1 = y$, $x_2 = \dot{y}$ and carrying out the reduction process of Section 15.4, we obtain the matrix state equation

$$\begin{bmatrix} \dot{x}_1 \\ \dot{x}_2 \end{bmatrix} = \begin{bmatrix} 0 & 1 \\ -2 & -3 \end{bmatrix}\begin{bmatrix} x_1 \\ x_2 \end{bmatrix} + \begin{bmatrix} 0 \\ u(t) \end{bmatrix} \qquad \text{or} \qquad \dot{\mathbf{x}} = \mathbf{A}\mathbf{x} + \mathbf{f}$$

with the initial condition $\mathbf{x}(0^+) = [0, 1]^T$. Then,

$$(\mathbf{sI}-\mathbf{A})^{-1} = \begin{bmatrix} \mathbf{s} & -1 \\ 2 & \mathbf{s}+3 \end{bmatrix}^{-1} = \begin{bmatrix} \dfrac{\mathbf{s}+3}{(\mathbf{s}+2)(\mathbf{s}+1)} & \dfrac{1}{(\mathbf{s}+2)(\mathbf{s}+1)} \\ \dfrac{-2}{(\mathbf{s}+2)(\mathbf{s}+1)} & \dfrac{\mathbf{s}}{(\mathbf{s}+2)(\mathbf{s}+1)} \end{bmatrix}$$

and $$\boldsymbol{\phi}(t)=\mathscr{L}^{-1}[(s\mathbf{I}-\mathbf{A})^{-1}]=\begin{bmatrix}2e^{-t}-e^{-2t} & e^{-t}-e^{-2t}\\ 2e^{-2t}-2e^{-t} & 2e^{-2t}-e^{-t}\end{bmatrix}$$

The solution to the matrix state equation is now given by

$$\mathbf{x}(t)=\boldsymbol{\phi}(t)\mathbf{x}(0^+)+\int_0^t \boldsymbol{\phi}(t-\tau)\mathbf{f}(\tau)\,d\tau$$

Only the first component of $\mathbf{x}(t)$ is of interest; thus,

$$x_1(t)=y(t)=[2e^{-t}-e^{-2t},\ e^{-t}-e^{-2t}]\begin{bmatrix}0\\1\end{bmatrix}+\int_0^t[2e^{-(t-\tau)}-e^{-2(t-\tau)},\ e^{-(t-\tau)}-e^{-2(t-\tau)}]\begin{bmatrix}0\\u(\tau)\end{bmatrix}d\tau$$

$$=e^{-t}-e^{-2t}+\int_0^t\{e^{-(t-\tau)}-e^{-2(t-\tau)}\}(1)\,d\tau$$

$$=\frac{1}{2}-\frac{1}{2}e^{-2t}$$

15.10 Solve the state equations

$$\frac{di}{dt}=-4i+\tfrac{1}{2}e^{-t}\quad \text{(A/s)}$$

$$\frac{dv}{dt}=-3v+e^{-t}\quad \text{(V/s)}$$

subject to the initial conditions $i(0^+)=1$ A and $v(0^+)=2$ V.

The equations in matrix form are

$$\frac{d}{dt}\begin{bmatrix}i\\v\end{bmatrix}=\begin{bmatrix}-4 & 0\\0 & -3\end{bmatrix}\begin{bmatrix}i\\v\end{bmatrix}+\begin{bmatrix}\frac{1}{2}e^{-t}\\e^{-t}\end{bmatrix}\qquad\text{with}\qquad \begin{bmatrix}i(0^+)\\v(0^+)\end{bmatrix}=\begin{bmatrix}1\\2\end{bmatrix}$$

Here $$\mathbf{A}=\begin{bmatrix}-4 & 0\\0 & -3\end{bmatrix}\qquad (s\mathbf{I}-\mathbf{A})^{-1}=\begin{bmatrix}s+4 & 0\\0 & s+3\end{bmatrix}^{-1}=\begin{bmatrix}\frac{1}{s+4} & 0\\0 & \frac{1}{s+3}\end{bmatrix}$$

and $$\boldsymbol{\phi}(t)=\mathscr{L}^{-1}[(s\mathbf{I}-\mathbf{A})^{-1}]=\begin{bmatrix}e^{-4t} & 0\\0 & e^{-3t}\end{bmatrix}$$

Consequently,

$$\begin{bmatrix}i(t)\\v(t)\end{bmatrix}=\begin{bmatrix}e^{-4t} & 0\\0 & e^{-3t}\end{bmatrix}\begin{bmatrix}1\\2\end{bmatrix}+\int_0^t\begin{bmatrix}e^{-4(t-\tau)} & 0\\0 & e^{-3(t-\tau)}\end{bmatrix}\begin{bmatrix}\frac{1}{2}e^{-\tau}\\e^{-\tau}\end{bmatrix}d\tau$$

$$=\begin{bmatrix}\frac{1}{6}e^{-t}+\frac{5}{6}e^{-4t} & \text{(A)}\\ \frac{1}{2}e^{-t}+\frac{3}{2}e^{-3t} & \text{(V)}\end{bmatrix}$$

Supplementary Problems

15.11 Write the state matrix equation for the circuit of Fig. 15-11, employing the indicated state variables.

Ans. $$\frac{d}{dt}\begin{bmatrix}v_x\\v_y\\i_z\end{bmatrix}=\begin{bmatrix}-\frac{1}{2} & 0 & \frac{1}{2}\\0 & -\frac{1}{2} & -\frac{1}{2}\\-1 & 1 & 0\end{bmatrix}\begin{bmatrix}v_x\\v_y\\i_z\end{bmatrix}+\begin{bmatrix}0\\\frac{1}{2}v_s\\0\end{bmatrix}.$$

15.12 Solve for the inductor current $i_L(t)$ in the circuit of Fig. 15-12.

Ans. $i_L(t)=-3\left(1-\frac{20}{3}t-e^{-20t/3}\right)u(t)$ (A)

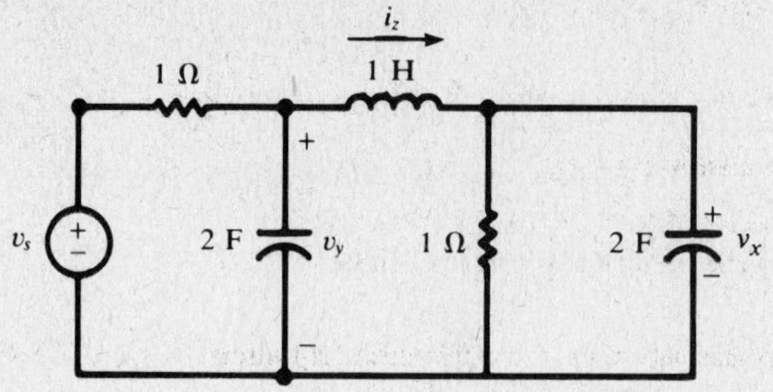

Fig. 15-11

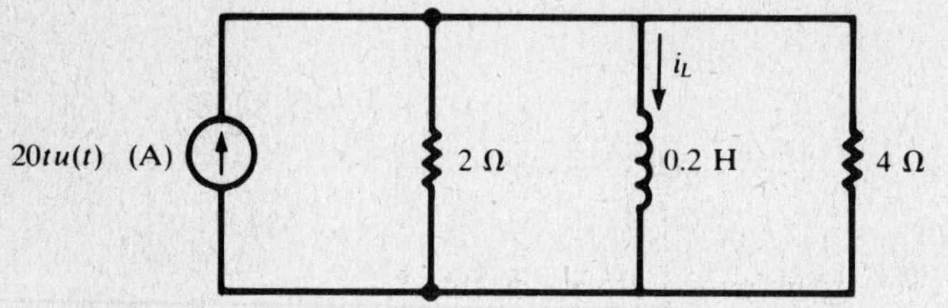

Fig. 15-12

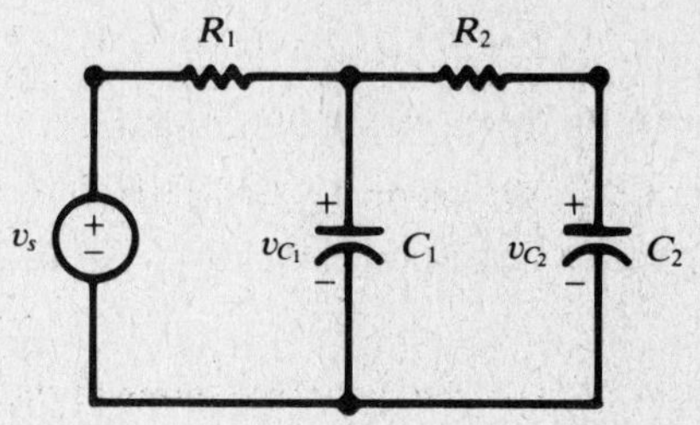

Fig. 15-13

15.13 For the circuit shown in Fig. 15-13, write the state equations in matrix form.

Ans.
$$\begin{bmatrix} \dot{v}_{C_1} \\ \dot{v}_{C_2} \end{bmatrix} = \begin{bmatrix} -\dfrac{1}{R_1C_1} - \dfrac{1}{R_2C_1} & \dfrac{1}{R_2C_1} \\ \dfrac{1}{R_2C_2} & -\dfrac{1}{R_2C_2} \end{bmatrix} \begin{bmatrix} v_{C_1} \\ v_{C_2} \end{bmatrix} + \begin{bmatrix} \dfrac{v_s}{R_1C_1} \\ 0 \end{bmatrix}$$

15.14 For the bridge network shown in Fig. 15-14, obtain the characteristic matrix and the forcing vector, using state variables $x_1 = i_L$, $x_2 = v_C$.

Ans.
$$\mathbf{A} = \begin{bmatrix} \dfrac{-2}{C(R_1+R_2)} & \dfrac{R_2 - R_1}{C(R_1+R_2)} \\ \dfrac{R_1 - R_2}{L(R_1+R_2)} & \dfrac{-2R_1R_2}{L(R_1+R_2)} \end{bmatrix}, \qquad \mathbf{f}(t) = \begin{bmatrix} \dfrac{v_1}{C(R_1+R_2)} + \dfrac{v_2}{C(R_1+R_2)} \\ \dfrac{R_2v_1}{L(R_1+R_2)} - \dfrac{R_1v_2}{L(R_1+R_2)} \end{bmatrix}$$

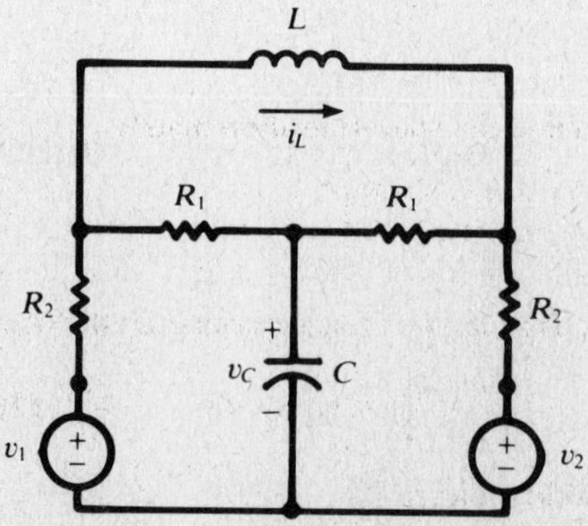

Fig. 15-14

15.15 A system matrix is given by

$$\mathbf{A} = \begin{bmatrix} -1/2 & -5/2 \\ 1/2 & -7/2 \end{bmatrix}$$

Determine the state transition matrix.

Ans. $\boldsymbol{\phi}(t) = \begin{bmatrix} \frac{5}{4}e^{-t} - \frac{1}{4}e^{-3t} & \frac{5}{4}e^{-3t} - \frac{5}{4}e^{-t} \\ \frac{1}{4}e^{-t} - \frac{1}{4}e^{-3t} & \frac{5}{4}e^{-3t} - \frac{1}{4}e^{-t} \end{bmatrix}$

15.16 A dynamical system is described by the differential equation

$$\dddot{y} + 4\ddot{y} + 5\dot{y} + 2y = v$$

Write the matrix state equation.

Ans. $\begin{bmatrix} \dot{x}_1 \\ \dot{x}_2 \\ \dot{x}_3 \end{bmatrix} = \begin{bmatrix} 0 & 1 & 0 \\ 0 & 0 & 1 \\ -2 & -5 & -4 \end{bmatrix} \begin{bmatrix} x_1 \\ x_2 \\ x_3 \end{bmatrix} + \begin{bmatrix} 0 \\ 0 \\ v \end{bmatrix}$

15.17 The characteristic matrix of a system is

$$\mathbf{A} = \begin{bmatrix} 0.2 & -0.05 \\ -0.1 & 0.12 \end{bmatrix}$$

Use the first three terms of the power-series expansion to calculate $\boldsymbol{\phi}(1) = e^{\mathbf{A}}$.

Ans. $e^{\mathbf{A}} \approx \begin{bmatrix} 1.245 & -0.066 \\ -0.132 & 1.1394 \end{bmatrix}$

15.18 The circuit of Fig. 15-15 has inputs v_x and v_y, and state variables v_C and i_L. Write the matrix state equation.

Ans. $\frac{d}{dt}\begin{bmatrix} v_C \\ i_L \end{bmatrix} = \begin{bmatrix} -8 & 3 \\ -5 & 0 \end{bmatrix} \begin{bmatrix} v_C \\ i_L \end{bmatrix} + \begin{bmatrix} 8v_y - 8v_x \\ 5v_y \end{bmatrix}$

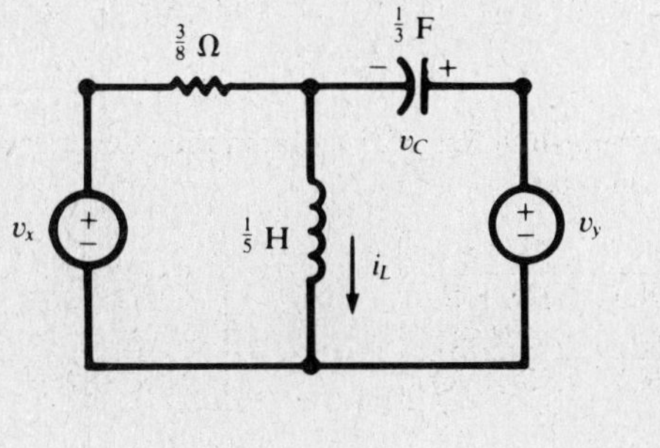

Fig. 15-15

15.19 For Problem 15.18, determine the state transition matrix.

Ans. $\boldsymbol{\phi}(t) = \begin{bmatrix} \frac{5}{2}e^{-5t} - \frac{3}{2}e^{-3t} & \frac{3}{2}e^{-3t} - \frac{3}{2}e^{-5t} \\ \frac{5}{2}e^{-5t} - \frac{5}{2}e^{-3t} & \frac{5}{2}e^{-3t} - \frac{3}{2}e^{-5t} \end{bmatrix}$

15.20 The natural response of a certain system is described by the homogeneous state equations

$$\frac{dy_1}{dt} + 7y_1 - y_2 = 0 \qquad \frac{dy_2}{dt} + 12y_1 = 0$$

Determine the state transition matrix.

Ans. $\boldsymbol{\phi}(t) = \begin{bmatrix} 4e^{-4t} - 3e^{-3t} & e^{-3t} - e^{-4t} \\ 12e^{-4t} - 12e^{-3t} & 4e^{-3t} - 3e^{-4t} \end{bmatrix}$

15.21 In Fig. 15-16, the switch is closed at $t=0$, when the instantaneous charge on the capacitor is zero. Use state-variable techniques to find the capacitor voltage $v_C(t)$.

Ans. $v_C(t)=\frac{50}{26}(5\sin 10t-\cos 10t+e^{-50t})$ (V)

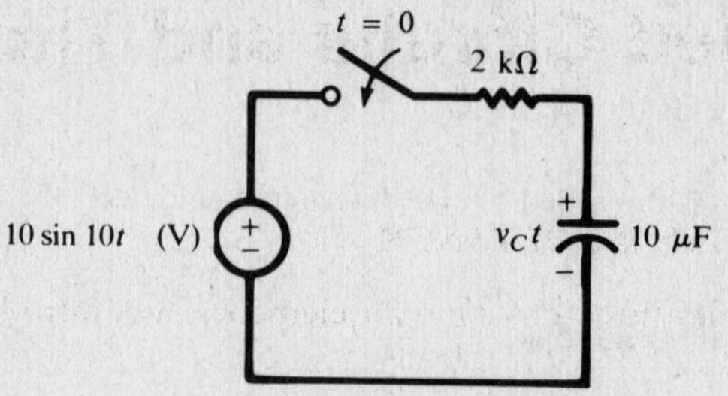

Fig. 15-16

15.22 Rewrite (*11a*), the solution of the state matrix equation, if the initial conditions are prescribed at $t=t_0$ instead of at $t=0$.

Ans. $\mathbf{x}(t)=e^{\mathbf{A}(t-t_0)}\mathbf{x}(t_0^+)+\int_{t_0}^{t} e^{\mathbf{A}(t-\tau)}f(\tau)\,d\tau \quad (t>t_0)$

Chapter 16

Coupled Circuits and Transformers

16.1 SELF- AND MUTUAL INDUCTANCE

Inductance, one of the three basic circuit elements, was introduced, in Table 2-1, via the voltage equation

$$v_L = L\frac{di}{dt}$$

where L is in H. Voltage v_L is the result of the time-variation of the magnetic flux ϕ which links one current path with another (Fig. 16-1). The effect is most pronounced when the conductors are arranged in series loops packed tightly together to enhance the linking of the flux, a construction called a *coil.* Resistance accumulates with each turn, so that in a practical sense pure inductance L cannot be achieved.

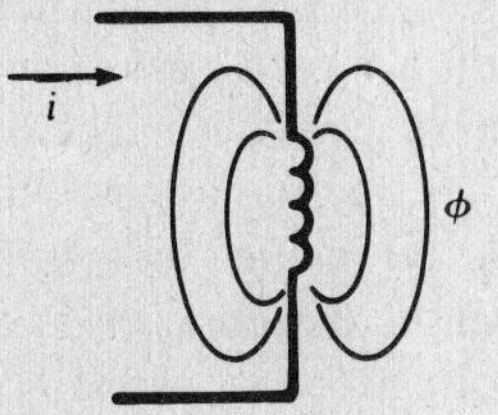

Fig. 16-1

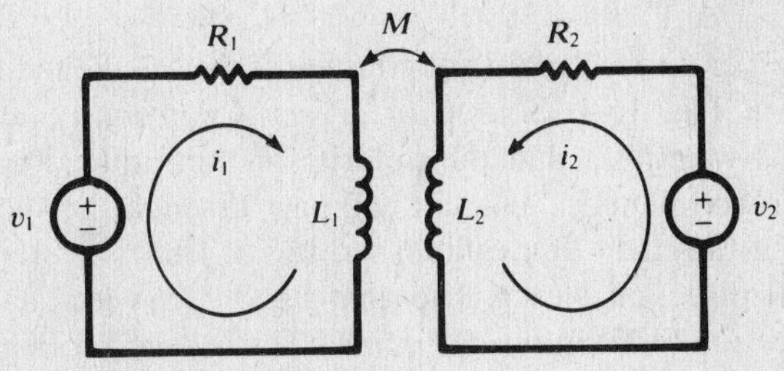

Fig. 16-2

Two conductors from different circuits in close proximity to each other are magnetically coupled to a degree that depends upon the physical arrangement and the rates of change of the currents. This coupling is increased when one coil is wound over another. If, in addition, a soft-iron core provides a path for the magnetic flux, the coupling is maximized. (However, the presence of iron can introduce nonlinearity.)

Figure 16-2 illustrates magnetic coupling of two circuits, which is measured by the *mutual inductance M.* Like L, M carries the units H. Because of the coupling, the equation of either circuit contains a term depending on the current change in the other circuit.

Time Domain	***Frequency Domain***
$R_1 i_1 + L_1\frac{di_1}{dt} \pm M\frac{di_2}{dt} = v_1$	$(R_1 + j\omega L_1)\mathbf{I}_1 \pm j\omega M\mathbf{I}_2 = \mathbf{V}_1$
$R_2 i_2 + L_2\frac{di_2}{dt} \pm M\frac{di_1}{dt} = v_2$	$(R_2 + j\omega L_2)\mathbf{I}_2 \pm j\omega M\mathbf{I}_1 = \mathbf{V}_2$ (*1*)

Section 16.2 will show how the signs on the M-terms are found. The frequency-domain equations have been written on the assumption that sources v_1 and v_2 are sinusoids, of the same frequency.

16.2 ANALYSIS OF COUPLED COILS

Polarities in Close Coupling

In Fig. 16-3, two coils are shown on a common core which channels the magnetic flux ϕ. This arrangement results in *close coupling*, of which more will be said in Section 16.4. To determine the proper signs on the voltages of mutual inductance, apply the right-hand rule to each coil: If the fingers wrap around in the direction of the assumed current, the thumb points in the direction of the flux. Resulting positive directions for ϕ_1 and ϕ_2 are shown on the figure. If fluxes ϕ_1 and ϕ_2 aid one another, then the signs on the voltages of mutual inductance are the same as the signs on the voltages of self-inductance. Thus, the plus sign would be written in all four equations (*1*). In Fig. 16-3, ϕ_1 and ϕ_2 oppose each other; consequently, the equations (*1*) would be written with the minus sign.

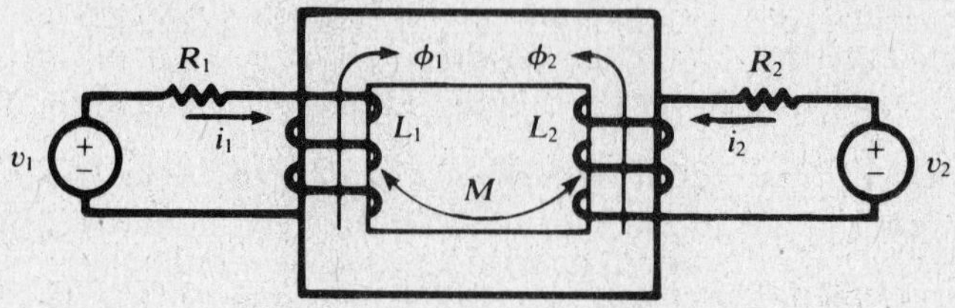

Fig. 16-3

Natural Current

Further understanding of coupled coils is achieved from consideration of a passive second loop as shown in Fig. 16-4. Source v_1 drives a current i_1, with a corresponding flux ϕ_1 as shown. Now, *Lenz's law* implies that the polarity of the induced voltage in the second circuit is such that if the circuit is completed, a current will pass through the second coil in such a direction as to create a flux opposing the main flux established by i_1. That is, when the switch is closed in Fig. 16-4, flux ϕ_2 will have the direction shown. The right-hand rule, with the thumb pointing in the direction of ϕ_2, provides the direction of the *natural current* i_2. The induced voltage is the driving voltage for the second circuit, as suggested in Fig. 16-5; this voltage is present whether or not the circuit is closed. When the switch is closed, current i_2 is established, with a positive direction as shown.

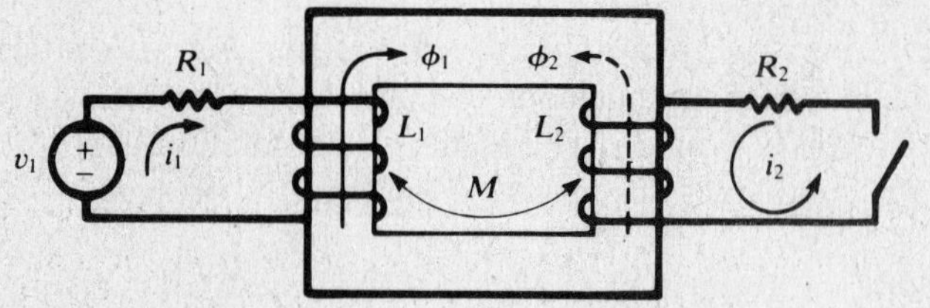

Fig. 16-4

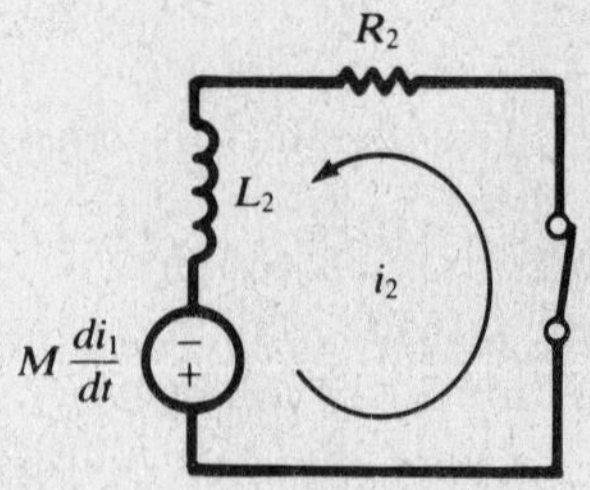

Fig. 16-5

EXAMPLE 16.1 Suppose the switch in the passive loop to be closed at an instant ($t = 0$) when $i_1 = 0$. For $t > 0$, the equation of the passive loop is (see Fig. 16-5)

$$R_2 i_2 + L_2 \frac{di_2}{dt} - M\frac{di_1}{dt} = 0$$

while that of the active loop is

$$R_1 i_1 + L_1 \frac{di_1}{dt} - M\frac{di_2}{dt} = v_1$$

Taking the Laplace transforms of the two equations, under the initial conditions $i_1(0^+) = i_2(0^+) = 0$, and eliminating $\mathbf{I}_1(\mathbf{s})$, we find

$$\mathbf{H}(\mathbf{s}) \equiv \frac{\text{response}}{\text{excitation}} = \frac{\mathbf{I}_2(\mathbf{s})}{\mathbf{V}_1(\mathbf{s})} = \frac{M\mathbf{s}}{(L_1L_2 - M^2)\mathbf{s}^2 + (R_1L_2 + R_2L_1)\mathbf{s} + R_1R_2}$$

and from the poles of $\mathbf{H}(\mathbf{s})$ we have the natural frequencies of i_2 (cf. Section 13.4).

Dot Rule

The sign on a voltage of mutual inductance can be determined if the winding sense is shown on the circuit diagram, as in Figs. 16-3 and 16-4. To simplify the problem of obtaining the correct sign, the coils are marked with dots at the terminals which are instantaneously of the same polarity.

To assign the dots to a pair of coupled coils, select a current direction in one coil and place a dot at the terminal where this current *enters* the winding. Determine the corresponding flux by application of the right-hand rule [see Fig. 16-6(*a*)]. The flux of the other winding, according to Lenz's law, opposes the first flux. Use the right-hand rule to find the natural current direction corresponding to this second flux [see Fig. 16-6(*b*)]. Now place a dot at the terminal of the second winding where the natural current *leaves* the winding. This terminal is positive simultaneously with the terminal of the first coil where the initial current entered. With the instantaneous polarity of the coupled coils given by the dots, the pictorial representation of the core with its winding sense is no longer needed, and the coupled coils may be illustrated as in Fig. 16-6(*c*). The following *dot rule* may now be used:

(1) when the assumed currents both enter or both leave a pair of coupled coils by the dotted terminals, the signs on the M-terms will be the same as the signs on the L-terms; but

(2) if one current enters by a dotted terminal while the other leaves by a dotted terminal, the signs on the M-terms will be opposite to the signs on the L-terms.

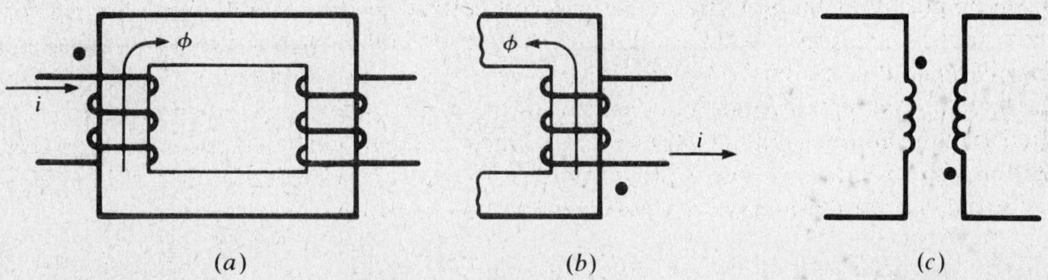

Fig. 16-6

EXAMPLE 16.2 The current directions chosen in Fig. 16-7(*a*) are such that the signs on the M-terms are opposite to the signs on the L-terms and the dots indicate the terminals with the same instantaneous polarity. Compare this to the conductively coupled circuit of Fig. 16-7(*b*), in which the two mesh currents pass through the common element in opposite directions, and in which the polarity markings are the same as the dots in the magnetically coupled circuit. The similarity becomes more apparent when we allow the shading to suggest two black boxes.

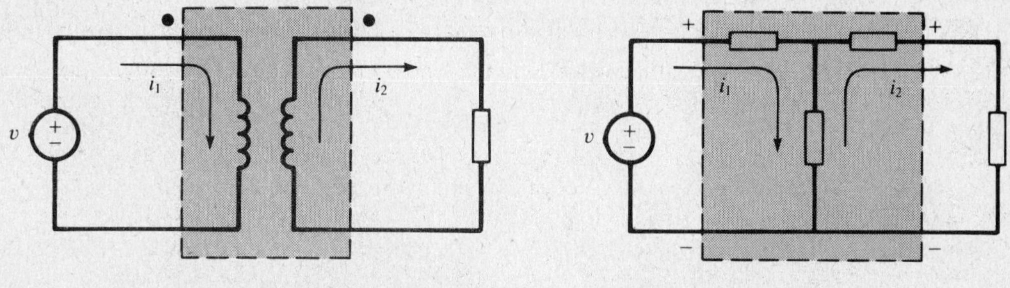

Fig. 16-7

16.3 CONDUCTIVELY COUPLED EQUIVALENT CIRCUITS

From the mesh current equations written for magnetically coupled coils, a conductively coupled equivalent circuit can be constructed. Consider the sinusoidal steady-state circuit of Fig. 16-8(*a*), with the mesh currents as shown. The corresponding equations in matrix form are

$$\begin{bmatrix} R_1 + j\omega L_1 & -j\omega M \\ -j\omega M & R_2 + j\omega L_2 \end{bmatrix} \begin{bmatrix} \mathbf{I}_1 \\ \mathbf{I}_2 \end{bmatrix} = \begin{bmatrix} \mathbf{V}_1 \\ 0 \end{bmatrix}$$

In Fig. 16-8(*b*) an inductive reactance, $X_M = \omega M$, carries the two mesh currents in opposite directions, whence

$$\mathbf{Z}_{12} = \mathbf{Z}_{21} = -j\omega M$$

in the **Z**-matrix. If now an inductance $L_1 - M$ be placed in the first loop, the mesh current equation for this loop will be

$$(R_1 + j\omega L_1)\mathbf{I}_1 - j\omega M \mathbf{I}_2 = \mathbf{V}_1$$

Similarly, $L_2 - M$ in the second loop results in the same mesh current equation as for the coupled-coil circuit. Thus, the two circuits are equivalent. The dot rule is not needed in the conductively coupled circuit, and familiar circuit techniques can be applied.

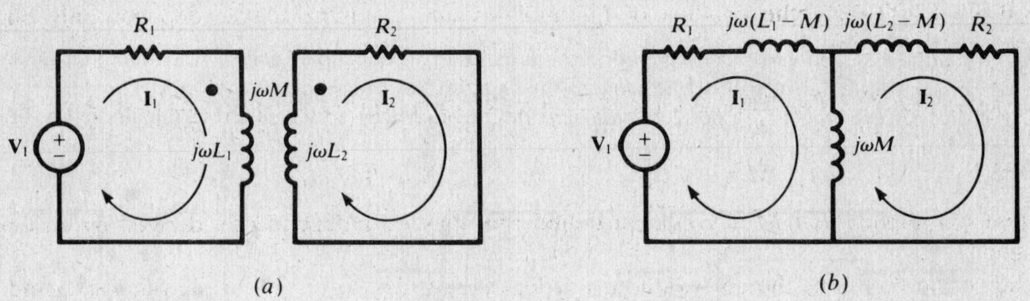

Fig. 16-8

16.4 COUPLING COEFFICIENT

By Faraday's law, a coil containing N turns, with magnetic flux ϕ linking each turn, has an induced emf (voltage) $e = N(d\phi/dt)$. A negative sign is frequently included in this equation to signal that the voltage polarity is established according to Lenz's law. By definition of self-inductance this voltage is also given by $L(di/dt)$; hence

$$L\frac{di}{dt} = N\frac{d\phi}{dt} \qquad \text{or} \qquad L = N\frac{d\phi}{di} \tag{2}$$

The unit of ϕ being the *weber*, where 1 Wb = 1 V·s, it follows from the above relation that 1 H = 1 Wb/A. Throughout this book it has been assumed that ϕ and i are proportional to each other, making

$$L = N\frac{\phi}{i} = \text{constant} \tag{2a}$$

In Fig. 16-9, the total flux ϕ_1 resulting from current i_1 through the turns N_1 consists of *leakage flux*, ϕ_{11}, and *coupling* or *linking flux*, ϕ_{12}. The induced emf in the coupled coil is given by $N_2(d\phi_{12}/dt)$. This same voltage can be written using the mutual inductance M:

$$e = M\frac{di_1}{dt} = N_2\frac{d\phi_{12}}{dt} \qquad \text{or} \qquad M = N_2\frac{d\phi_{12}}{di_1} \tag{3}$$

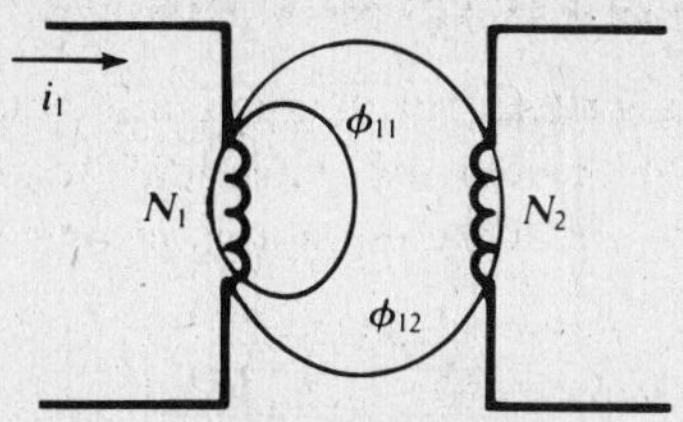

Fig. 16-9

Also, as the coupling is bilateral,

$$M = N_1 \frac{d\phi_{21}}{di_2} \tag{4}$$

The *coupling coefficient*, k, is defined as the ratio of linking flux to total flux:

$$k \equiv \frac{\phi_{12}}{\phi_1} = \frac{\phi_{21}}{\phi_2}$$

where $0 \le k \le 1$. Taking the product of (*3*) and (*4*), and assuming that k depends only on the geometry of the system,

$$M^2 = \left(N_2 \frac{d\phi_{12}}{di_1}\right)\left(N_1 \frac{d\phi_{21}}{di_2}\right) = \left(N_2 \frac{d(k\phi_1)}{di_1}\right)\left(N_1 \frac{d(k\phi_2)}{di_2}\right) = k^2\left(N_1 \frac{d\phi_1}{di_1}\right)\left(N_2 \frac{d\phi_2}{di_2}\right) = k^2 L_1 L_2$$

from which $$M = k\sqrt{L_1 L_2} \qquad \text{or} \qquad X_M = k\sqrt{X_1 X_2} \tag{5}$$

Note that (*5*) implies that $M \le \sqrt{L_1 L_2}$, a bound that may be independently derived by an energy argument.

If all of the flux links the coils without any leakage flux, then $k = 1$. On the other extreme, the coil axes may be oriented such that no flux from one can induce a voltage in the other, which results in $k = 0$. The term *close coupling* is used to describe the case where most of the flux links the coils, either by way of a magnetic core to contain the flux or by interleaving the turns of the coils directly over one another. Coils placed side-by-side without a core are loosely coupled and have correspondingly low values of k.

16.5 LINEAR TRANSFORMER

A *transformer* is a device for introducing mutual coupling between two or more electric circuits. The term *iron-core transformer* identifies the coupled coils which are wound on a magnetic core of laminated specialty steel to confine the flux and maximize the coupling. *Air-core* transformers are found in electronic and communications applications. A third group consists of coils wound over one another on a nonmetallic form, with a movable slug of magnetic material within the center for varying the coupling.

Attention here is directed to iron-core transformers where the permeability μ of the iron is assumed to be constant over the operating range of voltage and current. The development is restricted to two-winding transformers, although three and more windings on the same core are not uncommon.

In Fig. 16-10 the *primary winding*, of N_1 turns, is connected to the source voltage $\mathbf{V}_1$, and the *secondary winding*, of N_2 turns, is connected to the load impedance $\mathbf{Z}_L$. The coil resistances are shown by lumped parameters R_1 and R_2. Natural current $\mathbf{I}_2$ produces flux $\phi_2 = \phi_{21} + \phi_{22}$, while $\mathbf{I}_1$ produces $\phi_1 = \phi_{12} + \phi_{11}$. In terms of the coupling coefficient k,

$$\phi_{11} = (1-k)\phi_1 \qquad \phi_{22} = (1-k)\phi_2$$

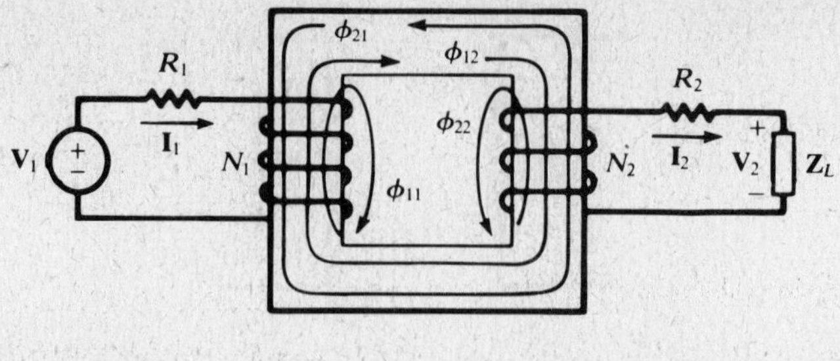

Fig. 16-10

From these flux relationships, *leakage inductances* can be related to the self-inductances:

$$L_{11} \equiv (1-k)L_1 \qquad L_{22} \equiv (1-k)L_2$$

The corresponding leakage reactances are:

$$X_{11} \equiv (1-k)X_1 \qquad X_{22} \equiv (1-k)X_2$$

It can be shown that the inductance L of an N-turn coil is proportional to N^2. Hence, for two coils wound on the same core,

$$\frac{L_1}{L_2} = \left(\frac{N_1}{N_2}\right)^2 \tag{6}$$

The flux common to both windings in Fig. 16-10 is the *mutual flux*, $\phi_m = \phi_{12} - \phi_{21}$. This flux induces the coil emfs by Faraday's law,

$$e_1 = N_1 \frac{d\phi_m}{dt} \qquad e_2 = N_2 \frac{d\phi_m}{dt}$$

Defining the *turns ratio*, $a \equiv N_1/N_2$, we obtain from these the basic equation of the linear transformer:

$$\frac{e_1}{e_2} = a \tag{7}$$

In the frequency domain, $\mathbf{E}_1/\mathbf{E}_2 = a$.

The relationship between the mutual flux and the mutual inductance can be developed by analysis of the secondary induced emf, as follows:

$$e_2 = N_2 \frac{d\phi_m}{dt} = N_2 \frac{d\phi_{12}}{dt} - N_2 \frac{d\phi_{21}}{dt} = N_2 \frac{d\phi_{12}}{dt} - N_2 \frac{d(k\phi_2)}{dt}$$

By use of (*3*) and (*2*), this may be rewritten as

$$e_2 = M \frac{di_1}{dt} - kL_2 \frac{di_2}{dt} = M \frac{di_1}{dt} - \frac{M}{a} \frac{di_2}{dt}$$

where the last step involved (*5*) and (*6*):

$$M = k\sqrt{(a^2L_2)(L_2)} = kaL_2$$

Now, defining the *magnetizing current* i_ϕ by the equation

$$i_1 = \frac{i_2}{a} + i_\phi \qquad \text{or} \qquad \mathbf{I}_1 = \frac{\mathbf{I}_2}{a} + \mathbf{I}_\phi \tag{8}$$

we have

$$e_2 = M \frac{di_\phi}{dt} \qquad \text{or} \qquad \mathbf{E}_2 = jX_M \mathbf{I}_\phi \tag{9}$$

According to (*9*), the magnetizing current may be considered to set up the mutual flux ϕ_m in the core.

In terms of coil emfs and leakage reactances, an equivalent circuit for the linear transformer may be drawn, in which the primary and secondary are effectively decoupled. This is shown in Fig. 16-11(a); for comparison, the dotted equivalent circuit is shown in Fig. 16-11(b).

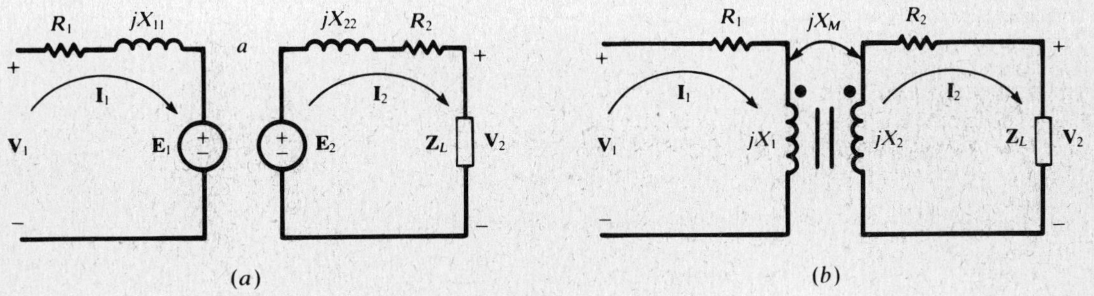

Fig. 16-11

EXAMPLE 16.3 Draw the voltage-current phasor diagram corresponding to Fig. 16-11(a), and from it derive the input impedance of the transformer.

The diagram is given in Fig. 16-12, in which θ_L denotes the phase angle of $\mathbf{Z}_L$. Note that, in accordance with (9), the induced emfs $\mathbf{E}_1$ and $\mathbf{E}_2$ lead the magnetizing current $\mathbf{I}_\phi$ by 90°. The diagram yields the three phasor equations

$$\mathbf{V}_1 = ajX_M\mathbf{I}_\phi + (R_1 + jX_{11})\mathbf{I}_1$$
$$jX_M\mathbf{I}_\phi = (\mathbf{Z}_L + R_2 + jX_{22})\mathbf{I}_2$$
$$\mathbf{I}_1 = \frac{1}{a}\mathbf{I}_2 + \mathbf{I}_\phi$$

Elimination of $\mathbf{I}_2$ and $\mathbf{I}_\phi$ among these equations results in

$$\frac{\mathbf{V}_1}{\mathbf{I}_1} \equiv \mathbf{Z}_{\text{in}} = (R_1 + jX_{11}) + a^2\frac{(jX_M/a)(R_2 + jX_{22} + \mathbf{Z}_L)}{(jX_M/a) + (R_2 + jX_{22} + \mathbf{Z}_L)} \tag{10}$$

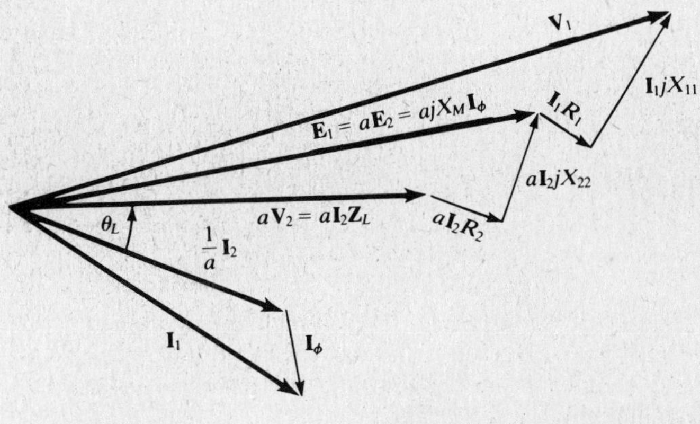

Fig. 16-12

If, instead, the mesh current equations for Fig. 16-11(b) are used to derive $\mathbf{Z}_{\text{in}}$, the result is

$$\mathbf{Z}_{\text{in}} = R_1 + jX_1 + \frac{X_M^2}{R_2 + jX_2 + \mathbf{Z}_L} \tag{10a}$$

The reader may verify the equivalence of (10) and (10a)—see Problem 16.34.

16.6 IDEAL TRANSFORMER

An *ideal transformer* is a hypothetical transformer in which there are no losses and the core has infinite permeability, resulting in perfect coupling with no leakage flux. In large power transformers the losses are so small relative to the power transferred that the relationships obtained from the ideal transformer can be very useful in engineering applications.

Referring to Fig. 16-13, the lossless condition is expressed by $\frac{1}{2}\mathbf{V}_1\mathbf{I}_1^* = \frac{1}{2}\mathbf{V}_2\mathbf{I}_2^*$ (see Section 9.3). But

$$\mathbf{V}_1 = \mathbf{E}_1 = a\mathbf{E}_2 = a\mathbf{V}_2$$

and so, a being real,

$$\frac{\mathbf{V}_1}{\mathbf{V}_2} = \frac{\mathbf{I}_2}{\mathbf{I}_1} = a \tag{11}$$

The input impedance is readily obtained from relations (*11*):

$$\mathbf{Z}_{\text{in}} = \frac{\mathbf{V}_1}{\mathbf{I}_1} = \frac{a\mathbf{V}_2}{\mathbf{I}_2/a} = a^2\frac{\mathbf{V}_2}{\mathbf{I}_2} = a^2\mathbf{Z}_L \tag{12}$$

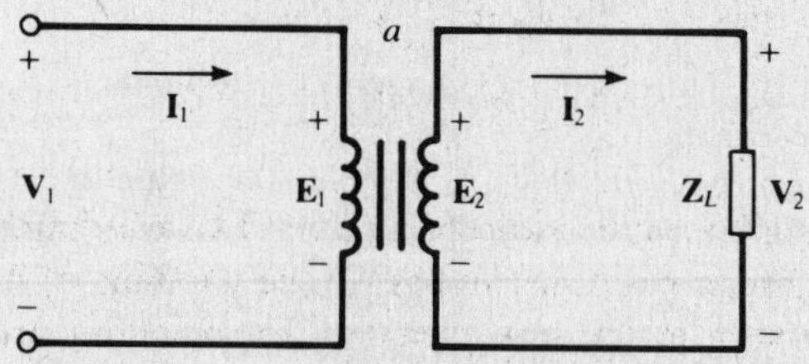

Fig. 16-13

EXAMPLE 16.4 The ideal transformer may be considered as the limiting case of the linear transformer of Section 16.5. Thus, in (*10*), set

$$R_1 = R_2 = X_{11} = X_{22} = 0$$

(no losses) and then let $X_M \to \infty$ (infinite core permeability), to obtain

$$\mathbf{Z}_{\text{in}} = \lim_{X_M \to \infty}\left[a^2 \frac{(jX_M/a)(\mathbf{Z}_L)}{(jX_M/a) + \mathbf{Z}_L}\right] = a^2\mathbf{Z}_L$$

in agreement with (*12*).

Ampere-Turn Dot Rule

Since $a = N_1/N_2$ in (*11*),

$$N_1\mathbf{I}_1 = N_2\mathbf{I}_2$$

i.e., the *ampere turns* of the primary equal the *ampere turns* of the secondary. A rule can be formulated which extends this result to transformers having more than two windings. A positive sign is applied to an ampere-turn product if the current enters the winding by the dotted terminal; a negative sign is applied if the current leaves by the dotted terminal. The *ampere-turn dot rule* then states that the algebraic sum of the ampere-turns for a transformer is zero.

EXAMPLE 16.5 The three-winding transformer shown in Fig. 16-14 has turns $N_1 = 20$, $N_2 = N_3 = 10$. Find $\mathbf{I}_1$ given that $\mathbf{I}_2 = 10.0\underline{/-53.13^\circ}$ A, $\mathbf{I}_3 = 10.0\underline{/-45^\circ}$ A.

With the dots and current directions as shown on the diagram,

$$N_1\mathbf{I}_1 - N_2\mathbf{I}_2 - N_3\mathbf{I}_3 = 0$$

from which

$$20\mathbf{I}_1 = 10(10.0\underline{/-53.13^\circ}) + 10(10.0\underline{/-45^\circ})$$
$$\mathbf{I}_1 = 6.54 - j7.54 = 9.98\underline{/-49.06^\circ}\ \text{A}$$

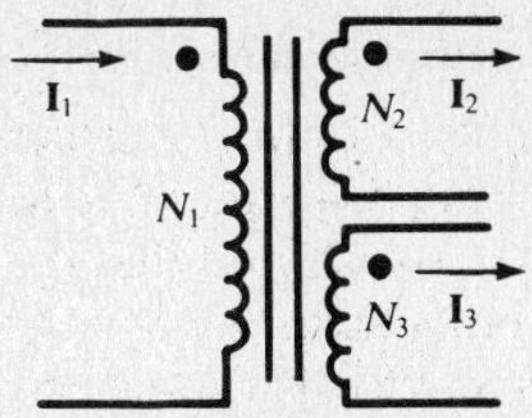

Fig. 16-14

Autotransformer

An *autotransformer* is an electrically continuous winding, with one or more taps, on a magnetic core. One circuit is connected to the end terminals, while the other is connected to one end terminal and to a tap, part way along the winding.

Referring to Fig. 16-15(a), the transformation ratio is

$$\frac{\mathbf{V}_1}{\mathbf{V}_2} = \frac{N_1 + N_2}{N_2} \equiv a + 1$$

which exceeds by unity the transformation ratio of an ideal two-winding transformer having the same turns ratio. Current $\mathbf{I}_1$ through the upper or series part of the winding, of N_1 turns, produces the flux ϕ_1. By Lenz's law the natural current in the lower part of the winding produces an opposing flux ϕ_2. Therefore, current $\mathbf{I}_n$ leaves the lower winding by the tap. The dots on the winding are as shown in Fig. 16-15(b). In an ideal autotransformer, as in an ideal transformer, the input and output complex powers must be equal.

$$\tfrac{1}{2}\mathbf{V}_1\mathbf{I}_1^* = \tfrac{1}{2}\mathbf{V}_1\mathbf{I}_{ab}^* = \tfrac{1}{2}\mathbf{V}_2\mathbf{I}_L^*$$

whence

$$\frac{\mathbf{I}_L}{\mathbf{I}_{ab}} = a + 1$$

That is, the currents also are in the transformation ratio.

Since $\mathbf{I}_L = \mathbf{I}_{ab} + \mathbf{I}_{cb}$, the output complex power consists of two parts:

$$\tfrac{1}{2}\mathbf{V}_2\mathbf{I}_L^* = \tfrac{1}{2}\mathbf{V}_2\mathbf{I}_{ab}^* + \tfrac{1}{2}\mathbf{V}_2\mathbf{I}_{cb}^* = \tfrac{1}{2}\mathbf{V}_2\mathbf{I}_{ab}^* + a(\tfrac{1}{2}\mathbf{V}_2\mathbf{I}_{ab}^*)$$

The first term on the right is attributed to conduction; the second, to induction. Thus, there exist both conductive and magnetic coupling between source and load in an autotransformer.

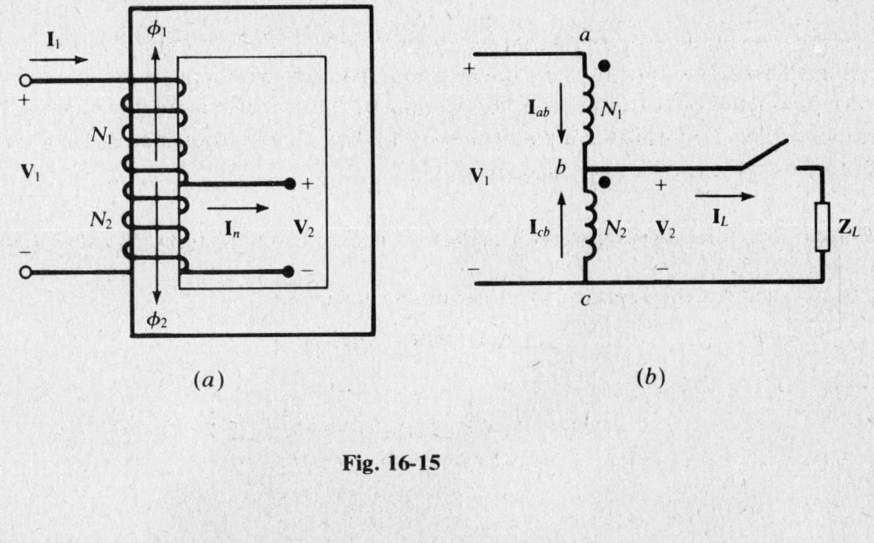

(a) (b)

Fig. 16-15

Solved Problems

16.1 When one coil of a magnetically coupled pair has a current 5.0 A, the resulting fluxes ϕ_{11} and ϕ_{12} are 0.2 mWb and 0.4 mWb, respectively. If the turns are $N_1 = 500$ and $N_2 = 1500$, find L_1, L_2, M, and the coefficient of coupling k.

$$\phi_1 = \phi_{11} + \phi_{12} = 0.6 \text{ mWb} \qquad L_1 = \frac{N_1\phi_1}{I_1} = \frac{500(0.6)}{5.0} = 60 \text{ mH}$$

$$M = \frac{N_2\phi_{12}}{I_1} = \frac{1500(0.4)}{5.0} = 120 \text{ mH} \qquad k = \frac{\phi_{12}}{\phi_1} = 0.667$$

Then, from $M = k\sqrt{L_1L_2}$, $L_2 = 540$ mH.

16.2 Two coupled coils have self-inductances $L_1 = 50$ mH and $L_2 = 200$ mH, and a coefficient of coupling $k = 0.50$. If coil 2 has 1000 turns, and $i_1 = 5.0 \sin 400t$ (A), find the voltage at coil 2 and the flux ϕ_1.

$$M = k\sqrt{L_1L_2} = 0.50\sqrt{(50)(200)} = 50 \text{ mH}$$

$$v_2 = M\frac{di_1}{dt} = 0.05\frac{d}{dt}(5.0 \sin 400t) = 100 \cos 400t \quad \text{(V)}$$

Assuming, as always, a linear magnetic circuit,

$$M = \frac{N_2\phi_{12}}{i_1} = \frac{N_2(k\phi_1)}{i_1} \qquad \text{or} \qquad \phi_1 = \left(\frac{M}{N_2k}\right)i_1 = 5.0 \times 10^{-4} \sin 400t \quad \text{(Wb)}$$

16.3 Apply KVL to the series circuit of Fig. 16-16.

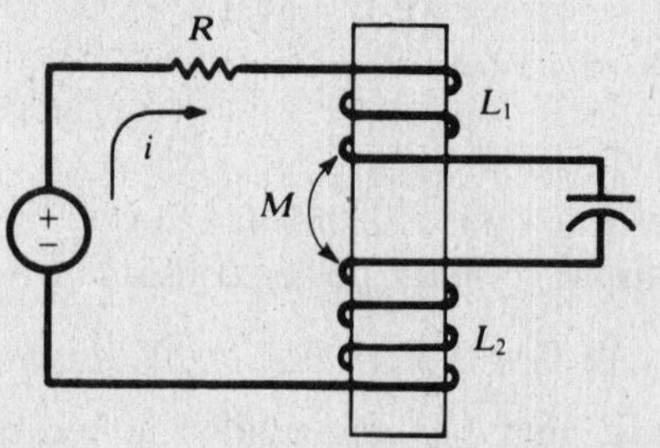

Fig. 16-16

Examination of the winding sense shows that the signs on the M-terms are opposite to the signs on the L-terms.

$$Ri + L_1\frac{di}{dt} - M\frac{di}{dt} + \frac{1}{C}\int i\,dt + L_2\frac{di}{dt} - M\frac{di}{dt} = v$$

or

$$Ri + L'\frac{di}{dt} + \frac{1}{C}\int i\,dt = v$$

where $L' \equiv L_1 + L_2 - 2M$. Because

$$M \leq \sqrt{L_1L_2} \leq \frac{L_1 + L_2}{2}$$

L' is nonnegative.

16.4 In a series aiding connection, two coupled coils have an equivalent inductance L_A; in a series opposing connection, L_B. Obtain an expression for M in terms of L_A and L_B.

As in Problem 16.3,

$$L_1 + L_2 + 2M = L_A \qquad L_1 + L_2 - 2M = L_B$$

which give

$$M = \frac{1}{4}(L_A - L_B)$$

This problem suggests a method by which M can be determined experimentally.

16.5 (*a*) Write the mesh current equations for the coupled coils with currents i_1 and i_2 shown in Fig. 16-17. (*b*) Repeat for i_2 as indicated by the dashed arrow.

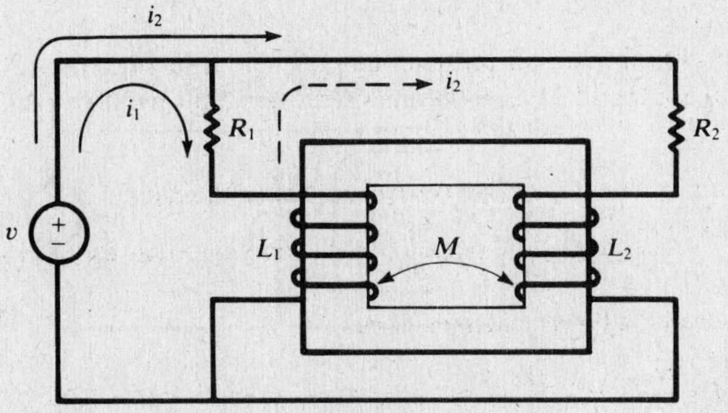

Fig. 16-17

(*a*) The winding sense and selected current directions result in signs on the M-terms as follows:

$$R_1 i_1 + L_1 \frac{di_1}{dt} + M \frac{di_2}{dt} = v$$

$$R_2 i_2 + L_2 \frac{di_2}{dt} + M \frac{di_1}{dt} = v$$

(*b*)

$$R_1(i_1 - i_2) + L_1 \frac{d}{dt}(i_1 - i_2) + M \frac{di_2}{dt} = v$$

$$R_1(i_2 - i_1) + R_2 i_2 + L_2 \frac{di_2}{dt} - M \frac{d}{dt}(i_2 - i_1) + L_1 \frac{d}{dt}(i_2 - i_1) - M \frac{di_2}{dt} = 0$$

16.6 Obtain the dotted equivalent circuit for the coupled circuit shown in Fig. 16-18, and use it to find the voltage **V** across the 10-Ω capacitive reactance.

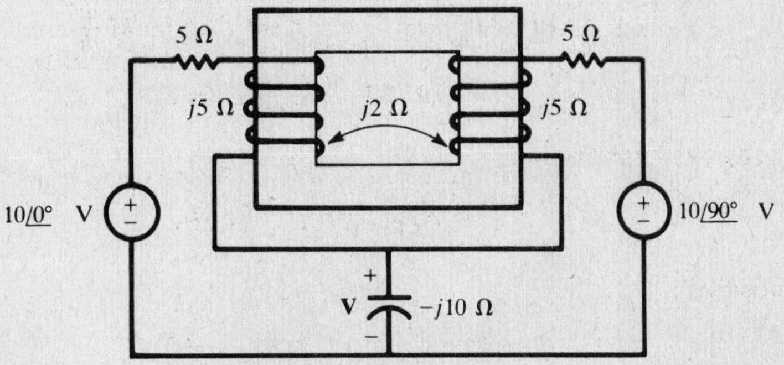

Fig. 16-18

To place the dots on the circuit, consider only the coils and their winding sense. Drive a current into the top of the left coil and place a dot at this terminal. The corresponding flux is upward. By Lenz's law, the flux at the right coil must be upward-directed to oppose the first flux. Then the natural current leaves this winding by the upper terminal, which is marked with a dot. See Fig. 16-19 for the complete dotted equivalent circuit, with currents $\mathbf{I}_1$ and $\mathbf{I}_2$ chosen for calculation of $\mathbf{V}$.

$$\begin{bmatrix} 5-j5 & 5+j3 \\ 5+j3 & 10+j6 \end{bmatrix}\begin{bmatrix} \mathbf{I}_1 \\ \mathbf{I}_2 \end{bmatrix} = \begin{bmatrix} 10\underline{/0^\circ} \\ 10-j10 \end{bmatrix}$$

$$\mathbf{I}_1 = \frac{\begin{vmatrix} 10 & 5+j3 \\ 10-j10 & 10+j6 \end{vmatrix}}{\Delta_\mathbf{z}} = 1.015\underline{/113.96^\circ} \quad \text{A}$$

and $\mathbf{V} = \mathbf{I}_1(-j10) = 10.15\underline{/23.96^\circ} \quad \text{V}$.

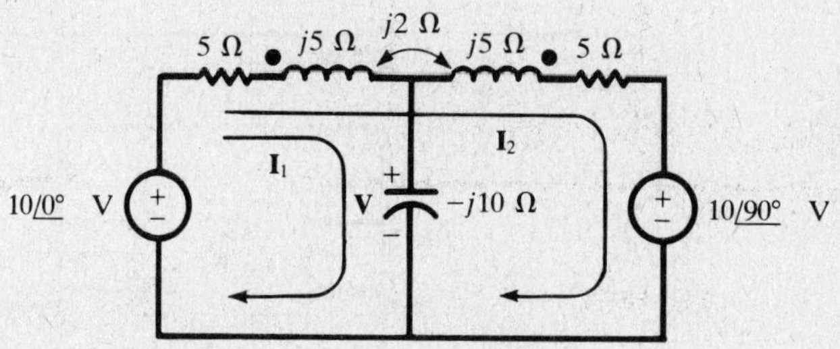

Fig. 16-19

16.7 Obtain the dotted equivalent for the circuit shown in Fig. 16-20 and use the equivalent to find the equivalent inductive reactance.

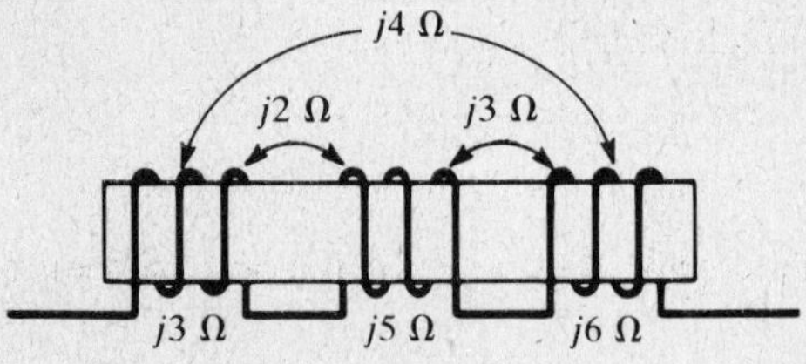

Fig. 16-20

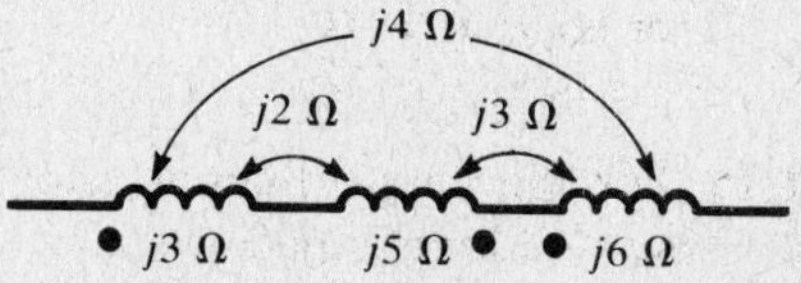

Fig. 16-21

Drive a current into the first coil and place a dot where this current enters. The natural current in both of the other windings establishes an opposing flux to that set up by the driven current. Place dots where the natural current leaves the windings. (Some confusion is eliminated if the series connections are ignored while determining the locations of the dots.) The result is Fig. 16-21.

$$\mathbf{Z} = j3 + j5 + j6 - 2(j2) + 2(j4) - 2(j3) = j12 \ \Omega$$

i.e., an inductive reactance of 12 Ω.

16.8 (*a*) Compute the voltage $\mathbf{V}$ for the coupled circuit shown in Fig. 16-22. (*b*) Repeat with the polarity of one coil reversed.

(*a*) $X_M = (0.8)\sqrt{5(10)} = 5.66 \ \Omega$, and so the $\mathbf{Z}$-matrix is

$$\begin{bmatrix} 3+j1 & -3-j1.66 \\ -3-j1.66 & 8+j6 \end{bmatrix}$$

Then
$$\mathbf{I}_2 = \frac{\begin{vmatrix} 3+j1 & 50 \\ -3-j1.66 & 0 \end{vmatrix}}{\Delta_\mathbf{z}} = 8.62\underline{/-24.79^\circ} \quad \text{A}$$

and $\mathbf{V} = \mathbf{I}_2(5) = 43.1\underline{/-24.79^\circ}$ V.

(*b*)

$$[\mathbf{Z}] = \begin{bmatrix} 3+j1 & -3+j9.66 \\ -3+j9.66 & 8+j6 \end{bmatrix}$$

$$\mathbf{I}_2 = \frac{\begin{vmatrix} 3+j1 & 50 \\ -3+j9.66 & 0 \end{vmatrix}}{\Delta_\mathbf{z}} = 3.82\underline{/-112.12^\circ} \text{ A}$$

and $\mathbf{V} = \mathbf{I}_2(5) = 19.1\underline{/-112.12^\circ}$ V.

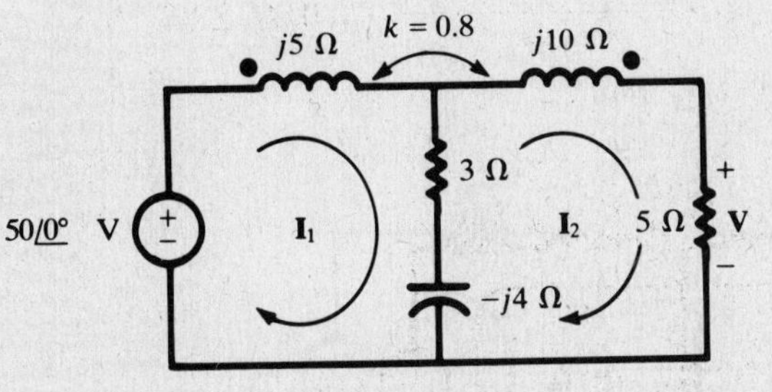

Fig. 16-22

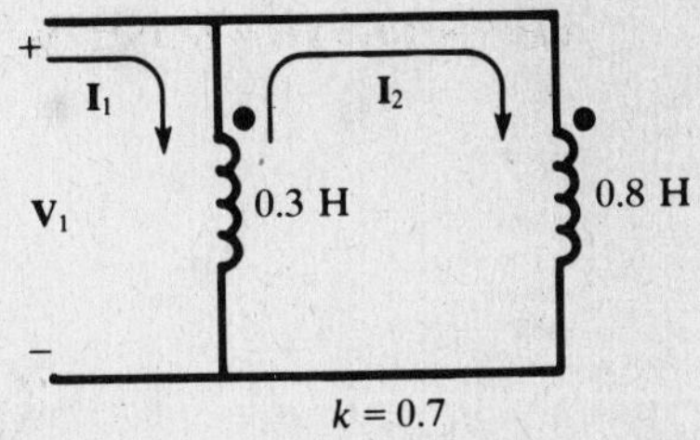

Fig. 16-23

16.9 Obtain the equivalent inductance of the parallel-connected, coupled coils shown in Fig. 16-23.

Currents $\mathbf{I}_1$ and $\mathbf{I}_2$ are selected as shown on the diagram; then $\mathbf{Z}_{in} = \mathbf{V}_1/\mathbf{I}_1$.

$$[\mathbf{Z}] = \begin{bmatrix} j\omega 0.3 & j\omega 0.043 \\ j\omega 0.043 & j\omega 0.414 \end{bmatrix}$$

and

$$\mathbf{Z}_{in} = \frac{\Delta_\mathbf{z}}{\Delta_{11}} = \frac{(j\omega 0.3)(j\omega 0.414) - (j\omega 0.043)^2}{j\omega 0.414} = j\omega 0.296$$

or L_{eq} is 0.296 H.

16.10 For the coupled circuit shown in Fig. 16-24, show that dots are not needed so long as the second loop is passive.

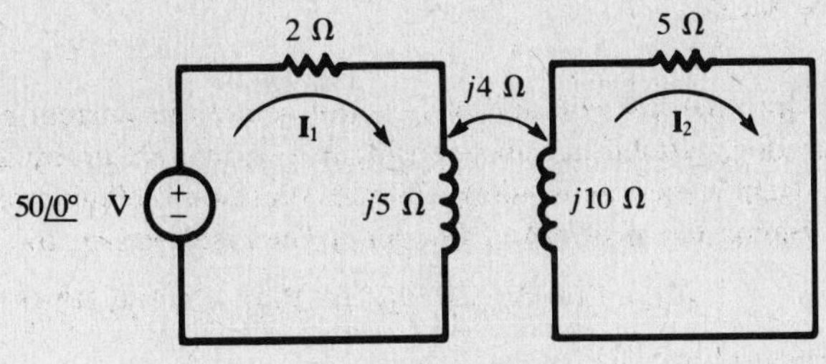

Fig. 16-24

Currents $\mathbf{I}_1$ and $\mathbf{I}_2$ are selected as shown.

$$\mathbf{I}_1 = \frac{\begin{vmatrix} 50 & \pm j4 \\ 0 & 5+j10 \end{vmatrix}}{\begin{vmatrix} 2+j5 & \pm j4 \\ \pm j4 & 5+j10 \end{vmatrix}} = \frac{250+j500}{-24+j45} = 10.96\underline{/-54.64^\circ} \text{ A}$$

$$\mathbf{I}_2 = \frac{\begin{vmatrix} 2+j5 & 50 \\ \pm j4 & 0 \end{vmatrix}}{\Delta_\mathbf{z}} = 3.92\underline{/-118.07 \mp 90^\circ} \text{ A}$$

The value of Δ_Z is unaffected by the sign on M. Since the numerator determinant for $\mathbf{I}_1$ does not involve the coupling impedance, $\mathbf{I}_1$ is also unaffected. The expression for $\mathbf{I}_2$ shows that a change in the coupling polarity results in a 180° phase shift. With no other phasor voltage present in the second loop, this change in phase is of no consequence.

16.11 For the coupled circuit shown in Fig. 16-25, find the ratio $\mathbf{V}_2/\mathbf{V}_1$ which results in zero current $\mathbf{I}_1$.

$$\mathbf{I}_1 = 0 = \frac{\begin{vmatrix} \mathbf{V}_1 & j2 \\ \mathbf{V}_2 & 2+j2 \end{vmatrix}}{\Delta_Z}$$

Then $\mathbf{V}_1(2+j2) - \mathbf{V}_2(j2) = 0$, from which $\mathbf{V}_2/\mathbf{V}_1 = 1 - j1$.

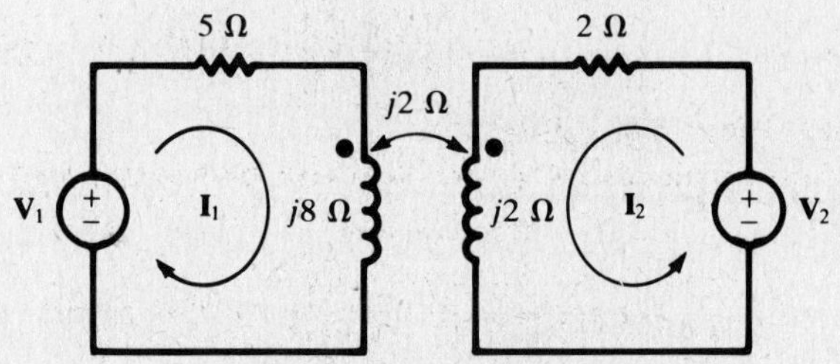

Fig. 16-25

16.12 In the circuit of Fig. 16-26, find the voltage across the 5 Ω reactance with the polarity shown.

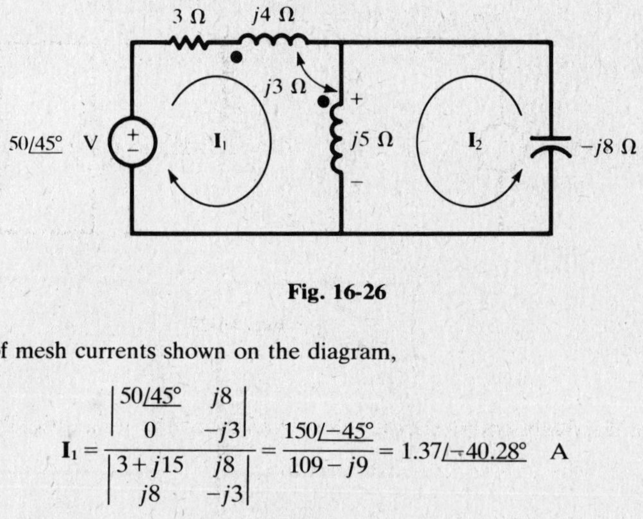

Fig. 16-26

For the choice of mesh currents shown on the diagram,

$$\mathbf{I}_1 = \frac{\begin{vmatrix} 50\underline{/45^\circ} & j8 \\ 0 & -j3 \end{vmatrix}}{\begin{vmatrix} 3+j15 & j8 \\ j8 & -j3 \end{vmatrix}} = \frac{150\underline{/-45^\circ}}{109-j9} = 1.37\underline{/-40.28^\circ} \quad \text{A}$$

Similarly, $\mathbf{I}_2 = 3.66\underline{/-40.28^\circ}$ A.

The voltage across the $j5$ is partly conductive, from the currents $\mathbf{I}_1$ and $\mathbf{I}_2$, and partly mutual, from current $\mathbf{I}_1$ in the 4 Ω reactance.

$$\mathbf{V} = (\mathbf{I}_1 + \mathbf{I}_2)(j5) + \mathbf{I}_1(j3) = 29.27\underline{/49.72^\circ} \quad \text{V}$$

Of course, the same voltage must exist across the capacitor:

$$\mathbf{V} = -\mathbf{I}_2(-j8) = 29.27\underline{/49.72^\circ} \quad \text{V}$$

16.13 Obtain Thévenin and Norton equivalent circuits at terminals ab for the coupled circuit shown in Fig. 16-27.

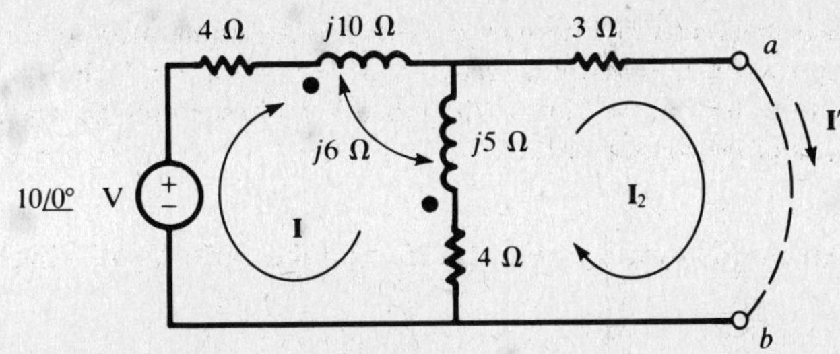

Fig. 16-27

In open circuit, a single clockwise loop current **I** is driven by the voltage source.

$$\mathbf{I} = \frac{10\angle 0^\circ}{8+j3} = 1.17\angle -20.56^\circ \quad \text{A}$$

Then $\mathbf{V}' = \mathbf{I}(j5+4) - \mathbf{I}(j6) = 4.82\angle -34.60^\circ \quad \text{V}$.

To find the short-circuit current $\mathbf{I}'$, two clockwise mesh currents are assumed, with $\mathbf{I}_2 = \mathbf{I}'$.

$$\mathbf{I}' = \frac{\begin{vmatrix} 8+j3 & 10 \\ -4+j1 & 0 \end{vmatrix}}{\begin{vmatrix} 8+j3 & -4+j1 \\ -4+j1 & 7+j5 \end{vmatrix}} = 0.559\angle -83.39^\circ \quad \text{A}$$

and

$$\mathbf{Z}' = \frac{\mathbf{V}'}{\mathbf{I}'} = \frac{4.82\angle -34.60^\circ}{0.559\angle -83.39^\circ} = 8.62\angle 48.79^\circ \quad \Omega$$

The equivalent circuits are pictured in Fig. 16-28.

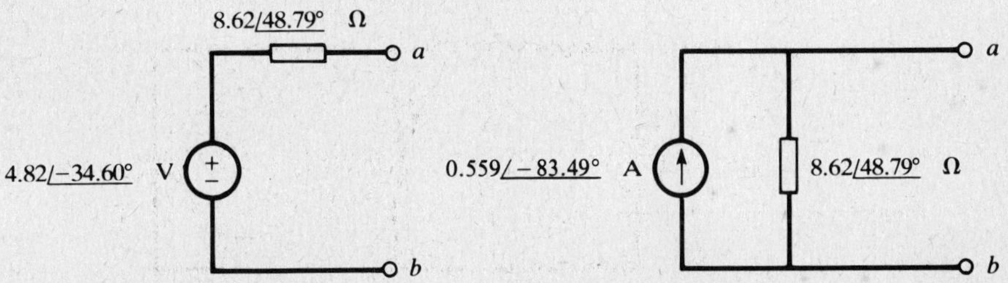

(a) (b)

Fig. 16-28

16.14 Obtain a conductively coupled equivalent circuit for the magnetically coupled circuit shown in Fig. 16-29.

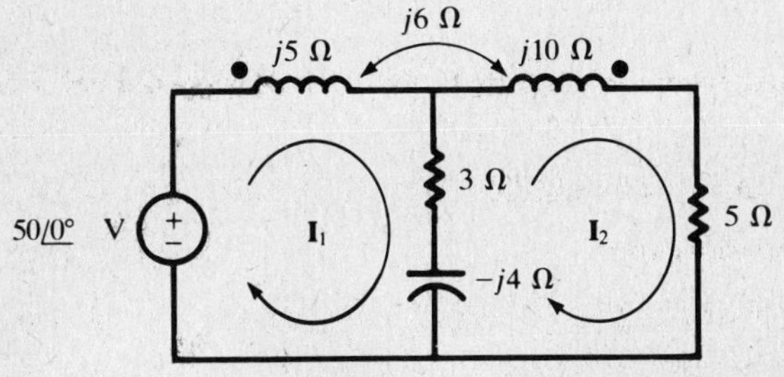

Fig. 16-29

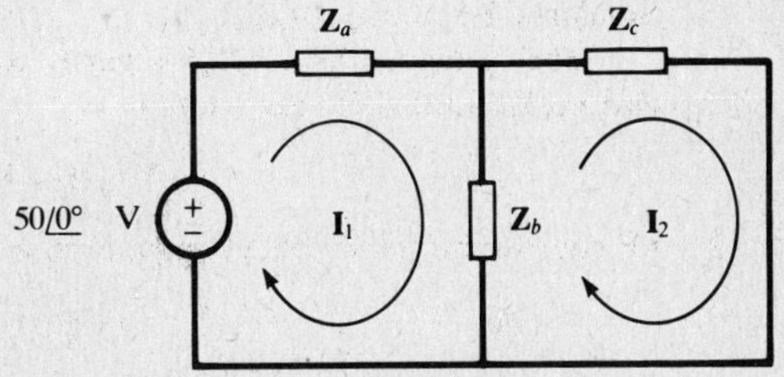

Fig. 16-30

Select mesh currents $\mathbf{I}_1$ and $\mathbf{I}_2$ as shown on the diagram and write the KVL equations in matrix form.

$$\begin{bmatrix} 3+j1 & -3-j2 \\ -3-j2 & 8+j6 \end{bmatrix}\begin{bmatrix} \mathbf{I}_1 \\ \mathbf{I}_2 \end{bmatrix} = \begin{bmatrix} 50\underline{/0^\circ} \\ 0 \end{bmatrix}$$

The impedances in Fig. 16-30 are selected to give the identical **Z**-matrix. Thus, since $\mathbf{I}_1$ and $\mathbf{I}_2$ pass through the common impedance, $\mathbf{Z}_b$, in opposite directions, $\mathbf{Z}_{12}$ in the matrix is $-\mathbf{Z}_b$. Then, $\mathbf{Z}_b = 3+j2\ \Omega$. Since $\mathbf{Z}_{11}$ is to include all impedances through which $\mathbf{I}_1$ passes,

$$3+j1 = \mathbf{Z}_a + (3+j2)$$

from which $\mathbf{Z}_a = -j1\ \Omega$. Similarly,

$$\mathbf{Z}_{22} = 8+j6 = \mathbf{Z}_b + \mathbf{Z}_c$$

and $\mathbf{Z}_c = 5+j4\ \ \Omega$.

16.15 For the transformer circuit of Fig. 16-11(*b*), $k=0.96$, $R_1 = 1.2\ \Omega$, $R_2 = 0.3\ \Omega$, $X_1 = 20\ \Omega$, $X_2 = 5\ \Omega$, $\mathbf{Z}_L = 5.0\underline{/36.87^\circ}\ \ \Omega$, and $\mathbf{V}_2 = 100\underline{/0^\circ}\ \ \text{V}$. Obtain the coil emfs $\mathbf{E}_1$ and $\mathbf{E}_2$, and the magnetizing current $\mathbf{I}_\phi$.

$$X_{11} = (1-k)X_1 = (1-0.96)(20) = 0.8\ \Omega \qquad X_{22} = (1-k)X_2 = 0.2\ \Omega$$

$$a = \sqrt{\frac{X_1}{X_2}} = 2 \qquad X_M = k\sqrt{X_1X_2} = 9.6\ \Omega$$

Now a circuit of the form Fig. 16-11(*a*) can be constructed, starting from the phasor voltage-current relationship at the load, and working back through $\mathbf{E}_2$ to $\mathbf{E}_1$.

$$\mathbf{I}_2 = \frac{\mathbf{V}_2}{\mathbf{Z}_L} = \frac{100\underline{/0^\circ}}{5.0\underline{/36.87^\circ}} = 20\underline{/-36.87^\circ}\ \ \text{A}$$

$$\mathbf{E}_2 = \mathbf{I}_2(R_2 + jX_{22}) + \mathbf{V}_2 = (20\underline{/-36.87^\circ})(0.3+j0.2) + 100\underline{/0^\circ} = 107.2 - j0.4\ \ \text{V}$$

$$\mathbf{E}_1 = a\mathbf{E}_2 = 214.4 - j0.8\ \ \text{V}$$

$$\mathbf{I}_\phi = \frac{\mathbf{E}_2}{jX_M} = -0.042 - j11.17\ \ \text{A}$$

16.16 For the linear transformer of Problem 16.15, calculate the input impedance at the terminals where $\mathbf{V}_1$ is applied.

Method 1

Completing the construction begun in Problem 16.15,

$$\mathbf{I}_1 = \mathbf{I}_\phi + \frac{1}{a}\mathbf{I}_2 = (-0.042 - j11.17) + 10\underline{/-36.87^\circ} = 18.93\underline{/-65.13^\circ}\ \ \text{A}$$

$$\mathbf{V}_1 = \mathbf{I}_1(R_1 + jX_{11}) + \mathbf{E}_1 = (18.93\underline{/-65.13^\circ})(1.2+j0.8) + (214.4 - j0.8)$$
$$= 238.2\underline{/-3.62^\circ}\ \ \text{V}$$

Therefore,

$$\mathbf{Z}_{\text{in}} = \frac{\mathbf{V}_1}{\mathbf{I}_1} = \frac{238.2\underline{/-3.62^\circ}}{18.93\underline{/-65.13^\circ}} = 12.58\underline{/61.51^\circ}\ \ \Omega$$

Method 2

By (*10*) of Example 16.3,

$$\mathbf{Z}_{\text{in}} = (1.2+j0.8) + 2^2\frac{(j4.8)(0.3+j0.2+5.0\underline{/36.87^\circ})}{0.3+j5.0+5.0\underline{/36.87^\circ}}$$

$$= \frac{114.3\underline{/123.25^\circ}}{9.082\underline{/61.75^\circ}} = 12.58\underline{/61.50^\circ}\ \ \Omega$$

Method 3

By (*10a*) of Example 16.3,

$$\mathbf{Z}_{\text{in}} = (1.2 + j20) + \frac{(9.6)^2}{0.3 + j5 + 5.0\underline{/36.87^\circ}}$$
$$= (1.2 + j20) + (4.80 - j8.94) = 12.58\underline{/61.53^\circ} \quad \Omega$$

16.17 In Fig. 16-31, three identical transformers are primary wye-connected and secondary delta-connected. A single load impedance carries current $\mathbf{I}_L = 30\underline{/0^\circ}$ A. Given

$$\mathbf{I}_{b2} = 20\underline{/0^\circ} \text{ A} \qquad \mathbf{I}_{a2} = \mathbf{I}_{c2} = 10\underline{/0^\circ} \text{ A}$$

and $N_1 = 10N_2 = 100$, find the primary currents $\mathbf{I}_{a1}$, $\mathbf{I}_{b1}$, $\mathbf{I}_{c1}$.

The ampere-turn dot rule is applied to each transformer.

$$N_1\mathbf{I}_{a1} + N_2\mathbf{I}_{a2} = 0 \quad \text{or} \quad \mathbf{I}_{a1} = -\frac{10}{100}(10\underline{/0^\circ}) = -1\underline{/0^\circ} \text{ A}$$

$$N_1\mathbf{I}_{b1} - N_2\mathbf{I}_{b2} = 0 \quad \text{or} \quad \mathbf{I}_{b1} = \frac{10}{100}(20\underline{/0^\circ}) = 2\underline{/0^\circ} \text{ A}$$

$$N_1\mathbf{I}_{c1} + N_2\mathbf{I}_{c2} = 0 \quad \text{or} \quad \mathbf{I}_{c1} = -\frac{10}{100}(10\underline{/0^\circ}) = -1\underline{/0^\circ} \text{ A}$$

The sum of the primary currents provides a check:

$$\mathbf{I}_{a1} + \mathbf{I}_{b1} + \mathbf{I}_{c1} = 0$$

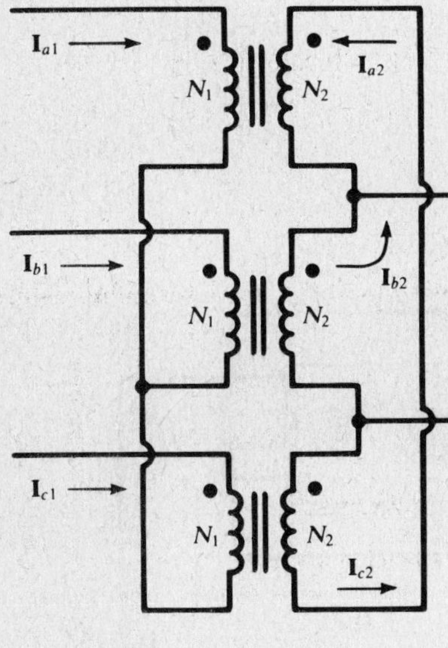

Fig. 16-31

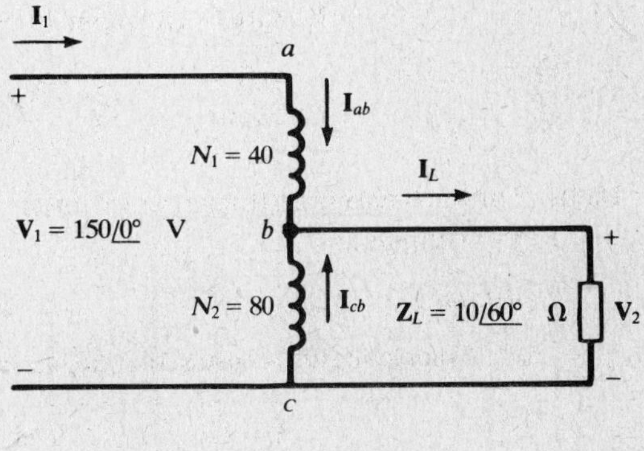

Fig. 16-32

16.18 For the ideal autotransformer shown in Fig. 16-32, find $\mathbf{V}_2$, $\mathbf{I}_{cb}$, and the input current $\mathbf{I}_1$.

$$a \equiv \frac{N_1}{N_2} = \frac{1}{2}$$

$$\mathbf{V}_2 = \frac{\mathbf{V}_1}{a+1} = 100\underline{/0^\circ} \text{ V} \qquad \mathbf{I}_L = \frac{\mathbf{V}_2}{\mathbf{Z}_L} = 10\underline{/-60^\circ} \text{ A}$$

$$\mathbf{I}_{cb} = \mathbf{I}_L - \mathbf{I}_{ab} = 3.33\underline{/-60^\circ} \text{ A} \qquad \mathbf{I}_{ab} = \frac{\mathbf{I}_L}{a+1} = 6.67\underline{/-60^\circ} \text{ A}$$

16.19 In Problem 16.18, find the apparent power delivered to the load by transformer action and that supplied by conduction.

$$\mathbf{S}_{\text{cond}} = \tfrac{1}{2}\mathbf{V}_2\mathbf{I}_{ab}^* = \tfrac{1}{2}(100\angle 0^\circ)(6.67\angle 60^\circ) = 333\angle 60^\circ \quad \text{VA}$$
$$\mathbf{S}_{\text{trans}} = a\mathbf{S}_{\text{cond}} = 167\angle 60^\circ \quad \text{VA}$$

Supplementary Problems

16.20 Two coupled coils, $L_1 = 0.8$ H and $L_2 = 0.2$ H, have a coefficient of coupling $k = 0.90$. Find the mutual inductance M and the turns ratio N_1/N_2. *Ans.* 0.36 H, 2

16.21 Two coupled coils, $N_1 = 100$ and $N_2 = 800$, have a coupling coefficient $k = 0.85$. With coil 1 open and a current of 5.0 A in coil 2, the flux is $\phi_2 = 0.35$ mWb. Find L_1, L_2, and M.
Ans. 0.875 mH, 56 mH, 5.95 mH

16.22 Two identical coupled coils have an equivalent inductance of 80 mH when connected series aiding, and 35 mH in series opposing. Find L_1, L_2, M, and k. *Ans.* 28.8 mH, 28.8 mH, 11.25 mH, 0.392

16.23 Two coupled coils, with $L_1 = 20$ mH, $L_2 = 10$ mH, and $k = 0.50$, are connected four different ways: series aiding, series opposing, and parallel with both arrangements of winding sense. Obtain the equivalent inductances of the four connections. *Ans.* 44.1 mH, 15.9 mH, 9.47 mH, 3.39 mH

16.24 Write the mesh current equations for the coupled circuit shown in Fig. 16-33. Obtain the dotted equivalent circuit and write the same equations.

Ans. $(R_1 + R_3)i_1 + L_1\dfrac{di_1}{dt} + R_3 i_2 + M\dfrac{di_2}{dt} = v$

$(R_2 + R_3)i_2 + L_2\dfrac{di_2}{dt} + R_3 i_1 + M\dfrac{di_1}{dt} = v$

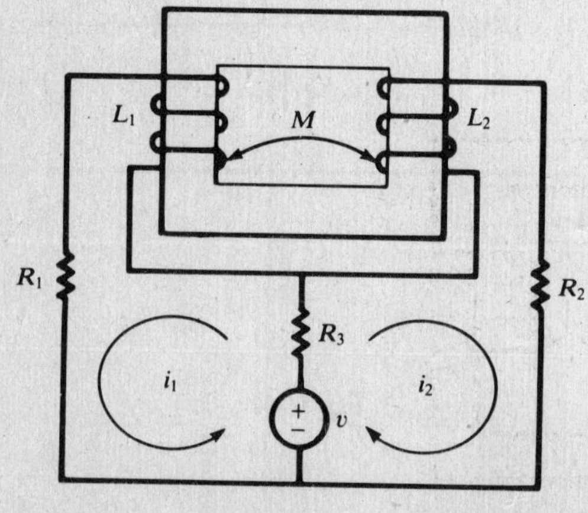

Fig. 16-33

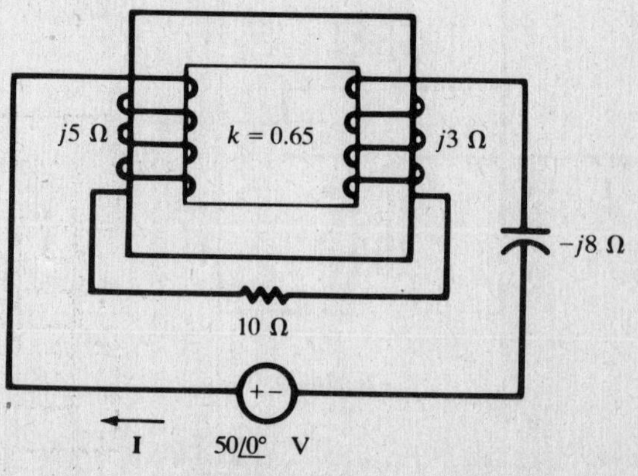

Fig. 16-34

16.25 Write the phasor equation for the single-loop, coupled circuit shown in Fig. 16-34.
Ans. $(j5 + j3 - j5.03 - j8 + 10)\mathbf{I} = 50\angle 0^\circ$

16.26 Obtain the dotted equivalent circuit for the coupled circuit of Fig. 16-34. *Ans.* See Fig. 16-35.

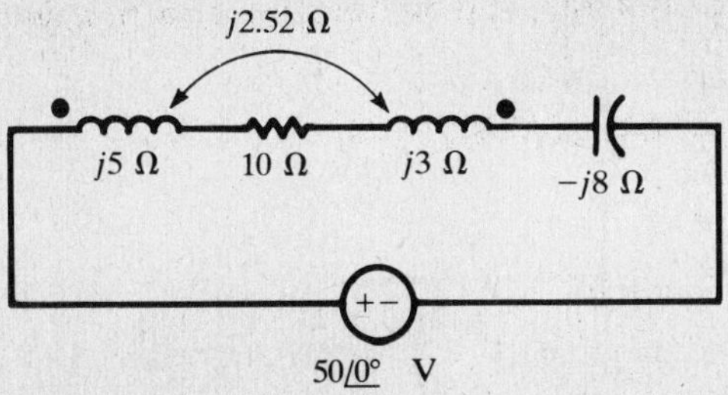

Fig. 16-35

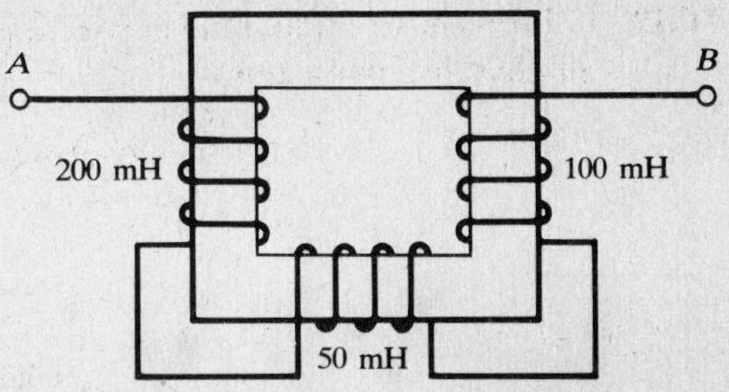

Fig. 16-36

16.27 The three coupled coils shown in Fig. 16-36 have coupling coefficients of 0.50. Obtain the equivalent inductance between the terminals AB. *Ans.* 239 mH

16.28 Obtain two forms of the dotted equivalent circuit for the coupled coils shown in Fig. 16-36. *Ans.* See Fig. 16-37.

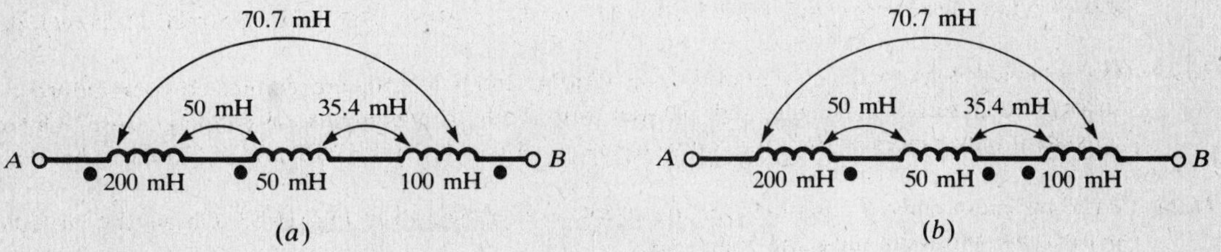

Fig. 16-37

16.29 (*a*) Obtain the equivalent impedance at terminals AB of the coupled circuit shown in Fig. 16-38. (*b*) Reverse the winding sense of one coil and repeat. *Ans.* (*a*) $3.40\angle 41.66^\circ$ Ω; (*b*) $2.54\angle 5.37^\circ$ Ω

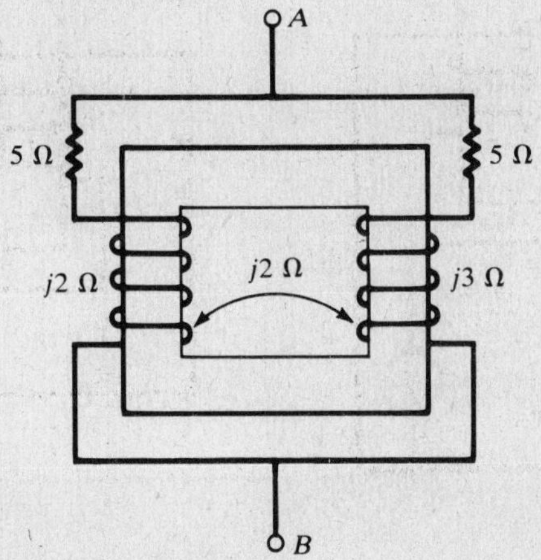

Fig. 16-38

16.30 In the coupled circuit shown in Fig. 16-39, find $\mathbf{V}_2$ for which $\mathbf{I}_1 = 0$. What voltage appears at the 8 Ω inductive reactance under this condition? *Ans.* $141.4\angle -45^\circ$ V, $100\angle 0^\circ$ V (+ at dot)

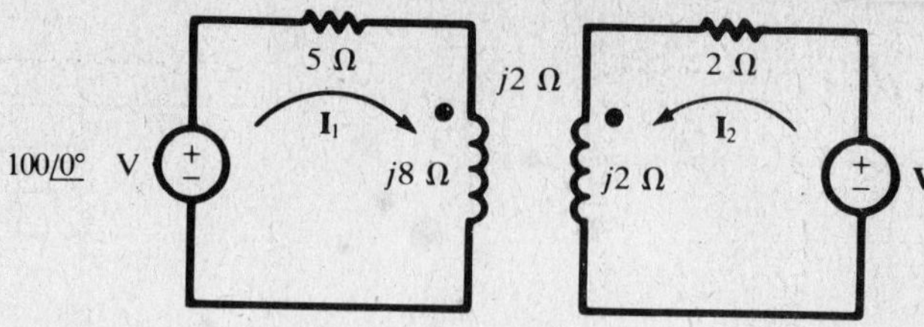

Fig. 16-39

16.31 Find the mutual reactance X_M for the coupled circuit of Fig. 16-40, if the average power in the 5-Ω resistor is 45.24 W. *Ans.* 4 Ω

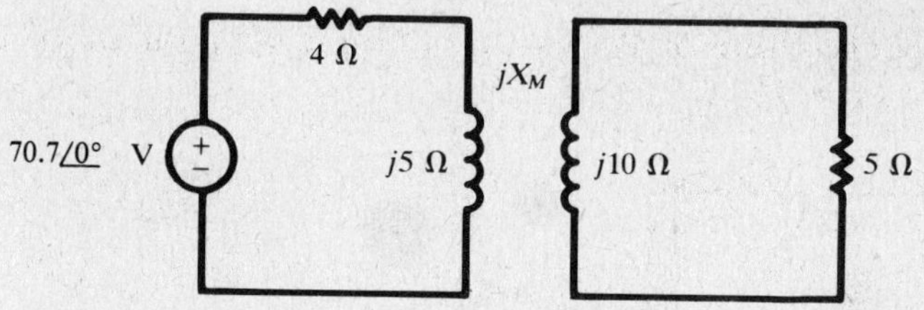

Fig. 16-40

16.32 For the coupled circuit shown in Fig. 16-41, find the components of the current $\mathbf{I}_2$ resulting from each source $\mathbf{V}_1$ and $\mathbf{V}_2$. *Ans.* $0.77\underline{/112.6^\circ}$ A, $1.72\underline{/86.05^\circ}$ A

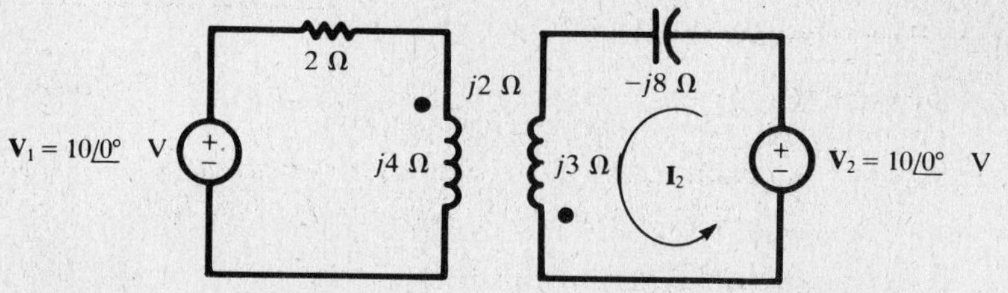

Fig. 16-41

16.33 Determine the coupling coefficient k in the circuit shown in Fig. 16-42, if the power in the 10-Ω resistor is 32 W. *Ans.* 0.791

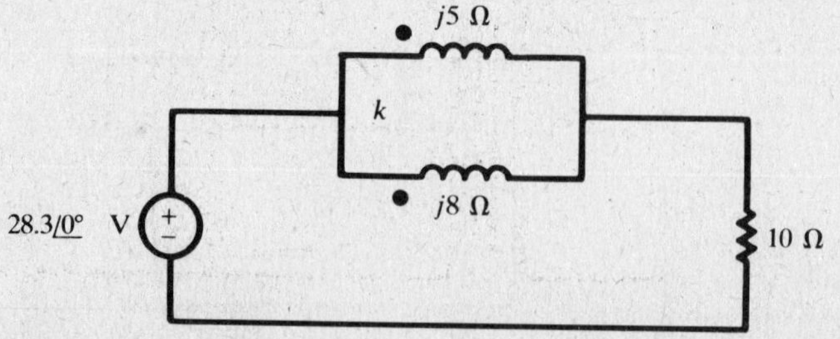

Fig. 16-42

16.34 In *(10)*, replace a, X_{11}, X_{22}, and X_M by their expressions in terms of X_1, X_2, and k, thereby obtaining *(10a)*.

16.35 For the coupled circuit shown in Fig. 16-43, find the input impedance at terminals ab. *Ans.* $3 + j36.3$ Ω

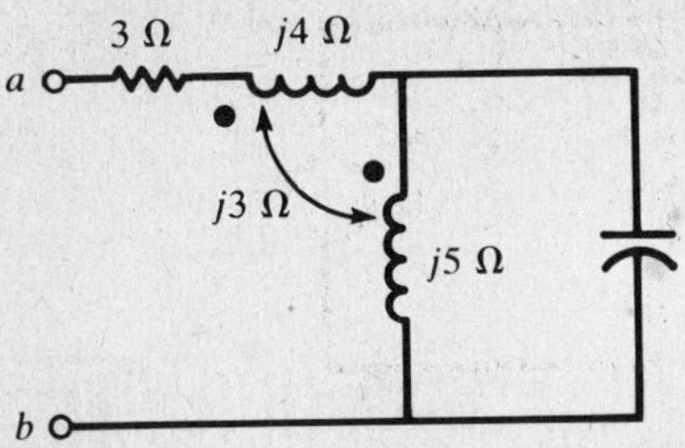

Fig. 16-43

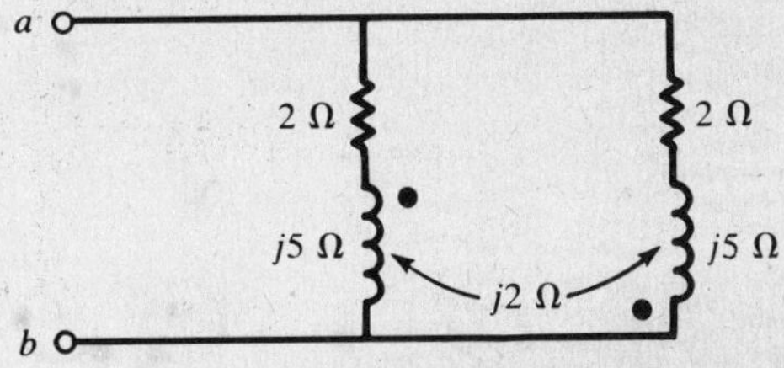

Fig. 16-44

16.36 Find the input impedance at terminals *ab* of the coupled circuit shown in Fig. 16-44.
Ans. $1 + j1.5 \quad \Omega$

16.37 Find the input impedance at terminals *ab* of the coupled circuit shown in Fig. 16-45.
Ans. $6.22 + j4.65 \quad \Omega$

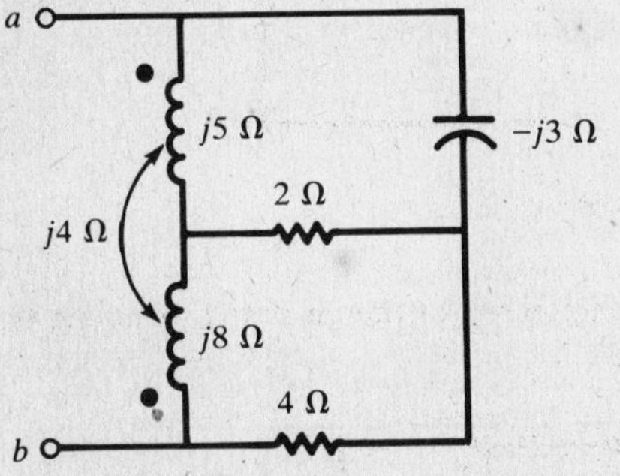

Fig. 16-45

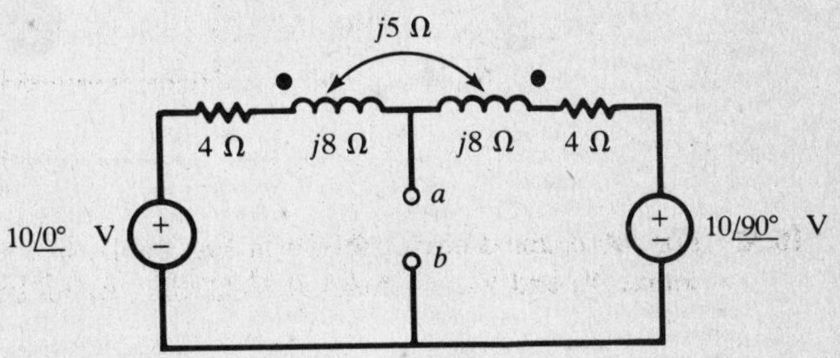

Fig. 16-46

16.38 Obtain Thévenin and Norton equivalent circuits at terminals *ab* of the coupled circuit shown in Fig. 16-46. *Ans.* $\mathbf{V}' = 7.07\angle 45^\circ$ V, $\mathbf{I}' = 1.04\angle -27.9^\circ$ A, $\mathbf{Z}' = 6.80\angle 72.9^\circ \quad \Omega$

16.39 For the ideal transformer shown in Fig. 16-47, find $\mathbf{I}_1$, given

$$\mathbf{I}_{L1} = 10.0\angle 0^\circ \text{ A} \qquad \mathbf{I}_{L2} = 10.0\angle -36.87^\circ \text{ A} \qquad \mathbf{I}_{L3} = 4.47\angle -26.57^\circ \text{ A}$$

Ans. $16.5\angle -14.04^\circ$ A

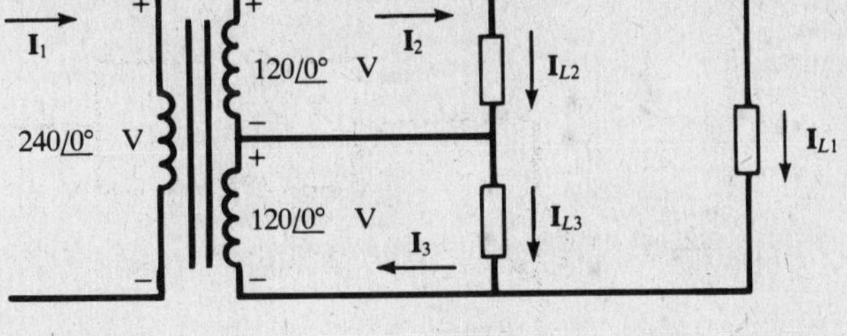

Fig. 16-47

16.40 When the secondary of the linear transformer shown in Fig. 16-48 is open-circuited, the primary current is $\mathbf{I}_1 = 4.0\angle -89.69^\circ$ A. Find the coefficient of coupling k. *Ans.* 0.983

16.41 For the ideal transformer shown in Fig. 16-49, find $\mathbf{I}_1$, given $\mathbf{I}_2 = 50\angle -36.87^\circ$ A and $\mathbf{I}_3 = 16\angle 0^\circ$ A.
Ans. $26.6\angle -34.29^\circ$ A

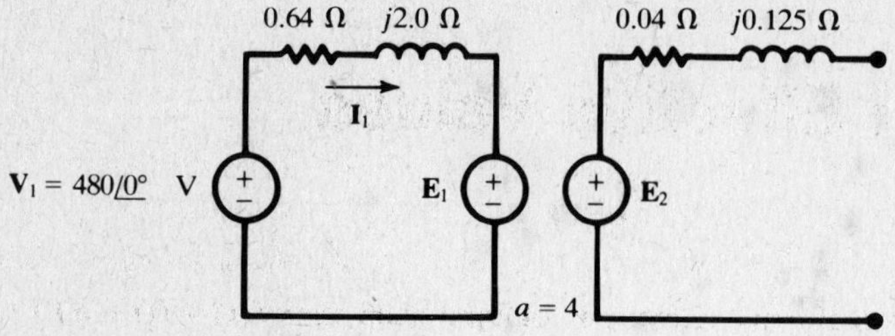

Fig. 16-48

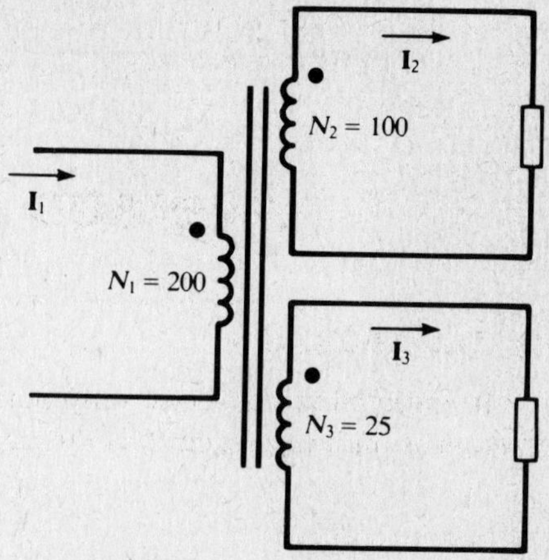

Fig. 16-49

16.42 Considering the autotransformer shown in Fig. 16-50 ideal, obtain the currents $\mathbf{I}_1$, $\mathbf{I}_{cb}$, and $\mathbf{I}_{dc}$.
Ans. $3.70\angle 22.5^\circ$ A, $2.12\angle 86.71^\circ$ A, $10.34\angle 11.83^\circ$ A

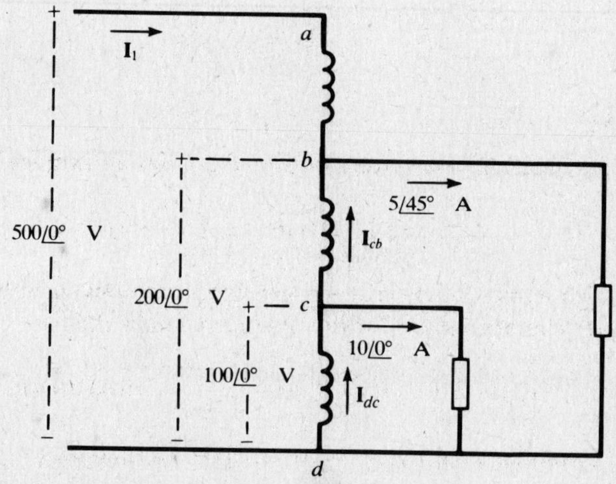

Fig. 16-50

Appendix A

Average and Effective Values

A1 WAVEFORMS

In basic circuit analysis, only periodic waveforms—functions of time such that $f(t) = f(t + nT)$, where n is an integer and T is the period—are studied in depth. See Fig. A-1 for examples.

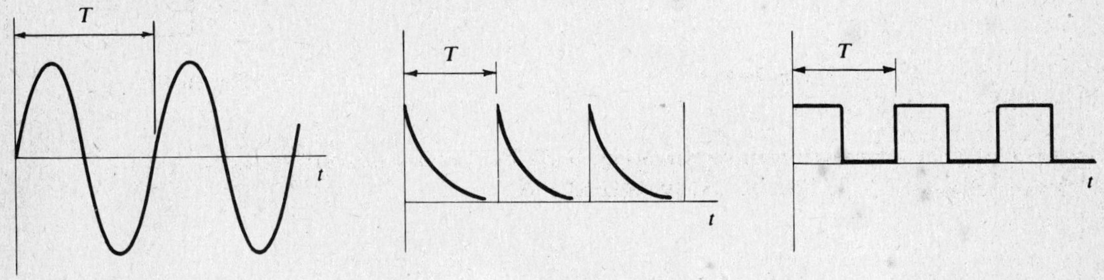

Fig. A-1

A2 AVERAGE VALUE

The general periodic function $y(t)$, with a period T (s), has an average value, Y_{avg}, given by

$$Y_{avg} = \frac{1}{T}\int_0^T y(t)\,dt = \frac{1}{T}\int_{t_0}^{t_0+T} y(t)\,dt$$

Some functions, of which $y = Y_m \sin(\omega t)$ is an example, are conveniently treated as functions of ωt (rad) rather than of t. The expression for the average value is then

$$Y_{avg} = \frac{1}{T}\int_0^T y(\omega t)\,d(\omega t) = \frac{1}{T}\int_{\omega t_0}^{\omega t_0+T} y(\omega t)\,d(\omega t)$$

where T is the period in radians; e.g., $T = 2\pi$ for $y(\omega t) = Y_m \sin(\omega t)$.

EXAMPLE A1 Determine the average value of the waveform shown in Fig. A-2; for $0 < t < 0.05$ s,

$$y(t) = 10e^{-200t}$$

Computing the average over the first period,

$$Y_{avg} = \frac{1}{T}\int_0^T y\,dt = \frac{1}{0.05}\int_0^{0.05} 10e^{-200t}\,dt = \frac{10}{0.05(-200)}\Big[e^{-200t}\Big]_0^{0.05}$$
$$= -1[e^{-10} - e^0] = 1.00$$

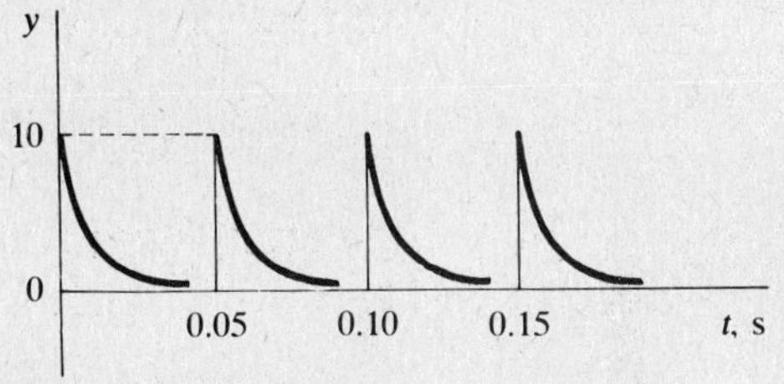

Fig. A-2

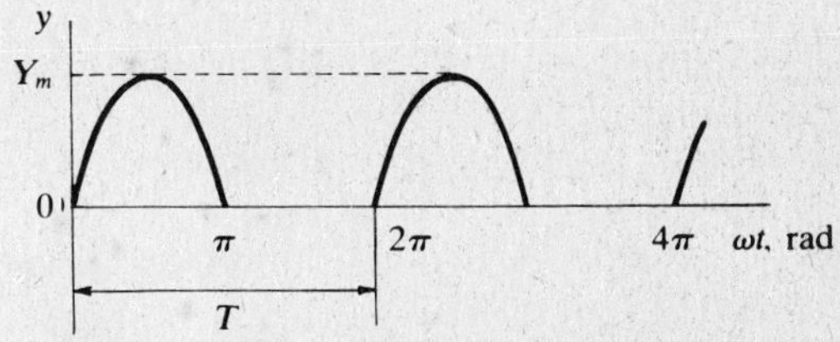

Fig. A-3

EXAMPLE A2 Determine the average value of the waveform shown in Fig. A-3.

For $0 < \omega t < \pi$, $y = Y_m \sin \omega t$; for $\pi < \omega t < 2\pi$, $y = 0$. The period is 2π.

$$Y_{avg} = \frac{1}{2\pi}\left\{\int_0^{\pi} Y_m \sin \omega t \, d(\omega t) + \int_{\pi}^{2\pi} 0 \, d(\omega t)\right\} = 0.318\, Y_m$$

A3 ROOT-MEAN-SQUARE OR EFFECTIVE VALUE

The instantaneous power in a resistance is given by $p(t) = [i(t)]^2 R$. The periodic current $i(t)$ is said to have the *effective* (or *rms*) value I_{eff} (or I_{rms}) if a constant current of that value would result in the same average power as the periodic current. Thus,

$$I_{rms}^2 R = \frac{1}{T}\int_0^T [i(t)]^2 R \, dt$$

from which

$$I_{rms} = \sqrt{\frac{1}{T}\int_0^T [i(t)]^2 \, dt}$$

The term *root-mean-square*, from which the letters rms are taken, serves as a reminder of how the rms value is obtained: take the **root** of the **mean** of the function **squared**. The general periodic function $y(t)$ has the effective value

$$Y_{eff} = \sqrt{\frac{1}{T}\int_0^T [y(t)]^2 \, dt} = \sqrt{\frac{1}{T}\int_{t_0}^{t_0+T} [y(t)]^2 \, dt}$$

EXAMPLE A3 Find the effective value of the waveform shown in Fig. A-2.

$$Y_{rms}^2 = \frac{1}{T}\int_0^T y^2 \, dt = \frac{1}{0.05}\int_0^{0.05} 100e^{-400t} \, dt = 5.00 \qquad \text{or} \qquad Y_{rms} = 2.24$$

EXAMPLE A4 Find the effective value of the waveform shown in Fig. A-3.

$$Y_{rms}^2 = \frac{1}{2\pi}\int_0^{\pi} (Y_m \sin \omega t)^2 \, d(\omega t) = \tfrac{1}{4} Y_m^2 \qquad \text{or} \qquad Y_{rms} = \tfrac{1}{2} Y_m$$

A4 THE EFFECTIVE VALUE OF A TRIGONOMETRIC SERIES

According to *Parseval's identity*, the function

$$y(t) = a_0 + (a_1 \cos \omega t + a_2 \cos 2\omega t + \cdots) + (b_1 \sin \omega t + b_2 \sin 2\omega t + \cdots)$$

has the effective value

$$Y_{rms} = \sqrt{a_0^2 + (\tfrac{1}{2}a_1^2 + \tfrac{1}{2}a_2^2 + \cdots) + \tfrac{1}{2}b_1^2 + \tfrac{1}{2}b_2^2 + \cdots)}$$
$$= \sqrt{a_0^2 + (A_1^2 + A_2^2 + \cdots) + (B_1^2 + B_2^2 + \cdots)}$$

where A_n is the effective value of $a_n \cos n\omega t$ and B_n is the effective value of $b_n \sin n\omega t$.

EXAMPLE A5 Find the rms value of $v = 50 + 141.4 \sin \omega t + 35.5 \sin 3\omega t$ (V).

$$V_{rms} = \sqrt{(50)^2 + \tfrac{1}{2}(141.4)^2 + \tfrac{1}{2}(35.5)^2} = 114.6 \text{ V}$$

Appendix B

Complex Number System

B1 COMPLEX NUMBERS

A *complex number* $\mathbf{z}$ is a number of the form $x + jy$, where x and y are real numbers and $j = \sqrt{-1}$. We write: $x = \text{Re}\,\mathbf{z}$, the *real part of* $\mathbf{z}$; $y = \text{Im}\,\mathbf{z}$, the *imaginary part of* $\mathbf{z}$. Two complex numbers are equal if and only if their real parts are equal and their imaginary parts are equal.

B2 COMPLEX PLANE

A pair of orthogonal axes, with the horizontal axis displaying $\text{Re}\,\mathbf{z}$ and the vertical axis $j\,\text{Im}\,\mathbf{z}$, determine a complex plane in which each complex number is a unique point. Refer to Fig. B-1, on which six complex numbers are shown. Equivalently, each complex number is represented by a unique vector from the origin of the complex plane, as illustrated for the complex number $\mathbf{z}_6$ in Fig. B-1.

$\mathbf{z}_1 = 6$
$\mathbf{z}_2 = 2 - j3$
$\mathbf{z}_3 = j4$
$\mathbf{z}_4 = -3 + j2$
$\mathbf{z}_5 = -4 - j4$
$\mathbf{z}_6 = 3 + j3$

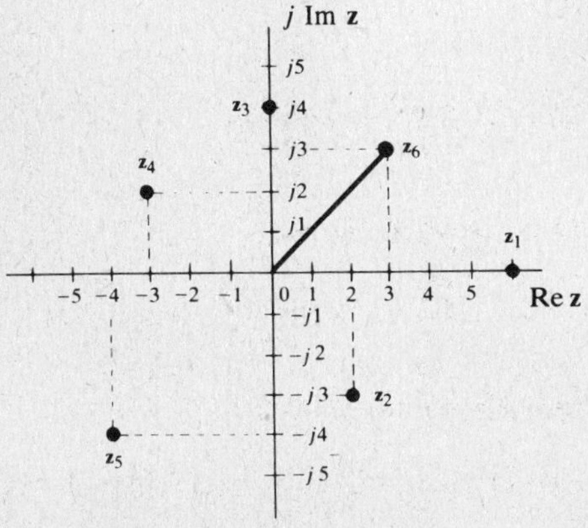

Fig. B-1

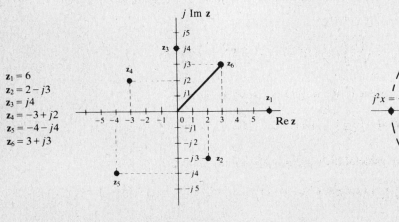

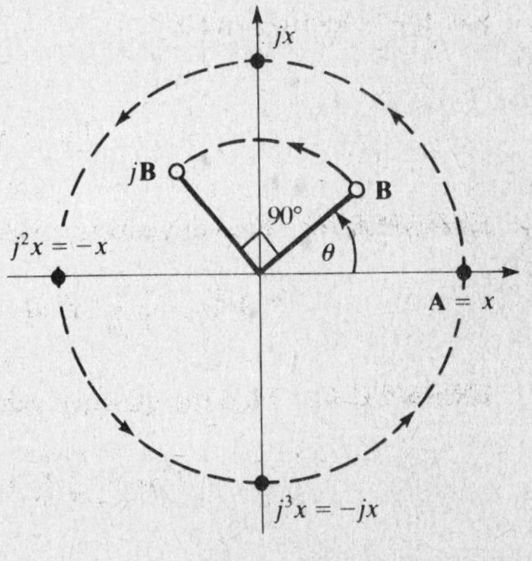

Fig. B-2

B3 VECTOR OPERATOR j

In addition to the definition of j given in Section B1, it may be viewed as an operator which rotates any complex number (vector) $\mathbf{A}$ 90° in the counterclockwise direction. The case where $\mathbf{A}$ is a pure real number, x, is illustrated in Fig. B-2. The rotation sends $\mathbf{A}$ into jx, on the positive imaginary axis. Continuing, j^2 advances $\mathbf{A}$ 180°; j^3, 270°; and j^4, 360°. Also shown in Fig. B-2 is a complex number $\mathbf{B}$ in the first quadrant, at angle θ. Note that $j\mathbf{B}$ is in the second quadrant, at angle $\theta + 90°$.

B4 OTHER REPRESENTATIONS OF COMPLEX NUMBERS

In Section B1 complex numbers were defined in *rectangular form*. In Fig. B-3, $x = r\cos\theta$, $y = r\sin\theta$, and the complex number $\mathbf{z}$ can be written in *trigonometric form* as

$$\mathbf{z} = x + jy = r(\cos\theta + j\sin\theta)$$

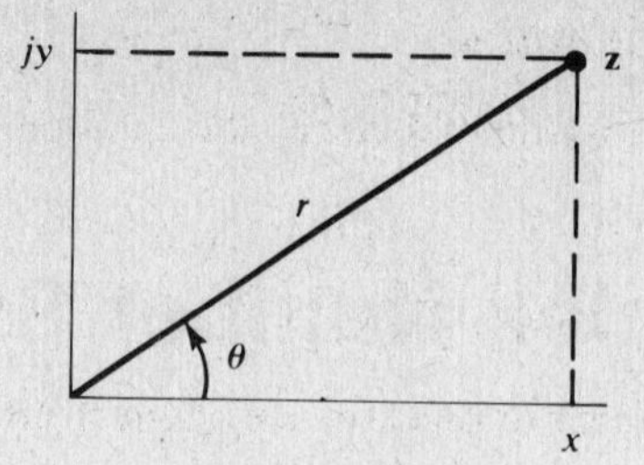

Fig. B-3

where r is the *modulus* or *absolute value* (the notation $r = |\mathbf{z}|$ is common), given by $r = \sqrt{x^2 + y^2}$, and the angle $\theta = \arctan(y/x)$ is the *argument* of $\mathbf{z}$.

Euler's formula, $e^{j\theta} = \cos\theta + j\sin\theta$, permits another representation of a complex number, called the *exponential form*:

$$\mathbf{z} = r\cos\theta + jr\sin\theta = re^{j\theta}$$

A third form, widely used in circuit analysis, is the *polar* or *Steinmetz form*, $\mathbf{z} = r\underline{/\theta}$, where θ is usually in degrees.

B5 SUM AND DIFFERENCE OF COMPLEX NUMBERS

To add two complex numbers, add the real parts and the imaginary parts separately. To subtract two complex numbers, subtract the real parts and the imaginary parts separately. From the practical standpoint, addition and subtraction of complex numbers can be performed conveniently only when both numbers are in the rectangular form.

EXAMPLE B1 Given $\mathbf{z}_1 = 5 - j2$ and $\mathbf{z}_2 = -3 - j8$,

$$\mathbf{z}_1 + \mathbf{z}_2 = (5 - 3) + j(-2 - 8) = 2 - j10$$
$$\mathbf{z}_2 - \mathbf{z}_1 = (-3 - 5) + j(-8 + 2) = -8 - j6$$

B6 MULTIPLICATION OF COMPLEX NUMBERS

The product of two complex numbers when both are in exponential form follows directly from the laws of exponents.

$$\mathbf{z}_1\mathbf{z}_2 = (r_1 e^{j\theta_1})(r_2 e^{j\theta_2}) = r_1 r_2 e^{j(\theta_1 + \theta_2)}$$

The polar or Steinmetz product is evident from reference to the exponential form.

$$\mathbf{z}_1\mathbf{z}_2 = (r_1\underline{/\theta_1})(r_2\underline{/\theta_2}) = r_1 r_2\underline{/\theta_1 + \theta_2}$$

The rectangular product can be found by treating the two complex numbers as binomials.

$$\mathbf{z}_1\mathbf{z}_2 = (x_1 + jy_1)(x_2 + jy_2) = x_1x_2 + jx_1y_2 + jy_1x_2 + j^2y_1y_2$$
$$= (x_1x_2 - y_1y_2) + j(x_1y_2 + y_1x_2)$$

EXAMPLE B2 If $\mathbf{z}_1 = 5e^{j\pi/3}$ and $\mathbf{z}_2 = 2e^{-j\pi/6}$, then $\mathbf{z}_1\mathbf{z}_2 = (5e^{j\pi/3})(2e^{-j\pi/6}) = 10e^{j\pi/6}$.

EXAMPLE B3 If $\mathbf{z}_1 = 2\underline{/30^\circ}$ and $\mathbf{z}_2 = 5\underline{/-45^\circ}$, then $\mathbf{z}_1\mathbf{z}_2 = (2\underline{/30^\circ})(5\underline{/-45^\circ}) = 10\underline{/-15^\circ}$.

EXAMPLE B4 If $\mathbf{z}_1 = 2 + j3$ and $\mathbf{z}_2 = -1 - j3$, then $\mathbf{z}_1\mathbf{z}_2 = (2 + j3)(-1 - j3) = 7 - j9$.

B7 DIVISION OF COMPLEX NUMBERS

For two complex numbers in exponential form, the quotient follows directly from the laws of exponents.

$$\frac{\mathbf{z}_1}{\mathbf{z}_2}=\frac{r_1e^{j\theta_1}}{r_2e^{j\theta_2}}=\frac{r_1}{r_2}e^{j(\theta_1-\theta_2)}$$

Again, the polar or Steinmetz form of division is evident from reference to the exponential form.

$$\frac{\mathbf{z}_1}{\mathbf{z}_2}=\frac{r_1\underline{/\theta_1}}{r_2\underline{/\theta_2}}=\frac{r_1}{r_2}\underline{/\theta_1-\theta_2}$$

Division of two complex numbers in the rectangular form is performed by multiplying the numerator and denominator by the conjugate of the denominator (see Section B8).

$$\frac{\mathbf{z}_1}{\mathbf{z}_2}=\frac{x_1+jy_1}{x_2+jy_2}\left(\frac{x_2-jy_2}{x_2-jy_2}\right)=\frac{(x_1x_2+y_1y_2)+j(y_1x_2-y_2x_1)}{x_2^2+y_2^2}=\frac{x_1x_2+y_1y_2}{x_2^2+y_2^2}+j\frac{y_1x_2-y_2x_1}{x_2^2+y_2^2}$$

EXAMPLE B5 Given $\mathbf{z}_1=4e^{j\pi/3}$ and $\mathbf{z}_2=2e^{j\pi/6}$,

$$\frac{\mathbf{z}_1}{\mathbf{z}_2}=\frac{4e^{j\pi/3}}{2e^{j\pi/6}}=2e^{j\pi/6}$$

EXAMPLE B6 Given $\mathbf{z}_1=8\underline{/-30^\circ}$ and $\mathbf{z}_2=2\underline{/-60^\circ}$,

$$\frac{\mathbf{z}_1}{\mathbf{z}_2}=\frac{8\underline{/-30^\circ}}{2\underline{/-60^\circ}}=4\underline{/30^\circ}$$

EXAMPLE B7 Given $\mathbf{z}_1=4-j5$ and $\mathbf{z}_2=1+j2$,

$$\frac{\mathbf{z}_1}{\mathbf{z}_2}=\frac{4-j5}{1+j2}\left(\frac{1-j2}{1-j2}\right)=-\frac{6}{5}-j\frac{13}{5}$$

B8 CONJUGATE OF A COMPLEX NUMBER

The *conjugate* of the complex number $\mathbf{z}=x+jy$ is the complex number $\mathbf{z}^*=x-jy$. Thus,

$$\text{Re}\,\mathbf{z}=\frac{\mathbf{z}+\mathbf{z}^*}{2}\qquad \text{Im}\,\mathbf{z}=\frac{\mathbf{z}-\mathbf{z}^*}{2j}\qquad |\mathbf{z}|=\sqrt{\mathbf{z}\mathbf{z}^*}$$

In the complex plane, the points $\mathbf{z}$ and $\mathbf{z}^*$ are mirror images in the axis of reals.

In exponential form: $\mathbf{z}=re^{j\theta}$, $\mathbf{z}^*=re^{-j\theta}$.

In polar form: $\mathbf{z}=r\underline{/\theta}$, $\mathbf{z}^*=r\underline{/-\theta}$.

In trigonometric form: $\mathbf{z}=r(\cos\theta+j\sin\theta)$, $\mathbf{z}^*=r(\cos\theta-j\sin\theta)$.

Conjugation has the following useful properties:

(i) $(\mathbf{z}^*)^*=\mathbf{z}$

(ii) $(\mathbf{z}_1\pm\mathbf{z}_2)^*=\mathbf{z}_1^*\pm\mathbf{z}_2^*$

(iii) $(\mathbf{z}_1\mathbf{z}_2)^*=\mathbf{z}_1^*\mathbf{z}_2^*$

(iv) $\left(\frac{\mathbf{z}_1}{\mathbf{z}_2}\right)^*=\frac{\mathbf{z}_1^*}{\mathbf{z}_2^*}$

Appendix C

Matrices and Determinants

C1 SIMULTANEOUS EQUATIONS AND THE CHARACTERISTIC MATRIX

Many engineering systems are described by a set of linearly independent simultaneous equations of the form,

$$\begin{aligned} y_1 &= a_{11}x_1 + a_{12}x_2 + a_{13}x_3 + \cdots + a_{1n}x_n \\ y_2 &= a_{21}x_1 + a_{22}x_2 + a_{23}x_3 + \cdots + a_{2n}x_n \\ &\cdots\cdots\cdots\cdots\cdots\cdots\cdots\cdots \\ y_m &= a_{m1}x_1 + a_{m2}x_2 + a_{m3}x_3 + \cdots + a_{mn}x_n \end{aligned}$$

where the x_j are the independent variables, the y_i the dependent variables, and the a_{ij} are the coefficients of the independent variables. The a_{ij} may be constants or functions of some parameter.

A more convenient form may be obtained for the above equations by expressing them in matrix form.

$$\begin{bmatrix} y_1 \\ y_2 \\ \cdots \\ y_m \end{bmatrix} = \begin{bmatrix} a_{11} & a_{12} & a_{13} & \cdots & a_{1n} \\ a_{21} & a_{22} & a_{23} & \cdots & a_{2n} \\ \cdots & \cdots & \cdots & \cdots & \cdots \\ a_{m1} & a_{m2} & a_{m3} & \cdots & a_{mn} \end{bmatrix} \begin{bmatrix} x_1 \\ x_2 \\ \cdots \\ x_n \end{bmatrix}$$

or $\mathbf{Y} = \mathbf{AX}$, by a suitable definition of the product $\mathbf{AX}$ (see Section C3). Matrix $\mathbf{A} \equiv [a_{ij}]$ is called the *characteristic matrix* of the system; its *order* or *dimension* is denoted as

$$d(\mathbf{A}) \equiv m \times n$$

where m is the number of rows and n is the number of columns.

C2 TYPES OF MATRICES

Row matrix. A matrix which may contain any number of columns but only one row; $d(\mathbf{A}) = 1 \times n$. Also called a *row vector.*

Column matrix. A matrix which may contain any number of rows but only one column; $d(\mathbf{A}) = m \times 1$. Also called a *column vector.*

Diagonal matrix. A matrix whose nonzero elements are all on the principal diagonal.

Unit matrix. A diagonal matrix having every diagonal element unity.

Null matrix. A matrix in which every element is zero.

Square matrix. A matrix in which the number of rows is equal to the number of columns; $d(\mathbf{A}) = n \times n$.

Symmetric matrix. Given

$$\mathbf{A} \equiv \begin{bmatrix} a_{11} & a_{12} & a_{13} & \cdots & a_{1n} \\ a_{21} & a_{22} & a_{23} & \cdots & a_{2n} \\ \cdots & \cdots & \cdots & \cdots & \cdots \\ a_{m1} & a_{m2} & a_{m3} & \cdots & a_{mn} \end{bmatrix} \qquad d(\mathbf{A}) \equiv m \times n$$

the *transpose* of **A** is

$$\mathbf{A}^T \equiv \begin{bmatrix} a_{11} & a_{21} & a_{31} & \dots & a_{m1} \\ a_{12} & a_{22} & a_{32} & \dots & a_{m2} \\ a_{13} & a_{23} & a_{33} & \dots & a_{m3} \\ \dots & \dots & \dots & \dots & \dots \\ a_{1n} & a_{2n} & a_{3n} & \dots & a_{mn} \end{bmatrix} \qquad d(\mathbf{A}^T) \equiv n \times m$$

Thus, the rows of **A** are the columns of $\mathbf{A}^T$, and vice versa. Matrix **A** is *symmetric* if $\mathbf{A} = \mathbf{A}^T$; a symmetric matrix must then be square.

Hermitian matrix. Given

$$\mathbf{A} \equiv \begin{bmatrix} a_{11} & a_{12} & a_{13} & \dots & a_{1n} \\ a_{21} & a_{22} & a_{23} & \dots & a_{2n} \\ \dots & \dots & \dots & \dots & \dots \\ a_{m1} & a_{m2} & a_{m3} & \dots & a_{mn} \end{bmatrix}$$

the *conjugate* of **A** is

$$\mathbf{A}^* \equiv \begin{bmatrix} a_{11}^* & a_{12}^* & a_{13}^* & \dots & a_{1n}^* \\ a_{21}^* & a_{22}^* & a_{23}^* & \dots & a_{2n}^* \\ \dots & \dots & \dots & \dots & \dots \\ a_{m1}^* & a_{m2}^* & a_{m3}^* & \dots & a_{mn}^* \end{bmatrix}$$

Matrix **A** is *hermitian* if $\mathbf{A} = (\mathbf{A}^*)^T$; that is, a hermitian matrix is a square matrix with real elements on the main diagonal and complex conjugate elements occupying positions that are mirror images in the main diagonal. Note that $(\mathbf{A}^*)^T = (\mathbf{A}^T)^*$.

Nonsingular matrix. An $n \times n$ square matrix **A** is *nonsingular* (or *invertible*) if there exists an $n \times n$ square matrix **B** such that

$$\mathbf{AB} = \mathbf{BA} = \mathbf{I}$$

where **I** is the $n \times n$ unit matrix. The matrix **B** is called the *inverse* of the nonsingular matrix **A**, and we write $\mathbf{B} = \mathbf{A}^{-1}$. If **A** is nonsingular, the matrix equation $\mathbf{Y} = \mathbf{AX}$ of Section C1 has, for any **Y**, the unique solution

$$\mathbf{X} = \mathbf{A}^{-1}\mathbf{Y}$$

C3 MATRIX ARITHMETIC

Addition and Subtraction of Matrices

Two matrices of the same order are conformable for addition or subtraction; two matrices of different orders cannot be added or subtracted.

The sum (difference) of two $m \times n$ matrices, $\mathbf{A} = [a_{ij}]$ and $\mathbf{B} = [b_{ij}]$, is the $m \times n$ matrix **C** of which each element is the sum (difference) of the corresponding elements of **A** and **B**. Thus, $\mathbf{A} \pm \mathbf{B} = [a_{ij} \pm b_{ij}]$.

EXAMPLE C1 If

$$\mathbf{A} = \begin{bmatrix} 1 & 4 & 0 \\ 2 & 7 & 3 \end{bmatrix} \qquad \mathbf{B} = \begin{bmatrix} 5 & 2 & 6 \\ 0 & 1 & 1 \end{bmatrix}$$

then

$$\mathbf{A} + \mathbf{B} = \begin{bmatrix} 1+5 & 4+2 & 0+6 \\ 2+0 & 7+1 & 3+1 \end{bmatrix} = \begin{bmatrix} 6 & 6 & 6 \\ 2 & 8 & 4 \end{bmatrix}$$

$$\mathbf{A} - \mathbf{B} = \begin{bmatrix} -4 & 2 & -6 \\ 2 & 6 & 2 \end{bmatrix}$$

The transpose of the sum (difference) of two matrices is the sum (difference) of the two transposes:

$$(\mathbf{A} \pm \mathbf{B})^T = \mathbf{A}^T \pm \mathbf{B}^T$$

Multiplication of Matrices

The product **AB**, in that order, of a $1 \times m$ matrix **A** and an $m \times 1$ matrix **B** is a 1×1 matrix $\mathbf{C} \equiv [c_{11}]$, where

$$\mathbf{C} = [a_{11} \ a_{12} \ a_{13} \ \dots \ a_{1m}] \begin{bmatrix} b_{11} \\ b_{21} \\ b_{31} \\ \dots \\ b_{m1} \end{bmatrix}$$

$$= [a_{11}b_{11} + a_{12}b_{21} + \cdots + a_{1m}b_{m1}] = \left[\sum_{k=1}^{m} a_{1k}b_{k1}\right]$$

Note that each element of the row matrix is multiplied into the corresponding element of the column matrix and then the products are summed. Usually, we identify **C** with the *scalar* c_{11}, treating it as an ordinary number drawn from the number field to which the elements of **A** and **B** belong.

The product **AB**, in that order, of the $m \times s$ matrix $\mathbf{A} = [a_{ij}]$ and the $s \times n$ matrix $\mathbf{B} = [b_{ij}]$ is the $m \times n$ matrix $\mathbf{C} = [c_{ij}]$, where

$$c_{ij} = \sum_{k=1}^{s} a_{ik}b_{kj} \qquad (i = 1, 2, \dots, m, \quad j = 1, 2, \dots, n)$$

EXAMPLE C2

$$\begin{bmatrix} a_{11} & a_{12} \\ a_{21} & a_{22} \\ a_{31} & a_{32} \end{bmatrix} \begin{bmatrix} b_{11} & b_{12} \\ b_{21} & b_{22} \end{bmatrix} = \begin{bmatrix} a_{11}b_{11} + a_{12}b_{21} & a_{11}b_{12} + a_{12}b_{22} \\ a_{21}b_{11} + a_{22}b_{21} & a_{21}b_{12} + a_{22}b_{22} \\ a_{31}b_{11} + a_{32}b_{21} & a_{31}b_{12} + a_{32}b_{22} \end{bmatrix}$$

$$\begin{bmatrix} 3 & 5 & -8 \\ 2 & 1 & 6 \\ 4 & -6 & 7 \end{bmatrix} \begin{bmatrix} I_1 \\ I_2 \\ I_3 \end{bmatrix} = \begin{bmatrix} 3I_1 + 5I_2 - 8I_3 \\ 2I_1 + 1I_2 + 6I_3 \\ 4I_1 - 6I_2 + 7I_3 \end{bmatrix}$$

$$\begin{bmatrix} 5 & -3 \\ 4 & 2 \end{bmatrix} \begin{bmatrix} 8 & -2 & 6 \\ 7 & 0 & 9 \end{bmatrix} = \begin{bmatrix} 5(8) + (-3)(7) & 5(-2) + (-3)(0) & 5(6) + (-3)(9) \\ 4(8) + 2(7) & 4(-2) + 2(0) & 4(6) + 2(9) \end{bmatrix} = \begin{bmatrix} 19 & -10 & 3 \\ 46 & -8 & 42 \end{bmatrix}$$

Matrix **A** is conformable to matrix **B** for multiplication, i.e. the product **AB** is defined, only when the number of columns of **A** is equal to the number of rows of **B**. Thus, if **A** is a 3×2 matrix and **B** is a 2×5 matrix, then the product **AB** is defined, but the product **BA** is not defined. If **D** and **E** are 3×3 matrices, both products **DE** and **ED** are defined. However, it is not necessarily true that **DE** = **ED**.

The transpose of the product of two matrices is the product of the two transposes *taken in reverse order*:

$$(\mathbf{AB})^T = \mathbf{B}^T\mathbf{A}^T$$

If **A** and **B** are nonsingular matrices of the same dimension, then **AB** is also nonsingular, with

$$(\mathbf{AB})^{-1} = \mathbf{B}^{-1}\mathbf{A}^{-1}$$

Multiplication of a Matrix by a Scalar

The product of a matrix $\mathbf{A} \equiv [a_{ij}]$ by a scalar k is defined by

$$k\mathbf{A} = \mathbf{A}k \equiv [ka_{ij}]$$

i.e., each element of **A** is multiplied by k. Note the properties

$$k(\mathbf{A} + \mathbf{B}) = k\mathbf{A} + k\mathbf{B} \qquad k(\mathbf{AB}) = (k\mathbf{A})\mathbf{B} = \mathbf{A}(k\mathbf{B}) \qquad (k\mathbf{A})^T = k\mathbf{A}^T$$

C4 DETERMINANT OF A SQUARE MATRIX

Attached to any $n \times n$ matrix $\mathbf{A} \equiv [a_{ij}]$ is a certain scalar function of the a_{ij}, called the *determinant of* $\mathbf{A}$. This number is denoted as

$$\det \mathbf{A} \quad \text{or} \quad |\mathbf{A}| \quad \text{or} \quad \Delta_{\mathbf{A}} \quad \text{or} \quad \begin{vmatrix} a_{11} & a_{12} & \dots & a_{1n} \\ a_{21} & a_{22} & \dots & a_{2n} \\ \dots & \dots & \dots & \dots \\ a_{n1} & a_{n2} & \dots & a_{nn} \end{vmatrix}$$

where the last form puts into evidence the elements of $\mathbf{A}$, upon which the number depends. For determinants of order $n = 1$ and $n = 2$, we have explicitly

$$|a_{11}| = a_{11} \qquad \begin{vmatrix} a_{11} & a_{12} \\ a_{21} & a_{22} \end{vmatrix} = a_{11}a_{22} - a_{12}a_{21}$$

For larger n, the analogous expressions become very cumbersome, and they are usually avoided by use of Laplace's expansion theorem (see below). What is important is that the determinant is defined in such a way that

$$\det \mathbf{AB} = (\det \mathbf{A})(\det \mathbf{B})$$

for any two $n \times n$ matrices $\mathbf{A}$ and $\mathbf{B}$. Two other basic properties are:

$$\det \mathbf{A}^T = \det \mathbf{A} \qquad \det k\mathbf{A} = k^n \det \mathbf{A}$$

Finally, $\det \mathbf{A} \neq 0$ if and only if $\mathbf{A}$ is nonsingular.

EXAMPLE C3 Verify the determinant multiplication rule for

$$\mathbf{A} = \begin{bmatrix} 1 & 4 \\ 3 & 2 \end{bmatrix} \qquad \mathbf{B} = \begin{bmatrix} -2 & 9 \\ 1 & \pi \end{bmatrix}$$

We have

$$\mathbf{AB} = \begin{bmatrix} 1 & 4 \\ 3 & 2 \end{bmatrix}\begin{bmatrix} -2 & 9 \\ 1 & \pi \end{bmatrix} = \begin{bmatrix} 2 & 9+4\pi \\ -4 & 27+2\pi \end{bmatrix}$$

and

$$\begin{vmatrix} 2 & 9+4\pi \\ -4 & 27+2\pi \end{vmatrix} = 2(27+2\pi) - (9+4\pi)(-4) = 90 + 20\pi$$

But

$$\begin{vmatrix} 1 & 4 \\ 3 & 2 \end{vmatrix} = 1(2) - 4(3) = -10$$

$$\begin{vmatrix} -2 & 9 \\ 1 & \pi \end{vmatrix} = -2(\pi) - 9(1) = -9 - 2\pi$$

and indeed $90 + 20\pi = (-10)(-9 - 2\pi)$.

Laplace's Expansion Theorem

The *minor*, M_{ij}, of the element a_{ij} of a determinant of order n is the determinant of order $n-1$ obtained by deleting the row and column containing a_{ij}. The *cofactor*, Δ_{ij}, of the element a_{ij} is defined as

$$\Delta_{ij} = (-1)^{i+j} M_{ij}$$

Laplace's theorem states: In the determinant of a square matrix $\mathbf{A}$, multiply each element in the pth row (column) by the cofactor of the corresponding element in the qth row (column), and sum the products. Then the result is 0, for $p \neq q$; and $\det \mathbf{A}$, for $p = q$.

It follows at once from Laplace's theorem that if $\mathbf{A}$ has two rows or two columns the same, then $\det \mathbf{A} = 0$ (and $\mathbf{A}$ must be a singular matrix).

Matrix Inversion by Determinants;
Cramer's Rule

Laplace's expansion theorem can be exhibited as a matrix multiplication, as follows:

$$\begin{bmatrix} a_{11} & a_{12} & a_{13} & \dots & a_{1n} \\ a_{21} & a_{22} & a_{23} & \dots & a_{2n} \\ \dots & \dots & \dots & \dots & \dots \\ a_{n1} & a_{n2} & a_{n3} & \dots & a_{nn} \end{bmatrix} \begin{bmatrix} \Delta_{11} & \Delta_{21} & \Delta_{31} & \dots & \Delta_{n1} \\ \Delta_{12} & \Delta_{22} & \Delta_{32} & \dots & \Delta_{n2} \\ \dots & \dots & \dots & \dots & \dots \\ \Delta_{1n} & \Delta_{2n} & \Delta_{3n} & \dots & \Delta_{nn} \end{bmatrix}$$

$$= \begin{bmatrix} \Delta_{11} & \Delta_{21} & \Delta_{31} & \dots & \Delta_{n1} \\ \Delta_{12} & \Delta_{22} & \Delta_{32} & \dots & \Delta_{n2} \\ \dots & \dots & \dots & \dots & \dots \\ \Delta_{1n} & \Delta_{2n} & \Delta_{3n} & \dots & \Delta_{nn} \end{bmatrix} \begin{bmatrix} a_{11} & a_{12} & a_{13} & \dots & a_{1n} \\ a_{21} & a_{22} & a_{23} & \dots & a_{2n} \\ \dots & \dots & \dots & \dots & \dots \\ a_{n1} & a_{n2} & a_{n3} & \dots & a_{nn} \end{bmatrix}$$

$$= \begin{bmatrix} \det \mathbf{A} & 0 & 0 & \dots & 0 \\ 0 & \det \mathbf{A} & 0 & \dots & 0 \\ \dots & \dots & \dots & \dots & \dots \\ 0 & 0 & 0 & \dots & \det \mathbf{A} \end{bmatrix}$$

or

$$\mathbf{A}(\operatorname{adj} \mathbf{A}) = (\operatorname{adj} \mathbf{A})\mathbf{A} = (\det \mathbf{A})\mathbf{I}$$

where $\operatorname{adj} \mathbf{A} \equiv [\Delta_{ji}]$ is the transposed matrix of the cofactors of the a_{ij} in the determinant of $\mathbf{A}$, and $\mathbf{I}$ is the $n \times n$ unit matrix.

If $\mathbf{A}$ is nonsingular, one may divide through by $\det \mathbf{A} \neq 0$, and infer that

$$\mathbf{A}^{-1} = \frac{1}{\det \mathbf{A}} \operatorname{adj} \mathbf{A}$$

This means that the unique solution of the linear system $\mathbf{Y} = \mathbf{AX}$ is

$$\mathbf{X} = \left(\frac{1}{\det \mathbf{A}} \operatorname{adj} \mathbf{A}\right)\mathbf{Y}$$

which is Cramer's Rule in matrix form. The ordinary, determinant form is obtained by considering the rth row ($r = 1, 2, \dots, n$) of the matrix solution. Since the rth row of $\operatorname{adj} \mathbf{A}$ is

$$[\Delta_{1r} \;\; \Delta_{2r} \;\; \Delta_{3r} \;\; \dots \;\; \Delta_{nr}]$$

we have:

$$x_r = \left(\frac{1}{\det \mathbf{A}}\right)[\Delta_{1r} \;\; \Delta_{2r} \;\; \Delta_{3r} \;\; \dots \;\; \Delta_{nr}] \begin{bmatrix} y_1 \\ y_2 \\ y_3 \\ \dots \\ y_n \end{bmatrix}$$

$$= \left(\frac{1}{\det \mathbf{A}}\right)(y_1\Delta_{1r} + y_2\Delta_{2r} + y_3\Delta_{3r} + \dots + y_n\Delta_{nr})$$

$$= \left(\frac{1}{\det \mathbf{A}}\right) \begin{vmatrix} a_{11} & \dots & a_{1(r-1)} & y_1 & a_{1(r+1)} & \dots & a_{1n} \\ a_{21} & \dots & a_{2(r-1)} & y_2 & a_{2(r+1)} & \dots & a_{2n} \\ \dots & \dots & \dots & \dots & \dots & \dots & \dots \\ a_{n1} & \dots & a_{n(r-1)} & y_n & a_{n(r+1)} & \dots & a_{nn} \end{vmatrix}$$

The last equality may be verified by applying Laplace's theorem to the rth column of the given determinant.

C5 EIGENVALUES OF A SQUARE MATRIX

For a linear system $\mathbf{Y} = \mathbf{AX}$, with $n \times n$ characteristic matrix $\mathbf{A}$, it is of particular importance to investigate the "excitations" $\mathbf{Y}$ that produce a proportionate "response" $\mathbf{X}$. Thus, letting $\mathbf{Y} = \lambda\mathbf{X}$, where λ is a scalar,

$$\lambda\mathbf{X} = \mathbf{AX} \qquad \text{or} \qquad (\lambda\mathbf{I} - \mathbf{A})\mathbf{X} = \mathbf{O}$$

where $\mathbf{O}$ is the $n \times 1$ null matrix. Now, if the matrix $\lambda\mathbf{I} - \mathbf{A}$ were nonsingular, only the trivial solution $\mathbf{X} = \mathbf{Y} = \mathbf{O}$ would exist. Hence, for a nontrivial solution, the value of λ must be such as to make $\lambda\mathbf{I} - \mathbf{A}$ a singular matrix; that is, we must have

$$\det(\lambda\mathbf{I} - \mathbf{A}) = \begin{vmatrix} \lambda - a_{11} & -a_{12} & -a_{13} & \dots & -a_{1n} \\ -a_{21} & \lambda - a_{22} & -a_{23} & \dots & -a_{2n} \\ \dots & \dots & \dots & \dots & \dots \\ -a_{n1} & -a_{n2} & -a_{n3} & \dots & \lambda - a_{nn} \end{vmatrix} = 0$$

The n roots of this polynomial equation in λ are the *eigenvalues* of matrix $\mathbf{A}$; the corresponding nontrivial solutions $\mathbf{X}$ are known as the *eigenvectors* of $\mathbf{A}$.

Setting $\lambda = 0$ in the left side of the above *characteristic equation*, we see that the constant term in the equation must be

$$\det(-\mathbf{A}) = \det[(-1)\mathbf{A}] = (-1)^n(\det \mathbf{A})$$

Since the coefficient of λ^n in the equation is obviously unity, the constant term is also equal to $(-1)^n$ times the product of all the roots. *The determinant of a square matrix is the product of all its eigenvalues*—an alternate, and very useful, definition of the determinant.

Index